V • O • L • U • M • E 2

METHODS IN CHEMICAL ECOLOGY

BIOASSAY METHODS

About the cover: The cover depicts a female bolas spider in her hunting posture. The spider aggressively mimics the sex pheromone blend of a few species of moths, and thus attracts only male moths. These prey are caught by the sticky ball (or bolas) at the end of the vertical line.

About the cover: The cover depicts a female bolas spider in her hunting posture. The spider aggressively mimics the sex pheromone blend of a few species of moths and thus attracts only male moths. These prey are caught by the sticky ball (or bolas) at the end of the vertical line.

VOLUME 2

METHODS IN CHEMICAL ECOLOGY

BIOASSAY METHODS

EDITED BY

KENNETH F. HAYNES

University of Kentucky, Department of Entomology, Lexington, KY

JOCELYN G. MILLAR

University of California, Department of Entomology, Riverside, CA

SPRINGER SCIENCE+BUSINESS MEDIA, LLC

Library of Congress Cataloging-in-Publication Data

Methods in chemical ecology
p. cm.
Includes bibliographical references and index.
Contents: v. 1. Chemical methods / edited by Jocelyn G. Millar and Kenneth F. Haynes-- v. 2. Bioassay methods / edited by Kenneth F. Haynes and Jocelyn G. Millar.
ISBN 978-1-4613-7471-8 ISBN 978-1-4615-5411-0 (eBook)
DOI 10.1007/978-1-4615-5411-0
1. Chemical ecology--Methodology. I. Millar, Jocelyn G. II. Haynes, K.F.
QH541.15.C44M48 1998
577'.01'54--dc21 97-39820
CIP

Originally published by Kluwer Academic Publishers in 1998
Softcover reprint of the hardcover 1st edition 1998

Printed on acid-free paper.

Dedicated to the memory
of Catharine Kennedy Haynes and William Warren Haynes

Contents

Preface xv

Contributors xix

1. Bioassays with marine microorganisms
Kelly M. Jenkins, Paul R. Jensen, and William Fenical 1

1.1. Chemical ecology of marine microorganisms 2
1.2. Ecological relevance of bioassays 3
1.3. Antimicrobial assays 5
1.3.1. Direct challenge/competitive interaction assays 6
1.3.2. Antibiotic disk susceptibility assays 7
1.3.3. Bioautography 10
1.3.4. Direct algal cell count assays 11
1.3.5. Spectrophotometric assays 11
1.3.6. Viral plaque formation assay 13
1.3.7. Agar plate assay for filamentous fungi 13
1.3.8. Fungal interference competition assay 14
1.3.9. Associational defense 15
1.4. Behavioral assays 16
1.4.1. Chemotaxis 16
1.4.1.1. Capillary assay 17
1.4.1.2. Modified capillary assays 18
1.4.1.3. Chamber method 20
1.4.1.4. Dual-well slide method 21
1.4.2. Gamete chemotaxis 21
1.4.2.1. Müller method 21
1.4.2.2. Miller method 22
1.4.3. Motility 23
1.4.4. Settlement assays—laboratory studies 24
1.4.4.1. Microscope slide assay 25
1.4.4.2. Agar matrix assay 26

1.4.4.3. Spectrophotometric assay 26
1.4.4.4. Microalgal attachment assay 27
1.4.5. Settlement assays—field studies 28
1.4.5.1. Tile assay 28
1.4.5.2. Microscope slide assay 29
1.4.5.3. Microalgal assays 29
1.5. Summary and conclusions 31
1.6. Acknowledgments 32
1.7. References 32

2. Bioassays with marine and freshwater macroorganisms
Mark E. Hay, John J. Stachowicz, Edwin Cruz-Rivera, Stephan Bullard, Michael S. Deal, and Niels Lindquist 39

2.1. Introduction 40
2.2. Foraging cues 42
2.2.1. Isolation and preparation of foraging cues 43
2.2.2. Vertebrates 45
2.2.3. Invertebrates 47
2.3. Feeding cues 51
2.3.1. Isolation and preparation of feeding cues 55
2.3.2. Feeding deterrents 56
2.3.2.1. Vertebrates 56
2.3.2.1.1. Field assays 56
2.3.2.1.2. Laboratory assays 59
2.3.2.2. Invertebrates 63
2.3.2.2.1. Field assays 63
2.3.2.2.2. Laboratory assays 64
2.3.3. Feeding stimulants 70
2.3.4. Multiple cues and defensive synergisms 72
2.4. Consequences of consuming defensive metabolites 74
2.4.1. Vertebrates 76
2.4.1.1. Short-term assays 76
2.4.1.2. Long-term assays 80
2.4.2. Invertebrates 82
2.4.2.1. Short-term assays 82
2.4.2.2. Long-term assays 85
2.5. Toxin-mediated prey capture 90
2.6. Chemically mediated detection of and responses to predators 92
2.6.1. Vertebrates 93
2.6.1.1. Chemical cues from conspecifics 93
2.6.1.2. Chemical cues from predators 95
2.6.2. Invertebrates 97
2.6.2.1. Behavioral responses 97
2.6.2.2. Induced responses 99
2.7. Intraspecific chemical communication 102
2.8. Chemically mediated homing behavior 106

2.8.1. Vertebrates 106
2.8.2. Invertebrates 108
2.9. Settlement cues 108
2.9.1. Vertebrates 108
2.9.2. Invertebrates 109
2.9.3. Recommendations regarding still water assays 114
2.10. Allelopathy and antifouling 117
2.10.1. Invertebrates 117
2.10.2. Seaweeds 119
2.11. Chemical ecology within a broader environmental context 120
2.12. Conclusions 121
2.13. Acknowledgments 122
2.14. References 122

3. Bioassay methods for fungi and oomycetes
James L. Kerwin and Melinda J. Semon 142

3.1. Introduction 142
3.2. Intraspecific interactions—reproduction 143
3.2.1. Chytridiomycetes 144
3.2.2. Oomycetes 145
3.2.3. Yeast 147
3.2.4. Basidiomycetes and ascomycetes 148
3.3. Intraspecific population interactions 148
3.4. Interspecific interactions 151
3.4.1. Exploitation of resources—saprophytes 151
3.4.2. Exploitation of resources—parasites and pathogens 151
3.4.2.1. Plant parasites and pathogens 151
3.4.2.2. Animal parasites and pathogens 157
3.4.2.3. Fungal and lichen parasites and pathogens 159
3.4.3. Mutualistic and symbiotic interactions 160
3.4.3.1. Vesicular-arbuscular mycorrhizal fungi 160
3.4.3.2. Ectomycorrhizal fungi 162
3.4.3.3. Fungal-insect associations 162
3.4.4. Competition—allelopathy 163
3.4.5. Defense against herbivores and pathogens 163
3.5. Conclusions 165
3.6. Acknowledgments 165
3.7. References 165

4. Bioassays for allelopathy in terrestrial plants
John T. Romeo and Jeffrey D. Weidenhamer 179

4.1. Introduction 179
4.2. Case studies illustrating appropriate bioassays 184
4.2.1. Bracken fern 184
4.2.1.1. Vegetation pattern 184
4.2.1.2. Environmental factors 185

4.2.1.3. Allelopathy 186
4.2.2. The California chaparral 188
4.2.2.1. Role of allelopathy 188
4.2.2.2. Physical and biotic factors 189
4.2.2.3. Bioassays to establish mechanism of allelochemical transport 189
4.2.3. Rye 191
4.2.3.1. Greenhouse experiments and bioassays 192
4.2.3.2. Chemical studies 193
4.2.4. The Florida scrub 195
4.2.4.1. Vegetation pattern and soil 195
4.2.4.2. Resource competition 195
4.2.4.3. Allelopathy 196
4.2.4.4. Chemical studies 196
4.2.5. Crowberry 198
4.2.5.1. Laboratory bioassays 198
4.2.5.2. Greenhouse and field studies 199
4.3. Density-dependent phytotoxicity 200
4.4. Practical considerations 202
4.5. Acknowledgments 205
4.6. References 205

5. Bioassay methods with terrestrial invertebrates
J. Daniel Hare 212

5.1. Introduction 213
5.1.1. Scope of the chapter 214
5.1.2. General considerations for laboratory bioassays 214
5.1.3. Statistical analysis of choices 215
5.2. Behavioral bioassays for odors, pheromones, and other volatile compounds 216
5.2.1. General considerations 216
5.2.2. Moving air bioassays (olfactometers and wind tunnels) 217
5.2.3. Still-air bioassays 223
5.2.4. Modifications for specific purposes 225
5.3. Bioassays for contact oviposition stimulants—two case studies 228
5.3.1. Butterflies 229
5.3.2. Parasitic Wasps 233
5.3.3. Data Analysis 237
5.4. Measurement of preference 237
5.4.1. Host-plant selection for feeding 237
5.4.1.1. Neutral substrates 238
5.4.1.2. Artificial diets 240
5.4.1.3. Plant tissue substrates 241
5.4.1.4. Leaf disks 243
5.4.1.5. Other natural substrates 243
5.4.2. Details of bioassay design 243

	5.4.3.	Data collection	244
	5.4.4.	Data analysis	247
5.5.	Postingestive bioassays		247
5.6.	Measurements in diet studies: growth rate, consumption rate, and efficiency of conversion of food to biomass		248
	5.6.1.	Data acquisition	252
		5.6.1.1. Consumption	252
		5.6.1.2. Weight gain	253
		5.6.1.3. Measurements of excreta	253
	5.6.2.	Alternatives to gravimetry	254
		5.6.2.1. Uric acid determination	254
		5.6.2.2. Calorimetry and respirometry	255
	5.6.3.	Summary	256
5.7.	Alternative methods to separate preingestive and postingestive effects		257
5.8.	Contact and volatile toxicity		259
5.9.	Conclusions		260
5.10.	Acknowledgments		260
5.11.	References		261

6. Bioassay methods for amphibians and reptiles
Robert T. Mason, Douglas P. Chivers, Alicia Mathis, and Andrew R. Blaustein 271

6.1.	Introduction		272
6.2.	Amphibians		273
	6.2.1.	Orientation and homing	273
		6.2.1.1. Frogs and toads	273
		6.2.1.2. Salamanders	277
	6.2.2.	Chemical cues in kin recognition	278
		6.2.2.1. Larval amphibians	278
		6.2.2.2. Post-metamorphosis	282
	6.2.3.	Territorial pheromones	282
		6.2.3.1. Source of the chemical signal	282
		6.2.2.2. Response variables	284
	6.2.4.	Marking behaviors	286
	6.2.5.	Reproductive pheromones and sexual advertisement	287
		6.2.5.1. Source of the chemical signal	287
		6.2.5.2. Response variables	287
		6.2.5.3. Species discrimination	288
		6.2.5.4. Male courtship pheromones	288
	6.2.6.	Predator/prey interactions	289
		6.2.6.1. Chemical alarm cues	289
		6.2.6.2. Antipredator responses	292
	6.2.7.	Noxiousness and toxicity	294
6.3.	Reptiles		296
	6.3.1.	Olfaction and vomerolfaction	296

6.3.2. Attractants and repellents 297
6.3.3. Aggregation 299
6.3.4. Trailing 302
6.3.5. Sexual behavior 305
6.3.6. Femoral pores 307
6.3.7. Copulatory plugs 307
6.3.8. Combat behavior 308
6.3.9. Predator/prey recognition 308
6.4. Conclusions 310
6.5. Acknowledgments 311
6.6. References 311

7. Bioassays for mammals and birds
Dale L. Nolte and J. Russell Mason 326

7.1. Introduction 327
7.2. Chemical senses 327
7.2.1. Olfaction 327
7.2.2. Vomeronasal chemoreception 328
7.2.3. Trigeminal chemoreception 328
7.2.4. Terminal nerve 329
7.2.5. Septal organ 329
7.2.6. Taste 329
7.3. Test paradigms 330
7.3.1. Naturalistic observations 331
7.3.2. Orientating responses 331
7.3.3. Choice tests 331
7.3.4. Flavor-avoidance learning 333
7.3.5. Operant conditioning 335
7.3.6. Single-subject designs 336
7.4. Experimental apparatus 337
7.4.1. Feed trays 337
7.4.2. Drinking tubes 337
7.4.3. Arena 338
7.4.4. Mazes 339
7.5. Intraspecific behaviors 339
7.5.1. Detection and recognition 342
7.5.1.1. Mammal bioassays 342
7.5.1.2. Bird bioassays 346
7.5.2. Reproduction 349
7.5.3. Dominance 351
7.5.4. Territorial behaviors 352
7.6. Interspecific behaviors 355
7.6.1. Resource exploitation 356
7.6.1.1. Mammal bioassays 356
7.6.1.2. Fetal mammal bioassays 360
7.6.1.3. Lactating mammal bioassays 362

7.6.1.4. Bird bioassays 363
7.6.2. Predator avoidance (defense) 365
7.6.3. Community or mutualistic interactions 368
7.7. Case studies 369
7.7.1. Chemical correlates and sensory perception 369
7.7.2. Responses of birds to aversive stimuli 370
7.7.3. Relating repellency and chemical structure 371
7.7.4. Conspecifics response to scent marks 373
7.7.5. Aversive components of predator odors 374
7.7.6. Chemically mediated foraging by black bears 376
7.8. Summary 378
7.9. Acknowledgments 378
7.10. References 378

Index 397

Preface

Identification of chemicals that affect the naturally occurring interactions between organisms requires sophisticated chemical techniques, such as those documented in volume 1, in combination with effective bioassays. Without an effective bioassay, the identification becomes akin to looking for a needle in a haystack, but without any idea of what a needle looks like. To a large extent semiochemical identifications must be driven by bioassays.

The design of bioassays for use in chemical ecology is governed by the sometimes conflicting objectives of ecological relevance and the need for simplicity. Bioassay design should be based on observations of the interactions between organisms in their natural context, a theme that appears throughout this volume. As a result, this volume is as much about ecology and behavior as it is about specific methods. It is impossible to design a relevant bioassay, whether it is simple or complex, without understanding at least the fundamentals of how chemical cues or signals mediate the interaction in nature. Thus, the development of bioassay methods must be driven by an understanding of ecology and a knowledge of the natural history of the organisms under study. Given such an understanding, it is often possible to design assays that are both ecologically relevant and easy to perform.

Because bioassays are developed (or should be) based on observations in natural habitats, they can be as diverse as the organisms that are involved in chemically mediated interactions. However, there is much to be gained by evaluating bioassays that have been developed by other investigators, especially if those assays involved similar types of interactions, occurred in similar habitats, *or* were carried out with closely related organisms. This is our central rationale for assembling the diverse chapters in this volume. With our invitations to authors, we have attempted to cover a broad range of taxa, but in a single volume we could not hope to be comprehensive. We have encouraged the authors to provide a wide range of coverage of types of interactions, but each author had to make a compromise between the depth and breadth of their coverage. While there is much to be gained by reading and using the information on a focal group of organisms, it is also instructive to learn how investigators working with different taxa approach analogous problems in chemical ecology and bioassay design. The field of chemical ecology can only benefit from such pooling of ideas and methods.

In addition to providing guidance for the development of specific bioassays, chapter authors have pointed to some important lessons learned through evaluation of a history of method development in their fields. Not infrequently, inappropriate bioassay design has led to misidentification of semiochemicals. Some of these historical lessons appear in several chapters. For example, concentration is a confounding factor in almost any bioassay. At some exposure level, the same compound may be attractive, then repellent, and finally toxic. Therefore it is crucial to relate the response to a biologically relevant concentration. This is often much more difficult than it sounds. Biologically active compounds may be sequestered in tissues, and macerating tissue may give an unrealistically high estimate of natural concentration. Conversely the compound may be actively synthesized at the moment of release, and a "snapshot" taken at the time of extraction may dramatically underestimate the effective concentration of the compound.

Another example of such a historical lesson is that translation from laboratory bioassay back to the natural context to reevaluate the results is a critical component of completing the identification of a semiochemical. If the need for simplicity has driven the investigator to laboratory bioassays, then returning to the organisms' habitat represents the ultimate test of the conclusions of laboratory bioassay-driven identification.

This volume is intended to be a sourcebook for bioassay design for problems in chemical ecology. However, it is not a "cookbook," and investigators will need to modify the bioassays described for the specific organisms with which they work. For example, in addition to the biological reasons for specificity in bioassays, investigators working with vertebrate animals will need to verify that their procedures satisfy the regulations and ethical standards for the humane use of animals in research. For historical completeness, this volume includes references to procedures that would not meet current animal care and use standards. These collected works are presented to show the development of ideas and methods, but readers must ensure that the historical procedures are modified to comply with current practices in the use of laboratory animals.

Because of the broad scope of this volume, it was not possible for the editors to provide the depth of evaluation that could come only from the authors' peers. We would like to thank the outside reviewers of the individual chapters, including Drs. James E. Adaskaveg, Frank A. Einhellig, Brent M. Graves, Frederick D. Provenza, Nancy M. Targett, Paul A. Weston, and other anonymous reviewers. These individuals made a very valuable contribution to this book through their insightful comments and suggestions. We would also like to thank a few individuals who commented on the overall organization of this volume before invitations to authors were made, including Drs. J. Daniel Hare, Nancy L. Reagan, Leslie A. Weston, and Paul A. Weston. We would like to thank our authors for their dedication to the very difficult task of covering broad fields of chemical ecology. They often had to describe and evaluate methods considerably beyond the confines of their own research interests. For this special effort we are very grateful. Finally, we would like to acknowledge our spouses, Elizabeth W. Haynes and Dorothy E. Hartley, for their continuing patience and support.

Contributors

Andrew R. Blaustein
Department of Zoology
Oregon State University
Corvallis, OR 97331 USA

Stephan Bullard
University of North Carolina
at Chapel Hill
Institute of Marine Sciences
3431 Arendell Street
Morehead City, NC 28557 USA

Douglas P. Chivers
Department of Biological Sciences
University of Maine
Orono, ME 04469 USA

Edwin Cruz-Rivera
University of North Carolina
at Chapel Hill
Institute of Marine Sciences
3431 Arendell Street
Morehead City, NC 28557 USA

Michael S. Deal
University of North Carolina
at Chapel Hill
Institute of Marine Sciences
3431 Arendell Street
Morehead City, NC 28557 USA

William Fenical
Scripps Institution of Oceanography
Center for Marine Biotechnology and
Biomedicine
University of California at San Diego
La Jolla, CA 92093 USA

J. Daniel Hare
Department of Entomology
University of California
Riverside, CA 92521 USA

Mark E. Hay
University of North Carolina
at Chapel Hill
Institute of Marine Sciences
3431 Arendell Street
Morehead City, NC 28557 USA

Kelly M. Jenkins
Scripps Institution of Oceanography
Center for Marine Biotechnology and
Biomedicine
University of California at San Diego
La Jolla, CA 92093 USA

Paul R. Jensen
Scripps Institution of Oceanography
Center for Marine Biotechnology and
Biomedicine
University of California at San Diego
La Jolla, CA 92093 USA

James L. Kerwin
Botany Department
Box 351330
University of Washington
Seattle, WA 98195 USA

Niels Lindquist
University of North Carolina
at Chapel Hill
Institute of Marine Sciences
3431 Arendell Street
Morehead City, NC 28557 USA

Alicia Mathis
Department of Biology
Southwest Missouri State University
Springfield, MO 65804 USA

J. Russell Mason
United States Department of Agriculture
Animal and Plant Health Inspection
Service
Wildlife Services
National Wildlife Research Center
Predation Ecology & Behavior Project
Utah State University
Logan, UT 84322 USA

Robert T. Mason
Department of Zoology
Oregon State University
Corvallis, OR 97331 USA

Dale L. Nolte
United States Department of Agriculture
Animal and Plant Health Inspection
Service
Wildlife Services
National Wildlife Research Center
9701 Blomberg Street
Olympia, WA 98512 USA

John T. Romeo
Department of Biology
University of South Florida
Tampa, FL 33620 USA

Melinda J. Semon
Department of Food Science
and Human Nutrition
Washington State University
Pullman, WA 99164 USA

John J. Stachowicz
University of North Carolina
at Chapel Hill
Institute of Marine Sciences
3431 Arendell Street
Morehead City, NC 28557 USA

Jeffrey D. Weidenhamer
Department of Chemistry and Geology
Ashland University
Ashland, OH 44805 USA

1

Bioassays with Marine Microorganisms

Kelly M. Jenkins, Paul R. Jensen, and William Fenical

1.1. Chemical ecology of marine microorganisms 2
1.2. Ecological relevance of bioassays 3
1.3. Antimicrobial assays 5
1.3.1. Direct challenge/competitive interaction assays 6
1.3.2. Antibiotic disk susceptibility assays 7
1.3.3. Bioautography 10
1.3.4. Direct algal cell count assays 11
1.3.5. Spectrophotometric assays 11
1.3.6. Viral plaque formation assay 13
1.3.7. Agar plate assay for filamentous fungi 13
1.3.8. Fungal interference competition assay 14
1.3.9. Associational defense 15
1.4. Behavioral assays 16
1.4.1. Chemotaxis 16
1.4.1.1. Capillary assay 17
1.4.1.2. Modified capillary assays 18
1.4.1.3. Chamber method 20
1.4.1.4. Dual-well slide method 21
1.4.2. Gamete chemotaxis 21
1.4.2.1. Müller method 21
1.4.2.2. Miller method 22
1.4.3. Motility 23
1.4.4. Settlement assays—laboratory studies 24
1.4.4.1. Microscope slide assay 25
1.4.4.2. Agar matrix assay 26
1.4.4.3. Spectrophotometric assay 26
1.4.4.4. Microalgal attachment assay 27
1.4.5. Settlement assays—field studies 28
1.4.5.1. Tile assay 28
1.4.5.2. Microscope slide assay 29
1.4.5.3. Microalgal assays 29

1.5. Summary and conclusions 31
1.6. Acknowledgments 32
1.7. References 32

1.1. Chemical Ecology of Marine Microorganisms

The aim of marine chemical ecology is to understand the significance of marine natural products in mediating the interactions between organisms in the marine environment. Our awareness of this subject has increased dramatically in recent years through the development of ecologically relevant bioassays (see Paul 1992; chapter 2) the results of which have provided us with a fundamental understanding of the effects of marine natural products on predators, larvae, and fouling macroorganisms. Marine natural products can also affect microorganisms, and although our understanding of these interactions is not as well developed, recent efforts to design new bioassays that integrate chemistry and microbiology have provided insight into their effects on taxonomically diverse marine microorganisms such as bacteria, microalgae, and fungi. With the advent of improved methodologies, we hope to learn the extent to which marine natural products influence co-occurring microbial populations and thereby obtain a better understanding of the evolutionary significance of marine secondary metabolite production.

Unraveling the microbiological effects of marine natural products is an enormously complex, potentially far-reaching, and experimentally difficult task. Microorganisms colonize all marine environments, from the deepest ocean trenches to the shallowest estuaries. In most cases, these populations are taxonomically diverse, e.g., 1 ml of seawater typically contains 1×10^7 viruses, 1×10^4 cyanobacteria, 1×10^6 bacteria, 1×10^3 yeasts, and 1×10^3 microalgae (Rheinheimer 1992), which display extraordinary levels of metabolic plasticity reflecting their ability to survive despite variable environmental conditions. Due to the ubiquitous nature of marine microbes, all plants and animals in the sea are in essence bathed in a "microbial soup," the taxonomy and physiology of which is not constant. This microbial mix includes potential pathogens and detrimental surface colonizing microorganisms as well as beneficial symbionts and benign associates. Despite constant contact with diverse microbes, the surfaces of most marine plants and animals remain relatively free from infection and are not colonized in the rapid and predictable manner characteristic of inanimate surfaces. These differences may be due to the maintenance of antimicrobial chemical defenses targeting pathogens and surface colonizing microbes. Conversely, chemical cues may be used to attract preferred microbial associates. The challenging task awaiting the microbial chemical ecologist is to determine if and to what extent marine natural products affect co-occurring and potentially deleterious microorganisms.

We can speculate that the potential for marine natural products to mediate microbial behavior is great. Marine plants and animals are a rich source of structurally novel secondary metabolites (Faulkner 1995 and references cited within), many of which display potent biological activities in pharmacological assays. These activities suggest a potential function in chemical defense, however, at present a correlation between pharmacological activity and ecological function has not been established (e.g., Bakus et al. 1985, Davis et al. 1989, Wahl et al. 1994). For the ecological significance of marine natural products to be elucidated at the microbial level, it is necessary to test these substances at realistic concentrations against microorganisms that have the potential to contact the producing organism in nature. Although both of these objectives represent significant challenges and are discussed in greater detail later, they are being met through the development of new bioassays and the adaptation of already existing assays to marine situations.

The purpose of this chapter is to provide an introduction to the types of bioassays that have been used to study the effects of marine natural products on marine microorganisms. Bioassays that test one living organism against another may detect activities that are not due to natural products chemistry, and therefore these assays are discussed to a limited extent. This chapter is not meant to serve as a review of the chemical ecology of marine microorganisms. Instead, our aim is to describe basic bioassay methods that have been used to monitor the responses of marine microorganisms to marine natural products and to discuss factors that must be taken into consideration when interpreting the ecological significance of assay results. We hope this review will assist researchers in choosing the type of bioassay that best suits their experimental objectives and encourage the continued development of new methodologies that will enhance our understanding of the chemical ecology of marine microorganisms. For the purpose of this review, microorganisms include viruses, bacteria (including cyanobacteria), fungi, microalgae, protozoa, and gametes. Marine invertebrate larvae are covered in chapter 2. Although to the best of our knowledge, no reviews of the chemical ecology of marine microbes have been published, the ecological roles of marine natural products have been reviewed (Paul 1992) and general texts on marine microbiology (Austin 1988; Rheinheimer 1992) and microbial ecology (Wood 1965) are available. To provide a concise and manageable amount of material in this chapter, subjects such as environmental toxicology and the development of commercial antifoulants have been excluded.

1.2. Ecological Relevance of Bioassays

In this review, we repeatedly mention the concept of ecological relevance in relation to bioassays. This warrants explanation, as the ability to derive ecological meaning from bioassay results is the foundation on which this chapter is written. There are two basic components to ecological relevance. The first is the selection of the microorganism(s) used in the bioassay. The effects of a marine natural product on a microorganism have more ecological significance if the microorganism is one that interacts with the producing organism in nature. The significance of these effects is further enhanced if there is an ecological basis for the producing organism either to attract or defend against a particular microbe, e.g., there is ecological rationale to attract a symbiont and deter a pathogen. In designing an effective bioassay, it is crucial that the microorganisms selected be given careful consideration. Unfortunately, and in most cases, we have insufficient information about the microbiology of the plants and invertebrates being studied to be certain that the best test microbe has been selected. In such cases, an educated best approximation must be made. If we are interested in examining the potential role of a natural product in antibacterial chemical defense, we might select as test microbes known marine pathogens (even if we do not know that they cause disease in the producing organism) and microbes isolated from the surrounding seawater (which have the potential to come in contact with the species being studied). Designing experiments that can be performed *in situ* and thereby testing against natural assemblages of microorganisms is an ideal but often experimentally more difficult method to obtain ecologically relevant results.

The second important variable that must be taken into consideration when interpreting the ecological relevance of bioassay results is the concentration at which an extract or purified material is tested (see Fenical & Pawlik 1991). For bioassay results to have ecological relevance, an attempt must be made to reproduce the concentrations potentially experienced by microbes in nature. It is relatively simple to estimate an appropriate test concentration when the organism receiving the chemical signal is a predator such as a fish and the chemical composition of its prey can be calculated. When a microorganism is being tested, however, it is difficult to estimate ecologically relevant concentrations as we do not know the microscale localization of metabolites and whether these substances are released into the surrounding seawater. Concerning the latter, at least two studies have demonstrated that secondary metabolites can be released from marine invertebrates into the surrounding seawater (Thompson 1985; Coll et al. 1982).

Because it is extraordinarily difficult to estimate the metabolite concentrations experienced by microbes in nature, it is often necessary to test extracts at a standard concentration. In the absence of more specific information, a logical standard is what can be considered the natural or whole tissue concentration. Natural concentrations are best approximated using a volumetric method (i.e., measure the volume of tissue extracted, determine the weight of extract obtained from that volume, and divide the extract weight by the volume extracted). Regardless of how the test concentration is calculated, it is fundamentally important for all experiments that the method of extraction, the concentrations at which materials are tested, and how these concentrations relate to those obtained from the organism collected be reported. We know that marine natural products can have significant ecological effects on invertebrate larvae and macropredators such as fish (see Paul 1992), and it is now in the hands of the microbial ecologist to expand our understanding of the ecological effects of these substances on marine microorganisms.

1.3. Antimicrobial Assays

Antibiotics have traditionally been defined as chemical substances produced by microorganisms that have the capacity, in dilute solutions, to inhibit the growth of or destroy other microorganisms. A current and more practical operational definition would also include substances produced by macroorganisms (i.e., plants and invertebrates) that show activity against microorganisms. Antimicrobial assays have for the most part been designed to test crude mixtures or purified substances for the presence of antibiotics. Antimicrobial activities detected by these assays, however, are not always due to what would traditionally be considered an antibiotic compound. For example, many substances become toxic when tested at high concentrations; consequently, antibiotic activities must be considered in the context of test concentration. Such nontraditional antibiotics should not be customarily ignored in the context of ecological investigation. Since we know so little about how marine organisms use chemical substances to regulate associated microbial populations, it remains possible that certain substances produced in high concentrations have ecologically significant effects on co-occurring microorganisms.

Antimicrobial assays have long been used by the pharmaceutical industry to screen the extracts of plants and microbial fermentations for the presence of organic metabolites that inhibit the growth of medically important microorganisms. It is logical, therefore, that these same biomedically oriented methods were applied when researchers expanded their search for new drugs to include plants, animals, and microorganisms from the sea. The expansion of biomedical studies to marine organisms has been highly productive, and it is now widely accepted that the marine environment is a rich source of novel organic metabolites possessing biomedically relevant antimicrobial activities (Attaway & Zaborsky 1993). But since there is no evidence that biomedically relevant antimicrobial activities are correlated to ecological function, few substantiated conclusions have been drawn about the ecological roles of antibiotic metabolites. This subject has received increasing interest in recent years, and as a result, standard pharmacological assays have been modified, and new methods developed, so that marine natural products can be tested against ecologically relevant marine microorganisms.

Antimicrobial assays represent a valuable method by which marine plant, animal, and microbial extracts can be tested for inhibitory effects against co-occurring and potentially deleterious microbes (e.g., pathogens, competitors, or fouling microbes). The ability of secondary metabolites to inhibit ecologically relevant microbes can then be used to assess the potential function of marine natural products in antimicrobial chemical defense, a subject for which we presently have essentially no knowledge. Although experimental data are lacking, a solid ecological rationale for the function of secondary metabolites in antimicrobial chemical defense comes from the well-documented occurrence of marine microbial pathogens. Marine plants and animals are susceptible to infection (Austin 1988; Sparks 1985) and, since they lack cell-based immune responses, antimicrobial secondary metabolites could provide an effective mechanism to combat infection or provide antimicrobial prophylaxis. It should be possible, through the use of antimicrobial assays, to detect constitutive antimicrobial chemical defenses that function via antibiotic activity. It should also be possible to use these assays to determine if marine extracts and purified materials are toxic to potential fouling microorganisms (see review by Clare 1996).

The following section describes various methods that have been used to test marine natural products for antimicrobial activity. These assays typically measure toxicity and/or growth inhibition. Antimicrobial assays are relatively easy to perform and are often characterized by simple design and implementation. Although the detection of antimicrobial activity is generally a straightforward process, the ecological interpretation of the results is complex. Generally, microbes are exposed to test compounds either by diffusion of the material in agar seeded with the test microbe or addition of the material to a liquid culture. For the effects of this exposure to have ecological relevance, there must be some evidence that microbes are exposed to similar chemical concentrations in nature. Since it is difficult to discern the chemical concentrations experienced by microbes in nature, ecological interpretations of antimicrobial activities must be made with caution. The antimicrobial assays that are described represent standard methods that, with simple modification, can be applied to a variety of marine microorganisms. These assays should be viewed with an open mind and modified as needed to meet specific experimental objectives. Although assays that have been used only in terrestrial studies have not been included, some of these methods, with minor modifications, could be applied to marine microorganisms.

1.3.1. Direct Challenge/Competitive Interaction Assays

It has been proposed that antibacterial compounds play an important part in the competitive interactions among terrestrial bacteria (Fredrickson & Stephanopoulos 1981). Given that marine bacteria have long been known to produce antimicrobial substances (Rosenfield & Zobell 1947), it seems appropriate to investigate the ecological role of these substances in the ocean. Most of the early work with marine bacteria tested for effects against terrestrial and biomedically important microbes and therefore is not discussed. One of the first studies to include marine microorganisms in the search for antibiotic activity examined the ability of marine bacteria to inhibit the growth of marine yeasts using a direct challenge assay (Buck et al. 1962). In this method, the test yeast was first inoculated onto the surface of an agar medium. This was followed by the addition of a sterile cellulose disk onto which one or two drops of a 48-h unfiltered bacterial culture was added. The marine bacteria formed a colony on the disk and antifungal activity was recorded as zones of inhibited growth around the disk. Although direct challenge assays are simple to perform, the activities detected are often not reproduced in cell-free culture filtrates. For example, active cell-free broths were not produced by any of the bacteria that exhibited antifungal activity in a strain-versus-strain direct challenge assay (Buck et al. 1962). These results suggest that some activities detected by this method are not due to the production of extracellular antibiotics but instead may be due to other factors such as competition for growth-limiting nutrients.

A direct challenge assay was employed by Lemos et al. (1991) to demonstrate that antibiotic-producing marine bacteria dominated non-antibiotic-producing strains when grown in mixed liquid culture. The rapid extinction of non-antibiotic-producing strains combined with growth rates of control cultures led these researchers to conclude that antibiotic production, not nutrient depletion, was responsible for the observed growth inhibition. These experiments showed a high degree of ecological relevance in that bacterial competition was tested in filtered and autoclaved unammended seawater. Individual strains were grown and adjusted to an equal optical density, and then equal volumes of each strain inoculated into the same flask. Growth of the individual strains was monitored in mixed culture over time by colony counts (this method requires that different strains have distinguishable colony characteristics). Competitive interaction assays have also been used to examine the role of toxic metabolites in competition among microalgae (Pratt 1966; Rijstenbil 1989) and marine fungi (Miller et al. 1985).

1.3.2. Antibiotic Disk Susceptibility Assays

The antibiotic disk susceptibility test or agar disk-diffusion assay (Acar 1980), as it is often termed, has been a mainstay of the modern pharmaceutical industry and is probably the simplest and most common method used to test for antimicrobial activity. In this method, a cell-free culture broth, culture extract, or purified compound is applied in solution to a small paper disk (usually 6–7 mm diameter). The disk is allowed to dry, it is placed on an agar plate that has been seeded with a test microorganism, and the plate is incubated at conditions appropriate for the microbe's growth. This assay relies on the diffusion of the test material through the agar where it contacts the test microbe and, if active, produces a zone of inhibited microbial growth (clear zone) around the disk (Fig. 1.1). Disk diffusion assays are simple to run, requiring only small amounts of material and no complex equipment. Their ability to rapidly identify active components makes them especially useful in the initial screening for antimicrobial activities and as a means of following activity during chemical purification.

To prepare for this assay, the medium is typically inoculated either by adding

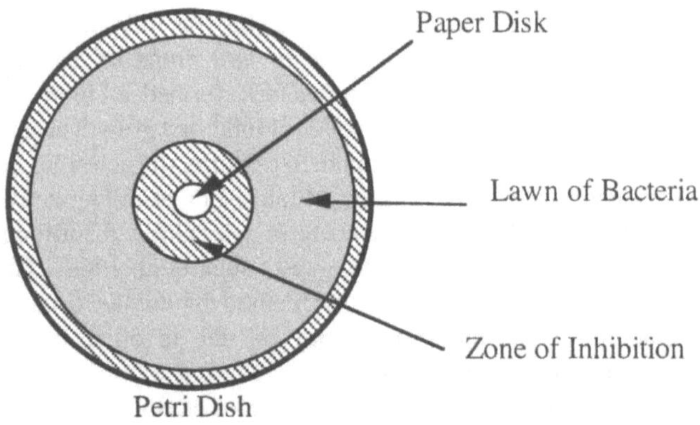

Figure 1.1. Representation of a typical disk susceptibility assay. A paper disk impregnated with a suspected antibiotic compound is applied to an agar plate seeded with a test microorganism and the area of inhibited microbial growth is indicated by a cleared zone surrounding the disk.

a liquid microbial culture to molten agar or by directly applying a culture to the surface of a solidified agar plate with a sterile swab or glass spreader. The objective of inoculation is to obtain sufficient and reproducible microbial growth, so that following incubation, the plate becomes turbid except for any area around the disk where growth is inhibited. Microbes that tend not to work well in the disk-diffusion assay are those that grow as clumps and therefore cannot be evenly seeded throughout or on top of the agar medium (e.g., filamentous fungi). The size and age of the inoculum, incubation temperature, growth medium, medium volume, and inoculation method must be modified depending on the microorganism tested. Once appropriate conditions have been selected, reproducibility will be maximized if a standard protocol is followed and solvent controls and standard antibiotic disks are run for each experiment. Antibiotic activities are generally reported as the diameter of the zone of inhibited microbial growth around a disk. The zone may not always be clearly demarcated due to various factors such as low potency and the presence of antibiotic-inactivating enzymes. For these reasons, the most complete and homogeneous zone of no growth should be measured to the nearest millimeter using a consistent method.

The size of zones of inhibition are dependent on a number of factors other than antibiotic potency including the concentration tested, the incubation temperature, the rate of compound diffusion, and the physical characteristics of the agar (Barry 1980). Some of these variables can be accounted for by ensuring that a standard antibiotic disk produces a consistent zone of inhibition. Although measuring zones of inhibited microbial growth is relatively simple, any ecological interpretation of the results must be made with caution. For example, the detection of antibiotic activity in an invertebrate extract is not sufficient evidence for the existence of an antimicrobial chemical defense. Activity does, however, suggest that an antimicrobial chemical defense may be present and that additional study is warranted. Likewise, the absence of antimicrobial activity is not conclusive evidence for the lack of antimicrobial chemical defense in that organism. Chemical defenses may be inducible or function by mechanisms other than toxicity, and therefore are not detected with the disk-diffusion assay.

Another reason for careful interpretation of the disk-diffusion assay is that it is subject to false-positive results that can be misinterpreted as antibiotic activity. For example, physical characteristics of the extract (viscosity, pH, etc.) can generate small zones of growth inhibition when bacteria are inoculated directly onto the surface of the agar plate. In addition, we have observed that some primary metabolites can inhibit growth when tested at high concentrations. It is also possible that simple molecules, or extract degradation products, can exhibit mild antibiotic properties. For these reasons, it is important that replicate extracts are tested and that small zones of inhibition are interpreted with caution. It is also important to clearly state the concentrations tested, even if naturally occurring concentrations are not known, so that activities can be reproduced and evaluated at a later time.

The disk-diffusion assay has been modified on numerous occasions for testing marine bacteria (e.g., Jensen et al. 1996; Kim 1994), yeasts (Amade et al. 1987; McClintock & Gauthier 1992), and microalgae (Chan et al. 1980), with strain selection depending on experimental objectives. Although there are currently no reports in the literature, zoosporic marine fungi such as thraustochytrids are also amenable to this method. Two of the earliest studies employing modifications of the disk-diffusion assay tested marine bacteria for sensitivity to sponges (Burkholder & Ruetzler 1969) and gorgonians (Burkholder & Burkholder 1958). These original studies were used in a descriptive sense to evaluate trends in antibacterial activity and did not include characterization of the active compounds. Subsequently, disk-diffusion assays were used for the isolation of novel marine natural products as well as for surveys of marine invertebrates (e.g., Amade et al. 1987) for activity against marine bacteria. In a few cases, bacteria were isolated from the same environment as the organisms being studied. For example, McCaffrey and Endean (1985) and Walls et al. (1993) examined antibacterial activity against microbes isolated from the surrounding seawater, and Becerro et al. (1994) used bacteria isolated from sediment. Other studies have been performed using bacteria isolated directly from the surface tissues of the invertebrates under study (Wahl et al. 1994) or using known marine invertebrate pathogens (Jensen et al. 1996).

The choice of assay organism is perhaps one of the most difficult aspects of ecologically motivated studies of antibiosis. This is primarily due to the fact that most marine microbes are not readily cultured by the current techniques. For example, one study has shown that the number of bacteria obtained by plating techniques can be three orders of magnitude lower than the actual number of bacteria as evidenced by microscopy (Simidu et al. 1983). Thus, when making statements about the ecological role of antibiotic activity in the marine environment, it must be acknowledged that the assay organism chosen may represent only a small subsample of the microbial community. Continued gains in our understanding of the culturability and composition of marine microbial communities will prove to be an asset to microbial chemical ecology.

In addition to marine bacteria, microalgae have also been used in disk-diffusion assays. This has especially been the case when the antifouling properties of marine natural products have been addressed (see review by Clare 1996), as microalgae can significantly contribute to marine fouling (Wahl 1989) and antifouling chemical defenses targeting microalgae have been proposed (Targett et al. 1983; Slattery et al. 1995). Disk-diffusion assays have also been used to examine the effects of extracts from the eelgrass *Zostera marina* on the growth of eight species of microalgae (Harrison & Chan 1980). In a separate study, the chemical basis for competition among marine microalgae was addressed (Chan et al. 1980) by studying the effects of microalgal extracts on the growth of the pennate diatom *Cylindrotheca fusiformis*. In this study, diatoms in late exponential or early stationary phase were plated onto an appropriate agar medium by placing 1×10^5 cells in the middle of a plate and then spreading them evenly across the surface with a sterile glass spreader. A paper disk impregnated with the extract or compound to be tested was placed in the center of the plate and the plates incubated upside down at 15–18°C with a total irradiance of 40 $\mu E \cdot m^{-2} \cdot s^{-1}$ provided by daylight fluorescent lamps. The authors acknowledged that the ecological significance of the assays should be interpreted carefully because: (1) pennate diatoms take up organic substances more readily than other marine microalgae and thus may be more susceptible to the organic compounds added and (2) the concentrations tested in the assay were likely to be abnormally high and thus do not represent ecologically relevant amounts of material.

1.3.3. Bioautography

While the disk-diffusion assay is the most prevalent method of testing extracts for antimicrobial activity, bioautography is useful as a follow-up method for the rapid recognition of the active component(s) in a crude mixture. Bioautography is performed by directly exposing a culture of the microorganism to a chromatogram of the extract (see review by Betina 1973). Bioautography begins with the separation of a crude extract on a thin-layer chromatography (TLC) plate. The solvent system must be tested before the assay is run to ensure adequate separation of the individual extract components (see Stahl 1969). TLC plates are then dried thoroughly to remove residual solvents and exposed to the test microorganism. Methods of exposure include transferring the compounds to a piece of moistened filter paper by placing it in contact with the chromatogram for 20 min, allowing the paper to dry, and then placing it in contact with an agar plate that has been inoculated with the test microbe (Bergquist & Bedford 1978). Alternatively (and more commonly), the TLC plate is placed directly on the inoculated agar for approximately 10 min, after which the chromatogram is removed and the plate is incubated at conditions sufficient for the development of confluent microbial growth (Walls et al. 1993). Zones of inhibited growth on the agar plate can then be compared to a second control chromatogram that has been developed with an appropriate indicator solution. While bioautography can aid in the identification of the active compound in a mixture, the application of this technique is limited to some degree in ecological studies because the concentration of the active metabolite is unknown. Thus, this technique is best used as a means of localizing the activity in an extract to a particular compound. Once the active compound is isolated, it can then by analyzed by a method such as the disk-diffusion assay in which the metabolite concentration can be more readily controlled.

1.3.4. Direct Algal Cell Count Assays

The antimicrobial activities of marine natural products have been tested against marine microalgae using direct cell count methods. For example, the muricins, a group of steroidal aminoglycosides isolated from the gorgonian *Muricea fruticosa*, have been shown to inhibit the growth of the marine diatom *Phaeodactylum tricornutum* (Bandurraga & Fenical 1985). In this study, 5 μl of a diatom suspension was used to inoculate tubes containing a seawater-based algal growth medium to which 100 μg of compound dissolved in 2.5 μl of ethanol had been added. Treatment and control cultures were grown for 4 days at 24°C, after which cell densities were estimated by direct cell counts using a hemacytometer and microscope. At 100 ppm, the muricins inhibited diatom growth by 60–80% relative to controls.

Slattery et al. (1995) used a similar assay to study the effects of soft coral extracts on the growth of the Antarctic benthic diatom *Navicula* sp. In this study, 20 g of freeze-dried soft coral tissue were extracted and the extract was dried and weighed. The extract was then tested at tissue-level concentration, although how these levels were calculated was not thoroughly discussed. The chloroform and aqueous methanol extracts of two of the Antarctic soft corals were found to significantly inhibit diatom growth at concentrations at or below those obtained from the animal tissues. Although the results of both of these assays were discussed in terms of potential antifouling properties, they measure microalgal toxicity and the correlation between toxicity and antifouling activity has not been established. Slattery et al. (1995) further investigated this subject using field settlement assays against a natural assemblage of fouling microorganisms. Field assays can provide a more accurate prediction of antifouling activity and are discussed in a later section.

1.3.5. Spectrophotometric Assays

In addition to the agar plate disk-diffusion assay, spectrophotometric methods have frequently been used to measure microbial growth inhibition. Spectrophotometric methods generally require that the test microorganism be grown in liquid culture. Growth of the culture is then monitored by changes in percent light transmittance or optical density. Microbial cultures are treated with chemical extracts and the growth of treatment relative to control cultures (solvent only) compared. These assays are useful for the rapid examination of microbial growth responses to chemical extracts and for bioassay-guided separation of active materials. The major disadvantage of this method is that the materials tested must be water soluble.

Spectrophotometric methods can readily be applied to a variety of marine microorganisms including bacteria, unicellular fungi, and microalgae. Strains that tend not to perform well in these assays, as with the disk-diffusion assay, are those that do not grow uniformly in liquid culture. Recently, spectrophotometric methods were used by Azumi et al. (1990) to demonstrate that halocyamine A, isolated from the ascidian *Halocynthia roretzi*, had an antimicrobial effect against the marine bacteria *Achromobacter aquamarinus* and *Pseudomonas perfectomarinus*. This study was notable in that the active compound was isolated from the hemocytes (blood cells) of the ascidian, and that these cells have been proposed to play an important role in defense.

Spectrophotometric methods have also been used to measure the growth of microalgae exposed to chemical extracts. McLachlan & Craigie (1964) examined the antialgal effects of the extracellular exudate of the macroalga *Fucus vesiculosus*. Cultures of seven different genera of marine microalgae were grown in an enriched seawater medium to which the exudate of *F. vesiculosus* was added. Microalgal growth was estimated by changes in absorbance and compared to untreated control cultures. In this study, the exudate of *F. vesiculosus* was found to inhibit the growth of all microalgae examined, however, there was considerable between-strain variation in sensitivity. Additional controls for these types of experiments should include examining the absorbance properties of the extract itself and thus perform adequate controls for the absorbance of highly colored extracts or extracts that decompose to give colored by-products.

Spectrophotometric methods were used in a more indirect manner by Targett et al. (1983) to examine the effects of gorgonian metabolites on the growth of the benthic marine diatom *Navicula salinicola*. In this study, compounds or extracts were dissolved in 1.0 ml of distilled water, mixed with 15 ml of a molten agar medium, and poured into petri plates. Following solidification, the plates were inoculated with *N. salinicola*, incubated for 8 days, and then extracted with 90% aqueous acetone. Growth was estimated by measuring extract chlorophyll *a* fluorescence. Using this method, extracts of the gorgonians *Leptogorgia setacea* and *L. virgulata* both inhibited diatom growth at 100 μg/ml and 500 μg/ml, respectively. The pure compound homarine, isolated from both gorgonians, inhibited the growth of *N. salinicola* at concentrations at or below those found in the gorgonian tissue. Similar experiments have been performed for the Caribbean gorgonians *Pseudopterogorgia americana* and *P. acerosa* (Targett 1988) and most recently for sponges and holothurians (Mokashe et al. 1994).

1.3.6. Viral Plaque Formation Assay

The importance of viruses in the marine environment has only recently been recognized. Viruses are involved in structuring phytoplankton communities and represent important effectors of bacterial mortality (Proctor & Fuhrman 1990; Suttle et al. 1991). Viruses are also associated with diseases in marine vertebrates and invertebrates (Austin 1988 and references therein). These recent discoveries indicate that viruses represent a potentially serious threat to the health of marine macro- and microorganisms and provide a clear ecological rationale for the maintenance of antiviral chemical defenses. The isolation of antiviral metabolites from marine invertebrates (Rinehart et al. 1981) and marine bacteria (Myouga et al. 1995) provides preliminary evidence that such defenses are maintained, however, additional studies are warranted.

One approach to examine the potential ecological function of antiviral marine natural products is to test for effects against marine viruses, or more specifically, viruses that naturally infect the producing organism. Although to the best of our knowledge, the later has never been attempted (this represents an important area for future research), there have been a few cases where marine viruses were tested for sensitivity to a marine natural product. One such case is that of the secondary metabolite halocyamine A, isolated from the ascidian *Halocynthia roretzi* (Azumi et al. 1990). This compound, which has already been discussed for its antibacterial activity, was tested in a whole cell plaque-formation assay against two fish RNA viruses using a method based on that of Kamei et al. (1987). In this assay, uninfected trout ovary cells were added to the wells of a microtiter plate and incubated at 15°C for 24 h. Following incubation, the medium was removed and 0.1 ml of a virus particle solution was added and incubated with the cells for 1 h. The wells were then washed twice with Hank's balanced salt solution (Gibco) and halocyamine A, dissolved in dimethylsulfoxide, was added to each well. The cells were then resuspended in culture medium and incubated at 15°C for an additional 7 days. The number of plaques formed was determined after fixation with 10% formalin and staining with 0.1% crystal violet. Halocyamine A was shown to inhibit the growth of both fish viruses, leading the authors to propose that halocyamine A may act as a defensive compound to inhibit viral infection in the ascidian *H. roretzi*.

1.3.7. Agar Plate Assay for Filamentous Fungi

Filamentous marine fungi are heterotrophic eukaryotes that for the most part function nutritionally as parasites or saprophytes. Marine fungi are well known to cause disease in marine algae, higher plants, invertebrates, and vertebrates (Kohlmeyer & Kohlmeyer 1979; Goff & Glasgow 1980; Austin 1988), and therefore there is an ecological rationale for susceptible organisms to maintain antifungal chemical defenses. Although secondary metabolites with antifungal activity have been isolated from marine plants and invertebrates, the ecological role of these substances in antifungal chemical defense remains virtually unknown. As with antimicrobial chemical defenses in general, this absence of ecological information is largely due to a lack of experimental methods by which such interactions can be documented.

Standard antibiotic disk-diffusion assays, as already described in a previous section, are often not readily amenable to filamentous fungi due to difficulties in obtaining uniform growth on the agar plate. In some cases, depending on the species tested, fungal spores can be collected and used to seed plates as per assays using unicellular microorganisms such as bacteria. This inoculation method generally results in a uniform lawn of fungal growth within which clear zones of antibiotic activity surrounding a disk can be readily visualized. Some strains, however, may not produce spores and an alternative method is to inoculate the test plate with an agar block transferred from a stock culture. The fungal mycelium will then grow radially out from the block and toward extract impregnated paper disks that have been placed around the inoculum. Antifungal extracts will inhibit growth of the spreading mycelium around the disk and activities can be recorded as the diameter of zones of inhibition.

An alternative antifungal assay was developed by Thompson et al. (1985) to test sponge extracts for activity against the filamentous marine fungi *Lindra thalassiae*, *Lulworthia* sp., and *Alternaria* sp. In this assay, compounds or extracts were incorporated into a glucose-yeast extract medium and added to plastic, multiwell trays at a final concentration of 10 μg/ml. Liquid fungal cultures were then added to each well and the trays were sealed with plastic wrap and incubated for 7 days at room temperature. The presence or absence of fungal growth was determined visually for each well relative to a control. The concentrations used in this assay were two to three orders of magnitude lower than the natural concentrations estimated to occur in the sponge tissues. The conservative concentrations at which extracts were tested and the fact that only substances that completely inhibited growth were scored as active may account for the detection of antifungal activity in only one compound (heteronemin). The study by Thompson et al. (1985) was one of the first to use marine fungi to investigate antifungal activity in marine organisms and to take into consideration the concentrations at which the compounds occur in the producing organisms. For these reasons, this study represents a significant advancement in the evaluation of ecologically relevant antifungal chemical defenses.

1.3.8. Fungal Interference Competition Assay

Interference competition between fungi can include antibiotic production as a method of excluding competitors from available resources, a subject investigated for marine fungi by Strongman et al. (1987). In this study, the marine fungus *Monodictys pelagica* was grown in 1-ml cultures containing a medium composed of glucose (5 g/L), yeast extract (0.5 g/L), and 28% artificial seawater inoculated with 0.5 mg of fungal mycelium. Chemical extracts of 27 marine fungi were dissolved in 80% ethanol/water and aliquots of each extract (10 μl) were added to the tubes containing *M. pelagica* to give a final concentration of 1–2 mg/ml. The tubes were then incubated for 10 days at 25°C. If an extract caused visible growth inhibition, the fungal mycelium was filtered (0.45 μm), washed with distilled water, dried, and weighed. Comparisons were then made between the dry weights of treatments, solvent controls, and nystatin antibiotic controls (1 mg/ml). Active extracts were subjected to bioassay-guided chemical purification. This led to the identification of the antifungal compound culmorin from the marine fungus *Leptosphaeria oraemaris*. In addition, the fungus *Arenariomyces trifurcatus* produced an active fraction that contained a mixture of unidentified triglycerides. The study by Strongman et al. (1987) was the first to fully characterize a marine fungal natural product in an ecological context. Due to the length of time required to perform this assay (10 days), it is not ideally suited for rapid bioassay guided fractionation.

1.3.9. Associational Defense

Antimicrobial chemical defenses can be extraordinarily complex. Not only can macroorganisms produce secondary metabolites targeting microbes, but microbial symbionts can produce antimicrobial metabolites that protect the host from infection by other microorganisms. Although few examples of marine antimicrobial chemical defense have been documented, an interesting associational chemical defense has been reported (Gil-Turnes et al. 1989). In this example, the antifungal compound 2,3-indolinedione (isatin) was isolated from the organic extract of a bacterial culture obtained from the embryos of the crustacean *Palaemon macrodactylus*. An agar plate disk-diffusion assay with the crustacean pathogen *Lagenidium callinectes* was used to isolate this compound.

Following isolation, experiments were performed to demonstrate that isatin chemically defends the embryos against infection by the pathogenic fungus *L. callinectes*. In these experiments, three groups of embryos were treated with penicillin-G to remove the isatin-producing bacteria and a fourth control group was left untreated. The three treated groups were then either (1) dipped in isatin and exposed to *L. callinectes*, (2) reinfected with the associated isatin-producing bacterium and exposed to *L. callinectes*, or (3) exposed directly to *L. callinectes*. The control group (not treated with penicillin-G) was also exposed to *L. callinectes*. Microscopic examination of the embryos following *L. callinectes* exposure revealed that 60% of the bacteria-free (penicillin-treated) embryos exposed to either 2,3-indolinedione or reinfected with the associated bacterium survived, while only 5% of the untreated bacteria-free embryos survived. The control group had an 80% survival rate. Based on these results, Gil-Turnes et al. (1989) proposed that the associated bacterium (*Alteromonas* sp.) chemically defends the embryos of *P. macrodactylus* from the pathogenic fungus *L. callinectes* by production of the antifungal compound 2,3-indolinedione.

1.4. Behavioral Assays

Behavioral bioassays are those in which the nonphysiological responses of microorganisms to marine natural products or extracts are monitored. The types of responses examined depend on experimental objectives and include chemotaxis, effects on motility, and attachment to a surface. Behavioral assays can be used to detect positive and negative effects, general toxicity, and microbial responses to nontoxic chemical signals, making these types of assays more sensitive and informative than the toxicity assays discussed in the previous sections. Behavioral assays are often more difficult to design, implement, and quantify than toxicity assays, and this may be one reason why they have not been extensively used in marine chemical ecology. Bioassays designed to help elucidate the behavioral effects of marine natural products on ecologically relevant microorganisms are discussed below and can provide important insight into such ecological phenomena as fouling, reproduction, and symbiosis.

1.4.1. Chemotaxis

Bioassays designed to measure microbial chemotactic responses are among the most extensively used behavioral assays. Chemotaxis in terrestrial and aquatic microorganisms has been reviewed by Chet & Mitchell (1976) and implicated in bacterial predator-prey relationships, symbioses, pathogenic associations, and associations with both corals and algae. Although to a lesser extent, chemotaxis has also been examined in marine fungi (Muehlstein et al. 1988), cyanobacteria (Willey & Waterbury 1989), and microalgae (Sjoblad et al. 1978; Cooksey & Cooksey 1988). Chemotaxis assays have generally been used to study the responses of microbes to nutrients and inorganic salts, and therefore our understanding of the ecological significance of chemotaxis to secondary metabolites is limited. These assays have been included in this review, however, since they could be readily applied to studies of natural products. In the context of chemical defense, one significant piece of ecological information that has come from the study of chemotaxis is that microorganisms can display a negative chemotactic response to sublethal chemical concentrations, indicating that extracts or purified materials that have important ecological effects on microorganisms may in some cases not be detected in assays that measure only toxicity. As with all assays, the ecological interpretation of chemotaxis assay results is challenging and attempts have been made to develop more realistic assay methods to show that chemotaxis is an effective method by which aquatic microbes can orient to a substance in the presence of realistic flow conditions (Walsh & Mitchell 1978).

1.4.1.1. Capillary Assay

Most studies of bacterial chemotaxis have used some form of the capillary tube assay originally described by Pfeffer (1884) and later modified by Adler (1973). This method, often referred to as the Adler method, measures the movement of bacteria into a capillary that has been filled with a test chemical. While the original method examined chemotaxis in *Escherichia coli*, this method, with minor modifications, has proven effective for measuring chemotaxis in marine microorganisms.

In a generalized Adler method, one end of a capillary tube (1 μl disposable micropipette, 3 cm long with an internal diameter of 0.2 mm) is flame sealed. The entire capillary tube is then quickly passed through a flame, and while warm, the open end is plunged into a reservoir containing the test chemical dissolved in chemotaxis medium. The liquid is drawn up into the capillary as it cools and the filled capillary is then withdrawn from the reservoir and inserted into a chemotaxis chamber, which is constructed by placing a U-shaped melting point capillary tube between a microscope slide and a coverslip (Fig. 1.2). The chamber is filled with an appropriate chemotaxis medium and inoculated with bacteria so that the final concentration is approximately 6×10^7 cells/ml. After a 1-h incubation, the capillary is removed from the chamber and the exterior rinsed with sterile water. The sealed end of the capillary is then broken and the contents are

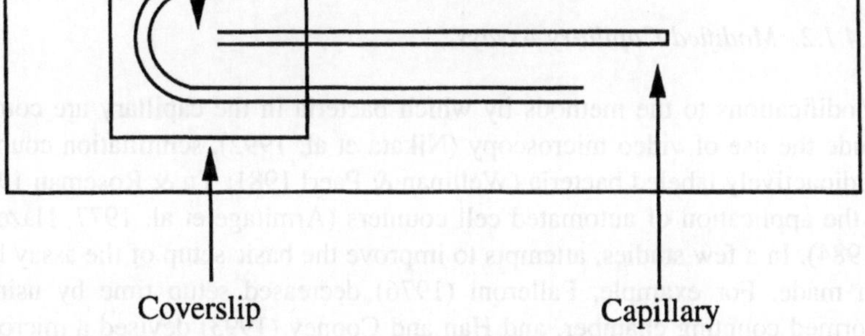

Figure 1.2. Diagram of a typical apparatus used in bacterial chemotaxis assays. The U-tube and cover glass form a chamber into which a bacterial suspension is added. Capillaries containing potential chemoattractants are then immersed into the chamber and the number of bacteria that move into the capillary are counted relative to a control. Modified from Adler (1973).

dispensed into a test tube containing a medium suitable for the growth of the test microbe. The inoculated broth is then diluted and plated onto a suitable agar medium, and following incubation, colonies are counted and treatment and controls are compared. Alternatively, bacteria in the capillary tube can be directly counted with a microscope.

The advantages of the capillary assay are its simplicity, quantitative nature, and high sensitivity. Alternative methods for studying chemotaxis such as the swarm plate method of Adler (1966) require that the chemoattractant be metabolized. This is not necessary in the standard capillary assay. In addition, due to the small size of the chemotaxis chamber, only small amounts of compound are required to perform the experiments. The main disadvantage of this method is that the compound tested must be soluble in the chemotaxis medium.

The Adler capillary assay has been used with marine bacteria to demonstrate chemotactic responses to a variety of extracellular products and crude extracts. Chet et al. (1971) showed a chemotactic response in predatory marine bacteria (*Pseudomonas* spp.) to exudates of their respective prey organisms. Bell & Mitchell (1972) demonstrated that marine bacteria are chemotactic to algal culture filtrates and proposed the existence of a "phycosphere" similar in ecological characteristics to the rhizosphere of terrestrial ecosystems. Furthermore, *Vibrio alginolyticus* has been shown to exhibit positive chemotaxis to mucus obtained from the surface of the coral on which it grows (Ducklow & Mitchell 1979). It should be noted, however, that most studies examining bacterial chemotaxis have focused on water-soluble nutrients (see review in Chet & Mitchell 1976) and not bioactive secondary metabolites. In this context, it has been observed that there is no correlation between the metabolism of a substance and its ability to induce a positive chemotactic response in bacteria (Weibull 1960), indicating that nonmetabolizable secondary metabolites may have significant effects on bacterial chemotactic behavior.

1.4.1.2. Modified Capillary Assays

Modifications to the methods by which bacteria in the capillary are counted include the use of video microscopy (Nikata et al. 1992), scintillation counting of radioactively labeled bacteria (Wellman & Paerl 1981; Yu & Roseman 1993), and the application of automated cell counters (Armitage et al. 1977; Hazen et al. 1984). In a few studies, attempts to improve the basic setup of the assay have been made. For example, Palleroni (1976) decreased setup time by using a preformed counting chamber, and Han and Cooney (1993) devised a microtiter plate and lid to replace the original U-shaped chamber. The later authors examined chemotaxis in two species of marine bacteria, *Serratia* sp. and *Pseudomonas fluorescens*, by passing capillary tubes through holes in a microtiter plate lid and into wells of the plate (Fig. 1.3). One advantage of this method is that the 96-well plate allows for the simultaneous testing of large numbers of samples.

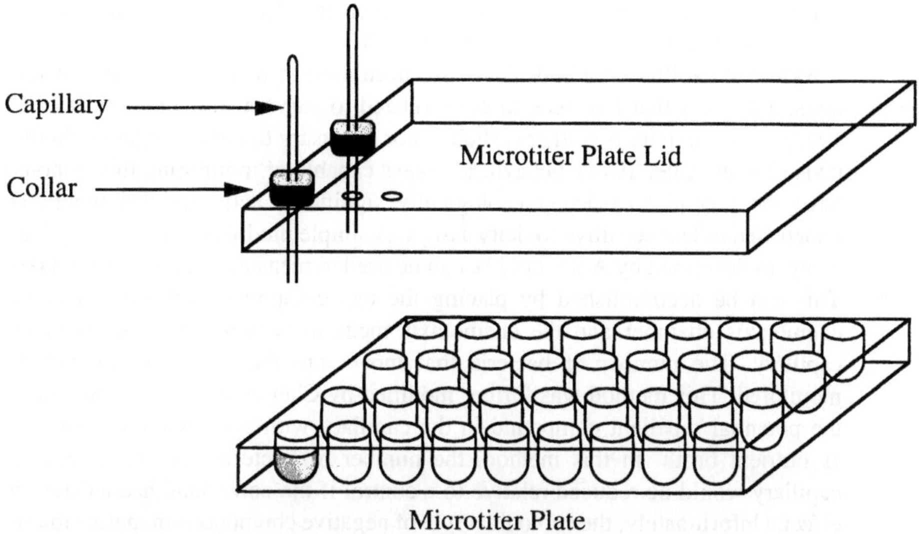

Figure 1.3. Diagram of a modified capillary assay. The wells in a microtiter plate are filled with a suspension of microbes and then capillaries containing a putative chemoattractant are immersed into the wells. The number of microorganisms that move into the capillary are then counted, indicating the chemotactic potential of the compound in question.

Although most chemotaxis experiments involved bacteria, a few studies have examined other marine microorganisms. One of the first studies to look at chemotaxis in marine microalgae was performed by Sjoblad et al. (1978). They examined the chemotactic response of the marine unicellular alga *Dunaliella tertiolecta* to varying concentrations of ammonium ion (NH_4^+). Essentially, the Adler method was used with modifications in the size of the glass capillary. Muehlstein et al. (1988) demonstrated a positive chemotactic response by zoospores of the marine fungus *Rhizophydium littoreum* to environmental compounds. A simple modification of the Adler method was used in this study (0.4-mm-diameter capillaries) to accommodate the larger size of the fungal zoospores. In addition to positive chemotaxis to some primary metabolites, zoospores also showed positive chemotactic responses to extracts of two algae: *Bryopsis* sp. and *Codium* sp. This study demonstrated the important observation that compounds supporting growth were not necessarily chemoattractant and chemoattractants did not always support growth. This is also the first study to examine chemotaxis of marine fungal zoospores. The role of chemotaxis in the chemical ecology of marine fungi is especially important because many of these microorganisms are pathogens. Fungal chemotaxis has been shown in terrestrial studies to be important in the attraction of pathogenic fungi to host plants (Chet & Mitchell 1976). However very little is known about the role of chemotaxis in the interactions between marine pathogenic fungi and their hosts, even though many such interactions have been described (Goff and Glasgow 1980).

Modified capillary methods have also been used to measure negative chemotaxis, a process that has been directly related to survival (Seymour & Doetsch 1973). Since toxicity is neither sufficient nor necessary to induce negative chemotaxis (Tso & Adler 1974), behavioral assays capable of monitoring this response have the potential to detect ecologically significant activities that would go undetected in less sensitive toxicity assays. A simple modification of the capillary assay, as described by Adler (1973), can be used to measure negative chemotaxis. This can be accomplished by placing the test compound and bacteria in the chemotaxis chamber and the chemotaxis medium (without compound) in the capillary. The number of bacteria that move into the capillary can then be monitored. This method was further modified by Chet et al. (1975), who placed the potential repellent compound in the capillary with a known attractant such as nutrient broth. In this method, the number of bacteria that move into the capillary would be reduced relative to a control if the compound had a deterrent effect. Unfortunately, the ecological role of negative chemotaxis in marine microorganisms is not well described, with most studies concentrating on antifouling research. For example, Mitchell (1975) examined negative chemotaxis in microalgae, however, these experiments were aimed at the commercial development of antifoulant products, and thus the materials tested were not necessarily ecologically relevant.

1.4.1.3. Chamber Method

Willey and Waterbury (1989) used a substantially different method to study chemotaxis in the marine cyanobacterium *Synechococcus* sp. This experiment involved the use of blind-well chemotaxis chambers (Neuroprobe, Inc., Cabin John, MD) that consisted of an upper (800-μl) and lower (200-μl) acrylic chamber separated by a polycarbonate filter (3.0 μm). A cell suspension (165 μl) of the cyanobacterium was placed in the lower chamber over which the polycarbonate filter was placed. An air space was left between the cell suspension and the filter to control the starting time of the experiment. The upper chamber was filled with sterile seawater containing the compound to be tested and then inverted, allowing the cell suspension to contact the polycarbonate filter and the seawater/compound solution. The experiments were run for 65 min, after which time the chambers were inverted to stop the experiment. The number of cells crossing the filter into the seawater chamber was determined by direct cell counts using epifluorescence microscopy. The motile strain of *Synechococcus* sp. tested in this assay elicited positive chemotaxis to compounds such as ammonia, nitrate, urea, glycine, and β-alanine. Control chambers with the same concentration of chemoattractant in both the upper and lower chambers failed to elicit a chemotactic response. While the compounds tested in this study were relatively simple metabolites, one could envision using this method to test more complex marine natural products. This method, like the capillary assay, is limited in that only water-soluble compounds can be tested.

1.4.1.4. Dual-Well Slide Method

Cooksey and Cooksey (1988) examined the chemotactic responses of the marine diatom *Amphora coffeaeformis* to nutrient gradients using a method originally developed by Zigmond (1977). Chemotaxis was measured visually using a specialized dual-well microscope slide assay. The slide was composed of two rectangular wells (1 mm deep × 4 mm wide) separated by a 1-mm bridge. To perform the assay, one well was filled with a chemotaxis medium containing the potential chemoattractant, and the other well was filled with only the chemotaxis medium. Thus, a chemical gradient was established across the bridge of the microscope slide. A coverslip inoculated with diatoms was placed across the wells such that the diatoms were exposed to the chemical gradient. Diatom chemotactic responses were then monitored by video microscopy. The advantage of the dual-well system is that both positive and negative chemotaxis can be measured by observing the movement of cells in either direction across the bridge. The major limitations of this assay are the specialized nature of the equipment and the time required if the objective is to screen a large number of pure compounds or extracts. The capillary assay avoids these problems by being rapid, relatively simple to implement, and inexpensive.

1.4.2. Gamete Chemotaxis

The importance of chemotaxis for the attraction of marine gametes has been established for many organisms (Miller 1973). The chemotaxis of algal gametes has been investigated (see reviews Maier & Müller 1986; Boland 1996), with the most commonly used assay being that developed by Müller (1976). In the scleractinian hard coral *Montipora digitata*, Coll et al. (1994) showed that sperm chemotaxis was mediated by a mixture of unsaturated fatty alcohols. Furthermore, Coll et al. (1995) studied sperm chemotaxis in *Lobophytum crassum* and demonstrated a chemotactic response to *epi*-thunbergol isolated from the soft coral. Sperm chemotaxis has also been studied in a wide range of marine organisms using an assay originally developed by Miller (1966). Both of these assays will be described in the following sections.

1.4.2.1. Müller Method

The Müller assay (Müller 1976) was developed to study chemotaxis in algal gametes. For this assay, a well (12 mm diameter, 1 mm height) was created by attaching a plastic ring to the bottom of a polystyrene petri dish. The well was then filled with 0.1 ml of an algal culture medium. Test compounds were incorporated into an inert, water insoluble, fluorinated hydrocarbon (FC-78, 3M Company, Düsseldorf) and four drops (~ 0.1 μl/drop) of compound treated FC-78 or FC-78 alone (control) were placed in different wells. Because FC-78 is more dense than the medium, it sinks and spreads out as a thin lens on the bottom of the well. Male gametes were then added to each well and the entire chamber placed in a black box fitted to a Zeiss microscope equipped with dark-field optics focused before covering the box. The gametes were incubated in the dark for 4 min, after which time the cover was removed and a photomicrograph was taken of the entire chamber. The assay was analyzed by counting the number of gametes in a field of predetermined size. The field size was chosen such that an average of 20–150 cells could be seen in the control well. By comparing treatment to control well counts, it was possible to demonstrate that algal gametes were attracted to substances isolated from their own species with threshold concentrations as low as 3×10^{-6} M. This assay has proven to be a very effective method of studying gamete chemotaxis.

1.4.2.2. Miller Method

While the Müller method has found widespread use, a second and substantially different assay (Miller 1966) has also been used to measure gamete chemotaxis. The Miller method as originally described was difficult to quantify and for this reason was modified (Miller 1979) so that minimum active concentrations could be calculated. In this assay, egg extracts were dissolved in seawater and injected through a microcapillary (50 μm diameter) into a solution of actively swimming sperm (Miller 1973). The behavior of sperm surrounding the tip of the capillary was then observed with a microscope fitted with dark-field optics. Photographs of each experiment were made with a 16-mm motion picture camera and were analyzed using an LW Motion Analyzer to project each frame individually such that sperm locations could be plotted. Extract activities were quantitatively evaluated by performing 1:1 serial dilutions until chemotactic responses were no longer observed.

In this assay, the typical swimming patterns of sperm in response to conspecific egg extracts included rapid looping and subsequent clustering around the pipette tip. Seawater controls and extracts of other species did not induce equivalent sperm behavior. Using this assay, Miller (1979) was able to show that the sperm of all 32 species of hydromedusae exhibited chemotactic responses to conspecific egg extracts. This assay has also been used to show chemotactic responses for the sperm of echinoderms (Miller 1985), molluscs (Miller 1977), and ascidians (Miller 1982). In similar studies, the same basic assay was made more quantitative by measuring the diameter of sperm trajectories (Cosson et al., 1984) and by measuring the bend angle and rotation frequency of sperm flagella (Miller & Brokaw 1970). Recently, computer software has been used to measure sperm trajectories and thus study chemotaxis in the ascidians *Ciona savignyi* and *C. intestinalis* (Yoshida et al. 1993).

1.4.3. Motility

Bioassays have been designed to examine the effects of crude extracts and pure compounds on the motility of marine microorganisms. These assays can be used to ask questions about diverse ecological phenomena such as symbiotic associations, predator-prey interactions, and chemical defense. One such assay that is described below was developed by Ciereszko and Guillard (1989) to investigate the possibility that gorgonian secondary metabolites (terpenoids) aid in the capture of dinoflagellates as a source of food and symbionts (zooxanthellae). To address these questions, gorgonian metabolites were tested against a variety of marine flagellates including isolates of *Symbiodinium* sp. whose nonmotile forms are known to occur as symbiotic zooxanthellae. In this motility assay, flagellates were grown at 20°C with 14 h of light per day provided by fluorescent illumination (~ 1.5×10^{16} photons $\cdot$ s^{-1} $\cdot$ cm^{-2}). Test compounds were dissolved in 95% aqueous ethanol to achieve a concentration of 1 mg/ml and 1 ml of these solutions were added to 200 ml of algal culture medium (final test concentration ~ 5 ppm). Assays were carried out in plastic microtiter plates (12 wells, each 2 cm in diameter) by adding 1 ml of treated media or control to different wells followed by one drop of algal culture. The effect of compounds on flagellate motility was estimated visually for 1–2 days at irregular intervals with the aid of a stereomicroscope. Most of the flagellates showed decreased motility in response to the cembranolide eunicin. The major limitations to this type of assay are that (1) it is not clear if the immersion of cells in test compounds accurately mimics natural conditions and (2) it is not always possible to distinguish between toxicity and a reduction in motility. The possibility that toxicity caused reduced motility in this study (Ciereszko & Guillard 1989) is supported by the observation that in some cases the cells disintegrated after exposure to the extract.

The ability of toxic dinoflagellates to chemically defend themselves from predatory ciliates was investigated by Hansen (1989). In this study, exudates of the toxic red tide dinoflagellate *Alexandrium tamarense* and a nontoxic control species, *Heterocapsa triquetra*, were added to chambers containing the marine ciliate *Favella ehrenbergii*. Responses to the exudates were observed with a video microscopy system and judged based on changes in swimming behavior (forward, backward, or stationary) at specific time intervals. Dinoflagellate toxin levels were determined by extraction of replicate cultures. Using these methods, the cellular exudate of *A. tamarense* was shown to cause ciliary reversal in *F. ehrenbergii*. Additionally, the filtrate caused swelling and subsequent cell lysis. The nontoxic dinoflagellate *H. triquetra* did not induce behavioral or cellular responses in *F. ehrenbergii*. The conclusion drawn by Hansen (1989) from this study was that while grazing by *F. ehrenbergii* may control low-density populations of toxic dinoflagellates, the significant amounts of toxins produced by high cell densities preclude grazing.

1.4.4. Settlement Assays—Laboratory Studies

Most benthic, sessile marine organisms maintain clean surfaces despite constant exposure to potential fouling organisms. There are a number of possible mechanisms by which sessile organisms prevent microorganisms from settling on their surfaces (epibiosis) including the use of marine natural products (see review by Davis et al. 1989). Natural products may reduce settlement and the subsequent development of a fouling community by inducing negative chemotaxis, affecting the attachment process, or via toxicity or growth inhibition. Because settlement assays have the potential to measure all of these processes collectively, they are inherently more sensitive than antimicrobial or toxicity assays. Behavioral bioassays have been developed to assess the effects of marine natural products on the attachment of marine microorganisms to a substrate and thereby provide an indirect evaluation of the potential role of these substances in antifouling chemical defense.

Some of the first discussions associating marine natural products with reduced fouling were based on correlations between extract antimicrobial activity and the extent to which an organism was visibly fouled (e.g., Thompson et al. 1985). Additional studies have compared extract antibacterial activity to epibacterial abundance (e.g., Burkholder 1973; McCaffrey & Endean 1985). Although toxic natural products may play a role in antifouling chemical defense, there is mounting evidence that extract toxicity and antifouling activity are not correlated (e.g., Wahl et al. 1994; Bryan et al. 1994) and that nontoxic substances can affect microbial attachment to a surface. These results support the contention that behavioral assays are better predictors of ecological effects than assays that only detect toxicity.

Laboratory settlement assays are typically performed by coating an extract or purified material onto an artificial settlement surface (e.g., a glass microscope slide) or incorporating the substance into a three-dimensional settlement matrix (e.g., agar). The later method of stimulus presentation has advantages in that (1) tissue level concentrations can be reproduced and (2) leaching of water soluble materials and physical effects of the extract are reduced. When materials are coated onto a planar surface such as a microscope slide, even if the concentration of the starting material is known, it is virtually impossible to calculate the final test concentration after the solvent has been removed. Furthermore, the physical effects of the test substance on the glass slide can not be accommodated for by the control. The use of three-dimensional matrices overcomes this problem, however, this method also has complications in that there is no evidence that whole animal concentrations provide an accurate prediction of what microbes experience in nature. In addition, agar is not an entirely inert matrix and could interact with compounds or chemical extract and thus give rise to inaccurate results. The advantages of the microscope slide assay are that it requires small sample sizes, can be used effectively for bioassay-guided chemical isolation, and due to its simplicity can be used to screen large numbers of extracts.

As mentioned throughout this chapter, one of the most important experimental variables that must be considered in microbial chemical ecology is the concentration at which substances are tested. Efforts to approximate natural concentrations have been made in a few instances by extracting only surface tissues (Wahl et al. 1994) or by measuring the concentrations at which compounds are released from the organism into the surrounding seawater (Thompson 1985). Although settlement assays have primarily been used to assess the effects of chemical extracts on the attachment of marine bacteria, these methods are also generally amenable to other types of microbes.

1.4.4.1. Microscope Slide Assay

The antifouling activity of extracts of the seagrass *Zostera marina* were examined for effects on the settlement of a marine bacterium isolated from its surface (Todd et al. 1993). In this assay, extracts and pure compounds dissolved in methanol were coated onto the frosted ends of glass microscope slides. Controls consisted of slides coated with methanol. The solvent was then evaporated and the slides placed into 50-ml plastic tubes containing a suspension of *Acinetobacter* sp. (10^6 cells/ml). The tubes were incubated on a rotary shaker and every 20 min for 4 h slides were removed, stained with Hoechst (no. 2287, Sigma Co.) and the number of cells attached to surfaces at various extract concentrations counted by epifluorescence microscopy. This study was important in that it established a connection between antifouling activity and chemical compounds isolated from the seagrass.

A similar assay was used to examine the effects of extracts from three soft corals: *Alcyonium paessleri*, *Clavularia frankliniana*, and *Gersemia antarctica*, against three species of marine bacteria (Slattery et al. 1995). In this assay, replicate 20-g soft coral samples were extracted in 200 ml of solvent and then diluted to tissue level concentrations. Twenty-five microliters of each soft coral extract or a solvent control were then spotted onto glass microscope slide coverslips. The coverslips were placed into individual wells of tissue culture plates that contained 2 ml of sterile-filtered seawater seeded with 50 µl of a bacterial suspension (7.5×10^6 cells/ml). Following a 48-h incubation, the coverslips were washed with sterile-filtered seawater to remove unattached cells and stained with DAPI (4,6-diamidino-2-phenylindole). The number of cells in each of five replicate microscope fields per coverslip were counted using epifluorescence microscopy. Bacterial attachment was found in this study to be inhibited by the chloroform extracts of both *A. paessleri* and *G. antarctica* and the methanol extract of *G. antarctica*. The active chloroform extracts did not display antimicro-

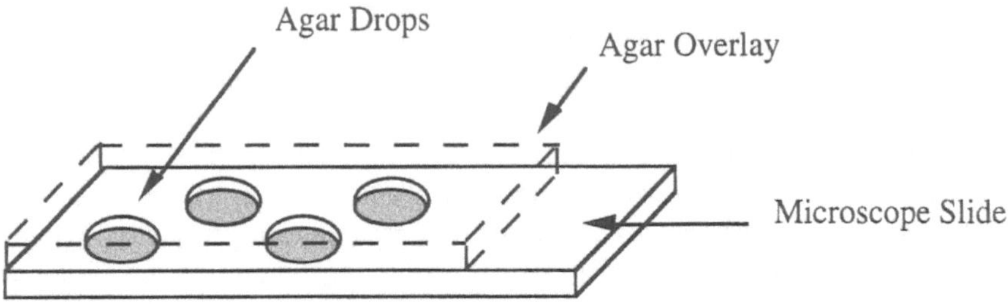

Figure 1.4. Diagram of a microscope slide assay. Agar drops containing a suspected antifouling compound are applied to a microscope slide and covered with an agar overlay. The microscope slide is then placed into a suspension of the test microbe, and after a predetermined incubation time, the slide is removed and the number of attached microorganisms counted by microscopy. Modified from Wahl et al. (1994).

bial activity supporting the observation that nontoxic extracts can reduce attachment and possibly function in antifouling chemical defense.

1.4.4.2. Agar Matrix Assay

Wahl et al. (1994) described an assay that measures bacterial attachment to an extract impregnated agar surface. Using this method, tissue level concentrations from two species of ascidians were reproduced by incorporating extracts into volumes of molten agar equal to the volumes of tissue extracted. Drops of extract impregnated agar were then placed on the surface of microscope slides and after cooling covered with an agar overlay to create a smooth surface (Fig. 1.4).

The settlement slides were then incubated in a microscope slide staining dish containing a suspension of bacteria isolated from the ascidians. To ensure a homogeneous distribution of bacteria, the dish was agitated on a rotary shaker at 60 rpm. The plates were incubated for approximately 15 min or until settlement density on extract-free agar-coated slides reached 10–100 cells per microscope field. Following incubation, the slides were removed from the bacterial suspension, rinsed to remove unattached cells, and stained with DAPI. The numbers of bacteria attached to the agar surface were then counted by epifluorescence microscopy and the number of bacteria on treatment versus control sections of the slide was compared. Using this method, it was possible to demonstrate a correlation between epibacterial abundance and the effects of extracts on attachment. It was also noted that antibiotic activities were not correlated with antisettlement effects, suggesting that nontoxic metabolites can deter surface fouling.

1.4.4.3. Spectrophotometric Assay

Spectrophotometric methods have been used to indirectly estimate the numbers of bacteria attaching to a surface (e.g., Shea & Williamson 1990). Bryan et al.

(1996) used a modification of this original assay to examine echinoderm body-wall extracts for effects on bacterial attachment. In this assay, stationary phase cultures of two marine bacteria, *Alteromonas luteo violacea*, which was isolated from the surrounding seawater, and *Deleya marina*, were transferred to a microtiter plate and centrifuged at 1000 rpm for 10 min. The culture broth was removed and 100 µl of sterile seawater (control) or sterile seawater plus echinoderm body-wall extract was added to the cells. The extract concentrations tested (3.0 mg/ml seawater) reflected natural concentrations (2.1–3.5 mg/g body tissue) following the assumption that 1 g of body tissue has the same volume as 1 ml of water. In addition, four half-log dilutions of the extracts were tested (0.6, 0.12, 0.024, and 0.0048 mg/ml seawater). Following an incubation period, the seawater was removed and the cells stained for 5 min with 100 µl of a 0.1% crystal violet solution. Unattached cells and excess dye were then removed by rinsing with deionized water.

The numbers of bacteria attaching to the plates were estimated by extracting the stain from the cells by addition of 100 µl of desoxycholate to each well and measuring absorbance at 595 nm. Absorbance was converted to cell numbers by comparison to a standard curve that was generated by incubating polystyrene coverslips for 1, 2, 4, 8, and 16 h with the test bacteria. Bacteria attaching to the coverslip were stained with crystal violet, counted with a Zeiss compound microscope and Petroff-Hauser counter, and the total number of cells on each coverslip extrapolated from ten (0.0025-mm^2) grids. The stain was then extracted with sodium desoxycholate and 100-µl aliquots transferred into wells of a new microtiter plate. Absorbance was measured as before and a standard curve generated relating absorbance to cell number. Using this method, it was possible to demonstrate that the extracts of three species of echinoderms caused significant reductions in bacterial attachment, while eight increased attachment.

1.4.4.4. Microalgal Attachment Assay

The importance of microalgae in conditioning surfaces for settlement by marine macroorganisms (larvae, macroalgae, etc.) has been stressed (Mitchell & Kirchman 1984; Wahl 1989). Although most studies addressing the effects of marine natural products on microalgal attachment have employed field experiments (see below), Hauser et al. (1975) devised a laboratory assay to monitor the responses of the colorless marine dinoflagellate *Crypthecodinium cohnii* to various compounds impregnated in agar. For this assay, diagonally opposed quadrants of a four-section petri dish were filled with agar containing a test compound or agar alone as a control. The dishes were then filled with 20 ml of a dinoflagellate suspension and following a 3-h incubation the solution was decanted and the plates stained with Lugol's iodine. Following staining, the plates were rinsed with distilled water and the number of cells imbedded in the agar surface counted with the aid of a dissecting microscope. In this assay, only dinoflagellates attached to the agar surface were counted, those killed or immobilized by toxic compounds were washed away. This behavioral assay is similar to that used by other researchers to monitor bacterial attachment and demonstrates the broad applicability of this type of assay to diverse microbes. By adding test compounds to a solid matrix, rather than to a liquid, microbes have the opportunity to respond to a chemical gradient. This then avoids the problems associated with immersing microbes in unrealistic chemical concentrations, provided that experiments are run for a sufficiently short duration and test materials do not leach quickly from the matrix.

1.4.5. Settlement Assays—Field Studies

The importance of testing compounds against natural populations of organisms has been stressed (Gerhart 1984; Harvell et al. 1988; Paul 1992) and certainly provides a more accurate picture of the ecological effects of marine natural products than laboratory-based experiments. In the case of antifouling studies, it has been proposed that laboratory assays may be biased due to excessive metabolite concentration, unrealistic flow conditions, or artifacts of extract preparation (Pawlik 1992). Microbiological field assays are inherently more complex than laboratory assays, and this may explain why few methods have been developed. One complication is that natural populations of microorganisms generally occur at lower densities than those used in controlled laboratory experiments, and as a result, field experiments must be deployed for longer periods of time. Long-term deployment runs the risk of loss of test material due to leaching, a subject recently addressed by Henrikson and Pawlik (1995). Mixed populations of microbes are also much more difficult to quantify than pure cultures. Naturally occurring microbes come in unusual shapes and sizes and subjective decisions have to be made about what organisms should be counted when visual quantification is required. Indirect quantification methods such as protein analysis or staining avoid this problem, however, they too must be interpreted carefully as one large individual can have more impact on the results than many small individuals.

1.4.5.1. Tile Assay

Compounds and extracts from the sponge *Aplysilla glacialis* collected in the Bahamas were tested for antifouling activity in San Diego, CA, using an *in situ* settlement assay developed by Bobzin and Faulkner (1992). In this assay, materials were applied to unglazed ceramic tiles by immersing the tiles in known concentrations of compound or crude extract. The tiles were found to absorb 0.3 ml of solvent, and thus the amount of material applied could be estimated. Control tiles were prepared by immersion in untreated solvent. Ten experiments were deployed by suspending treated and control tiles in the water column with Dacron® fishing line for a period of 2 weeks. Following seawater incubation, the tiles were retrieved, rinsed with sterile seawater, and stored at – 10°C until they were analyzed for total protein using the bicinchoninic acid (BCA) protein assay (Pierce, Rockford, IL) as an indirect estimation of the magnitude of surface fouling. Visual inspection of the plates at the termination of the experiment indicated heavy fouling by bacteria, fungi, and unicellular algae. Results of the protein analysis indicated that the tiles coated with the compounds manoöl and, to a lesser extent, cholesterol endoperoxide, were more heavily fouled than the controls.

To determine the extent to which extracts were retained on the tiles, compounds were back extracted after they had been exposed to seawater for 24 h. Loss of the pure compounds manoöl and cholesterol endoperoxide during seawater immersion was found to be slight with approximately 90% of both compounds being recovered. However, crude extracts showed much lower retention with values ranging from 3 to 21% recovered after immersion. This is one of first experiments to demonstrate that metabolite leaching could be a substantial problem in the design of settlement assays. Thus, as was acknowledged by the authors, this assay may be ineffective for crude extracts and water-soluble compounds that are readily desorbed from tile surfaces. The authors also propose that the pure compounds tested may have been responsible for increased fouling by acting as nutrients for primary colonizing bacteria, which in turn attracted secondary fouling organisms. Although Bobzin and Faulkner tested a natural assemblage of fouling organisms, the ecological validity of testing tropical organisms in temperate environments has been questioned (Bakus et al. 1985).

1.4.5.2. Microscope Slide Assay

The extracts of gorgonians, sponges and soft corals from Fiji were tested with an *in situ* antifouling assay by Wright (1989). Extracts were dissolved in abietic acid ($C_{20}H_{30}O_2$) and resin and the mixtures applied to preweighed glass microscope slides. Controls consisted of abietic acid resin and solvent. Coated slides were allowed to air-dry, placed in a holding device, and lowered to a water depth of 23 m. Following 38 days of incubation, the slides were collected, preserved in formalin, and brought back to the laboratory where they were reweighed. Changes in slide mass relative to controls were used to estimate antifouling activities of the extracts. This study was one of the first to investigate the effects of marine invertebrate extracts on fouling by co-occurring microorganisms. Antifouling activity was not observed for any of the organisms tested, while increased microbial film formation was seen for two extracts.

1.4.5.3. Microalgal Assays

Two different field assays were developed by Slattery et al. (1995) to examine microalgal settlement. In the first experiment, Antarctic soft coral extracts were incorporated into a 5% agar solution at dry weight concentrations equivalent to those found in the tissues extracted. Fifteen milliliters of each agar-extract solution was then added to each of four random wells in 12 six-well (35 × 10 mm; 17 ml) tissue culture plates. Control wells consisting of agar alone, agar containing solvent, or agar containing a known settlement inhibitor were also prepared. The tissue culture plates were then attached to settlement platforms constructed from PVC pipe and Plexiglas® and lowered to a water depth of 25 m for 1 year off McMurdo Station, Antarctica. On collection, the plates were rinsed to remove unattached epibionts and the settling organisms were visually identified. Settlement on the tissue culture plates was intense with approximately 90% of the agar surface covered. The assemblage of microorganisms included five genera of diatoms and three species of foraminiferans. There were, however, no significant differences among the treatment and control wells except for the extract of *Alcyonium paessleri* on which all but one genus of diatom was inhibited. Slattery et al. (1995) proposed that the lack of antisettlement effects in the agar-well experiment was possibly due to diffusion of the extract from the agar over the 1 year incubation and not a functional lack of bioactivity. Furthermore, they proposed that the settlement of only one diatom species (*Coscinodiscus* sp.) in response to the methanol extract of *A. paessleri* was possibly due to the loss of extract following changes in the composition of the fouling community. The loss of extract experienced in this assay could perhaps be reduced by the use of a matrix that is less susceptible to leaching such as Phytagel™ (Henrikson & Pawlik 1995). The problems associated with extract diffusion have been addressed in at least one study in which compound impregnated gels were back extracted after a 21-day exposure to seawater and the rate of diffusion for the extract was determined (Henrikson & Pawlik 1995). Long-term experiments such as this one could also be complicated by biological modification of metabolites by processes such as microbial degradation. The problem of chemical modification of the extract could be studied by examining controls in which the compounds are back extracted from the matrix and analyzed to determine the degree to which the metabolites have been chemically modified.

Slattery et al. (1995) designed a second assay to study the effects of metabolite leaching using glass microscope slides. In this assay, 1 ml of resolubilized soft coral extracts (or the appropriate solvent controls) were spotted onto glass microscope slides. Slides were then attached to a Plexiglas™ settling platform and deployed for a 4-week period. Using this assay, the chloroform extracts of two species of Antarctic soft corals, *Alcyonium paessleri* and *Gersemia antarctica* were determined to be effective deterrents to diatom settlement. While the compounds responsible for this activity were not chemically defined, the ability of chemical extracts to deter fouling by natural assemblages of microorganisms was demonstrated. The results for the more polar extracts are difficult to interpret due to extract diffusion following 3–5 days of deployment. The rapid loss of water-soluble materials from glass microscope slides once again illustrates the advantage of incorporating test materials into a three-dimensional matrix.

1.5. Summary and Conclusions

Microorganisms are ubiquitous in the sea and are in a constant state of interaction with marine plants, animals, and each other. These interactions can be beneficial (e.g., symbioses) or detrimental (e.g., pathogenesis, fouling), and depending on the outcome can provide incentive for the interacting organisms to either maximize or reduce the extent of the interaction. Although we do not have a clear understanding of the factors that control microbial interactions, there is mounting evidence implicating the secondary metabolites of marine plants and animals. We know that marine organisms are a rich source of biologically active secondary metabolites, and that many of these substances have ecologically significant effects on macroorganisms. The question that remains is to what extent do these metabolites also have important microbiological effects? For this question to be answered, a conscientious effort must be made to develop new bioassays that test compounds under realistic conditions against ecologically relevant microbes.

Developing bioassays to study the chemical ecology of marine microorganisms is an extraordinarily difficult task. The first major challenge that must be addressed is the choice of assay microbe(s). Clearly, experiments performed with natural assemblages of microorganisms provide a more realistic picture of the ecological effects of marine natural products than do controlled laboratory experiments. However, these methods are not always ideal for initial studies, as the results can be complex to interpret. The choice of test microbes for laboratory studies, although highly controlled, is also not straightforward. We know very little about microbial associations in the ocean and therefore, depending on the questions being addressed, it is often difficult to select the appropriate microbe. In all cases where the investigator hopes to obtain inferences about naturally occurring interactions, careful consideration must be given to the choice of test microorganism.

The second major challenge that must be addressed when attempting to assess the ecological significance of marine natural products is the concentration at which extracts or metabolites are tested. In most cases, it is difficult to estimate the concentrations experienced by microbes in nature, or even if they have the potential to contact these metabolites *in situ*, although this is certainly an area in which great progress can be made. One rational starting point is to test materials at whole tissue concentrations, i.e., the yield of material per volume tissue extracted. Active materials can then be diluted to determine minimum active concentrations, and these concentrations can be compared to those obtained from whole organisms and specific tissue extracts. Of course, if the microbe tested never contacts the active materials in nature, the effect has little meaning. Because of this, efforts to determine if metabolites are released from the producing organism, or located in tissues that are accessible to microbes, will shed important light on the ecological significance of experimentally demonstrated microbiological effects.

We have attempted in this chapter to describe bioassays in which marine secondary metabolites were tested for their effects on marine microorganisms. We have concentrated on assays that were developed to address ecological questions, e.g., do marine metabolites function in antimicrobial chemical defense? As can be seen from the contents of this chapter, the methods used to study the chemical ecology of marine microorganisms are not extensive, and as a result, we are only beginning to grasp the extent to which marine metabolites produce ecologically significant microbiological effects. Certainly for this field to continue to develop, new ecologically relevant bioassays will have to be developed and employed in a rational and thoughtful manner.

1.6. Acknowledgments

We gratefully acknowledge the continued financial support of the National Science Foundation, Chemistry Division, which allowed our studies of marine chemical ecology to be performed.

1.7. References

Acar, J.F. 1980. The disc susceptibility test. *In*: Antibiotics in Laboratory Medicine, ed. V. Lorian, pp. 24–54, Williams and Wilkins, Baltimore.

Adler, J. 1966. Chemotaxis in bacteria. Science **153**:708–716.

Adler, J. 1973. A method for measuring chemotaxis and use of the method to determine optimum conditions for chemotaxis by *Escherichia coli*. J. Gen. Microbiol. **74**:77–91.

Amade, P., C. Charroin, C. Baby & J. Vacelet. 1987. Antimicrobial activities of marine sponges from the Mediterranean Sea. Mar. Biol. **94**:271–275.

Armitage, J.P., D.P. Josey & D.G. Smith. 1977. A simple quantitative method for measuring chemotaxis and motility in bacteria. J. Gen. Microbiol. **102**:199–202.

Attaway, D.H. & O.R. Zaborsky. eds. 1993. Marine Biotechnology. Vol. 1. Pharmaceutical and Bioactive Natural Products. Plenum Press, New York.

Austin, B. 1988. Marine Microbiology, Cambridge University Press, New York.

Azumi, K., M. Yoshimizu, S. Suzuki, Y. Ezura & H. Yokosawa. 1990. Inhibitory effect of halocyamine, an antimicrobial substance from ascidian hemocytes, on the growth of fish viruses and marine bacteria. Experientia **46**:1066–1068.

Bakus, G.J., B. Schulte, S. Jhu, M. Wright, G. Green & P. Gomez. 1985. Antibiosis and antifouling in marine sponges: laboratory versus field studies. *In*: New Perspectives in Sponge Biology, ed. K. Rützler, pp. 102–108, Smithsonian Institution Press, Washington, DC.

Bandurraga, M.M. & W. Fenical. 1985. Evidence of a chemical adaptation against fouling in the marine octocoral *Muricea fruticosa* (Gorgonacea). Tetrahedron **41**:1057–1065.

Barry, A.L. 1980. Procedures and theoretical considerations for testing antimicrobial agents in agar media. *In*: Antibiotics in Laboratory Medicine, ed. V. Lorian, pp. 1–16, Williams and Wilkins, Baltimore.

Becerro, M.A., N.I. Lopez, X. Turon & M.J. Uriz. 1994. Antimicrobial activity and surface bacterial film in marine sponges. J. Exp. Mar. Biol. Ecol. **179**:195–205.

Bell, W. & R. Mitchell. 1972. Chemotactic and growth responses of marine bacteria to algal extracellular products. Biol. Bull. **143**:265–77.

Bergquist, P.R. & J.J. Bedford. 1978. The incidence of antibacterial activity in marine demospongiae; systematic and geographic considerations. Mar. Biol. **46**:215–221.

Betina, V. 1973. Bioautography in paper and thin-layer chromatography and its scope in the antibiotic field. J. Chromatogr. **78**:41–51.

Bobzin, S.C. & D.J. Faulkner. 1992. Chemistry and chemical ecology of the Bahamian sponge *Aplysilla glacialis*. J. Chem. Ecol. **18**:309–332.

Boland, W. 1996. The chemistry of gamete attraction: chemical structures, biosynthesis, and (a)biotic degradation of algal pheromones *In*: Chemical Ecology: The Chemistry of Biotic Interaction, eds. T. Eisner & J. Meinwald, pp. 87–102, National Academy Press, Washington, DC.

Bryan, P.J., J.B. McClintock, S.A. Watts, K. Marion & T.S. Hopkins. 1994. Antimicrobial activity of ethanolic body-wall extracts of echinoderms from the northern Gulf of Mexico. *In*: Echinoderms Through Time, Dijon, eds. B. David, A. Guille, J-P. Feral & M. Roux, pp. 17–23, Balkema, Rotterdam.

Bryan, P.J., D. Rittschof & J.B. McClintock. 1996. Bioactivity of echinoderm ethanolic body-wall extracts: an assessment of marine bacterial attachment and macroinvertebrate larval settlement. J. Exp. Mar. Biol. Ecol. **196**:79–96.

Buck, J.D., S.P. Meyers & K.M. Kamp. 1962. Marine bacteria with antiyeast activity. Science **138**:1339–1340.

Burkholder, P.R. 1973. The ecology of marine antibiotics and coral reefs. *In*: Biology and Geology of Coral Reefs, Vol. II, Biology, eds. O.A. Jones & R. Endean, pp. 117–182, Academic Press, New York.

Burkholder, P.R. & L.M. Burkholder. 1958. Antimicrobial activity of horny corals. Science **127**:1174–1175.

Burkholder, P.R. & K. Ruetzler. 1969. Antimicrobial activity of some marine sponges. Nature **222**:983–984.

Chan, A.T., R.J. Andersen, M.J. Le Blanc & P.J. Harrison. 1980. Algal plating as a tool for investigating allelopathy among marine microalgae. Mar. Biol. **59**:7–13.

Chet, I. & R. Mitchell. 1976. Ecological aspects of microbial chemotactic behavior. Annu. Rev. Microbiol. **30**:221–239.

Chet, I., S. Fogel & R. Mitchell. 1971. Chemical detection of microbial prey by bacterial predators. J. Bacteriol. **106**:863–867.

Chet, I., P. Asketh & R. Mitchell. 1975. Repulsion of bacteria from marine surfaces. Appl. Microbiol. **30**:1043–1045.

Ciereszko, L.S. & R.R.L. Guillard. 1989. The influence of some cembranolides from gorgonian corals on motility of marine flagellates. J. Exp. Mar. Biol. Ecol. **127**:205–210.

Clare, A.S. 1996. Marine natural product antifoulants: status and potential. Biofouling **9**:211–229.

Coll, J.C., B.F. Bowden, D.M. Tapiolas & W.C. Dunlap. 1982. In situ isolation of allelochemicals released from soft corals (Coelenterata: Octocorallia): a totally submersible sampling device. J. Exp. Mar. Biol. Ecol. **60**:293–299.

Coll, J.C., B.F. Bowden, G.V. Meehan, G.M. König, A.R. Carroll, D.M. Tapiolas, P.M. Aliño, A. Heaton, R. de Nys, P.A. Leone, M. Maida, T.L. Aceret, R.H. Willis, R.C. Babcock, B.L. Willis, Z. Florian, M.N. Clayton & R.L. Miller. 1994. Chemical aspects of mass spawning in corals. I. Sperm-attractant molecules in the eggs of the scleractinian coral *Montipora digitata.* Mar. Biol. **118**:177–182.

Coll, J.C., P.A. Leone, B.F. Bowden, A.R. Carroll, G.M. König, A. Heaton, R. de Nys, M. Maida, P.M. Aliño, R. H. Willis, R.C. Babcock, Z. Florian, M.N. Clayton, R.L. Miller & P.N. Aderslade. 1995. Chemical aspects of mass spawning in corals. II. (–)-Epi-thunbergol, the sperm attractant in the eggs of the soft coral *Lobophytum crassum* (Cnidaria: Octocorallia). Mar. Biol. **123**:137–143.

Cooksey, B. & K.E. Cooksey. 1988. Chemical signal-response in diatoms of the genus *Amphora.* J. Cell. Sci. **91**:523–529.

Cosson, M.P., D. Carré & J. Cosson. 1984. Sperm chemotaxis in siphonophores. II. Calcium-dependent asymmetrical movement of spermatozoa induces by the attractant. J. Cell Sci. **68**:163–181.

Davis, A.R., N.M. Targett, O.J. McConnell & C.M. Young. 1989. Epibiosis of marine algae and benthic invertebrates: natural products chemistry and other mechanisms inhibiting settlement and overgrowth. *In*: Bioorganic Marine Chemistry, Vol. 3, ed. P.J. Scheuer, pp. 85–113, Springer-Verlag, Berlin.

Ducklow, H.W. & R. Mitchell. 1979. Bacterial populations and adaptations in the mucus layers on living corals. Limnol. Oceanogr. **24**:715–725.

Faulkner, D.J. 1995. Marine Natural Products. Nat. Prod. Rep. **12**:223–269.

Fenical, W. & J.R. Pawlik. 1991. Defensive properties of secondary metabolites from the Caribbean gorgonian coral *Erythropodium caribaeorum.* Mar. Ecol. Prog. Ser. **75**:1–8.

Fredrickson, A.G. & G. Stephanopoulos. 1981. Microbial competition. Science **213**:972–979.

Gerhart, D.J. 1984. Prostaglandin A_2: an agent of chemical defense in the Caribbean gorgonian *Plexaura homomalla.* Mar. Ecol. Prog. Ser. **19**:181–187.

Gil-Turnes, M.S., M.E. Hay & W. Fenical. 1989. Symbiotic marine bacteria chemically defend crustacean embryos from a pathogenic fungus. Science **246**:116–118.

Goff, L.J. & J.C. Glasgow. 1980. Pathogens of Marine Plants, University of California, Santa Cruz.

Han, G. & J.J. Cooney. 1993. A modified capillary assay for chemotaxis. J. Ind. Microbiol. **12**:396–398.

Hansen, P.J. 1989. The red tide dinoflagellate *Alexandrium tamarense*: effects on behavior and growth of a tintinnid ciliate. Mar. Ecol. Prog. Ser. **53**:106–116.

Harrison, P.G. & A.T. Chan. 1980. Inhibition of the growth of micro-algae and bacteria by extracts of eelgrass (*Zostera marina*) leaves. Mar. Biol. **61**:21–26.

Harvell, C.D., W. Fenical & C.H. Greene. 1988. Chemical and structural defenses of Caribbean gorgonians (*Pseudopterogorgia* spp.). I. Development of an in situ feeding assay. Mar. Ecol. Prog. Ser. **49**:287–294.

Hauser, D.C.R., M. Levandowsky, S.H. Hutner, L. Chunosoff & J.S. Hollwitz. 1975. Chemosensory responses by the heterotrophic marine dinoflagellate *Crypthecodinium cohnii*. Microbial Ecol. **1**:246–54.

Hazen, T.C., R.V. Dimock, Jr., G.W. Esch, A. Mansfield & M.L. Raker. 1984. Chemotactic behavior of *Aeromonas hydrophila*. Cur. Microbiol. **10**:13–18.

Henrikson, A.A. & J.R. Pawlik. 1995. A new antifouling assay method: results from field experiments using extracts of four marine organisms. J. Exp. Mar. Biol. Ecol. **194**:157–165.

Jensen, P.R., C.D. Harvell, K. Wirtz & W. Fenical. 1996. Antimicrobial activity of extracts of Caribbean gorgonian corals. Mar. Biol. **125**:411–419.

Kamei, Y., M. Yoshimizu & T. Kimura. 1987. Plaque assay of *Oncorhynchus masou* virus (OMV). Fish Pathol. **22**:147–152.

Kim, K. 1994. Antimicrobial activity in gorgonian corals (Coelenterata, Octocorallia). Coral Reefs **13**:75–80.

Kohlmeyer, J. & E. Kohlmeyer. 1979. Marine Mycology: The Higher Fungi. Academic Press, New York.

Lemos, M.L., C.P. Dopazo, A.E. Toranzo & J.L. Barja. 1991. Competitive dominance of antibiotic-producing marine bacteria in mixed cultures. J. Appl. Bacteriol. **71**:228–232.

McCaffrey, E.J. & R. Endean. 1985. Antimicrobial activity of tropical and subtropical sponges. Mar. Biol. **89**:1–8.

McClintock, J.B. & J.J. Gauthier. 1992. Antimicrobial activities of Antarctic sponges. Antarct. Sci. **4**:179–183.

McLachlan, J. & J.S. Craigie. 1964. Algal inhibition by yellow ultraviolet-absorbing substances from *Fucus vesiculosus*. Can. J. Bot. **42**:287–292.

Maier, I. & D.G. Müller. 1986. Sexual pheromones in algae. Biol. Bull. **170**:145–175.

Miller, J.D., E.B.G. Jones, Y.E. Moharir & J.A. Findlay. 1985. Colonization of wood blocks by marine fungi in Langstone Harbour. Bot. Mar. **28**:251–257.

Miller, R.L. 1966. Chemotaxis during fertilization in the hydroid *Campanularia*. J. Exp. Zool. **162**:23–44.

Miller, R.L. 1973. Chemotaxis of animal sperm. *In*: Behavior of Microorganisms, ed. A. Perez-Miravette, pp. 31–47, Plenum Press, London.

Miller, R.L. 1977. Chemotactic behavior of chitons (Mollusca: Polyplacophora). J. Exp. Zool. **202**:203–212.

Miller, R.L. 1979. Sperm chemotaxis in the Hydromedusae. I. Species-specificity and sperm behavior. Mar. Biol. **53**:99–114.

Miller, R.L. 1982. Sperm chemotaxis in ascidians. Am. Zool. **22**:827–840.

Miller, R.L. 1985. Demonstration of sperm chemotaxis in Echinodermata: Asteroidea, Holothuroidea, Ophiuroidea. J. Exp. Zool. **234**:383–414.

Miller, R.L. & C.J. Brokaw. 1970. Chemotactic turning behavior of *Tubularia* spermatozoa. J. Exp. Biol. **52**:699–706.

Mitchell, R. 1975. Negative chemotaxis: a new approach to marine fouling control, Office of Naval Research Technical Report No. 1—Biofouling, Cambridge.

Mitchell, R. & D. Kirchman. 1984. The microbial ecology of marine surfaces. *In*: Marine Biodeterioration: An Interdisciplinary Study, eds. J.D. Costlow & R.C. Tipper, pp. 49–58, Naval Institute Press, Annapolis.

Mokashe, S.S., A. Garg, A.C. Anil & A.B. Wagh. 1994. Growth inhibition of periphytic diatoms by methanol extracts of sponges and holothurians. Indian J. Mar. Sci. **23**:57–58.

Muehlstein, L.K., J.P. Amon & D.L. Leffler. 1988. Chemotaxis in the marine fungus *Rhizophydium littoreum*. Appl. Environ. Microbiol. **54**:1668–1672.

Müller, D.G. 1976. Quantitative evaluation of sexual chemotaxis in two marine brown algae. Z. Pflanzenphysiol. **80**:120–130.

Myouga, H., M. Yoshimizu, K. Tajima & Y. Ezura. 1995. Purification of an antiviral substance produced by *Alteromonas* sp. and its virucidal activity against fish viruses. Fish Pathol. **30**:15–22.

Nikata, T., K. Sumida, J. Kato & H. Ohtake. 1992. Rapid method for analyzing bacterial behavioral responses to chemical stimuli. Appl. Environ. Microbiol. **58**:2250–2254.

Palleroni, N.J. 1976. Chamber for bacterial chemotaxis experiments. Appl. Environ. Microbiol. **32**:729–730.

Paul, V.J. 1992. Chemical defenses of benthic marine invertebrates. *In*: Ecological Roles of Marine Natural Products, ed. V.J. Paul, pp. 164–188, Comstock Publishing, Ithaca, NY.

Pawlik, J.R. 1992. Chemical ecology of the settlement of benthic marine invertebrates. Oceanogr. Mar. Biol. Annu. Rev. **30**:273–335.

Pfeffer, W. 1884. Locomotorische Richtungsbewegungen Durch Chemische Reize. Untersuchungen aus dem Botanischen Institut in Tübingen **1**:363–482.

Pratt, D.M. 1966. Competition between *Skeletonema costatum* and *Olisthodiscus luteus* in Narragansett Bay and in culture. Limnol. Oceanogr. **11**:447–455.

Proctor, L.M. & J.A. Fuhrman. 1990. Viral mortality of marine bacteria and cyanobacteria. Nature **343**:60–62.

Rheinheimer, G. 1992. Aquatic Microbiology, 4th ed. Wiley, New York.

Rijstenbil, J.W. 1989. Competitive interaction between *Ditylum brightwellii* and *Skeletonema costatum* by toxic metabolites. Neth. J. Sea Res. **23**:23–27.

Rinehart, K.L., J.B. Gloer, G.G. Hughes, H.E. Renis, J.P. McGovren, E.B. Swynenberg, D.A. Stringfellow, S.L. Kuentzel & L.H. Li. 1981. Didemnins: antiviral and antitumor depsipeptides from a Caribbean tunicate. Science **212**:933–935.

Rosenfeld, W.D. & C. Zobell. 1947. Antibiotic production by marine microorganisms. J. Bacteriol. **54**:393–398.

Seymour, F.W.K. & R.N. Doetsch. 1973. Chemotactic responses by motile bacteria. J. Gen. Microbiol. **78**:287–296.

Shea, C. & J.C. Williamson. 1990. Rapid analysis of bacterial adhesion in a microplate assay. Biotechniques **8**:610–611.

Simidu, U., W.J. Lee & K. Kogure. 1983. Comparison of different techniques for determining plate counts of marine bacteria. Bull. Jpn. Soc. Sci. Fish. **49**:1199–1203.

Sjoblad, R.D., I. Chet & R. Mitchell. 1978. Quantitative assay for algal chemotaxis. Appl. Environ. Microbiol. **36**:847–850.

Slattery, M., J.B. McClintock & J.N. Heine. 1995. Chemical defenses in Antarctic soft corals: evidence for antifouling compounds. J. Exp. Mar. Biol. Ecol. **190**:61–77.

Sparks, A.K. 1985. Synopsis of Invertebrate Pathology, Exclusive of Insects. Elsevier, New York.

Stahl, E. 1969. Thin Layer Chromatography, A Laboratory Handbook. Springer-Verlag, New York.

Strongman, D.B., J.D. Miller, L. Calhoun, J.A. Findlay & N.J. Whitney. 1987. The biochemical basis for interference competition among some lignicolous marine fungi. Bot. Mar. **30**:21–26.

Suttle, C.A., A.M. Chan & M.T. Cottrell. 1991. Use of ultrafiltration to isolate viruses from seawater which are pathogens of marine phytoplankton. Appl. Environ. Microbiol. **57**:721–726.

Targett, N.M. 1988. Allelochemistry in marine organisms: chemical fouling and antifouling strategies. *In*: Marine Biodeterioration. Advanced Techniques Applicable to the Indian Ocean, eds. M-F. Thompson, R. Sarojini & R. Nagabhushanam, pp. 609–617, Oxford and IBH Publishing, New Delhi.

Targett, N.M., S.S. Bishop, O.J. McConnell & J.A. Yoder. 1983. Antifouling agents against the benthic diatom, *Navicula salinicola*. Homarine from the gorgonians *Leptogorgia virgulata* and *setacea* and analogs. J. Chem. Ecol. **9**:817–829.

Thompson, J.E. 1985. Exudation of biologically-active metabolites in the sponge *Aplysina fistularis*. I. Biological evidence. Mar. Biol. **88**:23–26.

Thompson, J.E., R.P. Walker & D.J. Faulkner. 1985. Screening and bioassays for biologically-active substances from forty marine sponge species from San Diego, California, USA. Mar. Biol. **88**:11–21.

Todd, J.S., R.C. Zimmerman, P. Crews & R.S. Alberte. 1993. The antifouling activity of natural and synthetic phenolic acid sulphate esters. Phytochemistry **34**:401–404.

Tso, W.-W., & J. Adler. 1974. Negative chemotaxis in *Escherichia coli*. J. Bacteriol. **118**:560–576.

Wahl, M. 1989. Marine epibiosis. I. Fouling and antifouling: some basic aspects. Mar. Ecol. Prog. Ser. **58**:175–189.

Wahl, M., P.R. Jensen & W. Fenical. 1994. Chemical control of bacterial epibiosis on ascidians. Mar. Ecol. Prog. Ser. **110**:45–57.

Walls, J.T., D.A. Ritz & A.J. Blackman. 1993. Fouling, surface bacteria and antibacterial agents of four bryozoan species found in Tasmania, Australia. J. Exp. Mar. Biol. Ecol. **169**:1–13.

Walsh, F. & R. Mitchell. 1978. Bacterial chemotactic responses in flowing water. Microb. Ecol. **4**:165–168.

Weibull, C. 1960. Movement. *In*: The Bacteria, Vol. 1, eds. I.C. Gunsalus & R.Y. Stanier, pp. 153–206, Academic, New York.

Wellman, A.M. & H.W. Paerl. 1981. Rapid chemotaxis assay using radioactively labeled bacterial cells. Appl. Environ. Microbiol. **42**:216–221.

Willey, J.M. & J.B. Waterbury. 1989. Chemotaxis toward nitrogenous compounds by swimming strains of marine *Synechococcus* spp. Appl. Environ. Microbiol. **55**:1888–1894.

Wood, E.J.F. 1965. Marine Microbial Ecology. Chapman and Hall, London.

Wright, M. 1989. Antibiotic and antifouling properties of marine invertebrate extracts: comparative aspects of sponges and gorgonians. *In*: Bioactive Compounds from Marine Organisms: With Emphasis on the Indian Ocean: An Indo-United States Symposium, eds. M-F. Thompson, R. Sarojini & R. Nagabhushanam, pp. 351–356, Oxford and IBH Publishing, New Delhi.

Yoshida, M., K. Inaba & M. Morisawa. 1993. Sperm chemotaxis during the process of fertilization in the ascidians *Ciona savignyi* and *C. intestinalis*. Dev. Biol. **157**:497–506.

Yu, C. & S. Roseman. 1993. A modified capillary method for measuring bacterial chemotaxis. *In*: Lectins and Glycobiology, ed. H.-J. Gabius, pp. 441–444, Springer Verlag, Berlin.

Zigmond, S.H. 1977. Ability of polymorphonuclear leukocytes to orient in gradients of chemotactic factors. J. Cell Biol. **75**:606–616.

2

Bioassays with Marine and Freshwater Macroorganisms

Mark E. Hay, John J. Stachowicz, Edwin Cruz-Rivera, Stephan Bullard, Michael S. Deal, and Niels Lindquist

2.1. Introduction 40
2.2. Foraging cues 42
2.2.1. Isolation and preparation of foraging cues 43
2.2.2. Vertebrates 45
2.2.3. Invertebrates 47
2.3. Feeding cues 51
2.3.1. Isolation and preparation of feeding cues 55
2.3.2. Feeding deterrents 56
2.3.2.1. Vertebrates 56
2.3.2.1.1. Field assays 56
2.3.2.1.2. Laboratory assays 59
2.3.2.2. Invertebrates 63
2.3.2.2.1. Field assays 63
2.3.2.2.2. Laboratory assays 64
2.3.3. Feeding stimulants 70
2.3.4. Multiple cues and defensive synergisms 72
2.4. Consequences of consuming defensive metabolites 74
2.4.1. Vertebrates 76
2.4.1.1. Short-term assays 76
2.4.1.2. Long-term assays 80
2.4.2. Invertebrates 82
2.4.2.1. Short-term assays 82
2.4.2.2. Long-term assays 85
2.5. Toxin-mediated prey capture 90
2.6. Chemically mediated detection of and responses to predators 92
2.6.1. Vertebrates 93
2.6.1.1. Chemical cues from conspecifics 93
2.6.1.2. Chemical cues from predators 95
2.6.2. Invertebrates 97
2.6.2.1. Behavioral responses 97

2.6.2.2. Induced responses 99
2.7. Intraspecific chemical communication 102
2.8. Chemically mediated homing behavior 106
2.8.1. Vertebrates 106
2.8.2. Invertebrates 108
2.9. Settlement cues 108
2.9.1. Vertebrates 108
2.9.2. Invertebrates 109
2.9.3. Recommendations regarding still water assays 114
2.10. Allelopathy and antifouling 117
2.10.1. Invertebrates 117
2.10.2. Seaweeds 119
2.11. Chemical ecology within a broader environmental context 120
2.12. Conclusions 121
2.13. Acknowledgments 122
2.14. References 122

2.1. Introduction

The general methods for extraction, purification, and identification of marine secondary metabolites are similar to, or involve slight modifications of, those used with terrestrial organisms. Marine methodologies for small organic molecules have been reviewed by Norris and Fenical (1985), and those for phlorotannins by Ragan and Glombitza (1986). Many compounds, such as those involved in prey detection, predator detection, and home site selection are less well understood chemically, and general reviews for methodologies dealing with the chemistry of these types of compounds are not available. With the exception of the phlorotannins, the large majority of known secondary metabolites from marine organisms are lipid soluble (Faulkner 1996, and earlier reviews cited therein). This produces significant experimental advantages because known quantities of lipid-soluble metabolites can be dissolved in volatile organic solvents such as diethyl ether and coated at desired concentrations onto surfaces (for tests as settlement cues or antifouling agents); onto palatable prey that have been blotted dry; or onto finely powdered, dried algal or animal tissue that can then be imbedded in a gel matrix (for tests as defenses against consumers see McConnell et al. 1982; Targett et al. 1986; Hay et al. 1987a, 1987b, 1994; Paul 1987, 1992; Henrikson & Pawlik 1995; Schmitt et al. 1995). After the solvent evaporates, these hydrophobic compounds adhere to the treatment surfaces or foods and do not dissolve when immersed in seawater while being used in settlement or feeding assays. Comparisons of feeding or settlement on treatments versus controls (treated with solvent only) allow for a determination of the effects of the compound alone. Modifications of these basic procedures have been applied widely in both field and laboratory investigations.

One of the more significant ecological problems that marine biologists face when conducting bioassays is determining the natural concentration of a compound, and thus the concentration at which it should be tested. Natural concentrations of many marine secondary metabolites are not known because most of the available literature has been generated by natural products chemists interested in describing new compounds, rather than in carefully documenting their natural concentrations. These chemical studies rarely provide the yield of the compound in the organism being studied. When yields are given, they are conservative because (1) extraction is rarely, if ever, complete; (2) all isolation and purification techniques involve some loss of metabolites; and (3) handling, storage, and extraction techniques can result in compound degradation, and the occurrence and magnitude of these artifacts are rarely determined (Paul et al. 1987; Paul & Van Alstyne 1992; Cronin et al. 1995b).

Careful high-pressure liquid chromatography (HPLC) quantification of secondary metabolites from some marine organisms has now been conducted, and individual compounds range from <0.1% to about 9% of organism dry mass, with there sometimes being considerable variation between different populations or different portions of an individual (see Paul & Fenical 1986; Hay et al. 1988c; Meyer & Paul 1992, 1995; Bolser & Hay 1996; Cronin & Hay 1996a, 1996b, 1996c for examples). Geographic variation in chemical defenses also occurs. Tropical species of seaweeds and benthic invertebrates have often been claimed to have increased concentrations of secondary metabolites relative to similar temperate species (Hay & Fenical 1988). This general impression has rarely been tested directly, but recent tests with diverse groups of related tropical and temperate seaweeds did demonstrate increased chemical defenses among tropical populations and species, as well as dramatic within-region variance in some cases (Bolser & Hay 1996). Only investigations that carefully quantify amounts of compounds being released from organisms (in the case of waterborne cues) and the concentrations of compounds occurring in the populations being studied, or that extract known quantities of organisms and apply this extract to equivalent quantities of test organism, can be assured that they are using ecologically realistic concentrations. Knowing natural concentrations of metabolites occurring in whole organisms does not, however, guarantee ecologically meaningful results because the way the compound is applied and the test situations used in the assays are also critical (see below).

Marine chemical ecology requires collaboration between chemists and biologists. As marine chemical ecology began to emerge as a field, it went through an awkward stage where biologists did poor chemistry and chemists did poor biology. All fields suffer these types of growing pains at some stage in their development, but the increased productivity and scientific sophistication that has come with meaningful collaborations (both formal and informal) between chemists and ecologists argues that such collaborations should be strongly encouraged. If you are a biologist and want to investigate the ecological role of aquatic secondary metabolites, or primary metabolites with important ecological functions, then one of the most important methodological advances you can make may be to find a very good chemist that is willing to help. Chemists will be equally advantaged by working with experienced biologists.

All experiments require that investigators deal professionally with concerns of replication, independence, interspersion, and appropriate controls and statistical analyses. Because there are many excellent books and articles dealing with these topics, and because procedures often need to be designed to be specific for the types of species, compounds, and questions being investigated, we do not deal with specific aspects of design and statistical analyses. However, aquatic chemical ecology, like many areas in science, suffers from too many pseudoreplicated experiments where errors in experimental design make rigorous statistical analyses impossible. Hurlbert (1984) provides an excellent discussion of pseudoreplication that will help investigators avoid costly mistakes in designing their experiments.

We do not attempt to provide a complete and historic overview of all methods involved in aquatic chemical ecology, but rather hope to provide a sound conceptual discussion of various methods and approaches, and to describe recent procedures that meet present-day standards. Numerous excellent studies are not cited, usually because more recent studies provide similar access to the methods and greater access to other literature on the topic.

2.2. Foraging Cues

Many terrestrial studies have focused on how consumers use chemical cues to detect prey from a distance, to stimulate feeding on specific hosts, and to recognize appropriate hosts for oviposition (Rosenthal & Berenbaum 1992). In numerous instances, the specific chemicals cueing these behaviors have been isolated and identified. Understanding these chemical cues sometimes provides insights or tools that facilitate the suppression of pest populations or the preservation of desirable hosts or prey. Although many marine organisms clearly respond to chemical signals from their prey, the chemistry of these interactions is not well known.

In aquatic systems, the ability to detect and orient to food from a distance is potentially of considerable advantage, allowing consumers to detect food over an area larger than they can afford to search physically, given the constraints associated with predator avoidance and food acquisition as a function of searching effort. Additionally, vision can be severely restricted in aquatic habitats due to attenuation of light, particularly in deep or turbid waters. In contrast to the spatial limitation of visual cues, chemical odors can be carried over considerable distances. The possible role of distance chemoreception in mediating foraging behavior has been studied extensively in both freshwater and marine systems; however, most studies have been focused on physiological or behavioral questions rather than more ecological ones (e.g., see reviews by Hara 1982; Ache 1982; Derby & Atema 1988). The approaches of physiologists and ecologists commonly require different methodologies, yet many ecological studies have adopted methods used by physiologists and sometimes have drawn ecological conclusions based on inappropriate methods. While physiologist's methods may be appropriate for answering questions regarding neurological and physiological responses (Hara 1982), many ecological questions require that experiments be designed with adequate consideration given to the conditions under which animals forage in the field. Specifically, when testing for the presence and or effectiveness of chemical foraging cues, the researcher must consider (1) what constitutes an appropriate stimulus, (2) how to accurately and appropriately deliver the stimulus, and (3) how to quantify the response by the forager (i.e., what behaviors are appropriate indicators of an effect on foraging).

2.2.1. Isolation and Preparation of Foraging Cues

Methods of extracting and quantifying stimuli parallel those used in terrestrial studies (see other chapters in this book and volume 1). The brief discussion provided here highlights procedures that may provide the most ecologically appropriate foraging stimuli. Early studies gave little attention to the realism of the stimulus, and even some recent studies have used "off-the-shelf" pure chemicals as stimuli rather than identifying stimuli from the prey themselves—usually creating solutions in concentrations that have been found to occur in prey tissues without simultaneous testing of naturally occurring prey odors (e.g., Harpaz et al. 1987; Knutsen 1992). These studies are useful in determining the relationship of compound structure and function (Hara 1982), but extrapolation to foraging in the field is questionable. As an example, although amino acids can attract consumers, recent field measurements have shown that fluxes of amino acids from carrion are only occasionally above the threshold concentrations that allow detection by scavengers, and that fluxes from live organisms are much lower, and well below levels detectable by predators (Zimmer-Faust et al. 1996a). Additionally, natural stimuli can be complex mixtures of attractants, repellents, and neutral chemicals whose combined activity may differ widely from that of any component in isolation (e.g., Zimmer-Faust et al. 1984; Carr & Derby 1986a, 1986b, Zimmer-Faust 1987). Because of potential additive or interactive effects between components of a natural odor, even if the activity of a pure chemical matches that observed for natural odors, one should not conclude, based on this alone, that this chemical is responsible for the observed behavior in nature (Zimmer-Faust et al. 1984). Only bioassay-guided fractionation and factorial testing and recombination of various fractions for activity will allow true determination of the active component(s) of a signal.

A second approach is to use macerated prey flesh or chemical extracts of prey as the stimulus. This, however, has the disadvantage of releasing cues to the external environment that may normally be found only within prey tissue; such cues may not naturally be available as foraging cues (Rittschof et al. 1984; Luczkovich 1988). Despite these concerns, the activity of extracts or of ground up prey often parallels that of whole organisms or natural exudates (Rittschof et al. 1984; Daniel & Bayer 1987a; Anderson & Dale 1989). When this is the case, using prey extracts may have an advantage over using whole organisms or exudates because the concentration and composition of the stimulus can be precisely reproduced, allowing for confidence in replication and follow-up experiments (Zimmer-Faust 1989). However, the researcher must be aware that this match between the activity of an extract and that of prey rinses or whole organisms does not always occur (Atema et al. 1980), and compounds that are not normally released to the environment could interfere with or enhance the effect of a chemical signal in nonadditive ways (Zimmer-Faust et al. 1984; Carr & Derby 1986a, 1986b). For these reasons, chemical extracts or macerated tissues appear less appropriate for assaying the foraging phase of feeding, although they are valid stimuli in examining the consumption phase of feeding.

Ecological studies of chemoattraction should first demonstrate that consumers are attracted to whole, undisturbed prey in an ecologically realistic setting and that there are indications that this attraction could be chemically mediated (e.g., attraction happens without physical or visual cues). These initial studies can then be followed by bioassay-guided chemical investigations of signals released from intact prey. Given the difficulty of chemically characterizing the small quantities of compounds that are naturally released, investigators may find it necessary to use prey extracts or crushed prey. Once compounds are identified using these less ecologically realistic approaches, additional investigations should be conducted to determine that these compounds also function in ecologically realistic settings.

Natural exudates of prey often have been obtained by holding the prey in a volume of standing water for a period of time to allow accumulation of olfactory cues. Although this represents an improvement over crushing or extracting the prey, holding prey for extended periods in the absence of advection and diffusion (which are normally present in natural systems) can artificially increase stimulus concentrations. Additionally, holding prey in still water can cause the concentration and composition of prey odors to be altered by bacterial degradation or autolytic decomposition in seawater (Ogura 1975; Hollibaugh 1979; Ammerman & Azam 1985), uptake of dissolved organic matter by prey organisms (Manahan et al. 1983), prey nutritional history (Regnault 1981), and by prey physiological state (Bryant & Atema 1987). It is unlikely that any one method of stimulus preparation will be appropriate for all behavioral assays, although, for studies of foraging behavior, natural rinses appear to provide the most appropriate stimulus. However, if a stimulus must be reproduced precisely across many replicates, or bioassay-guided fractionation is desired, use of chemical extracts sometimes may be warranted (Zimmer-Faust 1989).

2.2.2. *Vertebrates*

Most investigators of chemically mediated foraging by marine vertebrates have used fishes as assay organisms. The extensive work on the chemosensory abilities of fishes has been reviewed many times (see reviews in Hara 1982; Atema et al. 1988), and studies of the toxicological impacts of waterborne chemicals on fish behavior also are common (Giattina et al. 1982). However, there have been far fewer studies focused on chemically mediated prey tracking and foraging. This is perhaps due more to the general perception that most fish are visual predators than to any inherent difficulties in performing realistic assays with marine vertebrates.

Unfortunately, bioassays designed for use in studies of toxicology or physiology (Cherry & Cairns 1982; Giattina & Garton 1983; Jones & Hara 1985) often have been adapted for the study of foraging behavior without modifications to make them more ecologically realistic. In some commonly used assays, fish are held in a small aquarium with filtered seawater and observed for some period of time to obtain data on baseline behavior. Then fish behavior in response to the localized injection of a chemical stimulus is compared to that of fish in a control aquarium with only filtered water injected. Quantification of behaviors usually involves recording activities thought to be indicative of foraging, such as changes in patterns of swimming (from figure 8 to circular), breakdown of schooling, rapid opening and closing of the mouth, "yawning," darting, or time spent at the end of the tank receiving the stimulus injection (Jones & Hara 1985; Steele et al. 1991; Knutsen 1992). Scoring this type of assay generally consists of sorting behaviors into categories, or ranking of responses on an arbitrary scale of intensity, rather than a strict quantification of behavior (Atema et al. 1980; Knutsen 1992). Although this method allows investigators to determine if the stimulus has elicited some response, it is often much less clear what that response means. We urge caution in inferring that stimuli are used as foraging cues based on the fact that they induce changes in behavior that may or may not be indicative of the ability of a predator to use that stimulus to track prey.

A more relevant behavior might be the time spent near the source of the stimulus (Jones & Hara 1985; Steele et al. 1991), but this requires a chemical gradient for the forager to follow rather than simply a discrete pulse of stimulus. Such a gradient may be produced by allowing a constant influx of stimulus-laden water (Atema et al. 1980; Giattina et al. 1982; Sola 1995). An observer or video camera can then record the position of the fish relative to the source of the stimulus, either relative to an inflow of a control solution in the same tank, or compared to a separate fish offered only a control. In these assays, fish behavior may be more clearly indicative of whether or not the cues can be used in tracking prey up a stimulus gradient. In at least one study where categorical behavioral observations like those described above indicated that fish were responding to a chemical signal, the fish were unable to localize the source of the signal, suggesting that chemically mediated foraging in some fishes may only provide a mechanism to guide fish to a general area where visual prey detection may then be employed (Atema et al. 1980).

Provided the odor plume contains chemical cues at concentrations that are ecologically realistic, assays that release an odor plume appear to mimic situations where cues are emitted from prey. But because these cues are waterborne, changing patterns of flow have important impacts on the structure and composition of odor plumes (Moore et al. 1994). Although the effects of changing flow regimes on chemically mediated foraging in fishes has not yet been investigated, ecologically realistic assays with invertebrates have shown that chemically mediated foraging may occur only under a limited range of field conditions (Weissburg & Zimmer-Faust 1993, 1994; Zimmer-Faust et al. 1995). Investigators should be aware that hydrodynamics can have important effects on foraging, and experiments should be conducted under conditions that approximate those occurring in the field.

Field studies of chemically mediated foraging of fishes are generally lacking. Perhaps the most ecologically realistic studies to date involve the use of live whole prey organisms as targets in experimental arenas that mimic the natural environment. Use of whole prey items addresses concerns over the appropriateness of the stimulus composition and concentration, because only naturally released materials are available as foraging cues. By manipulating the presence of visual and chemical cues to the pinfish, *Lagodon rhomboides*, Luczkovich (1988) was able to assay the relative importance of each in affecting prey detection. A single prey species was offered in an array that consisted of a tube with a sand bottom and containing a blade of seagrass as habitat for the prey. Tubes were manipulated to restrict visual, chemical, or both types of stimuli for both sensile and mobile prey items. A complete array consisted of five tubes (Fig. 2.1): a clear tube with a mesh top (visual + chemical stimuli), a sealed clear tube (visual only stimuli), an opaque tube with a mesh top (chemical only stimuli), an opaque, sealed tube (no stimulus), and an open clear tube with no prey inside as a control for the stimulatory effects of a clear glass tube. The experimental arena (a glass aquarium) had a natural sand bottom and a sea grass refuge area, and an apparatus allowed for diffuse flow over the glass tubes. The results provide evidence for a complementary role of visual and chemical cues in foraging; tubes offering either chemical or visual stimuli alone were not attacked more frequently than the no-prey control tube, but tubes offering both types of stimuli were attacked significantly more frequently than controls for all types of prey used (Luczkovich 1988).

Although assays that use whole organisms are perhaps the most ecologically realistic, it would be difficult to isolate the chemical cue(s) or mechanisms responsible for the observed behavior in this assay. As with any ecological study, there are trade-offs between realism and ability to manipulate the system, and combinations of approaches using whole organisms in ecologically realistic set-

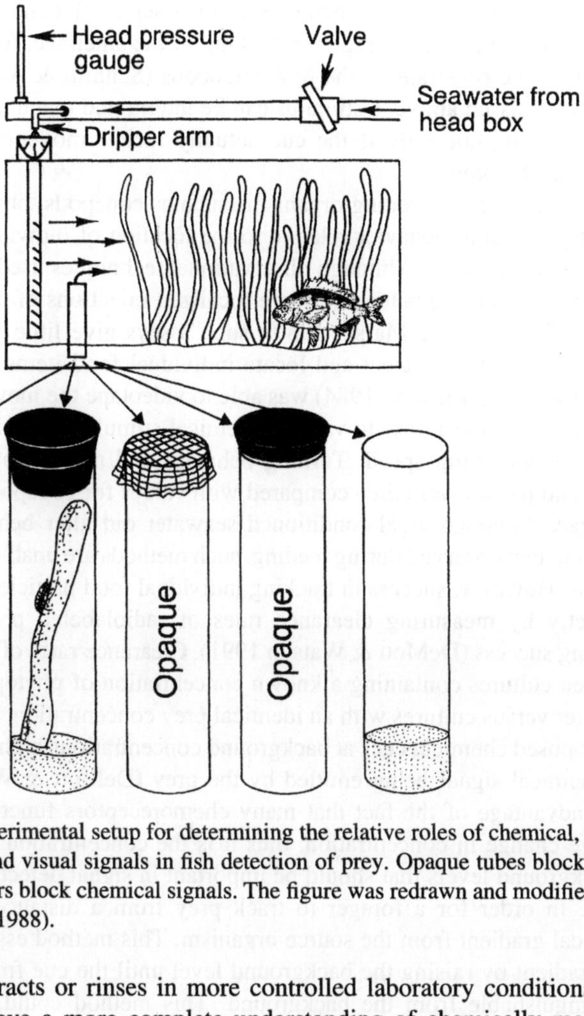

Figure 2.1. An experimental setup for determining the relative roles of chemical, visual, and both chemical and visual signals in fish detection of prey. Opaque tubes block visual signals, while stoppers block chemical signals. The figure was redrawn and modified from one in Luczkovich (1988).

tings and prey extracts or rinses in more controlled laboratory conditions may be needed to achieve a more complete understanding of chemically mediated foraging in fishes.

2.2.3. *Invertebrates*

Because invertebrates are easy to manipulate experimentally and have relatively simple nervous systems, invertebrates (particularly decapod crustaceans and gastropod mollusks) have been the subject of far more research on chemically mediated foraging than vertebrates. As with vertebrates, many workers have taken the approach of delivering a stimulus in a single pulse of odor injected into standing water by a syringe or pipette, and the behavioral response is compared to that for a control injection of water. For crustaceans, this response is usually quantified as the completion of a series of behaviors that resemble foraging activities, beginning with flicking of the antennae, followed perhaps by probing of the substrate by the first pereopods (Schmitt & Ache 1979). But, for reasons discussed above, a stimulus can be interpreted unambiguously as a long-range foraging cue only if the cue actually incites movement and guides the forager to the prey.

For smaller, filter-feeding organisms such as copepods, observing chemically mediated foraging behavior relative to the location of individual phytoplankton cells may be difficult. Although observational techniques used for larger animals can detect swarming behavior where localized injections of prey exudates have been made (Poulet & Marsot 1980), such assays give little information on the copepod's ability to detect and locate individual food items (algal cells) using chemoreception. Buskey (1984) was able to videotape the movements of individual copepods in response to various chemical stimuli and digitize their tracks to calculate swimming speed. Turning behavior and number of swimming speed bursts and pauses were then compared with values for copepods actively feeding on algae. Although algal-conditioned seawater did alter behavior in a fashion similar to that observed during feeding, such methods are unable to assess foraging success. However, success in tracking individual food particles has been assayed indirectly by measuring clearance rates of radiolabeled prey as a proxy for foraging success (DeMott & Watson 1991). Clearance rates of prey are compared between cultures containing a known concentration of phytoplankton in filtered seawater versus cultures with an identical prey concentration, but also containing the proposed chemical cues at background concentrations high enough to obscure the chemical signal being emitted by the prey (DeMott & Watson 1991). This takes advantage of the fact that many chemoreceptors function as detectors of relative change in concentration, thus it is the concentration of the cue relative to background levels that should be important in signal detection (Zimmer-Faust 1991). In order for a forager to track prey from a distance, there must be a chemical gradient from the source organism. This method essentially eliminates this gradient by raising the background level until the cue from the organism is indistinguishable from the background. This method could, however, lead to ambiguous results if feeding is altered not because of obscured chemical gradients, but because the proposed chemical cues for feeding have physiological effects on consumers at these artificially enhanced concentrations.

In contrast to the indirect methods used for smaller organisms, the effects of chemical gradients on foraging by larger consumers can be assayed more directly. A chemical gradient can be created by continuous infusion of a properly prepared test solution through a small tube into a tank holding potential predators (Jones & Hara 1985), or by allowing a test solution to diffuse from a dialysis bag (Rittschof et al. 1984) or a piece of filter paper (Daldorph & Thomas 1988). If these tests are conducted in standing water, then the potential predator needs to respond quickly or the chemical cue may diffuse evenly throughout the tank producing little, if any, gradient. This is less of a problem if open, flow-through systems are used. The concentration gradient of the proposed chemical cue can be viewed qualitatively using a colored dye tracer and a video camera. Alternatively, plume distribution can be quantified using fluorescent dye and a photodiode detector to resolve the structure of the odor plume on spatial scales of <1 cm, and on temporal scales <10 ms (Zimmer-Faust et al. 1988). An observer (or video camera) can track movement of the test animal relative to the source of the treatment versus control solution.

Frequently, behavioral response to a foraging cue is quantified by allowing the forager to select between two arms of a Y-shaped trough (Y-maze). Water flows into a central chamber from both arms; flow from one arm is laden with the hypothesized stimulus while the other arm serves as a control. Frequency of moving upstream (toward the Y) and choice between the stimulus-emitting arm versus the control arm are analyzed to determine if cues impact distance chemoreception (Carte & Faulkner 1986; Allmon & Sebens 1988; Boudreau et al. 1993). Provided stimuli are ecologically realistic, these methods can give an indication of the effect of a prey odor on foragers. However, in all of the above assays care must be taken in labeling compounds as attractants or repellents as a result of experiments over a narrow range of concentrations. Compounds that attract at low levels can be repellent at higher levels (Carte & Faulkner 1986) and vice versa (Giattina et al. 1982).

Despite the attractiveness of these simple approaches, these studies cannot accurately determine the concentration of a stimulant that provokes activity because concentrations of test solutions are known only at the instant they are released to the water column. By the time they reach the target organism, the concentration of the stimulus may be reduced to 1/1000 of its initial concentration (Harpaz et al. 1987), so precise active concentrations are difficult to determine using this method. To determine the precise concentration required to elicit a response, the stimulus must be delivered directly to the sensory organs. Although this is commonly done by dissection in neurophysiological studies (see review in Derby & Atema 1988), ecologists generally prefer to keep foragers intact when investigating their natural behaviors. Precise concentrations of assay metabolites can be determined using intact animals such as lobsters and crabs by strapping to the forager a backpack that holds two reservoirs with a tube leading from each reservoir to one of the forager's antennae (e.g., Zimmer-Faust 1993). One reservoir contains a solution with a stimulus, while the other contains a control seawater solution. When activated by remote control, each tube releases an equal volume of the solution contained in its reservoir, and stimulation or inhibition is inferred if the test organism moves toward or away from the side where the stimulus is released. To date, this method has been used only with solutions of pure compounds; however, it should also work with chemical extracts, prey-rinses, or even whole organisms that could be put in the reservoirs. This innovative technique provides the possibility of conducting well-controlled and exacting experiments under ecologically realistic conditions in the field.

Thus far, we have considered assays conducted without much attempt to mimic the natural environment in which consumers forage. All of the above experiments were conducted in arenas with little if any water movement and smooth sides that produce little or no turbulence. These conditions produce an unnatural distribution of stimulus patterns (see discussion in Zimmer-Faust 1989; Weissburg & Zimmer-Faust 1993). The environments in which marine organisms typically forage are characterized by at least some water flow (tidal currents or waves) and topographic complexity (even sand is rougher than a glass aquarium), both of which can increase turbulence. Increasing turbulence transforms an easily discernible gradient of odor from prey to predator into a patchy assortment of odor pulses that vary in strength and frequency with distance from the source (Moore & Atema 1991; Moore et al. 1994). Thus, the quality and quantity of chemical information available to foragers in the field may differ greatly from that available in simplified laboratory experiments.

Realistic field studies would greatly increase our understanding of how chemoreception mediates foraging in nature. Most previous efforts have consisted of simple assays in which a packet of food is placed in a cage some distance away from a group of nonvisual foragers (e.g., sea urchins). Successful location of the food source is compared for individuals located upstream, downstream, and cross stream from the food at various initial distances (Himmelman 1988; Young et al. 1993). Even if significantly more downstream foragers locate the food source, results are potentially confounded by the lack of an appropriate control (no scent). Movement that appears to be chemically mediated by the food could instead be explained by a simple tendency for upstream movement by foragers, by attractive properties of the cage containing the food, or by attraction to the disturbance created by placement of the cage or the food source.

More realistic and better controlled field assays have been conducted to investigate distance chemoreception (e.g., Rittschof et al. 1992), but even in these assays, relevant flow parameters are rarely quantified, and controlled manipulation of stimulus concentration and flow regime may be difficult to achieve. However, recent flume experiments manipulating both flow and bottom characteristics across a range of ecologically relevant levels have tested the effects of changing hydrodynamic regimes on chemoreception (Weissburg & Zimmer-Faust 1993, 1994). (Flume design and construction are critical considerations for conducting such experiments; this topic has been extensively reviewed by Nowell & Jumars 1987). These trials consisted of challenging blue crabs (*Callinectes sapidus*) to locate a live and intact clam (*Mercenaria mercenaria*) or clam extract. Current speeds in the flume were 0.0, 1.0, 3.8, or 14.4 cm/s and bottom roughness was manipulated at the 1.0 cm/s flow speed by changing the substrate from sand to gravel. Changes in these hydrodynamic parameters altered the strength and depth of penetration of turbulence in the boundary layer, which affected the structure of odor plumes (Moore & Atema 1991; Moore et al. 1994). Results suggest dramatic effects of flow on the ability of an organism to use chemical signals. In the absence of flow, predators were unable to locate prey. Slow flow (3 cm/s) resulted in efficient search paths and tracking success of nearly 100%, while fast flow (14 cm/s) resulted in convoluted search paths and low to moderate tracking success. Additional work across a range of slower flows (1–3 cm/s) suggests that search efficiency may actually increase with flow speed due to the streamlining of the odor plume (Scholz & Atema 1991). When bottom composition was altered from sand to gravel (increasing bottom roughness, and leading to greater turbulence and dispersion of the odor plume), but flow was held constant at 1.0 cm/s, tracking success decreased significantly (Weissburg & Zimmer-Faust 1993). These studies suggest that the use of chemical cues in locating distant prey may function only under a narrow range of conditions in nature, an idea that researchers should give strong consideration when designing and interpreting experiments involving the use of waterborne cues. In shallow estuaries, however, slow unidirectional flow over a period of several hours may occur; thus, chemoreception could be an important foraging tool in these habitats. Recent field experiments have shown that over 80% of crabs tested followed odor plumes emanating from injured bivalve prey or artificial plumes of clam bivalve mantle fluid to their source (Zimmer-Faust et al. 1995). Because this study took place in the field under natural flow conditions, it suggests that foragers can use chemical signals to locate natural prey items in the field.

All of the methods discussed above have been useful in advancing our knowledge of chemically mediated foraging; however, methodologies incorporating ecological realism provide much more relevant information on those processes affecting foragers in the field. Experiments that attempt to draw conclusions about natural behaviors of the foragers should consider that the behavior of animals toward chemical stimuli will vary depending on the animal's activity state (Zimmer-Faust & Case 1983; Zimmer-Faust et al. 1996b), hunger level (Atema et al. 1980; Costero & Meyers 1993; Cronin & Hay 1996c), feeding history (Daniel & Bayer 1987b), and the presence of conspecifics (Steele et al. 1991) or predators (see discussion in Zimmer-Faust 1989).

2.3. Feeding Cues

The assays described above assess how foragers use compounds to detect and move toward prey. Once prey have been encountered, feeding stimulants or deterrents affect the probability that prey will actually be consumed. Large numbers of investigations have demonstrated, in both field and laboratory assays, that secondary metabolites from seaweeds and benthic invertebrates serve as deterrents to feeding by marine consumers (reviewed by Hay & Fenical 1988;

Hay & Steinberg 1992; Paul 1992; Pawlik 1993; Pawlik et al. 1995). Alternatively, some secondary metabolites, and numerous primary metabolites, can serve as feeding stimulants. Because consumer pressure in marine systems can be extremely high and because these systems are often more experimentally tractable and less disturbed than terrestrial systems, it has been possible to clearly identify important consumers that compounds should be tested against, or to test compounds in the field against the diverse assemblage of natural consumers that occur there (reviewed by Hay & Steinberg 1992; Paul 1992; Pawlik 1993). Thus, marine investigators can apply purified metabolites at natural concentrations to the surfaces of otherwise palatable organisms or can imbed the metabolites in gel-based foods placed on natural reefs, and determine in a short period of time (minutes to hours) whether or not the compounds would affect predation under natural field conditions where prey are being attacked by a diverse assemblage of consumers (Hay et al. 1987a; Hay 1991; Pawlik & Fenical 1989; Chanas & Pawlik 1996 as examples). This ability to rapidly conduct ecologically realistic bioassays in the field appears to be an advantage that marine communities provide to chemical ecologists; these types of field bioassays of prey chemical defenses have rarely been conducted in terrestrial communities.

When nonpolar, lipid-soluble extracts from the green alga *Cymopolia* or the red alga *Gracilaria* were coated onto seaweeds that were then placed in seawater for 2–3 h, McConnell et al. (1982) reported that 100% of the *Gracilaria* extracts and 88% of the *Cymopolia* extracts remained on the coated plants. Similar experiments with the diterpene alcohol pachydictyol A showed that 93% of the compound remained on the coated plant after 24 h in seawater (Hay & Fenical 1988). Thin-layer chromatography of plants coated with many different lipid-soluble compounds and placed in the field on coral reefs for several hours show that the plants retain the compounds, but the total amount remaining was not assessed in these particular studies (Hay et al. 1987a, 1988a, 1988b; Paul 1987). Given the long retention of the few compounds or extracts that have been evaluated and the similar solubility characteristics of many marine secondary metabolites, this general methodology can probably be used with most nonpolar, lipid-soluble metabolites.

Assays involving phlorotannins represent much more of an experimental challenge for both marine biologists and chemists because they are large, water-soluble molecules that often occur in complex mixtures. The most common methods of estimating phlorotannin concentrations involve variations of the Folin-Denis procedure. This methodology has been discussed by numerous authors (Ragan & Glombitza 1986; Waterman & Mole 1994; Van Alstyne 1995). The Folin-Denis method does not directly measure phlorotannins but rather the concentration of phenolic hydroxyl groups, irrespective of the particular compound in which they occur. Unfortunately, several other compounds, without consistent structural similarities, like a phenolic hydroxyl group, also oxidize the reagents used in this assay and produce colored forms of the reagent that confound phenolic readings. These interfering compounds include urea, ascorbic acid, polypeptides, detergents, nontannic phenols, and other compounds. It is unfortunate that this 80-year-old technique that was developed for measuring phenols in urine (Folin & Denis 1915) has been so broadly used, and so poorly understood, by biologists that are attracted to this simple colorimetric method. When not subject to significant interference, this method might provide a general measure of total phenolics, but it does not differentiate among the many different phenolics that may be present. Although there appears to have been a general assumption among many biologists that total phenolics provide a good measure of bioactive, herbivore-deterrent phenolics, this assumption is subject to error. Boettcher and Targett (1993) demonstrated that different sized phlorotannins from brown algae have differential effects on herbivore digestive efficiency, and Clausen et al. (1990) demonstrated for tannins from terrestrial sources that very small structural differences between very similar tannins could lead to pronounced differences in their effects on herbivores. Additionally, the broadly held assumption that tannins act as digestibility reducers may also be in error in that, when tannins are active, they may act as toxins or gut lining irritants (Bernays et al. 1989). More importantly, there are now several investigations showing that both feeding and digestion of generalist herbivores may often be unaffected by even very high levels of marine phlorotannins (Steinberg & van Altena 1992; Targett et al. 1995).

This large variance in the ecological effects of phlorotannins mandates that ecologists assay the activities of phlorotannins from their test organisms rather than just measure total phlorotannins and assume that the particular phlorotannins they are studying have important ecological functions (e.g., as herbivore deterrents). Recent work is also indicating that deterrence previously attributed to algal phlorotannins could have been generated, at least in part, by compounds that are nonphenolic but that will commonly be impurities in the crude phenolic extracts that biologists often assay (M. Deal, M.E. Hay, and W. Fenical work in progress). Given all of these limitations, biologists should be extremely cautious about using these simple colorimetric techniques and assuming that these procedures provide meaningful chemical insights. Close collaboration between expert chemists and ecologists has proven to be critical in developing our understanding of the chemical ecology of organisms that produce smaller and more chemically tractable compounds. Thinking that professional chemical insights and talents are unnecessary when dealing with these more chemically difficult phlorotannins is wishful thinking that will impede development of rigorous understanding within this subfield of chemical ecology.

Many of the early investigations of prey chemical defenses were conducted by placing potential consumers into water containing secondary metabolites that had been extracted from potential prey. These assays lacked ecological realism in that (1) many of these compounds are not naturally leached into seawater; (2) if they were, they would disperse rapidly and not be present at significant concentrations; and (3) the potential predators would not normally encounter the compounds in this way. Although these types of assays are still used by some investigators, they have significant limitations for evaluating ecological phenomena and should be discontinued. More importantly, comparative assays have repeatedly shown that there is no predictive relationship between the effect of a compound placed in an organism's water and the effect of that compound if it is eaten by the organism (Hay & Fenical 1988; Sammarco & Coll 1992; Pawlik et al. 1995). As an example, low dosages of several algal metabolites killed omnivorous fishes in 1–3 min when placed in their water, but no negative effects of these compounds were noted when fish consumed this same amount of compound every day for 2–3 weeks (M.E. Hay & L. Matthews unpublished data). Unless prey normally release compounds into the water in situations that would allow for these compounds to be retained at high concentrations near the prey (e.g., small tide pools at low tide), these types of assays may be no more realistic than concluding that compounds in hops are toxic because humans die if immersed in beer. Potential feeding deterrents should be tested by *feeding* them to consumers; holding consumers in water containing compounds, injecting consumers with compounds, etc., are not acceptable methodologies.

An additional methodological problem is that some investigators commonly test marine natural products using unrealistic assay organisms (e.g., freshwater fish). Because there can be tremendous variance in the effect of a compound on feeding by different groups of consumers, or even different species within the same consumer group or genus (Paul et al. 1988; Wylie & Paul 1989; Hay & Steinberg 1992; Steinberg & van Altena 1992; Duffy & Hay 1994; Meyer et al. 1994; Pennings et al. 1996), data gathered using ecologically unrealistic consumers provide no firm insight into the function of the compounds in nature. Tests of metabolites as feeding deterrents or stimulants should thus be conducted using co-occurring consumers that have the potential to be ecologically important. This will often require considerable insight into the ecology and evolution of the communities and organisms being studied.

Additionally, assay foods should mimic as closely as possible the nutritional characteristics of the prey from which the test compounds or extracts have been obtained. Several experiments have shown that compounds that effectively deter feeding on lower quality prey may be ineffective deterrents if placed in higher quality prey (Duffy & Paul 1992; Hay et al. 1994; Pennings et al. 1994). The findings from some initial studies that extracted deterrent metabolites from nutritionally rich prey and tested these by adding them to gel-based foods containing feeding stimulants, but very little nutritional quality, have been called into question. Proving that compounds deter feeding on flavored packets of water may not be equivalent to showing that they would deter feeding on a nutritionally valuable prey organism. Given our limited understanding of the physiological and nutritional needs of most marine consumers (e.g., Horn 1989), it is often unclear how to best measure the nutritional value of prey or test foods. As a minimal requirement, test foods should approximate prey in ash-free dry mass and protein content per volume of food when possible.

2.3.1. *Isolation and Preparation of Feeding Cues*

Most feeding deterrents tested to date have been noticed, isolated, and identified by natural products chemists looking for unusual secondary metabolites. Once the chemists had these compounds, biologists then investigated their potential ecological effects on prey susceptibility to consumers. These studies have been important in building our understanding of marine chemical ecology and in demonstrating some of the natural functions of marine secondary metabolites; however, there are more ecologically realistic and intellectually satisfying approaches that should now be used. The traditional approach of testing pure and previously identified metabolites will not detect synergistic or additive effects of co-occurring metabolites and may not produce a full understanding of prey defenses because prey may possess additional, but unknown, metabolites that also affect feeding. Bioassays should be used to guide chemical discoveries of ecological importance.

Low palatability of intact prey could result from chemical, structural, morphological, nutritional, or other characteristic of a prey species. The structural and morphological traits can be destroyed by macerating the prey tissues and incorporating them into gel-based foods. If these foods are still unpalatable, then chemical deterrents are suggested. The chemical basis of deterrence can then be tested by incorporating water-soluble and lipid-soluble prey extracts into gel-based foods that are nutritionally similar to the prey and running feeding assays (Bolser & Hay 1996). If these assays demonstrate effects on feeding, then extracts can be partitioned by polarity or molecular size, feeding assays repeated, and the deterrent compounds detected and isolated by following their ecological effects. This approach allows for the detection of new bioactive metabolites or for finding that well-known metabolites have unexpected ecological effects. Examples of this general approach can be found in Van Alstyne and Paul (1992), Van Alstyne et al. (1994), Cronin et al. (1995a), and Bolser and Hay (1996).

Using the above approach will sometimes result in bioactivity being lost following repeated fractionation or when certain extraction procedures are used. This may result from metabolite instability, from active metabolites being separated among several different fractions (thus decreasing the concentration in each fraction), from bioactivities that are due to interactions of multiple metabolites that have been separated into different fractions, from compounds leaching from test foods, or from volatile metabolites being lost when solvents are removed by rotary evaporation or freeze-drying. Some of these possibilities can be evaluated by recombining fractions or by looking for obvious chemical degradation. Often even very simple chemical techniques such as thin-layer chromatography (TLC) are adequate for documenting this degradation. Because chemical conversions and instabilities can vary as a function of how samples are stored, extracted, partitioned, etc. (see Paul & Van Alstyne 1992; Cronin et al. 1995b as examples), the probability of solving such problems by developing more appropriate procedures will increase with increasing chemical expertise and experience. We thus, yet again, suggest that biologists seek input from experienced natural products chemists.

The types of bioassays suggested above can be conducted at ecologically realistic concentrations by placing crude extracts or fractions into test foods at the same concentration per food dry mass or per volume (if test foods mimic prey in nutritional value per volume) that occurs in the organisms from which they were extracted. For this to be valid, the nutritional quality of the test food should approximate the nutritional value of the organism from which the extract was obtained.

2.3.2. Feeding Deterrents

2.3.2.1. Vertebrates

A broad range of secondary metabolites from seaweeds, marsh plants, and benthic invertebrates have been shown to deter feeding by vertebrate consumers (Hay & Fenical 1988; Hay & Steinberg 1992; Paul 1992; Pawlik 1993). The vast majority of these studies have been conducted on fishes and have focused on feeding deterrents rather than stimulants (Hay 1991; Paul 1992). The very few studies conducted with other aquatic vertebrates such as sea otters and birds suggest that the basic methods used for fishes may also work with these vertebrates (Buschbaum et al. 1984; Kviteck 1991; Kviteck et al. 1991).

2.3.2.1.1. Field Assays

Initial field assays of marine secondary metabolites as feeding deterrents against fishes were based on modifications of laboratory methodologies developed by McConnell et al. (1982) and Targett et al. (1986). Because reef fishes rapidly graze on palatable prey transplanted onto consumer-rich areas of coral reefs (Hay 1984; Paul & Hay 1986), palatable prey can be used to conduct field assays of secondary metabolites or crude lipid-soluble extracts against the natural mix of fishes that occur on coral reefs (Hay et al. 1987a; 1988a, 1988c; Paul 1987; Hay 1991). These types of field assays can rarely be used to determine how specific species of fishes react to these metabolites, but they do provide a powerful way to address the advantages of feeding deterrents from the perspective of the prey. These procedures allow one to ask, if a particular prey species produced this compound at this concentration, would it suppress overall feeding on this prey by the mixed species group of fishes that attack this organism in the field? Procedures have been developed for using either natural prey items or artificial gel-based diets in the bioassays. Advantages of using natural prey as test organisms are that fishes rapidly recognize and feed on them, that they are readily available at even the most remote field sites, and that assay organisms can be chosen so that they approximate the nutritional value of the prey from which the suspected chemical defenses have been isolated. Potential disadvantages are that fishes may use visual rather than chemical cues in selecting prey with which they are familiar (i.e., learning to avoid the treated prey may require overcoming a lifetime of experience with a normally palatable prey) and that compounds usually have to be coated onto the exterior of intact prey where they may be unusually easy to detect and avoid. Evidence available to date indicates that these potential disadvantages may not be serious. As examples (1) fishes from a wide variety of geographic areas have repeatedly been shown to rapidly diminish their feeding on treated prey that they visually recognize as edible (see tables in Hay 1991; Paul 1992) and (2) the few experiments where the same compounds have been tested coated onto prey versus evenly embedded internally throughout an artificial prey's tissues have produced similar results (Cronin & Hay 1996c).

Most field assays with seaweed secondary metabolites have used co-occurring palatable seaweeds as assay organisms, while most field assays with invertebrate secondary metabolites have used artificial gel-based diets. Each of these approaches is outlined below.

To produce treatment plants possessing test compounds, hearty, structurally simple, and palatable seaweeds such as the sea grass *Thalassia* (Hay et al. 1987a, 1987b, 1987c) or the red algae *Gracilaria* or *Acanthrophora* (Hay et al. 1987b, 1988a) are quickly blotted dry, coated with known concentrations of compounds that have been dissolved in a volatile solvent such as diethyl ether, and the solvent is allowed to evaporate, leaving the plant with a coating of the test metabolite. Control plants are blotted and coated with only ether. Pairs of treatment and control plants can then be placed between the strands of three-stranded ropes and these ropes anchored slightly above the bottom (this eliminates grazing by nonswimming herbivores) in the field. Losses to grazing can be measured by decreases in length, projected surface area, or mass. Preliminary trials using this method indicated that some fishes (e.g., parrotfishes) initially bit treatment and control blades with equal frequency but returned to bite control blades more frequently than treatment blades (i.e., they appeared to detect the deterrent compounds by taste rather than smell) (Hay et al. 1988c). When only one treatment and one control plant were placed on each rope, initial sampling by the fishes often removed most plant material from both treatments and controls and obscured effects of the compounds. Effects of compounds on feeding became both visually and statistically obvious when four treatment or four control plants were placed on separate ropes and these ropes paired within 50 cm of each other (Hay et al. 1988c; M.E. Hay personal observation). This later procedure produced a small patch of each plant type and allowed fishes a chance to learn the taste of each patch without sampling them to extinction. This learning can happen very rapidly;

large differences in feeding on treatment versus control patches can occur in as little as 2–3 h. For additional details on this general methodology see Hay et al. (1987a, 1987b, 1988a, 1988c; Paul 1987). Although these methods have been used primarily with seaweeds, they potentially could be modified for use with some benthic invertebrates.

Many lipid-soluble algal metabolites have been assayed using the above procedures, and TLC analyses indicate that these types of compounds are usually still present at the end of the experiment (Targett et al. 1986; Hay et al. 1987a, 1987b, 1988a, 1988c; Paul 1987). However, a very few secondary metabolites tested in this manner may oxidize, decompose, irrevocably bond to the surface of the test plant, or be lost from the assay food (Hay et al. 1987a, 1988a, 1988c, M.E. Hay personal observation); thus screening for retention of test compounds following bioassays is advisable.

Herbivory in the field can vary dramatically over small spatial scales (Hay et al. 1983; Hay 1985, 1991). This high variance will diminish the statistical power of field assays unless treatment and control plants are paired in space. Placing members of a pair within 0.5 m of each other and separating different pairs by 4 m or more has been a commonly used procedure. To acquire data rapidly and thus minimize loss or degradation of compounds, transplants should be placed in areas where fishes are abundant and actively feeding. On coral reefs, the best areas tend to be shallow (3–10 m depths) and topographically complex (Hay 1985).

Field assays also can be conducted using gel-based (e.g., agar, sodium alginate, etc.) foods containing finely ground animal or plant material that stimulates feeding by consumers. These types of foods allow test compounds to be more evenly distributed throughout the "prey" tissues rather than coated onto exterior surfaces of the prey. These foods can be formed into a variety of shapes or sizes, and can be used to test the effects of other prey characteristics like the presence of siliceous or carbonate spicules in invertebrates or the presence of calcite or aragonite in calcified seaweeds (Harvell et al. 1988; Pawlik & Fenical 1989; Pennings & Paul 1992; Van Alstyne & Paul 1992; Hay et al. 1994; Chanas & Pawlik 1995).

Artificial seaweeds or invertebrates are made by mixing freeze-dried and finely powdered algae (palatable green seaweeds such as *Enteromorpha*, *Cladophora*, and *Ulva* commonly have been used) or animal tissue (squid, fish, or crustacean tissues) into agar, carrageenan, sodium alginate, or a mixture of carrageenan and paraffin wax and forming this into small strips, blocks, or other shapes depending on the form into which the mix is poured before it gels (see Harvell et al. 1988; Lindquist et al. 1992; Pennings & Paul 1992; Van Alstyne et al. 1992; Van Alstyne & Paul 1992; Hay et al. 1994; Pawlik et al. 1995). Forms can be as simple as small ice cube trays (Van Alstyne et al. 1992) or can be specially made to produce foods of particular sizes and shapes. Most investigators have incorporated some anchoring item (string, a scuba tank O-ring, etc.) into these gels so that they could be more easily transplanted and attached in the field. These items are then attached to ropes or other anchoring devices and transplanted into the field as described above (see Fig. 3 in Van Alstyne et al. 1992 as an example).

Alternatively, this same methodology can be used to make small food pellets with and without compounds that can then be released in the water column where they can be attacked by reef fishes. Feeding observations (number of bites per unit time, mass of pellet eaten, or acceptance versus rejection of the pellet) can be made on paired treatment versus control pellets in either field or laboratory situations (Lindquist et al. 1992; Pawlik et al. 1995; Lindquist & Hay 1996).

Production of these types of foods can vary slightly depending on the gelling agent and extracts being used, but all methods are reasonably similar with the exception that sodium-alginate-based foods do not mandate heating. Using sodium alginate will thus be advantageous if compounds are thermally unstable. Expanded procedures for preparing foods from agar, sodium alginate, and carrageenan are described by Harvell et al. (1988), Lindquist et al. (1992), Pennings and Paul (1992), Van Alstyne et al. (1992), Van Alstyne and Paul (1992), Hay et al. (1994), and Lindquist and Hay (1995). As an example, an agar-based food is prepared by mixing 20 ml of water with 0.72 g of agar and heating this in a microwave oven until it boils. The boiling water and agar are stirred and then poured into 16 ml of cold water containing 2–4 g of freeze-dried algae or animal matter (depending on the prey to be mimicked). This is stirred to assure uniform mixing and rapidly poured into a mold to harden. The rigidity of the food can be altered within useful limits by increasing or reducing the proportion of agar used.

To prepare treatment foods, natural concentrations of secondary metabolites are added to the food by placing the freeze-dried algal or animal tissue that is to be used in a small flask adding the desired amount of metabolite (i.e., the amount isolated from an equivalent volume or dry mass of the organism producing the metabolite) dissolved in diethyl ether, adding more ether until it covers all the food, and then removing the ether with a rotary evaporator or a stream of nitrogen (for more volatile compounds). This results in a uniform coating of metabolite on the food particles prior to their being added into the gelling agent. Control foods are treated identically but without addition of the metabolite. These foods can then be placed in the field as described above and loss of paired treatment and control foods measured as change in volume, mass, or length. If there is any autogenic change in these factors unrelated to consumer feeding, then the design and statistical concerns discussed by Peterson and Renaud (1989) need to be taken into account.

2.3.2.1.2. Laboratory Assays

Laboratory assays with vertebrates can use the same types of treatment and control foods as described above. However, laboratory assays allow modifications that may be advantageous under some circumstances. Two of the most commonly used ones are described below.

Because fishes in the laboratory can be isolated and closely observed, observations of whether they accept or reject a bite of treatment versus control food can be made much more reliably than in the field. Assays of invertebrate chemical defenses have commonly incorporated natural concentrations of secondary metabolites into small bite-sized food pellets that co-occurring predatory fishes had been trained to consume from a pipette. This procedure maximizes fish encounters with the food and thus rapidly allows for a determination of how a metabolite affects feeding once a fish has taken the potential prey into its mouth.

In these types of assays, food pellets are usually prepared by incorporating animal material (often macerated squid mantle) into a paste of 2% sodium alginate (by wet mass), squirting this from a syringe or small pipette into a solution of calcium chloride (which causes the alginate to harden into a strand of food with the consistency of cooked pasta), and then cutting this into small, bite-sized portions that can be offered individually to fish held in separate aquariums. The water and natural oils in this artificial diet are usually adequate to allow either water-soluble or lipid-soluble compounds to be mixed into the food before treating it with the calcium chloride. With some compounds, however, it may be advantageous to add a small quantity of ethanol to facilitate dispersal of lipid-soluble compounds (see Lindquist et al. 1992; Lindquist & Hay 1995 for examples of this approach). Alternatively, appropriate concentrations of compounds dissolved in solvent can be injected into a dried food, such as freeze-dried krill, and the solvent allowed to evaporate, producing a dried food particle containing the test compounds dispersed throughout the food (see Hay et al. 1989, 1990a, 1990b). Control foods are produced using solvent alone.

Feeding on these foods is recorded as the frequency of acceptance or rejection of treatment versus control pellets, with these pellets being offered alternately (see Hay et al. 1989, 1990a, 1990b; Lindquist et al. 1992; Pawlik et al. 1995; Lindquist & Hay 1996 as examples). Once fish become accustomed to this feeding methodology, they will very rapidly take pellets from the end of a pipette or from forceps each time these contact the water's surface. Because pellets are taken before they can leach their contents, this procedure allows the testing of water-soluble as well as lipid-soluble compounds.

For fishes that feed less rapidly or that will not feed when observers are nearby, some of the previously described field methods can be used or can be modified to better use the controlled conditions provided by the lab setting (see Targett et al. 1986; Hay et al. 1987a; Meyer & Paul 1992). As an example, Hay et al. (1994) modified food preparation procedures developed by Pennings and Paul (1992) to produce artificial algae in which rates of food removal would be especially easy to measure. These "seaweeds" were made by mixing freeze-dried and finely powdered green algae into an agar base and, using a methodological innovation developed by N. Lindquist, forming this onto fiberglass window screen

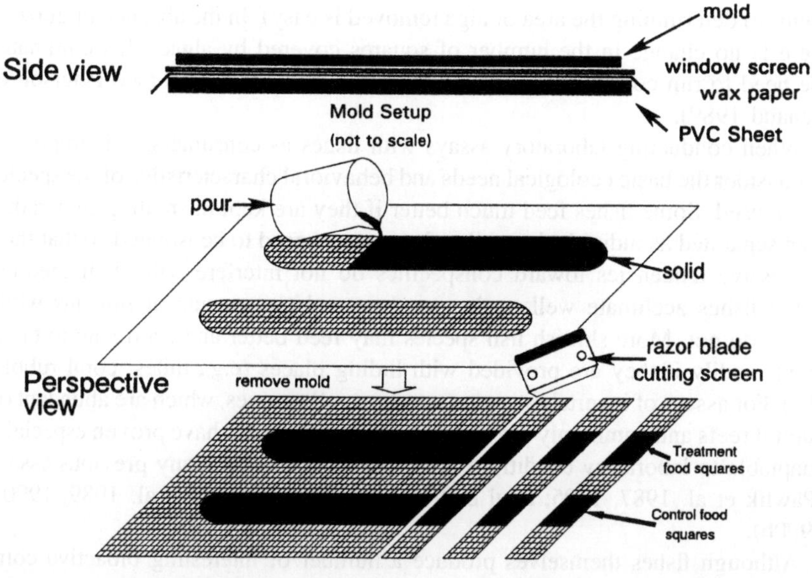

Figure 2.2. The mold used to form artificial foods onto window screen material and the resulting foods that are produced for assays with consumers. Consumption can be easily measured by counting the number of mesh squares cleared of food. The innovation of using window screening material for an imbedded matrix was developed by N. Lindquist. The figure was redrawn and modified from one in Hay et al. (1994).

mesh that provided support and a uniform internal grid (see Fig. 2.2) that allows quantification of amount eaten by counting the number of squares of the screen that have been cleared of algae.

This artificial seaweed was made by pouring the algal-agar mixture into the mold shown in Figure 2.2. A razor blade was used to smooth the food into the mold and assure a uniform thickness. The standard mold was made from a rectangular piece of 2-mm-thick Formica™ (a plastic laminate) with two 1.3- × 55.5-cm openings cut into it (this sizing can be altered to accommodate consumers that need more or less food). A piece of window screen was clamped between the mold and a smooth sheet of polyvinylchloride (PVC) plastic. Waxed paper placed beneath the screen prevented the food from sticking to the PVC sheet. For an assay, a control food (without secondary metabolite) is poured into one side of the mold and a treatment food (with metabolite) into the other. As the agar cools, it takes the shape of the mold and becomes firmly attached to the screen. Cutting the screens perpendicular to the food material produces strips of screen containing equal sized rectangles of both treatment and control foods (see Fig. 2.2) that can be presented to consumers in the lab. Feeding is measured by counting the number of screen squares from which algal material is completely removed (i.e., "algae" come with the equivalent of graph paper embedded in them, so determining the area of alga removed is easy). In the absence of grazers, there is no change in the number of squares covered by algae; this eliminates the need to run controls for changes unrelated to consumption (see Peterson & Renaud 1989).

When conducting laboratory assays with fishes as consumers, it is important to consider the basic ecological needs and behavioral characteristics of the species being used. Some fishes feed much better if they are kept in small groups rather than separated as individuals, while other species need to be isolated so that their aggressive tendencies toward conspecifics do not interfere with their feeding. Some fishes acclimate well to the presence and movements of humans while others do not. More skittish fish species may feed better and acclimate to tanks more readily if they are provided with hiding places (e.g., tubes, coral rubble, etc.). For assays of invertebrate compounds, small wrasses, which are abundant on natural reefs and commonly prey on benthic invertebrates, have proven especially adaptable to laboratory conditions and have been used in many previous assays (Pawlik et al. 1987, 1995; Paul & Van Alstyne 1992; Hay et al. 1989, 1990a, 1990b).

Although fishes themselves produce a number of interesting bioactive compounds that have been hypothesized to serve as feeding deterrents to larger fishes (Tachibana 1988), this assumption has not been tested directly using ecologically realistic procedures. Modification of some of the procedures outlined above should allow direct tests of this.

In a few cases, fishes have been used to assess the chemical noxiousness of larval, rather than adult, invertebrates. More investigators have extracted large numbers of larvae, incorporated larval extracts or pure compounds into an artificial food (usually squid paste), and presented this food and an equivalent control to fishes as small pellets (see Lindquist & Hay 1995, 1996 and references therein). In one case, fishes that did not normally co-occur with the chemically rich larvae consumed treated foods during the first 2 days of feeding, but regurgitated these foods within 2 h and refused to eat these foods in subsequent assays (Lindquist & Hay 1995). This assay points to the need to employ methods that allow for learned aversion in certain feeding assays. Chemically mediated learned aversion is relatively unstudied in aquatic systems, but a series of excellent studies using terrestrial vertebrates (Provenza & Malechek 1986; Provenza & Balph 1987; Burritt & Provenza 1989; Provenza et al. 1990; Distel & Provenza 1991) provide general methodologies and approaches that could serve as models for designing similar studies using aquatic organisms.

The squid-paste methodology used above allows assays of how compounds affect feeding; but the methodology does not assess the variability among individual larvae in the repellence of their chemical extracts. While investigating the chemical deterrence of ascidian larvae, Lindquist et al. (1992) developed a methodology for extracting individual larvae and infusing this extract into a similar sized food particle so that each food item would be representative of the individual

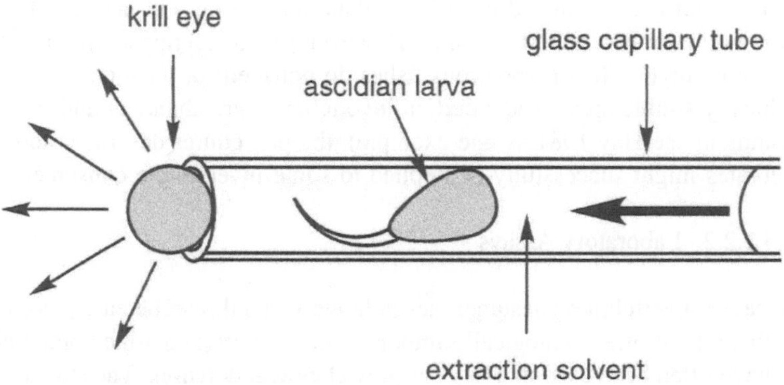

Figure 2.3. The "micromethod" used to extract organic soluble metabolites from an individual ascidian larva and impregnate that extract into an individual krill eye. After inserting the larva about 2.5 mm into a capillary tube, the end containing the larva (left end in this figure) was dipped into methanol, which was drawn about 2.5 mm past the larva by capillary action. The right end of the capillary tube was then sealed with parafilm to prevent evaporation from that end, and a single krill eye was placed on the open end of the tube. Evaporation of the solvent from the krill eye drew the solvent past the stationary larva, extracted the organic-soluble compounds, and impregnated them into the krill eye. As the last of the solvent evaporated, the eye would fall from the capillary tube. Controls consisted of krill eyes treated in the same manner, but without a larva being placed into the capillary tube. Figure and methods are from Lindquist et al. (1992).

larvae used to prepare it. This "micromethod" consisted of inserting an individual larvae about 0.25 cm into a capillary tube. This end of the capillary tube was then dipped into the extraction solvent (methanol) allowing capillary action to draw solvent into the tube past the larvae until the solvent was about 2 cm past the larva. The far end of the capillary tube was then sealed off with parafilm and a single freeze-dried eye from a krill (used because it approximated the size of a single larva) was placed on the other end of the capillary tube. Evaporation of the solvent from the krill eye drew the solvent past the stationary larva, extracting the organic-soluble compounds, and impregnating them into the krill eye. As the last of the solvent evaporated, the eye would fall from the end of the tube (see Fig. 2.3). Control eyes were prepared in the same manner but without a larva in the capillary tube. Paired sets of treatment and control eyes were then offered to fishes in the types of laboratory assays described previously.

2.3.2.2. *Invertebrates*

2.3.2.2.1. Field Assays

Field assays of chemical defenses against invertebrate consumers have rarely been conducted, possibly because palatable foods placed in the field are more rapidly found and consumed by fishes—thus, interfering with an unambiguous determination of invertebrate feeding. However, by conducting assays at specific times (e.g., night when herbivorous fishes do not feed) or locations (e.g., topographically simple areas where certain invertebrates are abundant and fishes are uncommon; see Hay 1981 as one example), the procedures described above for vertebrates might successfully be applied to some invertebrate consumers.

2.3.2.2.2. Laboratory Assays

Because invertebrate consumers are relatively small, can be easily housed in the lab, and are often ecologically important consumers in marine communities, they have often been used in assays of prey chemical defenses. The slow feeding responses of certain marine invertebrates (e.g., sea stars and some gastropods) have encouraged some investigators to use behavioral tests (tube foot or tentacle retractions) as proxies for feeding deterrence. We provide one widely used example of this below, but concentrate most of our efforts on assays that measure feeding more directly.

Barnes et al. (1970), Collins (1974), Sloan (1980), and other investigators have noted that the sensory tube feet of sea stars respond in a stereotypic manner when presented chemical stimuli extracted from preferred prey, low preference prey, and potential predators. Because sea stars use their tube feet both to sense chemical cues from a distance and as taste receptors, it is unclear if these reactions are best used as indicating distance chemoreception or palatability. We include them in this section on feeding simply because numerous previous investigators have interpreted their assays as indicative of feeding cues. This procedure assumes (McClintock et al. 1992) that sustained sensory tube feet retraction and arm withdrawal are indicative of a negative response to a given stimulant. Although using sensory tube foot behavior to predict a feeding response in sea stars seems reasonable, this assumption has not been tested by correlating tube foot response to an extract with sea star feeding behavior on a normally palatable food containing that extract. The following description of the "tube foot retraction" assay is summarized from McClintock et al. (1994). These assays are typically conducted by placing sea stars on their aboral side in a shallow bowl and covering them with sea water. Prey extracts or purified metabolites are mixed in a 1:1 ratio (wt/wt) with an inert silicone grease. The tip of a glass rod thinly coated with the silicone-extract mixture is presented to an extended tube foot on the tip of an arm such that the motion of the tube foot creates contact with the rod. The response of the tube foot to contact with the rod is recorded for a period of 60 s.

Three different controls are usually conducted simultaneously with experimental treatments: (1) a glass rod coated with a 1:1 ratio (wt/wt) of fish muscle extract and silicone (positive control), (2) a glass rod alone (mechanical control), and (3) a glass rod coated with silicone alone (silicone control). Trials of each of the four treatments (test extract and three controls) are conducted with the sequence of presentation being randomized. Retraction of the sensory tube feet after contact with a chemical treatment for a period significantly greater than that caused by the controls is assumed to indicate a feeding deterrent. In the data presented by McClintock et al. (1994), the values for all three types of controls (see above) were very similar, and all appear to differ substantially from the values for several of the tests using prey extracts. However, it should be noted that the only true control (i.e., differing from the treatment in only one factor) among the three used was the glass rod coated with silicone alone. Because the assumption in these assays is that the deterrent extracts should cause the avoidance of a food, not just the avoidance of silicone, it would be appropriate to use the fish muscle extract in silicone as a control and test it against chemical extracts added to this same silicone plus fish muscle treatment. This would indicate more clearly whether the positive response to fish extract was diminished by compounds from extracts of prey that are avoided by the sea star. Although these types of assays can produce considerable information in a short period of time, one needs to be cautious in unambiguously interpreting the activity of tube feet as a clear indication of effect on feeding. With the majority of invertebrate consumers, direct tests on feeding should be possible.

Many direct tests for chemical feeding deterrents against amphipods and sea urchins have been conducted by dissolving compounds in diethyl ether and coating the extracts or compounds onto the surface of an alga that is palatable. As an early example, Paul et al. (1987) coated secondary metabolites from the red seaweed *Ochtodes* onto uniform diameter disks of the green alga *Ulva* sp. A cork borer was used to produce the uniform sized algal disks. The compound coated disks were offered to amphipods along with solvent-coated control disks for 36–48 h. Because treatment and control disks were indistinguishable when placed together with the amphipods, a length of thread was pulled half-way through each disk and tied so that the position of a small knot indicated treatments versus controls. The amount of each disk consumed by the amphipods was determined by a point-intercept method using a dissecting microscope fitted with an ocular grid. Modifications of this general methodology have been used in several other studies with amphipods, urchins, or polychaetes (McConnell et al. 1982; Hay et al. 1987a, 1987b, 1988b, 1988d, 1990a; Duffy & Hay 1994).

The above procedure produces potential problems associated with coating compounds onto food surfaces rather imbedding them throughout the foods. These problems can be avoided by incorporating metabolites directly into gel-based foods. As examples, artificial seaweeds made from powdered seaweeds and gelling agents such as agar have been used in feeding studies with amphipods, urchins, and other small invertebrates. Hay et al. (1994) present a modification of the assay method that appears particularly good for use with small mesograzers because it allows for measurement of small amounts of total consumption. To form the artificial food for amphipods (Fig. 2.2) the algal-agar matrix is poured over window screen lying on a piece of waxed paper, a second piece of waxed paper is placed over the screen and a heavy flat object is placed on top to force the food into a thin uniform layer that fills openings in the screen. After the food has cooled and solidified, the wax paper is peeled away leaving the food affixed between the screen fibers. Small sections of the screen (commonly 7 × 7 mesh units) are cut out and treatment and control sections are simultaneously offered to the amphipods. A small notch cut in the corner of one of the treatments allows treatment and control foods to be easily identified. Replicates are monitored periodically and the algal foods removed (1) whenever one half or more of either the treatment or the control food has been eaten or (2) at the termination of the experiment (for replicates with low feeding rates). The amount eaten is determined by counting the number of screen squares from which the algal material has been completely removed. This methodology also works well with other small crustaceans such as crabs (Stachowicz & Hay 1996). In a small number of cases, a compound or extract added to these agar-based foods interferes with the normal gelling process, making the treatment food softer and more easily dislodged from the mesh backing than the control food (Bolser & Hay 1996). This problem can usually be remedied by increasing the concentration of agar, decreasing the concentration of test extract, or using an alternative gelling agent, such as carrageenan, sodium alginate, or Phytagel™.

Coating (Hay et al. 1987a, 1987b, 1988b, 1988d) or adsorbing (Steinberg 1984) compounds onto the surface of palatable seaweeds or presenting artificial seaweeds formed on window screen (Hay et al. 1994; Cronin & Hay 1996b, 1996c, 1996d; Bolser & Hay 1996) have both been used with larger mobile herbivores such as sea urchins or gastropods. Additionally, Steinberg (1988) demonstrated that most algal phlorotannins would remain in artificial foods made from agar for long enough periods of time to directly test the effects of these compounds on urchin or gastropod feeding. We (M. Deal and M.E. Hay work in progress) have also found that some nonphenolic compounds that are water soluble can be assayed in agar-based foods so long as the assays are completed within 1–2 h after the foods are immersed in water. The rates at which particular water soluble metabolites are lost from agar-based foods have rarely been measured, but they can be appreciable for some compounds (Steinberg 1988). Better assay procedures (perhaps microencapsulation) need to be developed for testing water-soluble compounds.

For omnivorous and carnivorous benthic invertebrates, Lindquist et al. (1992) used a squid-based artificial food (preparation described in the above section on vertebrates) in a study of ascidian chemical defenses. To present this food to benthic invertebrates, pellets of treatment and control foods were formed in brass or plastic rings placed over a small piece of the "hook" side of Velcro® glued to a heavy PVC disk. As they gelled, the pellets became securely attached to the PVC disk. The size of the pellets could be varied according to the size of the consumer by varying the diameter of the ring used as a mold. Treatment and control pellets were formed on opposite edges on the upper side of each disk.

One PVC disk and one test animal were placed in glass beakers filled with seawater. Each beaker was covered with fiberglass screen to prevent the animals from escaping and then placed back into flowing seawater. Individual animals were allowed to feed until about 50% of either pellet on a disk was consumed. Remaining portions of treatment and control pellets were cut away from the disk at the Velcro interface and weighed to determine loss to consumers. Controls to account for variance in pellet mass unrelated to the activity of the consumers were also conducted; the mass of treatment and control pellets not exposed to consumers did not differ. This method was used with hermit crabs, shrimp, and a gastropod.

Assays using invertebrates that are sedentary and do not move to their foods may require different methods than those used with mobile invertebrates. Sedentary consumers, unlike most mobile species, capture and consume food brought to them by passing water currents. The feeding mechanisms of these sensile invertebrates are diverse. Corals, anemones, and other cnidarians generally capture and consume prey individually; thus these kinds of consumers may make choices regarding the suitability of each food particle. In contrast, filter feeders such as bivalves and tunicates drive water through a netlike filter and then into the gut. These consumers may have less ability to discriminate between individual food particles (but see Tamburri & Zimmer-Faust 1996). Feeding assays using sessile benthic invertebrates as model consumers need to consider these differences.

Relatively few studies of prey chemical defenses have used sessile invertebrates as consumers. Lindquist (1996) used several species of Caribbean corals to examine the palatability of various invertebrate larvae relative to a palatable control food. Although not a test of extract or compound palatability, the assay methodologies that were used are applicable to such studies. Most corals are colonial, making it possible to offer separate polyps on the same colony a treatment and a control food simultaneously. This characteristic of clonal organisms enables the investigator to control for genetic or phenotypic variance among individual consumers. Lindquist (1996) tested larval palatability by simultaneously offering a piece of a squid based control food to one polyp and offering a larva to a nearby polyp. Frequency of consumption and subsequent regurgitation of each food type was noted and analyzed.

Sea anemones and some corals exist as solitary polyps. A single polyp should not be fed both a treatment and control food because response to unpalatable foods is not always immediate; they often regurgitate noxious foods a few minutes after consumption. A study by Lindquist & Hay (1995) illustrates this point. This study used the sea anemone *Aiptasia pallida* as a model larval predator to examine the long-term consequences of consuming prey chemical defenses. In this experiment, some anemones were fed a small squid-based food pellet that contained didemnin cyclic peptides (isolated from a Caribbean asicidian) while others were fed a similar pellet lacking didemnins. The assay food was made from squid and sodium alginate as described previously. In this investigation, anemones adjacent to one another in the field were used as paired treatment and control consumers. Because *A. pallida* reproduces asexually by budding of its adhesive papilla, adjacent anemones are likely to be clones, thus some degree of genetic and phenotypic similarity between treatment and control anemones can be achieved. Those anemones offered the pellets with didemnins would almost always ingest them, but within 1–60 min, 60–100% of the treatment anemones would regurgitate their pellets; control anemones rarely regurgitated. Similar assays with corals consuming live invertebrate larvae demonstrated that numerous larvae survived this short-term ingestion (Lindquist 1996).

As prey items become smaller, ecologically realistic assays of their chemical defenses often become increasingly difficult to conduct. However, the tremendous effects that compounds from marine microbes, such as toxic phytoplankton, can have on entire ecosystems (Paerl 1988; Hay & Fenical 1996) suggest that investigating the natural function of microbial metabolites should be a priority of marine chemical ecologists. A variety of planktonic organisms, including diatoms and dinoflagellates, produce bioactive secondary metabolites; however, the ecological function of these metabolites is not well known. One of the major impediments to advances in this general field is the difficulty of conducting ecologically realistic bioassays on the effects of metabolites produced by microbes. As an example, Shaw et al. (1995a, 1995b) found that the diatoms *Phaeodactylum tricornutum* and *Thalassiosira pseudonana* produced apo-fucoxanthinoids that appeared to deter feeding by copepods. Even though these compounds are held in the diatom cells and not released into the water, they tested the activity of these compounds by dissolving them into the water holding the copepods and measuring fecal production as an indication of feeding suppression. A decrease in fecal production was interpreted as a result of feeding deterrence even though the effect could have been caused by physiological stress, a narcotic effect that diminished activity levels, or an effect on gut passage rate instead of feeding. Although these conditions may affect copepod feeding, compounds that produce these effects are not necessarily feeding deterrent—altered temperature or salinity could produce these same effects.

These types of ecologically unrealistic assays were commonly used on microorganisms 20 years ago (and are still used by some investigators) but they have important limitations for evaluating most ecological phenomena. Numerous investigators have demonstrated that there is no predictive relationship between the effect of a compound placed in an organism's water and the effect of that compound if it is eaten by the organism (Hay & Fenical 1988; Sammarco & Coll 1992; Pawlik et al. 1995). Attempts to assess the physiological effects of most prey metabolites on consumers by dissolving metabolites into water holding the consumer have minimal ecological relevance and should be discontinued. The possible exception to this general caution involves microorganisms that both occur in dense blooms and release metabolites into the surrounding water. If specific metabolites are known to be released under natural conditions, then bioassays of these metabolites dispersed into the water may be justified.

A few investigators have developed more direct assays of how microbial metabolites affect invertebrate feeding. Ward & Targett (1989) examined the palatability of reverse-phase and normal-phase chromatography particles containing adsorbed algal metabolites that had been exuded into culture medium by six species of microalgae. Their procedures allowed treated particles to be presented to mussels in a manner that mimics how mussels would encounter whole microalgal cells. However, because most feeders are unlikely to disrupt prey surfaces on initial contact, particles coated with an extract or compound might provide stimuli that are not normally associated with the prey unless the defensive compounds are exuded or present on prey surfaces (the entire question of where compounds are located in prey and how they are sensed by consumers is relatively unstudied and provides a rich area for productive investigation). To prepare treatment particles, each type of chromatography material (silica, silica amino, alumina, octadecyl silane [ODS-2], and polystyrene—all with a diameter of about 10 μm) was separately placed on the bottom of a sintered-glass frit of a Buchner funnel. Culture filtrate (900 ml) or an equal amount of control media was drawn over the particles using vacuum filtration to yield the treated particle for the feeding assays. Treated spheres were resuspended in 200–220 ml of filtered seawater, kept cold, and used within 1 h. The ODS-2 spheres were dispersed in the seawater by adding about 15 μl of 20% sodium dodecyl sulfate. Three plastic dishes containing one mussel each (starved for 24 h prior to beginning an assay so that their guts were empty) were used for each treatment. The total concentration of chromatography spheres offered to each mussel was 1×10^4 spheres/ml. Red polystyrene spheres treated with culture medium were used as the control. Treatment and red control spheres were kept suspended using a magnetic stirrer, and the total concentration of spheres (treatment and reference) was 2×10^4 spheres/ml. As mussels fed on the mixture of treatment and control beads, they produced feces (from particles that have been digested) and pseudofeces (from particles that have been ingested, but then passed through without being digested). It was assumed that if no selection occurred between particles, then the ratio of red (control) beads to white (treatment) beads would not differ between feces and pseudofeces. If selection occurred there would be a difference. Mussels were allowed to feed for 1 h, then removed from the dish, washed with clean sea water and placed in another dish filled with aged sea water. Pseudofeces and feces were collected from the start of the experiment for 45 h. Selective feeding by the mussels was evaluated by determining the ratio of treatment to control beads.

Within planktonic communities, probably the most well studied trophic interactions are between microalgae and herbivorous zooplankton. Targett and Ward (1991) reviewed the literature regarding chemical mediation of feeding (both positive and negative) by phytophagous suspension-feeding invertebrates, including zooplankters. In their review, they discuss the implications of contact versus distance chemoreception in microalga-zooplankter interactions and assay methodologies for examining microalgal chemistry using phytophagous suspension feeders. Previous approaches involve assaying compounds from (1) microalgal cell homogenates or extracts, (2) cell-free culture media, and (3) pure microalgal metabolites.

In summary (Targett & Ward 1991), the use of homogenates from microalgal cells may not be relevant to grazer's chemical recognition of prey because grazers rarely, if ever, encounter whole cell homogenates under natural conditions (except perhaps when microalgal blooms senesce). The uncertainty regarding (1) the role of bacteria in modifying the composition of microalgal homogenates and cell-free extracts and (2) the composition of exudates from cultured versus natural microalgal populations are problematic for assays using homogenates or cell-free extracts. Assaying purified microalgal metabolites can provide information on the relationship between compound structure and ecological function, provided that compounds are offered to grazers in ways that mimic how the compounds are distributed within, on, or around the microalgal cells.

2.3.3. Feeding Stimulants

Most ecological research on chemically mediated feeding has focused on defenses against consumers (Hay & Fenical 1988; Hay & Steinberg 1992; Paul 1992; Pawlik 1993). Feeding stimulants and compounds that consumers use to search for food are of equal importance but are less thoroughly studied within an ecological context. Sakata (1989) reviews feeding attractants and stimulants for marine gastropods, but much of the methodology he describes has been used successfully with other invertebrates as well. His review shows that specific amino acids, sugars, carbohydrates, glycerolipids, etc. can attract or repel gastropods. Interest in phagostimulants has been driven more by the economic benefits of maximizing feeding rates of cultured gastropods than by ecological and evolutionary implications, and most investigations in this area have tested commercially available compounds rather than conducted bioassay-guided discovery of the metabolites that affect feeding in nature. However, Sakata and coworkers (Sakata et al. 1984, 1986) have developed methods for bioassay guided discovery of natural feeding stimulants. As a test substrate, they use crystalline cellulose (Avicel SF) coated on a glass plate. Circles (2.5 cm diameter) are drawn on this plate, a sample solution (extract or purified compound) is absorbed onto the cellulose within each circle, and the plate is placed on the bottom of a test tank containing abalone that have been starved for 24 h. After being left in the tank overnight, the density of radular feeding scar within the various circles (including control circles) are compared (Fig. 2.4). Observations made of assays in progress suggested abalone creep along the bottom, periodically sampling the substrate, but stop and feed within one area when a phagostimulant is encountered. Thus,

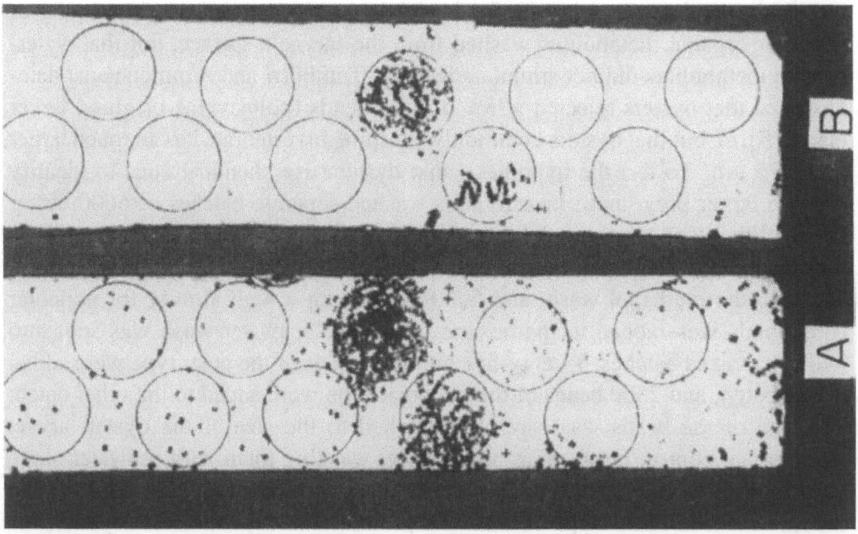

Figure 2.4. Avicel plates showing abalone grazing scars on circles containing different fractions from a Sephadex LH-20 column chromatography of the MeOH extracts from a prey organism. Aliquots of each fraction are absorbed into one of the marked circles, the solvent is allowed to evaporate, and the plates are left overnight on the bottom of aquariums where they can be grazed by abalone or other gastropods. Plates *A* and *B* were placed in separate aquaria. In both aquariums the fourth and fifth fractions from the bottom of the plate stimulated feeding. Overall feeding rate was obviously greater in aquarium A than in aquarium B. Methods and figure (slightly modified) are from Sakata (1989).

Sakata and coworkers suggest that this assay is useful for identifying phagostimulants but not attractants.

Sakata and coworkers have used the Avicel® plate bioassay procedure to isolate and characterize several phagostimulants from algae that are highly preferred by abalone and other creeping grazers. To date only lipid-soluble compounds have been identified as phagostimulants by this methods, whereas many water-soluble compounds have been identified previously as feeding attractants and stimulants for other marine consumers. This dichotomy probably results from the different bioassay methodologies used to investigate feeding attractants and stimulants in these different groups. Highly water-soluble compounds are unlikely to remain absorbed to Avicel for appreciable amounts of time, and thus unlikely to be identified as phagostimulants. Many other studies of feeding attractants or stimulants have assayed "off-the-shelf" compounds (carbohydrates, amino acids, and nucleoside) that tended to be water soluble.

Using a separate methodology, Tamburri and Zimmer-Faust (1996) found that oysters consumed numerous species of invertebrate larvae and that their recognition and acceptance of a food particle was stimulated by either polar or nonpolar organic metabolites washed from the larvae's surface, but that water-soluble metabolites did not stimulate feeding. Tamburri and Zimmer-Faust demonstrated that oysters rejected >75% of inert beads (polystyrene or glass) larger than 275 μm, but that oysters commonly consume invertebrate larvae much larger than 275 μm. To test the hypothesis that oysters use chemical cues to identify suitable larger prey, these investigators washed separate batches of 5000 oyster larvae with either seawater, 9:1 methanol/seawater, or 9:1 hexane/seawater. Five thousand chromatography beads with a high affinity for polar compounds were added to the methanol wash, and 5000 beads with a high affinity for nonpolar compounds were added to the hexane wash. The seawater wash was split into two equal-sized batches. Twenty-five hundred beads of the polar type were added to one batch, and 2500 beads of the nonpolar type were added to the other batch. The size of the beads was carefully matched to the size of the oyster larvae. Appropriate control beads were prepared by washing them with seawater alone (both polar and nonpolar beads), with methanol (polar beads), or with hexane (nonpolar beads). Each type of bead was presented to eight different oysters, but only one bead type was presented to each oyster. The beads were gently pipetted into the mantle cavity of a gaping oyster, and the percentage of beads in the feces versus pseudofeces was determined over a 24-h period. Beads conditioned with either the methanol or the hexane wash from the oyster larvae were consumed significantly more than were beads treated with solvent alone. Additionally, no significant difference in ingestion was found between beads treated with seawater alone versus the seawater rinse of the larvae. These results indicate that oysters are using lipid-soluble chemical cues to identify larger particles that are appropriate foods.

Because most investigations of feeding stimulants have been focused on aquacultural or similar needs, we do not consider this topic further because few studies provide examples that should be employed when ecological realism is desired. Many of the feeding deterrent assays discussed above could be modified to investigate feeding stimulants.

2.3.4. Multiple Cues and Defensive Synergisms

It is becoming increasingly apparent that consumers commonly make feeding choices based on a wide variety of prey characteristics (prey nutritional value, toughness, the presence of feeding stimulants versus deterrents, etc.). Although secondary metabolites can have strong direct effects on consumer feeding preferences, they do not act in isolation from other plant traits or from the environmental context in which consumer-prey interactions occur. An adequate understanding of chemical defenses will not be achieved until we more fully appreciate the full range of defenses that prey species can use, and how these defenses are integrated.

Because most studies of marine chemical defenses have manipulated individual compounds or structural traits rather than multiple prey traits, more investigation and development of appropriate methodology is needed.

The few studies that have simultaneously manipulated multiple prey characteristics indicate that prey nutritional quality and prey chemical defenses may work in concert to affect prey susceptibility to consumers. When Duffy and Paul (1992) and Pennings et al. (1994) evaluated how marine secondary metabolites affected feeding by reef consumers using foods that differed in their ratios of protein to carbohydrate, they found that some compounds were effective at reducing feeding on low-protein foods but ineffective at reducing feeding on higher-protein foods. Hay et al. (1994) noted a similar pattern when they varied algal concentrations in chemically defended agar-based foods being consumed by a sea urchin.

The defensive value of secondary metabolites in foods of variable nutritional quality should change as a function of the specific physiological effect of the metabolite on consumers. If compounds lower growth rates or digestive efficiencies rather than survivorship or fecundity, then a consumer might choose a defended high-quality food over an undefended lower-quality food because the consumer's net income from the defended, but nutritionally rich, prey could be higher. Alternatively, if a compound is toxic and strongly affects survivorship or reproduction, then consumers may be selected to avoid such a compound regardless of the nutritional value of the prey.

Chemically defended organisms often produce multiple secondary metabolites, several of which act as feeding deterrents. This provides the possibility for synergistic or additive effects among the multiple metabolites. Investigations are needed on the interactive effects of multiple metabolites; however, few of the studies available at present have found synergistic interactions among secondary metabolites. As an example, when Lumbang and Paul (1996) tested multiple brominated sesquiterpenes produced by the green algae *Neomeris annulata* against two reef fishes and an urchin, they found that each compound was significantly deterrent at or below its natural concentration and that all of the compounds together were not more deterrent than each alone. Stachowitz and Lindquist (1997) also have tested for, but not found, interactive effects among three structurally related secondary metabolites produced by the hydroid *Tridentata marginata*. All feeding deterrent effects documented when the three compounds were combined appeared to be due to the activity of one compound alone. More studies like this will be needed before the frequency of synergistic or additive effects can be assessed.

Several studies have indicated that the chemical and structural defenses (e.g., $CaCO_3$) that commonly co-occur in both marine plants (calcification) and benthic invertebrates (spicules, calcification) can function either additively or synergistically to reduce susceptibility to consumers (Hay et al. 1994; Schupp & Paul 1994; Meyer & Paul 1995; Pennings et al. 1996). In one of the more elaborate examples, the calcified alga *Halimeda goreaui* contains both an unusual secondary metabolite and a heavily calcified thallus. When the effects of the metabolite and the $CaCO_3$ on urchin grazing were tested separately using a nutritionally valuable agar-based food, neither had any deterrent effect; however, when these traits were combined they interacted synergistically to deter feeding (Hay et al. 1994). When this assay was repeated using a test food of lower quality, this relationship changed, indicating that food value, $CaCO_3$, and secondary metabolites may interact to affect consumer food choices. These assays used the agar-based window screen methods (Hay et al. 1994) that are modifications of those developed by Pennings and Paul (1992). Either of these procedures provide ways to manipulate multiple prey traits and assess their potential additive or synergistic effects. The statistical procedures that are most appropriate for assessing data from these types of experiments are debated (Hay 1996a; Pennings 1996) and in need of further refinement. Marine chemical ecologists have recently focused more attention on trying to understand prey chemical defenses within these broader and more ecologically realistic contexts; these efforts are well founded and need to continue. The rate of progress here will depend, in part, on the development of novel and more ecologically realistic methods.

2.4. Consequences of Consuming Defensive Metabolites

The investigations described above document how prey metabolites affect consumer behavior, but provide no information on why consumers would evolve to detect and avoid certain metabolites. To assess this, one needs to determine how consuming these metabolites affects consumer physiology and fitness. Assays on these topics consist of short-term assays that determine immediate effects (regurgitation, effects on behavior, effects on assimilation efficiency, etc.) and longer-term assays that measure effects on growth, survivorship, and fecundity. The time scale over which assays should be run will depend, to some extent, on the mode of action of the secondary metabolite being tested and the response variables being measured. Toxins may have quick modes of action, whereas other compounds may cause regurgitation (possibly limiting the toxic effect) or decrease growth or fecundity over the life cycle of the consumer. In long-term experiments that require the storage of compounds, compounds could degrade, necessitating periodic checks for purity and concentration. Water-soluble compounds could leach out of experimental foods, reducing the oral dosage received by test organisms as a function of time (Steinberg 1988), and potentially increasing levels of the compounds in the surrounding water. If treatment and control consumers are not kept in the same water, then ecologically unrealistic effects of compounds in the water can be misinterpreted as effects of consumption.

Because compounds that have no detectable effects when they are consumed can kill organisms in seconds to minutes when placed in their water (discussed in Hay 1996b), this confounding of effects is potentially serious. Finally, interpreting the data from animals fed *ad libitum* will be more difficult than interpreting experiments where consumption of treatment and control food is equalized because the effects of amounts of food eaten can be confounded with the direct physiological effects of the compounds on consumers (see the examples for insects in the review by Slansky 1992).

Although it seems reasonable to assume that consumers avoid prey containing certain secondary metabolites because those metabolites negatively affect consumer fitness, this has rarely been demonstrated, especially for marine consumers. Data from some terrestrial studies suggest that numerous deterrent plant compounds have no negative biological effects when consumed by herbivorous insects (Bernays & Chapman 1987; Bernays & Cornelius 1992). However, even for these well-studied cases of chemically mediated plant-insect interactions, it is often unclear whether consumers limit ingestion of chemically rich prey due to the negative postingestive effects or whether prey are simply producing compounds that consumers find distasteful but that have little effect on their fitness (Bernays 1991; Bernays & Cornelius 1992).

Studies on the consequences for marine consumers of ingesting prey secondary metabolites are rare, and several of the studies that have been conducted use methodologies that lack ecological realism or statistical rigor. As examples, some investigators have raised consumers on diets containing different secondary metabolites, but all individuals experiencing a given treatment diet were housed in the same culture vessel, thus making the experiment pseudoreplicated (see Hurlbert 1984 for a discussion of pseudoreplication). Studies that do not suffer these design limitations may, nonetheless, lack ecological realism. As examples, Hay et al. (1987b) and Steinberg and van Altena (1992) conducted adequately designed, longer-term studies that fed consumers diets containing algal metabolites at concentrations naturally occurring in some seaweeds. Although this seems reasonable, these generalist consumers that naturally feed on many different foods would never eat one species of seaweed for an extended period under natural conditions and thus would not consume this quantity of metabolite per unit time or per unit of total ingested food. These compound concentrations were therefore unnaturally high as a proportion of the total diet (see discussion in Lindquist & Hay 1995). As an additional consideration, the quality of food in which a compound is delivered can not only influence how a consumer reacts to the compound (Duffy & Paul 1992; Hay et al. 1994; Pennings et al. 1994), but it may also alter the capacity of an organism to detoxify a compound (Yang et al. 1992). Thus, experimental diets should approximate the nutritional content of natural prey. Below, we summarize a broad variety of approaches in hopes of stimulating more work in this important, but inadequately investigated area.

2.4.1. Vertebrates

2.4.1.1. Short-Term Assays

Rejection of a defensive chemical is not always instantaneous. Prostaglandins from gorgonian corals and didemnins from ascidians each induce regurgitation in fishes following ingestion. Fishes learn to avoid the compounds, or the types of foods containing the compounds, after successive ingestion and regurgitation of food treated with these metabolites. Regurgitation and its effects on learned aversion to foods occur in both the lab and field (Gerhart 1984; Pennings 1990, Lindquist & Hay 1995).

Gerhart (1984) demonstrated that fish usually ingested food pellets containing prostaglandins the first time they were offered, but rejected them in later trials. Following these observations, fish were fed pellets either with or without prostaglandins and watched closely for 25 min. Fish regurgitated about 7 min after ingesting a treated pellet, but no regurgitation occurred in fish that consumed control pellets. Fish held in water containing prostaglandins did not regurgitate, indicating that the response was not caused by prostaglandins leaching from the pellets into the fish's water. Field experiments were also conducted by feeding fishes pieces of boiled fish treated with prostaglandins or lacking prostaglandins. After ingestion of either a treatment or a control food, a fish was observed for 10 min. Fish regurgitated about 2 min after ingestion of the treated food; no regurgitation was observed in fish that ate the control food.

One problem encountered in Gerhart's study needs to be mentioned, especially to biologists who may not have the experience or equipment to follow chemical alterations in the compounds they are testing. The prostaglandins in the coral studied by Gerhart occur naturally as methyl diesters. The coral enzymes transform these molecules to the acid form of the prostaglandins when the coral tissue is ground, but this conversion is not rapid, so fishes biting corals would initially encounter the methyl diesters, rather than the acid form of the prostaglandins. Thus, the compounds tested against fishes were not those that actually occur in the coral's tissue, and the naturally occurring metabolite does not deter feeding by fishes (Pawlik & Fenical 1989).

Determining the appropriate "natural" product to test can, however, be a matter of perspective. Feeding by fish involves mechanical disruption of cells, which presumably releases enzymes, so enzymatic transformation of the prostaglandins in the coral could be occurring in the fishes gut. Enzymatic conversion of marine natural compounds is not limited to this coral (Paul & Van Alstyne 1992), so these types of questions will occur in other instances as well. Again, the best general advice we can offer on dealing with these types of problems is to make sure that investigations are conducted rigorously from both the chemical and biological perspectives. This will often mandate collaborative interactions between chemists and biologists.

A study by Pennings (1990) provides a novel and experimentally desirable approach for studying effects of prey metabolites on consumers. He raised sea hares (*Aplysia*) on red algae from which they sequester secondary metabolites and on green algae from which they do not sequester metabolites. This produced sea hares with high and low concentrations of secondary metabolites, respectively. When these were fed, or force-fed, to consumers, some fishes regurgitated the metabolite-rich sea hares more frequently than the metabolite-poor sea hares. Additionally, for regurgitated sea hares, those that were metabolite rich were digested less completely than those that were metabolite poor, although all were killed by the fish.

Regurgitation and its effect on learned aversion also have been studied in glaucous-winged gull chicks that were fed clams that had or had not accumulated paralytic shellfish poisoning (PSP) toxins from feeding on toxic dinoflagellates (Kvitek 1991). This study was particularly novel and valuable in that it was conducted in the field. Gull chicks are fed by their parents, so toxic and nontoxic clams of the same species (collected from two different locations) were made available inside territories of parents with inexperienced offspring. Because gulls open clams by flying away and dropping the clams to crack them open (thus, making it impossible to keep track of toxic versus nontoxic clams), shucked clams were placed in territories. Chicks ingesting the clams were observed and followed for 15 min (or until regurgitation occurred). Regurgitation in the adults was not followed because this is a common behavior involved in feeding chicks. One day after the first exposure to toxic clams, the onset of aversion was analyzed. This was accomplished by identifying chicks that consumed toxic clams the day before, and offering them a shucked nontoxic clam from the same species previously offered, followed by a shucked nontoxic clam of a different species. Offering a different species of clam determined if rejection was caused by gut fullness, rather than by learned aversion to the toxic species.

Dinoflagellate toxins, like those responsible for the paralytic shellfish poisoning mentioned above, and their effect on consumers have received considerable attention because of their devastating effects on ecosystems and marine fisheries, and because their transmission between trophic levels poses a danger to public health. The physiological effects of ciguatoxins, and paralytic and diarrhetic shellfish toxins on man have been reviewed with more of a biomedical focus and will not be discussed (Halstead 1978; Meier & White 1995). However, long-term ecological experiments requiring storage of dinoflagellate toxins have productively used methodologies from the medical field (e.g., using mice to score toxicity) to ensure that stored metabolites retained their bioactivities (Magnelia et al. 1992).

Early studies on the effects of dinoflagellate toxins were made by culturing dinoflagellates in vessels, and then placing fish in water taken from these cultures (Gates & Wilson 1960; Ray & Aldrich 1967; Sievers 1969). Although an ecologically unrealistic approach for most systems and species, these experiments are probably more realistic than the intraperitoneal injection of toxins used in several early studies (see collected papers in Russell & Saunders 1967). Injections that bypass the digestive system and various membrane barriers ignore the possible role of digestion or other processes in detoxifying these compounds. White (1981) provides an assessment of the effects of oral versus intraperitoneal delivery of metabolites. Ten fish from each of five species were starved for a day and then force-fed with extracts from the dinoflagellate *Gonyaulax excavata* (in 0.1-N HCl), using a syringe carrying a 14-gauge, 10-cm needle. Ten additional fish from each species were injected intraperitoneally with a similar solution. The dosage for both oral and intraperitoneal injections was adjusted by weight of the fish, and was equivalent for both methods of delivery. The same number of control fish received equivalent oral or intraperitoneal dosages of 0.1-N HCL (the carrier solvent for extracts used above). After recording mortality for 6 and 24 h, the LD_{50} (the amount of toxin needed to kill half the population) was calculated. The oral LD_{50} values were between 62 and 100 times higher than the intraperitoneal LD_{50} values, indicating that data from intraperitoneal injections should not be used to draw ecological conclusions about predator-prey interactions that do not normally involve injection of the compound (i.e., some fishes with toxins that are injected via hollow spines). Clearly, if an investigator wants to determine the physiological consequences of consuming a secondary metabolite, then the test organism should consume the compound. Investigations focused on ecological, rather than pharmacological, questions concerning prey chemical defenses should not inject their organisms or immerse them in solutions of metabolites.

Davin et al. (1986) used a more appropriate method when they fed bluehead wrasses (*Thalassoma bifasciatum*) on treatment diets prepared by sprinkling weighed amounts of freeze-dried dinoflagellate cells on an artificial food. Because some species of toxic dinoflagellates grow primarily on benthic surfaces, this could be an ecologically justifiable assay for those particular species. Control fish received the same food, but without the dinoflagellates. The base diet was composed mainly of shrimp, gizzard shad, and salmon feed. Fish were fed *ad libitum* on a daily basis, and their behavior was observed every 12 h for 4 days. In this study, color change, inactivity, loss of equilibrium, erratic swimming, jerky feeding movements, loss of orientation, and the time required to capture control and treatment fish with a dip net were used as indications of behavioral differences among treatments. Because natural densities of dinoflagellates on benthic surfaces and the density of cells used in feeding assays were both known, realistic doses could be achieved. The investigators also tested for toxicity of the stored material at three intervals during the course of the experiment, thus ensuring that the activity of the toxins had not decreased during storage. However, most response variables were subjective. Although behavioral changes are important to note and study, quantification of these changes can be influenced by the observer's experience. If these types of subjective data are to be collected, they should be recorded by observers that do not know which replicates are treatments and which are controls.

Secondary metabolites can have effects on secondary consumers if they accumulate in, or have greater effects on, higher-order predators. Davin et al. (1988) tested whether ciguatoxins could be transferred to higher trophic levels by offering predatory fish smaller prey fish that had either received an injection of dinoflagellate toxins extracted in methanol (= treatment fish) or received an injection of methanol only (= control fish). Levels of toxins in the predatory fish were then determined after they had fed on the control versus treatment prey. Gosselin et al. (1989) used more ecologically realistic methods to assess this same question. Natural assemblages of zooplankton were allowed to graze on either toxic or nontoxic strains of a dinoflagellate and then offered at various concentrations to different age classes of two species of larval fishes. The authors then monitored survival of the larval fishes.

The tri-trophic effects of paralytic shellfish poisoning toxins also have been studied on other vertebrates such as otters (Kvitek et al. 1991). This study observed behavioral differences when otters were feeding on butter clams containing PSP toxins (i.e., clams that had fed on PSP containing dinoflagellates) versus clams that were free of these toxins (i.e., clams that had not fed on PSP containing dinoflagellates). The otters were placed in floating cages and offered clams *ad libitum.* Otters were started on clams containing low levels of toxins, then switched to highly toxic clams (flown in from a different location), and then switched back to low toxicity clams. Otter feeding was monitored using scuba and surface observations via telescopes. When fed clams containing dinoflagellate toxins, otters twitched and screamed more often, tended to reject clam parts where toxins accumulated, and dove more often without successfully harvesting prey.

In addition to looking for behavioral modifications caused by consuming secondary metabolites, there have been several investigations of how marine secondary metabolites affect the assimilation efficiency of vertebrate consumers (Targett & Targett 1990; Ireland & Horn 1991; Boettcher & Targett 1993; Targett et al. 1995). Ireland and Horn (1991) coated pieces of a palatable green algae with water-soluble extracts from two temperate algae (treatments) or with only water (= control) and force fed these to a temperate herbivorous fish. The fish in these assays had to be force-fed because they would not willingly consume the treatment foods. Force-feeding was accomplished by anesthetizing fish in 1:10,000 MS222 in seawater, using soft plastic forceps to force pieces of control or treated algae (~1/g) into each fish's mouth, and pushing this food into the stomach with a smooth glass rod. Fish were then placed in fresh seawater until they recovered normal movement patterns; they were then returned to their holding tanks. Fish were observed for 0.5 h after feeding in case regurgitation required refeeding. Feces were collected at least twice a day, for 1–11 days, until the force-fed thalli were defecated. Feces and whole pieces of the palatable green alga used for the food were then dried and divided into triplicate subsamples for ashing at 500°C for 5 h. Two additional subsamples from food and feces were used for C and N analysis. Values from these determinations were then used to compute assimilation efficiency for both C and N of treatment versus control fish. The extract of one of the two algal species tested reduced the assimilation of nitrogen, but not of carbon, clearly demonstrating the potential importance of analyzing for multiple nutrients in such assays.

Boettcher and Targett (1993), also force-fed fish in a similar experiment to test how polyphenolics that differed in molecular weight affected the assimilation efficiency of a temperate fish. However, in addition to force-feeding fish an alga with and without added secondary metabolites, this study also included another set of fish that were allowed to feed freely on the same alga that was untreated. This treatment served as a control for the effects of force-feeding. The fish were starved for a day prior to any assay, then anesthetized with MS222 (60–80 mg/l adjusted to pH 8), and force fed on days 1, 3, and 5 of the experiment. Treatment foods consisted of algal disks treated with one of their polyphenolic fractions; control disks were treated with water alone. Feces were collected and analyzed to determine assimilation efficiency. Preparing some of the polyphenolic fractions at the desired concentrations increased the concentration of salts in the fraction. Assays testing the effects of salt on assimilation efficiency were performed by force-feeding fish with algal disks treated with either deionized water, normal seawater (which approximated the concentration of salt in the polyphenolic samples), or salt-saturated seawater. Salinity did not affect assimilation efficiency, but they found large differences in the effects of different sized phlorotannins on digestive efficiency. Although the force-feeding required for these assays is tedious and may stress the fish, these methods allow careful control of ingestion rate, and they permit tests of water-soluble metabolites. A possible disadvantage is that this method bypasses mechanical and digestive processes that may occur in the mouth of the animal; some terrestrial animals secrete compounds in their mouths that minimize the activity of prey secondary metabolites (Robbins et al. 1987).

2.4.1.2. Long-Term Assays

Long-term feeding trials assessing the effects of deterrent secondary metabolites on marine vertebrates are difficult to conduct because fishes that co-occur with chemically defended prey often innately avoid the compounds (Hay 1991; Paul 1992), and fishes that have no evolutionary experience with the defended prey (and thus do not initially recognize or avoid the compounds) may rapidly learn to avoid consuming foods containing the compounds (Gerhart 1984; Lindquist & Hay 1995). Thus, the challenge for conducting longer-term feeding assays has been to induce and sustain feeding. Attempts to solve this problem have involved adding additional feeding stimulants to the experimental foods, force-feeding, giving treatment fish no other food options and then feeding control fish only as much as the treatment fish eat, or using neurologically simple invertebrates (see section 2.4.2) that appear incapable of learning. Examples of these approaches are provided below.

Hay et al. (1987b) investigated how consuming the brown algal diterpene pachydictyol A as 1% of the total diet for a period of up to 20 days affected the temperate herbivorous fish *Diplodus holbrooki*. Because control fish feeding on untreated food ate more readily than treatment fish, the investigators limited food consumption by each control fish to the amount consumed by its paired treatment fish, which was kept in the same aquarium but separated by a mesh barrier. This ensured that any differences in growth or survivorship within fish pairs were due to effects of the compound when it was consumed, not due to differences in amounts of food eaten or due to effects of compounds leaching into the water. Fifteen sets of fish were paired by standard length, and growth was determined by measuring fish at the end of the experiment or whenever either member of a pair died. Each day, the treatment fish within a pair was offered food pellets containing 1% dry mass of pachydictyol A until no more pellets were consumed. Because treatment fish were reluctant to feed, pellets were dipped in a slurry of tuna flavored cat food before being offered to each fish. Initial feedings demonstrated that treatment fish would quit eating without this added feeding stimulant. The paired control fish was then fed the same number of control food pellets (also dipped into cat food) as had been eaten by the treatment fish. This compound suppressed fish growth by about 50%. However, Hay et al. (1987b) tested the effects of their algal metabolite by placing it into a commercial fish food that would have been much more nutritionally valuable than the seaweed from which the compound was obtained. If this procedure is repeated, investigators should use a food that closely approximates the natural prey organism.

Because the Caribbean fishes that were tested used taste to recognize and avoid defensive didemnins produced by tropical ascidians (Lindquist et al. 1992), Lindquist & Hay (1995) attempted to conduct long-term feeding studies using generalist temperate consumers that were unlikely to have had significant associations with these compounds. The omnivorous fish *Lagodon rhomboides* and later the sea anemone *Aiptasia pallida* (see section 2.4.2.2) were chosen for their investigation. Fish were held individually in flow through containers, with treatment and control fish interspersed. To acclimate fish to laboratory conditions and to train them to feed on artificial foods, every morning (0800–0900 h) for 10 days all fish were offered 10 pellets of an untreated commercial fish food having a total dry mass of 168 mg. On days 8–10 at 1200–1300 h, all fish also were offered nine small food pellets (each = 0.056 mg dry mass) that fish could swallow whole; these pellets were made from the squid-based assay food described previously (see section 2.3.2). Both the commercial fish food and the small squid pellets were readily consumed. Each afternoon (1400–1500 h) for 3 consecutive days beginning on day 8, control fish were offered an additional feeding of two larger squid-pellets (3.5 μl = 0.3 mg dry mass each) without didemnins; treatment fish were offered equivalent squid pellets without didemnins followed several minutes later by a squid pellet that contained 7 μg of a naturally occurring didemnin mixture. Because by day 10 a significant number of fish would not ingest the didemnin-containing food pellet, the experiment was terminated. Data analyzed for this assay were the number of treatment versus control fish that consumed the second pellet offered during the second afternoon feeding (with and without didemnins, respectively) and the number of fish regurgitating the small squid pellets within 2 h of the final feeding each day. Because these fish rapidly learned to avoid treatment foods by associating compound containing foods with the physiological stress they experienced following their ingestion, it was not possible to conduct longer assays using these consumers and methodologies.

2.4.2. Invertebrates

Studying the effects of secondary metabolites on invertebrate consumers involves the same considerations outlined for vertebrates; however, invertebrates generally offer the advantages of requiring less space (because of smaller size or lower mobility) and having shorter life cycles than vertebrates. Some invertebrates also produce eggs, gonads, or asexual clones that can be measured without histological sectioning, allowing an easier estimate of fitness.

However, the diversity of invertebrate feeding habits pose some challenges because of our limited ability to produce experimental foods that some invertebrates will eat. As one example, many ascoglossans, nudibranchs, and pycnogonids feed via sucking, rather than chewing, mouthparts. These animals will not accept foods requiring chewing and no attempts have been made to produce foods that would be suitable for animals using this feeding mode. Another cautionary remark involves invertebrates producing gametes or eggs in the laboratory. Such events can be an indication of stress under poor laboratory conditions, rather than a sign of good health and high fitness. Experiments that measure fitness using egg or gonad production must allow for comparisons among treatments under equal stresses or results may be confounded by design induced artifacts.

2.4.2.1. Short-Term Assays

Several investigations have focused on the immediate effects of dinoflagellate metabolites on invertebrates. Modes of action of these toxins have been investigated in bioassays using isolated tissues from invertebrates (see references in Ives 1987), but these biomedical-type approaches provide little information about the effects of metabolites under natural conditions. More ecologically focused bioassays have used filtrates from cultures, or cell homogenates of dinoflagellates suspected to be toxic. Animals are placed in jars containing filtered seawater to which different extracts are added, and the behavior and mortality rate are recorded (Sievers 1969; Huntley et al. 1986, 1987; Ives 1987; Uye & Takamatsu 1990; reviewed by Targett & Ward 1991). Many of these experiments have suffered from limited replication, inadequate controls, and an absence of ecologically realistic test conditions. In many instances, filtrates or extracts were acquired from cultures maintained at unnaturally high densities, or compounds that are rarely, if ever, leached to the water were extracted and tested in the water. Although some dinoflagellates can reach very high densities during bloom conditions and compounds can be released into the water as cells lyse, artificial cultures introduce many potential artifacts that may not resemble field conditions (Targett & Ward 1991). For example, the production of a secondary metabolite may vary with the age, density, and nutritional status of the phytoplankters; rarely have any of these factors been measured in the field. Also, compounds that are released in nature are subjected to flow, which may decrease their impact on surrounding animals. In most culture chambers, both flow and turnover of the water are minimal, potentially leading to artificially high concentrations of exudates. Finally, the production of phytoplankton metabolites may be influenced by other microbial associates. Thus, sterile, single-species cultures may not represent cells as they occur in nature. Some of these problems can be solved by improving replication, establishing proper controls, and using flow-through systems. Others will frustrate researchers until natural histories are better known and additional technologies are developed. The gross effects of toxic dinoflagellate blooms on marine animals have been well documented, and provide evidence that phytoplankton secondary metabolites can play important roles under natural conditions (Paerl 1988), but an adequate understanding of the importance and effects of released versus ingested toxins, and how this varies among species, requires further investigation.

Several investigations have evaluated the effects of consuming, or living in association with, unpalatable (toxic?) versus palatable phytoplankters (Huntley et al. 1986, 1987; Uye & Takamatsu 1990), but have not been able to unambiguously assess the role of chemical deterrents or toxins in producing the patterns they documented.

One approach involves comparing the effects of exudates from toxic and nontoxic clones of the same dinoflagellate species. Different strains of the same microalga can have different levels of toxicity (Ives 1985, 1987; Gosselin et al. 1989) that induce different levels of rejection by copepods. Using the nontoxic strain as base food, Ives (1987) cultured copepods in jars containing either filtrates from the nontoxic strain or filtrates of the most toxic strain. After running additional assays to determine the grazing rates and gut filling rates of copepods feeding on each of these strains, and how switching from toxic to nontoxic strains, and vice versa, affected the copepods feeding rates, Ives (1987) proposed that copepods were physiologically unable to differentiate among palatable and unpalatable foods prior to ingestion. It appeared that copepods initially consumed both cell types at equivalent rates but that animals feeding on the toxic cells then lowered their rate of consumption. Ives hypothesized that feeding declined because the toxin that treatment copepods had consumed caused neurological impairment that interfered with coordinated movement of their feeding appendages.

Microalgal strains that occur in toxic and nontoxic forms offer an important tool for studies of this sort. However, strains that differ in toxicity might also differ in other characteristics. If this approach is used, potential among-clone differences in other traits such as nutritional quality, cell morphology, etc., should also be evaluated.

The short-term behavioral and physiological responses of copepods to microbial compounds also have been observed directly using microvideo techniques (Sykes & Huntley 1987; see other studies using microvideo in Targett & Ward 1991). These studies involve (1) tethering copepods, (2) allowing them to feed on toxic versus nontoxic phytoplankton inside a chamber, (3) recording their activities using a high-speed video camera attached to a microscope, and (4) analyzing their responses by playing this film back at lower film speeds that allow clear viewing of even very rapidly occurring behaviors.

Copepods are tethered by placing them ventrally on a fine mesh overlying absorbent paper soaked in seawater, and gluing the dorsal side to a thin glass fiber or an animal hair, using cyanoacrylate glue. The glue is allowed to partially dry (until it has a gummy consistency) before being attached to the copepod so that the amount can be minimized, and so it does not run down to the ventral area of the copepod (Sykes & Huntley 1987). Tethered copepods can be maintained on palatable microalgae until used. In Sykes and Huntley (1987), copepods were starved prior to use, and then placed in spectrophotometer cuvettes or 100-ml chambers containing either seawater, a suspension of a palatable species of dinoflagellate, or one of four test dinoflagellates. After 15 min of acclimation to the chambers the copepods were observed for feeding behavior, fecal pellet production, heart rate, and mouthpart movements.

Tethering studies revealed that certain dinoflagellate species induced strong physiological responses in the copepods—such as elevated heart rates, regurgitation, loss of motor control, and twitching of the mouthparts. Details about the microvideo equipment and analysis of the tape can be found in Alcaraz et al. (1980), Price et al. (1983), and Sykes and Huntley (1987). Feeding by pelagic tunicates and bivalve larvae, also has been observed using microvideo (Deibel & Paffenhöfer 1988; Gallager 1988). Tethering of these soft animals for video feeding studies can be accomplished by placing a drop of Vaseline gel on a fine glass rod (e.g., a capillary tube) and affixing the organism to this gel.

The most direct approach for testing the effects of a metabolite on a consumer is to isolate the compound, or a bioactive chemical fraction, incorporate it into a palatable food, and feed it (versus an equivalent control food without the metabolite) to consumers. Harpacticoid copepods have been fed on fish food with and without lipophilic extracts of diatoms (Shaw et al. 1994). This direct approach is a significant improvement over many of the previous indirect approaches, but testing algal compounds on a high-protein food that may be nutritionally very different from the alga that produced the metabolite may limit the ecological interpretation of the results. The few studies that have manipulated both defensive compounds and the nutritional rewards of the assay foods have found that specific metabolites or extracts may be very effective deterrents when added to low-quality foods but ineffective when placed in foods of higher quality (Duffy & Paul 1992; Hay et al. 1994; Pennings et al. 1994). The use of freeze-dried phytoplankton or possibly finely ground seaweeds as basic foods for such experiments might avoid this potential problem. The development of algal based diets, such as *Spirulina* powder, for aquacultural purposes may provide more suitable foods for testing microalgal metabolites.

Some investigators have developed procedures for feeding copepods on microbeads (Van Alstyne 1986; Decho & Moriarty 1990). Incorporating microbial metabolites onto these beads might allow certain types of tests of their short-term or long-term effects on micrograzers, however, microbeads are not nutritious, and their application to the study of physiological responses therefore faces obvious limitations. Clearing rates of the copepods on each type of microbead must be measured to confirm ingestion, and controls having treated and untreated beads without copepods (for quantifying measurement error) and having copepods without microbeads (for starvation) should be run simultaneously (see Huntley et al. 1986, 1987 for a discussion of these types of assays). Microbeads have two additional limitations; they can not be used with compounds that readily dissolve in water, and their utility depends on whether or not the animals ingest them at all.

Microencapsulation can overcome some of the problems of microbeads because microcapsules could be made that contained both nutritive materials and secondary metabolites. Poulet and Marsot (1978) used microcapsules containing either albumin or albumin with phytoplankton homogenates to test food selectivity in copepods. Potentially, this technique can be used to feed micrograzers on diets with and without secondary compounds, possibly even water soluble metabolites. Microcapsules have been used to maintain micrograzers for extended periods (Jones et al. 1974, 1979), so it should be possible to produce microcapsules with nutritional traits that mimic natural prey and add test metabolites to these foods. Microcapsules appear to offer a significant advantage over the use of microbeads. The main disadvantage of microcapsules is that they are permeable to small molecules; this might compromise the dosage of certain compounds.

2.4.2.2. Long-Term Assays

Few studies have assessed how long-term consumption of prey secondary metabolites affects the physiology of invertebrate consumers. Two basic approaches have been used: (1) long-term experiments using foods that naturally contain or lack various secondary metabolites or (2) studies using artificially prepared diets with and without metabolites. The first kinds of experiments cultured consumers on natural foods known to contain secondary metabolites and compared growth on these foods to growth of consumers fed on natural foods that lack known secondary chemistry (e.g., Duffy & Hay 1991; Steinberg & van Altena 1992); or compare the effects of a natural diet in which the content of the metabolites is known to vary, for example old, new, and decomposing leaves of the same plant (Bärlocher & Newell 1994). These studies have been able to correlate metabolite presence with certain effects on consumers, but the effects of secondary metabolites are confounded by many other traits that may also vary among the different foods.

A more direct approach is to compare consumers feeding on a test food containing the metabolite or extract of interest versus consumers feeding on that same food without the added metabolite or extract. The basic design is similar to that outlined above for fish. We focus on this approach because it is more powerful than the less direct approaches.

In one of the earliest experiments on long-term effects, Paul and Fenical (1986) examined the survivorship of juvenile queen conch *Strombus costatus*) over a 14-day period when limited to consuming the palatable green alga *Enteromorpha* sp. coated with one of three green algal metabolites versus a solvent control. For this pseudoreplicated assay (all conch within a treatment were maintained and fed communally), the amount of food consumed by each treatment group of snails was not determined; thus the relative effects of reduced food consumption versus compound effect could not be adequately assessed. However, if altered to provide independent replicates and to quantify feeding, the methodology developed in this initial study of long-term effects could provide a useful template for future studies. This approach is tractable, inexpensive, and the delivery of the compounds to the animals is realistic. However, it can only be used for testing lipid soluble compounds because more polar metabolites will dissolve into the water.

If water-soluble metabolites are to be tested, artificial diets need to be made using a matrix that minimizes leaching of the compounds into the water. Although some polar metabolites rapidly leach from agar, agar disks enriched with algal particles or extracts (to provide sustenance for the animals) have been used to test the effects of some algal polyphenolics on consumers (Steinberg & van Altena 1992; Winter & Estes 1992). Larger algal phlorotannins are retained relatively well by this matrix, but small phlorotannins like phloroglucinol leach fairly rapidly (Steinberg 1988). In our own work (M. Deal and M.E. Hay work in progress), we have used agar-based foods (like those described in section 3) to demonstrate that several water-soluble metabolites deter feeding by urchins. This required running assays over short periods of time and monitoring effects of the metabolites as a function of how rapidly individual urchins fed. In several cases, we found treatment foods to be significantly deterrent over the first 1–12 h of the experiment, but to lose deterrence if run for longer than 12 h; suggesting that deterrent metabolites were leaching from these foods. Thus, one might be able to run long-term assays if these types of foods were prepared daily and placed with consumers for a short period during each day. If known metabolites are used, then they can be monitored over time and their rates of loss determined to validate methods for longer-term experiments. When doing bioassays on extracts or fractions containing unknown metabolites, this is not possible. In these cases, failure to detect significant effects could result from the extracts being inactive or from them rapidly dissolving out of the foods. Better methods for retaining water-soluble metabolites in submerged foods need to be developed.

Winter and Estes (1992) tested the effects of polyphenolics on abalone by feeding animals for 88 days and measuring growth as both length and mass changes. Eight tanks were set up, each containing previously measured abalone, and algal enriched agar disks with or without polyphenolics were offered to the replicates. Food disks were changed every 12–14 days, since preliminary assays showed that leaching of polyphenolics was negligible. The amount consumed by the abalone was calculated by taking the mass of the food disks at the time of introduction into the tanks, and after the period allowed for feeding. An advantage of their design is that calculation of the amount consumed in these long-term assays gives information regarding the dose the animals consumed, as well as whether the observed variation in growth was caused by starvation or by the physiological effects of the compounds. Winter and Estes (1992), however, tested polyphenolic concentrations that were above the natural concentrations found in the algae they studied, and they used artificial foods that were lower in nutritive value than natural seaweeds; both of these facts complicate interpretation of their results. Additionally, leaving food disks for 12–14 days might allow the establishment of microbial communities on the food. This possibility should not be underestimated because microbes can alter the value of a food by decomposing particular components or by enhancing the foods through their own nutritional value. Although these particulars complicate interpretation of the above study, its basic methods and approach are sound.

Steinberg & van Altena (1992) used agar based artificial diets to test the effects of phlorotannins on the snail *Turbo undulata* and the urchin *Tripneustes gratilla*. They also conducted parallel experiments using monospecific diets of natural seaweeds known to contain different types of metabolites, or lacking known defensive chemistry. The basic agar diets were enriched by adding dry algae to mimic the dry mass/wet mass ratio of natural seaweeds. These assays were run in large mesocosms, or in the field, by placing the animals inside crates fixed to the bottom. The mesocosms had running seawater, and combined with their large volume, the effect of any compound released from the diets to the water would be greatly diminished. Additionally, the rate of leaching of the phlorotannins was calculated before the experiments to provide a baseline for comparing dose-dependent artifacts, and the compounds were tested at concentrations comparable to their natural occurrence at certain times of the year. The feeding rates of the animals were measured during these long-term assays, and the authors also ran starvation controls. This investigation was thorough, well designed, and could serve as a model for similar studies.

When Lindquist and Hay (1995) tried to determine how periodic, low-level consumption of ascidian secondary metabolites (didemnins) affected the growth and survivorship of fish, fish rapidly learned to avoid the treatment food and prevented the study from running long enough to acquire meaningful data (discussed in section 2.4.1.2). These authors then used the anemone *Aiptasia pallida*, a consumer of invertebrate larvae that might be less capable of learning feeding aversion. Three adjacent anemones were collected from each of separate 20 locations and used as 20 blocks for three separate treatments. This anemone commonly reproduces asexually by breaking off small portions of its base that develop into daughter clones; thus the blocking procedure minimizes experimental variance due to genetic or phenotypic differences among anemones. Once removed from the substrate, the anemones were placed on a slightly inclined surface for 5 min to drain off expelled water, and then weighed. The consistent dry mass to wet mass ratio measured for 15 anemones (r^2 = 0.88) indicated that the measured wet mass of anemones provided a good approximation of their size. Anemones were then placed in separate 100-ml plastic bowls arranged in 20 rows of three bowls each. Each row of three bowls contained the anemones within a block. Seawater in each bowl was changed daily at 0800 and 1800 h. Each morning at 0800–0900 h, all anemones were fed a 16.8-mg pellet of a commercial fish food, accounting for >98% of total daily food intake. In addition to this major feeding, there was an additional feeding that comprised the various treatments. Treatments consisted of (1) anemones receiving a single squid pellet containing 7 μg of didemnin metabolites (this pellet accounted for only 1.8% of the total daily diet and was equivalent to consuming 15 ascidian larvae), (2) anemones receiving a squid pellet lacking didemnins (i.e., the control), and (3) anemones receiving no additional feeding. As with the fish, didemnins caused anemones to regurgitate (generally within 1 h), but they did not learn to avoid ingesting the didemnin-containing squid pellets. Because of the anemone's regurgitation of the treatment squid pellet, the additional control group (i.e., treatment 3 above) was added to account for the nutritional value lost due to regurgitation of the squid pellet.

After placing a squid pellet on a anemone's arm, investigators recorded whether or not the anemone ingested the pellet, and if ingested, whether the pellet was regurgitated within 1 h. Data were analyzed for significant differences between the number of treatment and control anemones (1) ingesting the squid pellet and (2) subsequently regurgitating it each day that squid pellets were offered. Periodically throughout the experiment, the number of daughter clones that had budded off each adult were also counted. On the final day of the experiment, the total number of clones present in each bowl was recorded and each adult and its pooled clones were weighed. Data were analyzed for differences between anemones ingesting squid pellets with didemnin versus anemones ingesting squid pellets lacking didemnins, and between anemones ingesting control squid pellets versus anemones not receiving any squid pellets. Traits compared in these analyses included (1) the change in wet mass of adults, (2) the total change in wet mass of adults plus clones, (3) the cumulative number of clones produced by each adult anemone, and (4) the average wet mass of daughter clones. These experiments indicated that consuming even very low amounts of chemically defended prey (<2% of the daily diet) could dramatically lower consumer growth and reproduction.

This methodology has several advantages. First, the consumption of treatment food (squid pellets) was monitored closely, so any effects of diet would not be confounded by differences in the amount of treatment food consumed by the animals. Second, the dosage of compounds delivered to the test organisms was realistic. The compounds in the pellets were used at natural concentrations, and the exposure to these compounds was designed to approximate the levels that a predator would encounter in nature. Because the test compounds are found in ascidian larvae and these larvae would comprise only a small part of the daily diet of most consumers, it would be unrealistic for the total diet of treatment animals to contain these compounds (a problem with earlier studies—Hay et al. 1987b; Steinberg & van Altena 1992). Finally, the authors were able to measure effects of metabolites on fitness (daughter colonies produced) as well as growth (usually assumed to be a proxy for fitness). One criticism pertaining to these assays is that, although the predator and the prey overlap in distribution, they rarely co-occur. However, as the authors note, it is difficult to study these sorts of physiological effects on predators that regulatory encounter the chemically defended prey, because these consumers often recognize and avoid consuming the metabolites. Using consumers that do not learn to reject the metabolites after repeated ingestions is one method for overcoming this problem.

Amphipods and isopods are the only crustaceans for which the long-term effects of consuming secondary metabolites have been studied. Newman et al. (1990) were the first to look at the effects of compounds from a freshwater macrophyte on the survivorship of the amphipod *Gammarus pseudolimnaeus*. However, their assays were ecologically unrealistic in that they tested how macrophyte compounds dissolved in water, rather than incorporated into foods, affected survivorship of the amphipods. More recently, Duffy and Hay (1991) described methods for raising amphipods from newly hatched juveniles to egg-bearing adults on different algal diets and determining how these diets affected growth, survivorship, fecundity, egg size, and age to first ovulation. Using this same basic approach coupled with the production of artificial, agar-based foods described in Hay et al. (1994), E. Cruz-Rivera and M.E. Hay (unpublished data) have been able to grow several species of mesograzers from newly hatched juveniles through two ovulations on diets differing in nutritional qualities, secondary metabolites, and various combinations thereof. Because some mesograzers (e.g., amphipods) are small, eat limited amounts of food, and grow from hatchlings to reproductive adults in as little as 23 days (Duffy & Hay 1991), they make exceptionally tractable subjects for studying the long-term effects of consuming secondary metabolites.

We recently tested the effects of the lipid-soluble secondary metabolites from the brown alga *Dictyota menstrualis* on the physiology of several marine amphipods, and an isopod (E. Cruz-Rivera & M.E. Hay unpublished data). In these experiments, newly hatched amphipods were obtained by placing egg-bearing females in individual petri dishes and collecting juvenile amphipods shortly after hatching. This approach minimized genetic variance and allowed an assessment of maternal effects, because offspring within a block (control and treatments) come from the same mother (see Duffy & Hay 1991). Individual juveniles from a mother were randomly assigned to either an artificial diet containing the algal metabolites, the same food without the metabolites but treated with solvent (= control), or a starvation treatment. Each amphipod was individually housed in a 60- × 15-mm petri dish. Artificial diets were made by incorporating a freeze-dried food into an agar base and forming this onto plastic screen mesh (described above and in Hay et al. 1994). Because the animals were fed *ad libitum,* separate experiments were run to measure consumption rates (under no choice conditions) of each amphipod species on each diet tested, so that the potential effects of reduced feeding on treatment foods could be evaluated relative to the physiological effects of consuming the compounds. The amphipods were cultured until maturity by replacing water and food daily, and scrubbing the dishes every 2 days to minimize microbial growth. The duration of these experiments required storage of the foods, therefore, samples of the foods containing the compounds were periodically taken to confirm the presence of the metabolites. Because amphipods are direct developers, several measurements of fitness could be obtained (age at first and second ovulation, eggs per clutch, size of eggs, and clutch volume). The same protocols were used to successfully culture the marine isopod *Sphaeroma quadridentata*, but the young were collected from the field and not obtained from captured females. Using this general approach, we also have been able to address questions involving the interactions of secondary chemistry with nutritional quality of the food by altering the composition of the dried ingredients added to the artificial diets.

2.5. Toxin-Mediated Prey Capture

Toxins delivered in the bites of aquatic organisms have been discussed in both folklore and modern medicine (Tamiya 1975; Halstead 1978; Russell 1984; Meier & White 1995). Despite improved understanding of the chemistry, toxicology, and pharmacology of many of these substances, studies on their ecological roles are rare.

It is often assumed that venomous mouthparts or stinging tentacles play a role in defense or prey capture; however, the ecological mechanisms underlying these hypotheses have rarely been tested directly. In the case of prey capture, the use of secondary metabolites has been inferred from (1) observations of foraging behavior and reactions of the attacked prey, (2) the existence of structures apparently adapted to deliver toxins, (3) isolation of metabolites associated with these structures, and (4) assessment of their toxicity by injecting the chemicals into animals or dissolving the chemicals in water and then exposing animals to the solution. As a consequence, many of these ecological studies have been descriptive or have not established clear relations between the occurrence of suspected metabolites and prey capture. Nonetheless, indirect evidence for the use of toxins by predators has been acquired for marine worms (Kem 1988), snails (Olivera et al. 1988; Kobayashi et al. 1989), cephalopods (Wells 1978; Hanlon & Messenger 1996), chaetognaths (Thuessen et al. 1988), cnidarians (Van-Praët 1985), and sea snakes (McCosker 1975, Tamiya 1975).

The mechanisms of envenomation are not always known. For example, early accounts of prey capture by octopi claimed that crabs were killed when the octopi released toxins into the water without the need of biting the prey. Later studies, however, failed to show envenomation by toxins dissolved in the water (see references in Wells 1978). Other studies have shown that crabs trapped by an octopus (*Eledone* spp.) exhibited signs of paralysis before the octopus had time to drill through the carapace of the crab (see references in Hanlon & Messenger 1996), but the site of entry for the toxin remains unknown.

The fangs, stylets, teeth, and beaks used by poisonous animals cause some degree of direct tissue damage to prey. Because the effects of this tissue trauma have not been separated from the effects of injected toxins, the relative importance of toxins in prey capture cannot be unambiguously evaluated. In sea snakes, as an example, researchers have assumed that the reduced size of the fangs renders the snakes incapable of capturing prey without the use of toxins (McCosker 1975). However, the diet of sea snakes is so poorly understood that generalizations, even ones that seem very reasonable, could be premature. For some predators, toxins may serve as a secondary rather than primary mode of capturing prey. For example, Olivera et al. (1988) note that certain cone snails capture fish by stinging them with their poisonous proboscis and then engulfing them, while others distend their rostrum and sting the fish only after it has been at least partially trapped. There may be more variance and subtlety in toxin-mediated prey capture than is presently appreciated.

Cnidarians pose a particularly difficult challenge. The toxins produced by jellyfish and some hydrozoans are among the most powerful known, and have been relatively well characterized (Russell 1984). However, because the toxins are delivered through diverse kinds of nematocysts, experimental manipulations may be difficult. The role of toxins in prey capture has been inferred from the isolation of nematocyst-associated toxins, the behavior of exposed prey, and extensive work on the mechanics of nematocyst discharge and adhesion (Hessinger & Lenhoff 1988). The difficulties involved in manipulating prey capture at a cellular level give ecologists little choice but to accept this circumstantial evidence as proof of toxin-mediated predation. Nonetheless, useful techniques for manipulating nematocysts and their associated components have been developed (Lenhoff 1983). If complemented with ecological studies on prey capture and handling, these physiological and pharmacological assays could provide additional tools for studying the natural roles of cnidarian toxins.

As guidelines, we suggest that researchers interested in the ecology of toxins as means of prey capture consider the following:

1. Foraging behavior and the steps involved in prey capture must be well documented. Animals that use piercing organs produce tissue trauma in their prey, and this damage should be assessed so that its effects are not confounded with those of the suspected toxin. Exposing prey to predators in which the toxins have been minimized or eliminated (perhaps by offering several prey right before the assay or by draining the venom in the case of sea snakes; Tamiya 1975), and comparing the results to those from predators with full complements of toxin may help solve some of these complications.
2. If toxins are used to capture food, then the presence of the toxin in the prey tissues should be confirmed once prey have been attacked. No studies have assayed for the presence of metabolites from the attackers inside the prey, even in cases where the toxins are known.
3. Once a presumed toxin has been isolated, direct testing on prey animals should be performed in ecologically realistic ways. When the suspected mode of toxin delivery is through piercing parts, injections should follow realistic dosages. Information on the dosage that a prey receives when attacked is nonexistent. Again, quantifying the toxin amount from previously attacked prey will help solve these uncertainties.
4. Prey used in experiments should be those that co-occur with the predator in question because toxins may be targeted to certain types of common prey. Kobayashi et al. (1989) note that venoms from fish eating cone snails are active against laboratory mice, whereas those from cone snails that feed on mollusks or worms are not.
5. As a final point, care should be exercised by researchers interested in these topics. Some marine animals possess extremely toxic venoms, and carelessness could lead to dramatically shortened careers.

2.6. Chemically Mediated Detection of and Responses to Predators

Just as chemical cues from prey are used as foraging cues by predators, prey also use chemical cues as a warning that predators are near or that conspecifics are being attacked. There are three main types of chemical cues that prey use as warnings of the threat of predation: (1) those actively released by conspecifics that can serve as warning signals, (2) those released passively when tissue is damaged, and (3) odors released directly by predators.

2.6.1. Vertebrates

2.6.1.1. Chemical Cues from Conspecifics

Aquatic vertebrates can use chemical cues released by conspecifics injured by predators as an alarm signal and take appropriate predator avoidance measures such as hiding or reducing movement (fishes—reviews by Pfieffer 1977; Smith 1992; amphibians—Hews 1988; chapter 6). Much of this work with fishes focused on the identification of the alarm chemical, or "schreckstoff," and the physiological responses of this chemical on conspecifics. While the ecological function of the physiological or behavioral response in these studies was often assumed (either implicitly or explicitly) to be antipredator in nature, this was rarely quantified in an ecologically meaningful way (e.g., Pfeiffer 1977; Pfeiffer et al. 1985). Recent work has focused more on quantifying ways in which this alarm response might be adaptive. Laboratory and field assays with fishes (Mathis & Smith 1993a, 1993b; Mathis et al. 1995) and laboratory assays with larval amphibians (Hews 1988) show that chemical cues from damaged individuals cause conspecifics to behave in ways that should reduce susceptibility to consumers, such as hiding or schooling. In at least one instance this response has been demonstrated to increase chances of survivorship in the presence of predators (Mathis & Smith 1993a). Chemical stimuli are prepared in one of two ways: either by obtaining an aqueous extract of skin removed from a freshly killed individual, or by blocking all nonchemical stimuli from a live conspecific being consumed by a predator. Both methods likely provide ecologically realistic stimuli, however, the second method includes chemicals emitted by the predator as well as those released by breaking of the prey skin. The second approach, therefore, requires an additional control (a predator alone) so that direct cues from the predator will not be confounded with cues from damaged prey (Hews 1988).

The laboratory assay used by Mathis and Smith (1993a) demonstrates the delivery of cues derived from skin extract. A glass aquarium is partitioned into three sections. One of the two edge sections contains four prey (fathead minnows), and several artificial plants and broken clay pots for shelter; the middle section is empty, and the other edge section contains a predator (pike). Use of shelter and shoaling behavior is recorded for 8 min, then 10 ml of either the alarm substance extract or distilled water (control) is poured into the center section. After waiting 2 min to allow all fish in the tank to detect the alarm cue, barriers are removed and the pike is allowed to roam throughout the tank. Release of alarm substance into the tank resulted in increased shelter use and shoaling, and increased the time required for successful capture of the treatment prey relative to the control prey.

Similar assays have been performed using amphibian larvae as prey and dragon fly naiads as predators, with the exception that the stimulus used emanated from the consumption of a live larva (Hews 1988). Visual cues were eliminated by holding the predator and prey in a perforated, opaque cage at one end of the experimental area (Fig. 2.5). At the other end, an identical cage containing only a predator served as a control. To control for mechanical stimuli of prey movement and chemical cues released by undamaged prey, both of these predator cages were divided in half, with the predator (and prey larvae in the treatment cage) on one half and five larvae with no predators on the other half. The predator cages were separated from the center of the tank by mesh screens, the center section was divided lengthwise, and a single tadpole was placed in each half. The time the tadpole spent in the half of the tank where a conspecific was being consumed was reduced compared to the time spent in the half with just a predator and intact larvae. Additionally, activity levels of the larvae (measured as number of times the center line was crossed) were reduced when one of the predator

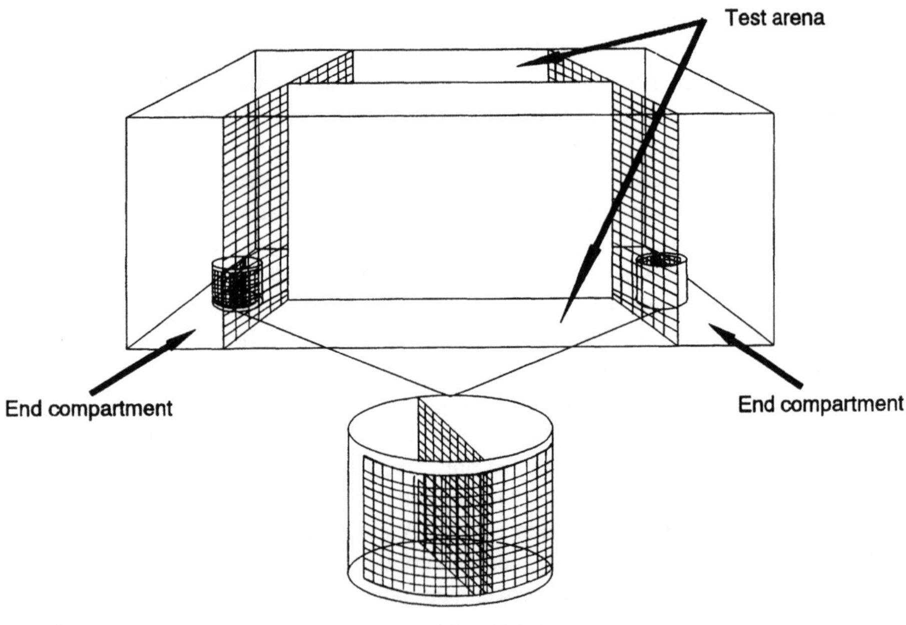

Figure 2.5. Test tank for alarm cue experiment. Circular cages (predator cages) contain combinations of predators and larvae as described in the text. These cages are placed in each end compartment, with the opaque side facing the test arena so that only chemical cues reach the test arena. The behavior of an individual larva in the test arena is recorded, including the location relative to each of the predator cages. The figure was redrawn and modified from one in Hews (1988).

cages held a conspecific that was being consumed relative to a cage holding a consumed heterospecific or a no-predator control larva.

These types of assays, while useful in demonstrating changes in prey behavior, cannot tell us whether or not organisms actually use these cues in the field. The few field experiments investigating alarm signals in fishes have followed similar methods (Mathis & Smith 1993b; Mathis et al. 1995). Alarm substance prepared from the skin extract of the fish (in these cases fathead minnows) was adsorbed into a cellulose sponge and placed in a minnow trap, while a separate sponge soaked with distilled water was placed in another trap as a control. Traps were placed in a line in the field with the ordering of traps done randomly. After several hours, the traps were checked, and less than 4% of the 849 fathead minnows caught were in traps treated with alarm chemicals, suggesting that fish avoid this extract under field conditions.

Care must be taken in constructing evolutionary arguments concerning the origin of the alarm signal based on these experiments. Although kin selection is sometimes invoked to explain how selection could favor release of an alarm cue by an individual that has already been injured, some evidence suggests that a more parsimonious (and experimentally testable) explanation may exist. Antifungal properties (see references in Smith 1992) of the alarm signal and the attraction of predators that interfere with the attacker, thus allowing the alarm-giver to escape (Mathis et al. 1995) have both been cited as possible direct benefits of the alarm chemical to the sender.

2.6.1.2. Chemical Cues from Predators

Just as prey avoid predators by reacting to alarm signals from conspecifics, they also may be able to detect chemicals from the predators themselves. Early efforts to explore this possibility tested the effects of predator-scented water on prey behavior in simple laboratory assays. Because prey would need to detect undamaged predators, natural exudates are a more appropriate stimulus than predator extracts. In one example (Rehnberg & Schreck 1987), the response of juvenile coho salmon to water that had held predators for the previous 12 h was evaluated using a Y-maze. Ten juvenile salmon were placed in the Y-maze just below the fork, and a test substance was introduced into one of the arms and a control (water only) was introduced into the other. Each arm had a funnel trap at its entrance, allowing fish to enter but not leave, and after 25 min, the number of fish that had entered each arm was recorded. Water scented by the predator was significantly avoided relative to the control, but water containing aqueous extract of predator skin did not affect prey behavior.

A similar laboratory assay was used to determine the effects of predator presence on activity levels and intraspecific aggression in salmon (Martel & Dill 1993). Treatments included water scented with conspecifics by holding fish in a 1-L tank of filtered water for 30 min, and "predator" water prepared similarly by holding a common merganser that was allowed to attack and partially consume a fish. Thus the "predator" odor was actually a mix of predator and injured conspecific odors. The authors recorded the distance juvenile salmon were willing to move to attack passing food (brine shrimp) and monitored salmon behavior toward mirrors placed in the experimental chamber before and after addition of treatment and control (filtered water) stimuli. Attack distance decreased dramatically relative to the control with the introduction of the predator scented water, as did the frequency, intensity, and duration of aggressive displays toward the mirrors. Water scented with only conspecifics had no effect on fish behavior, suggesting that predators, or some cue released by damaged conspecifics was responsible for eliciting this change in behavior. These assays are able to tell us about potential interactions between predator and prey, but problems associated with decomposition or concentration of the stimulus (see section 2.2.1) and a failure to incorporate elements of the natural environment such as flow, bottom topography, and vegetation into the experimental design hamper their extrapolation to the field.

Problems with the realism of stimulus composition have been addressed by holding the predator in a tank through which water is passed on its way into the tank where prey are held. The basic design for this assay consists of two sets of three connected tanks (Fig. 2.6). The first tank of each set is the source of water for the experiment; the second either contains a predator (treatment) or no predator (control); the third is the experimental arena where the behavior of the test organism is monitored, it contains an open area and a refuge provided by an opaque Plexiglas™ sheet (Petranka et al. 1987; Kats et al. 1988). Some species of amphibian larvae always used the refuge in tank three, regardless of whether predator scent was present or not. The design discussed above was modified for these species to funnel the predator and control water into different sides of a single experimental arena (tank 5; Fig. 2.6A). This new experimental arena was partitioned along three quarters of its length, and larvae were confined to the unpartitioned back one fourth by a mesh screen (Petranka et al. 1987; Kats et al. 1988). The partition functioned to keep predator-scented and control water on separate halves of the experimental arena, allowing larvae to choose between predator-scented and control water. These assays were scored based on percentage of time each larva spent in each half of the arena.

Although laboratory assays under controlled flow conditions can help work out the mechanism for predator avoidance, to unequivocally demonstrate that prey actually use the chemical scent of predators to avoid predation in nature, field experiments are required. In an attempt to investigate this behavior in the field, Sih et al. (1992) manipulated fishless stream pools to produce pools with and without fish. Fish were held in a perforated bucket on the bottom of treatment pools and covered with opaque mesh screening that blocked visual and mechanical indications of fish presence, but allowed diffusion of chemical cues. Behavior of laboratory-reared amphibian larvae (no prior exposure to fish) was quantified

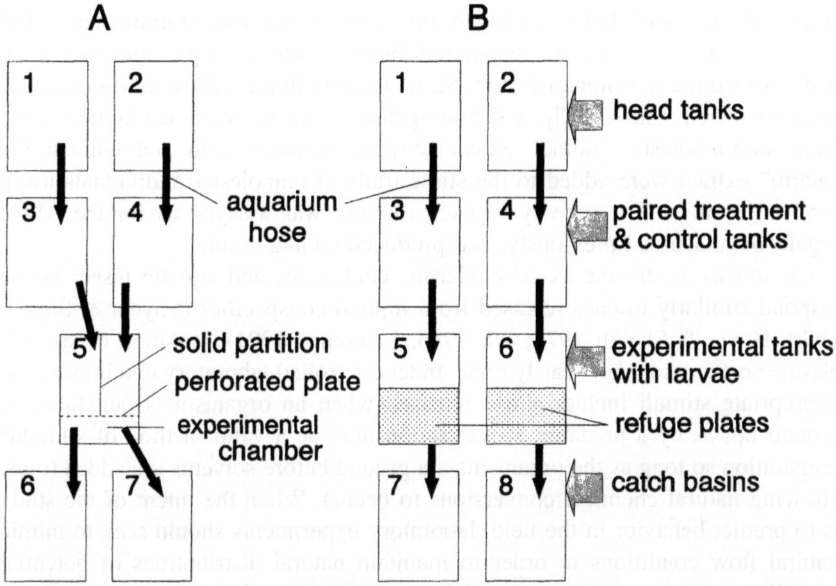

Figure 2.6. A top view of a gravitational flow-through system used in the study of how chemical cues from predators affect larval behavior. Tanks 3 and 4 hold either a predator (treatment) or are empty (control). In panel **A**, tank 5 is the experimental arena and larvae are given a simultaneous choice between predator-scented and control water. In panel **B**, Tanks 5 and 6 are separate experimental arenas for treatment and control stimuli (a no-choice test). The figure was redrawn and modified from one in Petranka et al. (1987).

after they were dropped from a height of 15 cm into fish and fishless pools at points of identical depth. In pools with fish, larvae entered refuges more frequently, and attempted to leave the pool more frequently than larvae in pools without fish (Sih et al. 1992), suggesting that these amphibians could detect and respond to chemical cues from predators in the field.

2.6.2. *Invertebrates*

2.6.2.1. *Behavioral Responses*

Chemicals released when organisms are attacked can serve to mark a location as dangerous to conspecifics. When sufficiently molested, the marine opisthobranch mollusk *Navanax inermis* secretes a mixture of bright yellow chemicals into its slime trail that causes an avoidance response in trail-following conspecifics (Sleeper et al. 1980). Individuals were allowed to lay elaborate slime trails of at least 100 cm length in an aquarium, then were prodded with forceps to induce the release of a bright yellow mixture of compounds (navenones A–C) into the slime trail. The animal was then carefully removed from the tank and another added. The new opisthobranch began following the trail almost immediately, but turned 90° or more when it encountered the place where the previous individual had released the navenone mixture. Slime trails with the yellow alarm chemical were extracted with $CHCl_3$, and homogenized into seawater containing slime from an unmolested animal. Seawater-slime mixtures with and without the "alarm" extract were added to the slime trails of unmolested individuals using a 50-ml syringe. The activity of each mixture was assayed as for the whole organisms discussed previously, and produced similar results.

Organisms as diverse as echinoderms, cnidarians, and aquatic insect larvae respond similarly to cues released from injured conspecifics (Snyder & Snyder 1970; Howe & Sheikh 1975; Sih 1986; Lawrence 1991). Isolation of specific alarm substances is most easily done under controlled laboratory conditions, and appropriate stimuli include those released when an organism is punctured or ground up, as by a predator, so extraction may be a valid method of stimulus preparation so long as the organisms are ground before solvents are added (thus, allowing natural chemical conversions to occur). When the intent of the study is to predict behavior in the field, laboratory experiments should seek to mimic natural flow conditions in order to maintain natural distributions of potential stimuli (see discussion in section 2.3). As in studies of foraging cues, a flume and predators or injured conspecifics can be used to present prey with a realistic distribution of odor cues. As an example, an experiment with an intertidal marine gastropod used a flume with some vertical drop to mimic the organism's intertidal habitat and tested the effects of the chemical scent of both predators and injured conspecifics on foraging behavior (Vadas et al. 1994). In this study, gastropods were placed just below the surface of the water in the tank, and flow moved from the top of the tank to the bottom where prey (mussels and barnacles) were located. Inflow water was allowed to pass over one of four tanks prior to entering the experimental arena: one tank contained crushed conspecifics, a second contained a predator (a decapod crab), a third contained both crabs and crushed conspecifics, and the fourth was a no-scent control. The number of active gastropods was decreased by odors from crushed conspecifics or from crushed conspecifics and crabs. Additionally, more snails sought refuge out of the water in the presence of these cues. However, when gastropods were starved, the predator and injured conspecific cues had no effect on snail behavior. There were also differences in responses to predator cues between well-fed snails raised on different diets, suggesting that physiological state and diet history of the prey organisms may alter their willingness to take risks. This demonstrated that intraspecific variability in responses to predator cues could influence results if the treatment of test organisms is not uniform prior to assays. These findings also suggest that the common procedure of starving test organisms before use in bioassays may significantly alter results, as has been demonstrated for some feeding assays (Cronin & Hay 1996c).

Although the apparatus used by Vadas et al. (1994) represents a substantial improvement in bioassay design, it does not reproduce the turbulent flow regime characteristic of many aquatic habitats. This is not a trivial point given that even moderate turbulence can have a substantial effect on the spatial distribution of an odor plume (Moore et al. 1994), and thus an organism's ability to locate an odor source (Weissburg & Zimmer-Faust 1993, 1994). As an alternative to increasingly complicated laboratory designs, behavioral responses might be better tested in the field, although there is admittedly a loss of precision in stimulus reproducibility. As an example of a simple field experiment, Snyder and Snyder (1970) placed between 6 and 20 individuals of the sea urchin *Diadema antillarum* in a clump and then smashed a single conspecific 25 cm upstream from the clump. Controls consisted of smashing bare ground. All of the urchins downstream from a crushed conspecific moved away from the area rapidly for 1 or 2 min before settling down, while none of the urchins in the bare sand control moved. *Diadema* responded specifically to crushed conspecifics, and exhibited no response to crushed heterospecifics, including other urchins, starfish, brittle stars, and holothurians (Snyder & Snyder 1970), although a pseudoreplicated design prevented legitimate statistical analyses.

Similar field experiments to determine if prey can detect the threat of predation from the scent of the predator alone have also been conducted (Peckarsky 1980). In these experiments, prey are placed in a field enclosure containing predators that are isolated in a small mesh cage. Changes in behavior due to the reception of chemical cues from the predator are inferred by measuring activity levels (frequency and distance of movement) or time spent in the region of highest odor concentration (as determined by tracking a plume of colored dye) before and after addition of the predator. Although these field studies demonstrate chemically mediated antipredator responses under natural conditions, they did not attempt to determine the chemicals responsible for the stimulus. This could be determined by extracting and fractionating predator-scented water or crushed conspecifics, releasing each fraction in the experimental apparatus used for whole organisms, and comparing behavioral responses to these fractions to an appropriate control. However, to determine precise concentrations affecting activity or to isolate the effects of flow regime on chemoreception, controlled laboratory studies likely would be needed (for discussion of appropriate method see section 2.3).

2.6.2.2. *Induced Responses*

Thus far we have considered only how chemical cues may help prey behaviorally avoid predators, but there is a large body of literature indicating that induction of physical or chemical defenses in response to predators is chemically mediated, and that this induction reduces the probability of prey being killed in an attack (reviewed by Harvell 1990). While much of the induction of these defenses requires physical damage to the prey, some rotifers, cladocerans, and bryozoans use chemicals emitted by predators or damaged conspecifics to cue the induction of protective spines (Gilbert 1966; Dodson 1989; Harvell 1992). Bioassays involve culturing of the prey animal in the presence and absence of the hypothesized warning cues. Treatments consisting of a starved predator, a predator fed daily, and homogenized prey are often employed to determine if induction is due to the scent of the predator, injured conspecifics, or some interaction between the two. The predators (or injured conspecifics) are held in a mesh bag that allows chemical cues to escape but blocks visual and mechanical cues. Because warning cues do not induce spine production in adult zooplankton, but rather in their offspring (Gilbert 1966; Krueger & Dodson 1981), induced responses for zooplankters are typically measured in the offspring. The length of spines is usually measured under a microscope and compared to spine lengths in individuals from a control culture that received no predator cues; however, a better measure of inducible defense would be to compare the susceptibility to consumers of the offspring from various predator cue treatments.

To be certain that it is chemical cues that are inducing increased defenses, culturing experiments should be repeated with predator-scented, predator and conspecific scented, and conspecific scented water to remove all nonchemical stimuli. Stimuli are commonly prepared by holding the above treatments in isolated small volumes of water to allow stimuli to accumulate (Gilbert 1966; Walls & Ketola 1989). This may allow the build up of unnaturally high levels of odors, and the alternative of using aqueous extracts of predators may release chemicals not included in predator odors in the field (see discussion in section 2.1). However, once chemically mediated induction has been demonstrated using whole organisms in ecologically realistic settings, bioassay guided fractionation of organism rinses or extracts could be used to determine the precise active component using methods similar to those discussed in previous sections. Using aqueous extracts may be more appropriate for assaying the effects of crushed conspecifics (see section 2.6.1), but such cues alone have not been implicated in inducing spine production in zooplankton (Walls & Ketola 1989), although the magnitude of a predator-induced response may be increased by the presence of crushed conspecifics (Grant & Bayly 1981).

Harvell (1986) demonstrated similar interactions while studying induction of defensive spines in an encrusting bryozoan. She manipulated bryozoan colonies and nudibranchs in the lab (using small mesh cages and separate culture containers) so that she had replicate bryozoans that were (1) being attacked by nudibranchs, (2) contacting and not contacting these attacked colonies (i.e., in the same containers, but not accessible to attack by the nudibranch), and (3) in similar containers but completely isolated from all nudibranchs. Nudibranchs were allowed to feed for 36 h. They were then removed and the spination of colonies censused 24 h later. She found that both attacked and unattacked bryozoans in containers with nudibranchs induced spines, while bryozoans in containers without nudibranchs did not produce spines. To determine whether chemical cues were coming from the predator or the attacked prey, she ground nudibranchs or spined (i.e., induced) bryozoan colonies in a tissue homogenizer and cultured bryozoan colonies for 36 h with supernatant from the macerated nudibranchs, supernatant from the macerated bryozoans, or water without added supernatant. After 36 h, 76% of colonies treated with nudibranch supernatant had produced spines while only 2–7% of control colonies or those treated with supernatant from spined bryozoan colonies had produced spines. Because she was able to induce spine formation without damaging colonies (i.e., by using nudibranch supernatant), she was also able to estimate the cost of induced defenses without having this cost confounded by previous damage due to attack. Induced colonies grew at 85% of the rate of control colonies, indicating a significant cost to induction of this morphological defense.

Theories of inducible defenses suggest that if defenses are only activated when danger is present, then there must be costs associated with the defense, and these costs are minimized by producing defenses only when needed (Harvell 1990). Measuring costs of inducible chemical, as opposed to morphological, defenses has been difficult and there is little direct evidence for such a cost among marine prey (Hay & Fenical 1988; Hay & Steinberg 1992). In both cladocerans and bryozoans, inducible morphological defenses that are chemically cued have been demonstrated to be costly in terms of reduced growth or clutch size of individuals and lower intrinsic growth rate of the population (Harvell 1986, Walls & Ketola 1989). Additionally, induced morphological changes in seaweeds that make them less susceptible to herbivores also appear to involve significant growth rate costs (Steneck & Adey 1976; Lubchenco & Cubit 1980; Hay 1981; Lewis et al. 1987). In contrast to these clear and consistent costs associated with induced morphological defenses, there are no equally clear demonstrations of costs associated with chemical defenses, although natural history observations and some relatively weak negative correlations between levels of chemical defenses and growth at some times of year suggest that such costs might occur (Hay & Fenical 1988; Hay & Steinberg 1992; Steinberg 1994, 1995; Peckol et al. 1996).

The potential difference in costs between induced chemical versus morphological defenses is interesting but may result from the limited number of chemical-induction studies that have been conducted on marine organisms. Additionally, some of the induction studies that have been conducted may have been carried out under stressful physical conditions that limited the ability of organisms to express defenses that they are capable of inducing under more natural conditions. As an illustration of this, Cronin and Hay (1996a) recently demonstrated that grazing by amphipods induced increased levels of terpenoid chemical defenses in a marine brown alga. Initial efforts to conduct this study in outside mesocosms with flow through seawater produced variable results; apparently because the physical stresses (temperature fluctuations, high levels of ultraviolet [UV] light, etc.) in this relatively benign culture facility sometimes compromised the alga's ability to produce defensive metabolites (Cronin & Hay 1996c; M.E. Hay,

S. Johnson & N. Lindquist unpublished data). However, when seaweeds were split in half and grown in the field with and without periodic amphipod grazing (accomplished by periodically enclosing all plants in small mesh cages and adding amphipods to treatment cages for 24 h; following removal of cages, local fishes appeared to remove most amphipods), those plant halves attacked by amphipods significantly increased production of chemical defenses relative to unattacked plant halves. We now have additional information (M.E. Hay, S. Johnson, and N. Lindquist unpublished data) indicating that chemically defended species often may be more affected by certain physical stresses than species that lack chemical defenses, and that this stress is often expressed by decreased production of chemical defenses. This suggests that experiments on induction would best be conducted in the field, or if they are conducted in the lab, they should be conducted with hearty species whose responses to laboratory conditions are well known.

2.7. Intraspecific Chemical Communication

In addition to the alarm signals described previously, chemical signaling can also be involved in a wide range of interactions where marine or aquatic animals identify conspecifics, search for mates, or synchronize reproductive behavior. Although there is considerable anecdotal evidence for such signals (e.g., see review for Crustacea by Dunham 1978), rigorous demonstrations of intraspecific chemical signaling in aquatic systems are scarce.

2.7.1. Vertebrates

One of the few studies to convincingly demonstrate chemical communication between fishes (other than the alarm signals discussed in section 2.6.1.1) involves the recognition of shoalmates (conspecifics collected from the same shoal) in the fathead minnow (*Pimephales promelas*). To see if minnows could recognize shoalmates, Brown and Smith (1994) used a system of barriers (Fig. 2.7) that transmitted either (1) visual cues only (solid, clear Plexiglas®); (2) chemical cues only (a system of two barriers each with opaque Plexiglas® across two thirds of the width and mesh across one-third, arranged such that the mesh portions were on opposite sides of the tank and blocked all visual cues), or (3) both visual and chemical cues (mesh screen). Barriers were placed 10 cm from each end of a $67 \times 35 \times 55$ cm high aquarium, dividing it into a stimulus chamber at each end of the aquarium and a test chamber in the middle. One stimulus chamber held a fish collected from the same shoal as the test fish, and in the other was a fish from a shoal approximately 200 m away. Test fish given only chemical cues or both chemical and visual cues behaved similarly, qualitatively and quantitatively, spending more time near shoalmates than unfamiliar fish. Visual cues alone produced no difference in time spent near shoalmates versus unfamiliar fish. Chemical recognition of shoalmates persisted even when fish had been isolated

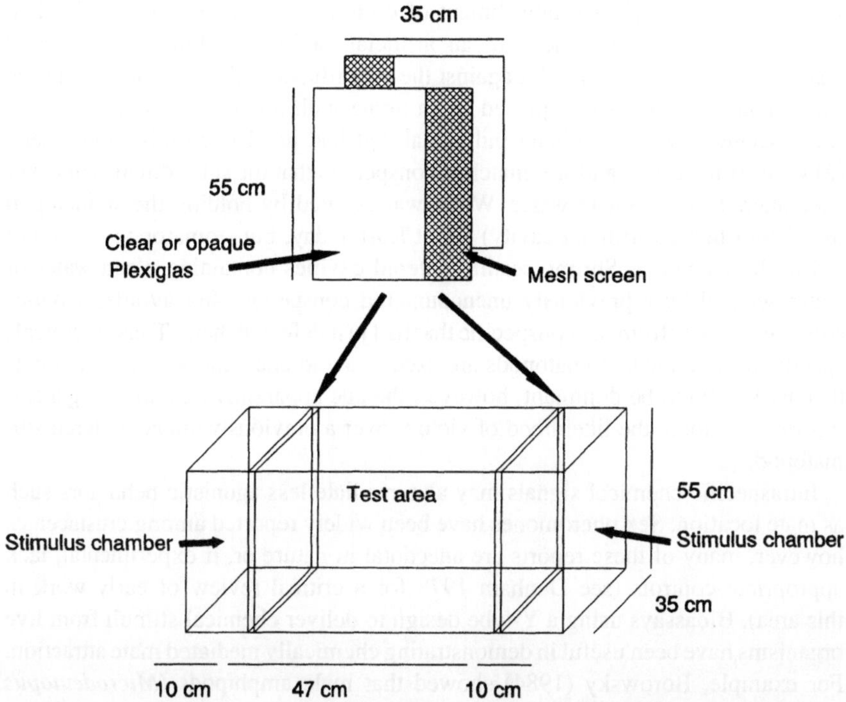

Figure 2.7. Diagram of a test tank and barriers used to separate visual from chemical cues involved in shoalmate recognition. Clear or opaque Plexiglas® was used depending on whether visual cues were being offered. Shoalmates or unfamiliar conspecifics were placed in stimulus chambers at the end of the test arena. Test fish tended to congregate near the stimulus chamber producing chemical cues from shoalmates. The figure was redrawn and modified from one in Brown and Smith (1994).

for 2 months (Brown & Smith 1994). It is hypothesized that this type of individual recognition may be advantageous because minnows that are familiar with one another exhibit enhanced antipredator behaviors (i.e., greater shoal cohesion) when exposed to predator stimuli than do minnows that are unfamiliar with each other (Chivers et al. 1995).

2.7.2. *Invertebrates*

Chemical cues also mediate interactions among conspecifics involved in aggressive disputes. Although little work has been conducted in this area, observations of staged interactions between stomatopod shrimp suggest that chemical signals allow recognition of individual conspecifics, and that these chemical signals mediate aggressive encounters (Caldwell 1979). Size and sex matched pairs of the stomatopod *Gonodactylus festai* were allowed to fight for 10 min, which was sufficient time for clear establishment of dominance. The dominant individual of each pair was then placed in an artificial cavity (a 30-ml bottle painted black) and allowed to defend it against the subordinate. Fifteen minutes later the subordinate individual was placed in an arena with a cavity holding either (1) water scented by the dominant individual that had just beaten this subordinate, (2) water from a size- and sex-matched conspecific that the subordinate had never encountered, or (3) clean water. Water was scented by holding the stomatopod in a 30-ml bottle (artificial cavity) for at least 1 day, but stomatopods were not fed in the container. Shrimp readily entered cavities containing clean water or water scented by a previously unencountered conspecific, but avoided cavities containing water from the conspecific that had just defeated them. Thus, chemicals specific to individual stomatopods are used to avoid encounters with individuals that are known to be dominant; however, the cue apparently contains no general information about the likelihood of victory over a previously unencountered stomatopod.

Intraspecific chemical signals may also mediate less agonistic behaviors such as mate location. Sex pheromones have been widely reported among crustaceans, however, many of these reports are anecdotal in nature or, if experimental, lack appropriate controls (see Dunham 1978 for a critical review of early work in this area). Bioassays using a Y-tube design to deliver chemical stimuli from live organisms have been useful in demonstrating chemically mediated mate attraction. For example, Borowsky (1984) showed that male amphipods (*Microdeutopus gryllotalpa*) are attracted to receptive females by a waterborne substance that is not released by other males or non-receptive females. The experimental apparatus consisted of a pair of 2-L aspirator bottles connected in series by plastic tubing, emptying into a single outflow, forming a Y-shaped chamber (Fig. 2.8). The aspirator bottles most distant from the Y contained 0, 30 or 60 male or female amphipods, depending on the treatment. Water was passed through these bottles at 100 ml/h, mixing equal volumes of water scented by each of the treatments at the junction of the Y. A test consisted of placing 15 male amphipods below the junction of the Y, and allowing them to move up either of the forks of the Y into an "arrival bottle" or downstream into the "departure bottle." Amphipods were prevented from moving into the bottle containing the stimulus producing organisms by a piece of mesh screening. A single piece of the green alga *Ulva lactuca* was placed in each of the arrival bottles to simulate natural conditions. After 5.5 h, the number of males in each of these bottles was counted and results compared by contingency table analysis. The entire apparatus was washed thoroughly in hot tap water between trials to remove residual signals. This methodology has been successfully combined with amphipod culture techniques (see section 2.4.2.2) to investigate the effects of diet on pheromone composition in the context of a potential mechanism promoting reproductive isolation (Stanhope et al. 1992).

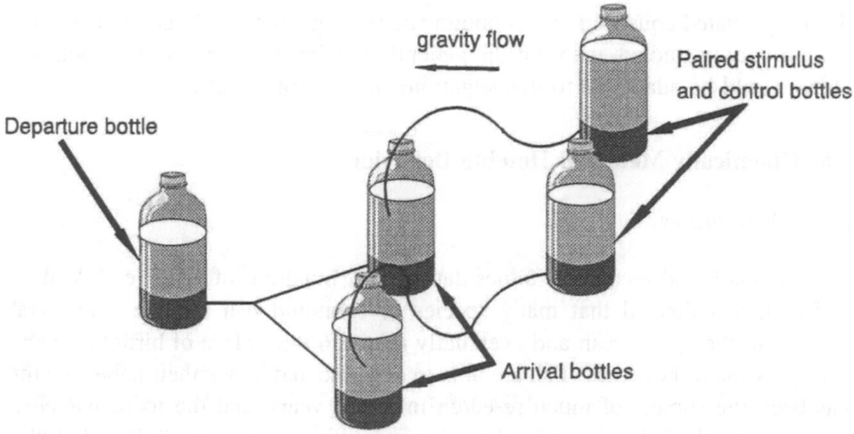

Figure 2.8. Y-maze apparatus used in the study of how amphipods use pheromones to locate mates. This is also a good example of the standard Y-maze technique used in many studies of chemoreception. The figure was redrawn and modified from one in Borowsky (1984).

The role of chemical cues in mediating other types of reproductive interactions are less well understood. Although many reproductive cycles appear to be on circadian or circatidal rhythms, some observations suggest that chemical cues also may play some role. Reiswig (1970) noted that sperm release by downcurrent sponges occurred only after release by upcurrent sponges, suggesting that a cue released by, or with, the sperm may synchronize downstream release. This does not explain what cues the upcurrent sponges to release, nor does it rule out tactile or circadian cues, but it does suggest that the role of chemical cues in synchronizing spawning cycles deserves attention.

A chemical basis for the initiation of larval release has been identified for decapod crabs (Forward & Lohmann 1983). These authors found that when exposed to water containing recently expelled larvae (concentration of 40 larvae/ml), the estuarine crab *Rhithropanopeus harissii* began to pump its abdomen, a behavior normally associated with the expulsion of larvae. Recently released larvae that were rinsed twice in filtered seawater did not elicit pumping any more frequently than a filtered seawater control, while crushed eggs (at the same 40 eggs/ml concentration) did elicit the pumping behavior. Additionally, eggs close to hatching (<1 day) elicited a pumping response in more females than eggs 8–10 days from hatching, suggesting that some cue is released from eggs at or near the time of hatching to signal expulsion by the mother.

Intraspecific chemical signaling could be involved in a wide range of interactions where marine and aquatic animals search for mates, avoid competing conspecifics, need to synchronize release of gametes or larvae, or identify near versus distantly related conspecifics. Although chemical mediation of such interactions is relatively unstudied, many of the general techniques and approaches outlined above could be adaptable to investigations in this general area.

2.8. Chemically Mediated Homing Behavior

2.8.1. Vertebrates

Various mark-and-recapture studies dating back hundreds of years (e.g., Walton 1653) have indicated that many species of salmonid fish migrate from natal streams to the open ocean and eventually return to their place of birth to spawn. The question of how these fish are able to orient to and locate their natal streams has been the subject of much research in recent years, and the focus has been primarily on the role of chemical cues (see reviews by Cooper & Hirsch 1982; Dittman & Quinn 1996). Initial assays in laboratory settings quickly showed that these fish were able to use olfaction to discriminate between water from their own stream and water from other sources (Idler et al. 1961; Fagerlund et al. 1963; Sutterlin & Gray 1973). However, detection of gradients of chemical cues over the small spatial scales characteristic of laboratory assays tells us little about the fishes' ability to use these chemical cues to locate their natal stream from distances of many kilometers. Wisby and Hasler (1954) conducted field tests of the homing ability of adult salmonids in a Y-shaped stream, of which each branch had its own population of coho salmon. Fish were collected from their home branches and the olfactory pits of half were plugged using some combination of Vaseline®, benzocaine ointment, and cotton plugs to prevent water from reaching the olfactory tissue, while the olfactory capabilities of the other half of the fish were left intact. Each fish was tagged and released 1.6 km below the fork of the stream, and allowed to repeat their upstream migration. Nearly all of the control fish returned to their home stream while the olfactory impaired fish split nearly equally between the two streams. Other species of salmonid have been demonstrated to exhibit similar homing behavior using these methods (Stasko 1971), but the interpretation of these types of studies can be questioned because there was no control for the nose-plugging procedure. It is possible that the trauma associated with the nose-plugging procedure, rather than the impaired olfaction, caused the alteration of behavior.

A more realistic approach is provided by experiments that collect developing eggs or larvae in one stream and transplant them to another or release hatchery reared fishes in nonnative streams (review by Ricker 1972). These transplant experiments have demonstrated that nonnative fishes are able to return to the stream from which they were released, suggesting that the fish are somehow "remembering" something about the stream as they leave it, and that these "remembered" cues allow them to return as adults. Further work has shown that this imprinting takes place within a very short time period during the transition from the parr to smolt stage, just prior to the downstream migration (Hasler & Scholz 1983; Dittman et al. 1996). Although these studies provide solid evidence that fish can cue on their natal stream, they are unable to say much about whether that cue is visual or chemical.

A more conclusive test of chemically mediated homing behavior is to allow fish to imprint on an artificial chemical cue, release them in the ocean, then lace a stream with the imprinting chemical and see if there are differences in return rates to that stream versus others in the area. In one of the more refined tests using this methodology, Scholz et al. (1976) divided hatchery-raised salmon into three groups, one was imprinted on morpholine at 5×10^{-5} mg/L, a second on phenethyl alcohol (PEA) at 10^{-3} mg/L, and the third was held in filtered water and served as a control (no imprinting chemical). Salmon were imprinted by holding individuals at the parr stage in large aquaria with the imprinting chemical (or filtered seawater) from April 1 to May 13. After the imprinting process, investigators released fish 5 km offshore in Lake Michigan, between two rivers that would be scented with the imprinting chemical during the next year's migration. During the spawning migration 18 months later, the investigators released morpholine into one stream to produce a streamwater concentration of 5×10^{-5} mg/L and PEA into another to produce a streamwater concentration of 5×10^{-3} mg/L. These concentrations were calculated on the basis of mean flow; actual values ranged between 3×10^{-4} and 1×10^{-5} mg/L for morpholine and 2×10^{-2} and 1×10^{-3} mg/L for PEA due to variations in flow. Fish were censused in these streams and in 16 other nearby nonscented ones during the spawning run by electrofishing, gill-netting, and creel surveys. The majority of the fish exposed to morpholine were recovered in the morpholine scented stream; the majority of PEA imprinted fish were recovered in the PEA scented stream. In contrast, significant numbers of the control fish were recovered in many of the different locations (scented and nonscented).

Further direct evidence for orientation by chemoreception was provided by tagging fish from the above experiment with ultrasonic transmitters, and charting the reaction of treatment and control fish to plumes of morpholine in open lake waters (Scholz 1980). Morpholine imprinted fish always stopped swimming when they encountered a morpholine plume, and remained until it dissipated (0.5–0.4 h); they never stopped for plumes of other chemicals (PEA). In contrast, fish never exposed to morpholine (control) swam directly through plumes of morpholine without stopping (Scholz 1980). The author concluded that morpholine was responsible for guiding the migration of fish imprinted with this chemical to the morpholine scented stream.

Testing the effectiveness of artificial homing cues in guiding migration by hatchery-reared and imprinted fishes has provided insights into the mechanisms involved in imprinting and homing; however, recent views suggest that they may present an oversimplified view of the process (Dittman & Quinn 1996). In experiments by Scholz et al. (1976), hatchery-reared fish were imprinted once and then brought directly out to open water, but stable environmental conditions in these rearings may not provide the stimuli required for elevation of hormone levels that could lead to additional imprinting (Dittmann et al. 1996). Peaks in thyroid hormone levels at the parr-smolt transformation are thought be responsible for imprinting, but smaller peaks occur before the transformation to smolt, and thyroid hormone levels are sensitive to a wide range of environmental factors such as exposure to novel water, temperature changes, and flow regime (Dittman & Quinn 1996 and references cited therein). Thus imprinting could potentially occur as a series of different imprints; migration upstream could then be guided by a series of olfactory waypoints learned at developmentally regulated periods and perhaps cued by environmental stimuli (Dittman & Quinn 1996). Most studies have imprinted smolts in filtered water using a single chemical cue under conditions of constant temperature and no flow. In nature, imprinting must occur over a wide range of physical conditions (temperature, flow), and the chemical composition of imprinting signals could well be a complex mixture of odors. The effects of flow on imprinting have not yet been considered in experiments, but some evidence suggests that wild salmon may learn and recognize complex mixtures and sequences of odorants in a fundamentally different manner than they imprint on a single chemical (Dodson & Bitterman 1989). Further research on homing using wild salmon imprinted under more natural conditions should provide valuable insights into the ecology of homing behavior.

2.8.2. Invertebrates

Many marine invertebrates have been noted to exhibit some form of homing behavior, including limpets (reviewed in Branch 1984), sea urchins (Carpenter 1984), and decapod crustaceans (reviewed in Vannini & Cannicci 1995). However, the chemical mechanisms, if any, by which these animals orient are largely unknown. Crustaceans (and to some degree mollusks) are well known for their ability to orient to food using chemical cues (see section 2.3). Although they might also use this same facility for chemoreception to direct homing activities or longer-distance migrations, direct evidence for this lacking (Vannini & Cannicci 1995). Methods discussed in this and other sections of this chapter could be modified to test the role of chemical cues in invertebrate homing.

2.9. Settlement Cues

2.9.1. Vertebrates

Although many investigators speculate that larvae of some coral reef fishes might use chemical cues to identify suitable settlement sites (Sale 1991), there has been almost no attempt to test this hypothesis. The single available test (Sweatman 1988) is, however, an excellent example of designing assays to achieve ecological realism and potential statistical rigor. The physical efforts and logistical complexities involved in this study may also demonstrate why so few investigations like this have been conducted. To assess the effect of resident-produced chemical cues on the settlement selectivity of coral-reef fish larvae, Sweatman created 0.5-m-diameter "standard coral units" (SCUs) on a 600-m broad sand plain using living branches of the coral *Pocillopora damicornis*. Each SCU received 20 transplanted adult damselfish of the species *Dascyllus aruanus*, 20 of the species *D. reticulatus*, or no resident adult fish (the control SCU). Twelve-volt submersible bilge pumps were then used to pump water, and any chemical cues produced by resident fish, from the source SCUs, through microirrigation hose, and onto 8–15 settlement SCUs located 20–30 m away. The pumping apparatus was designed so that the investigator could reposition the connections leading from the source SCUs to the settlement SCUs; flow was then randomized from the source to the settlement SCUs daily, thus ensuring that the settlement patterns observed for each treatment were caused by chemical cues from the source SCUs and were not confounded by positional effects of individual settlement SCUs. Experiments were performed for several weeks, with the newly settled juvenile fish being counted and removed from the settlement SCUs daily. Over the course of the experiment, significantly more *Pomacentrus* spp. settled on control settlement SCUs than on treatment settlement SCUs; all other fish species, however, settled independently of treatment type. These results suggest that some larval fishes use chemical cues in selecting appropriate settlement sites.

These methods are quite rigorous, but they are logistically difficult, time consuming, and subject to disruption by severe weather. In addition, the success of the experiment is dependent upon an adequate and continuous supply of competent larval settlers. Sweatman's (1988) was run over two settlement seasons. In the first, only damselfish were monitored and only six settled during the entire season. In the second season, all settling fishes were monitored with 14 damselfish and 557 total fish settling. Such risky experimental endeavors may be better suited to tenured faculty than to younger investigators, but few studies in chemical ecology, either terrestrial or marine, have achieved this degree of ecological realism.

2.9.2. *Invertebrates*

A dispersive larval stage occurs in the life history of most species of aquatic benthic invertebrates. Although these larvae can vary greatly in their morphological, physiological, and behavioral characteristics, all must eventually locate a suitable substratum for settlement and metamorphosis. Because acceptable substrata are often difficult to locate in vast, heterogeneous aquatic environments, there has been considerable interest in determining how larvae detect appropriate settlement locations; chemical cues (both settlement inducers and retardants) have been a major focus of these investigations (Burke 1986; Wahl 1989; Pawlik

1992). The following section presents a brief outline of historical and contemporary settlement assay methodologies.

Initial investigations into larval settlement concentrated on documenting how settlement varied as a function of adult organisms already occupying the settlement substratum (Knight-Jones 1951; Scheltema et al. 1981). These experiments usually were performed by placing known, or estimated, numbers of larvae into vessels containing substrata of various types. One substratum would be devoid of established organisms while others supported established organisms at various densities. Larvae settling on each substratum were counted. These studies served as useful springboards for later work, but generally did not address directly the role that chemistry played in settlement selection. They focused on demonstrating substratum selective settlement, but usually assumed, rather than tested, the role of chemical signals.

Somewhat similar investigations have focused on the chemically-mediated behavior of planktonic larvae before they settle. Marsden (1987; Marsden et al. 1990) developed methods for documenting the relative attractiveness of chemical cues to planktonic larvae. She constructed small experimental chambers that would hold organisms or pieces of substrata in one end, be open to planktonic larvae on the other end, but keep larvae separated from contact with the test organism or substrate by placing plankton netting across the middle of the chamber. After pieces of treatment organisms or substrata were placed in the chambers, the chambers were placed into larval cultures where soluble chemicals released from the treatment material could be detected by larvae. In most of these assays, one treatment and one control chamber (no organism or substrata present) were placed in each larval culture. After a period of time, the chambers were upended, and number of the larvae inside each recorded to determine if cues either attracted or repelled larvae. Care needs to be taken in interpreting these assays, because if larvae do not respond rapidly and the chemical cues are easily leached to the media, then cues could become evenly dispersed throughout the larval culture, obscuring the chemical gradient to which larvae might respond. Additionally, if larval responses are chemically cued by compounds that are bound to surfaces and do not leach into the water, then larvae will cue to these compounds only on contact. Surface bound cues may be relatively common and will not be detected by these types of assays. As an example, assays by Jensen and Morse (1984) allowed larvae to contact settlement surfaces, and demonstrated that larvae of a gregarious polychaete needed physical contact with a chemically active substrata before metamorphosis was induced.

The above methods can determine if chemical cues are present, but say little about the chemical composition of the cue. Methods facilitating this include the use of organismal washes, homogenates, or extracts, and the fractionation of these chemical mixtures along size or polarity gradients, followed by bioassays of the different fractions.

Organismal washes have usually been prepared by holding a certain number or volume of organisms (or substrata types) in standing water for a specified period of time (Standing et al. 1984; Hadfield & Scheuer 1985; Zimmer-Faust & Tamburi 1994), or by collecting the outflowing water from a running tank of water in which treatment organisms are being held (Thompson 1985). The major advantages to using organismal washes lie in the ease with which the washes can be prepared, and with the fact that the organism of interest does not need to be damaged to obtain them. However, in collecting washes, some investigators have held large masses of organisms in limited volumes of water for extended periods of time. The stresses involved in this could be considerable and could result in laboratory prepared washes differing, in both form and concentration (see section 2.2.1), from those that are present in the field (but see Hadfield & Scheuer 1985). Additionally, it is often difficult to isolate unknown active compounds from washes.

Organismal homogenates have been prepared by blending tissues (or whole organisms) of interest until the structural material has been degraded. This slurry is then used directly, or it is centrifuged and the supernatant presented to competent larvae in settlement assays (Standing et al. 1984; Durante 1991). The pros and cons of using homogenates are similar to those of the organismal washes, but homogenates are even less ecologically realistic (especially ones using slurries of whole organisms instead of the supernatant). Homogenizing organisms releases many compounds that might not be released at all from natural, undamaged organisms, and even for cues that are normally released, homogenizing may increase levels of these cues many fold over natural levels. Organismal homogenates may serve as a useful first step if they lead to more rigorous settlement investigations, and additional assay techniques; they have been used effectively as positive settlement controls (Toonen 1993).

Some investigators also have obtained lipid-soluble extracts by rinsing organisms, or isolated tissues, with organic solvents (Pawlik 1989; Davis et al. 1991; Bryan et al. 1996). In brief, the material of interest is serially bathed in organic solvents of differing polarities. Once the bathing process is complete, fractions may then be used separately or in combination with other fractions in settlement assays. Assays of this nature are generally performed by placing a small drop of extract onto a settlement substratum (often a microscope slide) and then allowing the solvent to evaporate to dryness; once the solvent has evaporated, the lipid-soluble portions of these extracts will stay on the slide when it is immersed in sea water. By using the results of bioassays to guide further fractionation, different and increasingly specific solvent mixtures or fractionation techniques can be used to help isolate active compounds.

There are two potentially major problems that must be considered when using the types of approaches discussed above. Unfortunately, in several previous studies, these potential problems may have been actual problems. First, investigators often extract all metabolites from a three-dimensional organism and concentrate these on a two-dimensional assay surface. This can produce a grossly inflated concentration of extract, an effect due to the oily nature of the extract rather than specific metabolites it contains (i.e., what is an appropriate control for effects on surface texture, charge, etc.?), and can result in internally bound metabolites that never occur on the organism's surface being incorrectly identified as metabolites that affect larval settlement in nature. Surficial chemistry, not internal chemistry, should be the focus of these assays. Second, larvae that settle move from the water column to the bottom of assay containers; not surprisingly, so do larvae that die. If investigators do not examine "settled" larvae to see if they have attached, metamorphosed, or started normal development, then compounds or extracts that kill larvae may be incorrectly identified as settlement cues. These errors in interpretation can be avoided if investigators acquire adequate information about the basic biology and natural history of their assay organisms.

Larval toxicity or antifouling assays have also received considerable attention (Thompson et al. 1985; Pawlik 1990; Slattery et al. 1995). These experiments are generally performed using crude extracts or purified compounds from chemically interesting organisms and then placing competent larvae into varying concentrations of these extracts or metabolites that have been dispersed into seawater (a procedure that is ecologically unjustified unless the compounds are known to be released to the water). After a period of time, the number of larvae surviving, settling, or metamorphosing in each extract is recorded and compared with similar data from control solutions. These types of assays can provide a quick screening for certain types of bioactivity, but they often have limited ecological relevance because many metabolites that are normally bound in organisms are dispersed into the media, thus contacting larvae in unrealistic ways and concentrations. Additionally, some of the solvents that have been used to carry these compounds may themselves have large effects on larval health and behavior. Pennington and Hadfield (1989) demonstrated that some larvae will settle and metamorphose in response to numerous organic solvents; most settlement, however, occurs in response to high concentrations, or to long periods of exposure.

Until recently, most studies dealing with the chemical mediation of larval settlement have been conducted in still water. Although results that are obtained from still-water assays can be useful, the ecological relevance of these types of studies can be questioned because they do not incorporate ecologically realistic flow regimes into the design (Eckman 1983; Butman 1987). While it may be possible for small, weakly swimming, larvae to respond directly to chemical stimuli while they are in small volumes of still water or in the surface boundary layer, it is unclear that larvae would be able to respond to similar stimuli under natural flow conditions in the field (Butman 1987).

Some of the most recent advances in larval settlement methodology involve using flumes to determine the response of larvae to chemical signals under various flow conditions (Butman et al. 1988; Butman & Grassle 1992; Pawlik et al. 1991; Turner et al. 1994). In general, flume experiments have been performed by placing larvae and larval mimics (often glass spheres) into a flume upstream from an array of experimental settlement substrata. Assays are often conducted for periods of 12–24 h at steady flow rates ranging between 2.0–6.0 cm/s, which is comparable to moderately slow natural flow rates (Butman & Grassle 1992). Following an assay, the number of settlers in control and treatment substrata are compared to the distribution of passive larval mimics to see if chemical induction of settlement has occurred. For an excellent discussion of flume structure, protocol for flume usage, and flume-related problems and biases, see Butman and Grassle (1992). Although flumes have been used most often for assaying substratum bound or substratum specific cues, such as those found in association with muds (Butman et al. 1988; Butman & Grassle 1992) or sands (Pawlik et al. 1991), they may also be used to assay water soluble chemical cues (Turner et al. 1994).

Because aquatic invertebrate larvae are extraordinarily diverse, it is impossible to describe a single methodological approach for field experimentation. Each project may need specific methodologies that account for particular characteristics of the larvae being investigated. However, with this limitation in mind, it is possible to describe some field methodologies that have worked in the past or that show promise for overcoming some of the problems mentioned above. Gerhart et al. (1988) placed a structure of vertical rods and horizontal surfaces subtidally in the field. Half of this structure had been treated with gorgonian extracts in methylene chloride while the other half had been treated with the organic solvent alone. After 24 h, the structure was recovered, and organisms that had settled on the different sides were enumerated. Although this initial experiment was pseudoreplicated, adding independent replicates will allow these procedures to provide statistically rigorous results.

In the Gulf of California, Raimondi (1988) treated Plexiglas™ settlement plates with homogenates from various intertidal organisms and deployed them in the *Cthamalus* settlement zone of the rocky intertidal. Experiments ran for 24 h, after which the plates were recovered and the number of barnacle settlers recorded. Several of the homogenates produced significant increases in barnacle settlement.

Henrikson and Pawlik (1995) recently developed a new methodological approach and demonstrated that organic crude extracts from test organisms could be incorporated into disks of Phytagel™ (an agarlike substance), these disks transplanted into the field for up to 6 weeks, and the disk surface itself monitored for inhibition or stimulation of larval settlement. Gels containing extracts from some organisms significantly deterred or stimulated settlement relative to control gels (containing no extracts) or Plexiglas™ plates (control gels and Plexiglas™ plates exhibited similar settlement patterns). This method is advantageous in that (1) compounds are *in* rather than *on* a surface, (2) they are slowly released to the surface over a considerable period of time (after 21 days in flowing seawater, 56% of the mass of a sponge crude extract was still in the gels), and (3) testing in the field allows assays in flow and against the natural mix of larvae that encounter surfaces. Potential, but as yet untested, limitations would include possible chemical or biological degradation of metabolites over extended assay periods and the possibility that some metabolites from crude extracts could leach at a much faster rate than others. This method also could be adapted to the lab and is in some ways a variation of the novel procedure developed by Wahl et al. (1994) for testing extract effects on microbial settlement and development.

2.9.3. Recommendations Regarding Still-Water Assays

Although numerous investigations have now demonstrated that settlement assays conducted in flumes, or in the field, have considerable advantages over still-water assays conducted in the lab, the limited availability of flumes, the ease of conducting still-water assays in the lab, and the limited funding needed to conduct laboratory assays all suggest that still-water assays will continue to be used—at least during initial stages of investigations. We therefore provide a few general suggestions regarding these methods even though we feel that field or flume assays would be preferable.

Investigations concerning the chemical ecology of larval settlement should be based on strong natural history and experimental field studies documenting patterns of larval settlement or adult abundance, and investigators should realize that many nonchemical factors can also affect these patterns—and therefore need to be considered as alternate hypotheses to chemical mediation of settlement. In many previous studies focused on the chemical ecology of larval settlement, ecologically meaningless bioassays were used to track novel chemistry that may have had no real effect on larval settlement in the field, or sound bioassays were used to suggest the importance of chemical signals, but the chemistry of these signals was rarely investigated rigorously (see Pawlik 1992 for a review of this general field). With the increasing maturity of the field and the increasing communication and collaboration between chemists and biologists, a more productive and intellectually satisfying approach now seems possible. Below we describe an investigation by Schmitt et al. (1995) as an example of how still-water laboratory assays, natural history observations, natural products chemistry, and tests of alternative hypotheses can be integrated to provide an increased understanding of chemically mediated settlement by invertebrate larvae.

In the field, these authors demonstrated that the brown alga *Dictyota menstrualis* was less frequently and less heavily fouled than any of the other common seaweeds with which it co-occurred. Although such patterns are commonly interpreted as evidence of chemically mediated antifouling activity, these patterns also could arise due to algal species being distributed in different microhabitats that favored or suppressed fouling organisms, due to some algal species having communal symbionts that clean their surfaces of foulers (Duffy 1990, Stachowicz & Hay 1996), or due to other unrecognized possibilities. Thus, before conducting extensive chemical studies, it would be prudent to determine if fouling larvae actually avoid this generally unfouled seaweed, or if their low numbers might result from postsettlement processes.

When assessing fouling of seaweeds in the field, the bryozoan *Bugula neritina* was one of the species most commonly found growing on seaweeds, and was the only macroorganism found on *Dictyota*. When larvae from this bryozoan were placed in petri dishes with equal-sized portions of seven different co-occurring seaweeds, larvae settled extensively on most seaweeds, but avoided settling on *Dictyota*. When larvae were placed into petri dishes with *Dictyota* and with an alga that they commonly fouled, larvae contacted *Dictyota* as frequently as they contacted the other alga; however, they settled and metamorphosed on the other alga, but not on *Dictyota*. When larvae were placed into petri dishes containing water collected from the middle of *Dictyota* plants versus water collected distant from *Dictyota* plants, there was no difference in rates of larval settlement or metamorphosis. All of the above experiments and observations suggested that some surface quality of *Dictyota* deterred settlement of *Bugula* larvae, and that this deterrent characteristic was not released into the water near *Dictyota* where it could repel larvae before they contacted the plant.

Measurements of surface wettability (a physical property of surfaces that can affect fouling) for seaweed species that were avoided or preferred by settling *Bugula* larvae suggested that wettability could not explain why *Bugula* larvae did not settle on *Dictyota*. However, when gentle surface swabbings of *Dictyota* were extracted with organic solvents and transferred in these solvents onto one half of the surface of numerous polystyrene settling plates (area of the plate coated was equal to the surface area of the alga that had been swabbed), *Bugula* larvae avoided settling on the treated half of the plates relative to the plate halves that had been coated only with the organic solvents. This indicated that a lipid-soluble chemical that deterred *Bugula* settlement occurred on the surface of *Dictyota*. Chemical investigations confirmed the presence of diterpene alcohols in surface swabbings of *Dictyota*, and bioassays of these compounds indicated that they negatively affected the survivorship, growth, and development of *Bugula* larvae when larvae were confined in water with these compounds and thus unable to avoid them by delaying settlement.

Although the investigation by Schmitt et al. (1995) has numerous limitations (e.g., surface swabbings may break surficial cells, producing extract concentrations that are too high, or they may simply smear rather than absorb surficial metabolites, producing extract concentrations that are too low), it demonstrates that still-water laboratory assays can provide useful insights into chemical mediation of invertebrate settlement. However, the recent methodologies developed by Wahl et al. (1994) and Henrikson and Pawlik (1995) should allow very similar investigations to be conducted almost entirely in the field. If these methods for field investigations prove to be as tractable as they seem, then most future studies should probably incorporate these methodologies.

When using either field or laboratory approaches, there are several other procedural questions that should be addressed. First, larval settlement often occurs more rapidly on surfaces that have been allowed to accumulate a natural biofilm of microbes by immersing them in flowing seawater for hours to days before an experiment (Brancato & Woollacott 1982). Using biofilmed surfaces for settling assays allows more rapid assays of metabolites that degrade with time or that slowly dissolve from experimental surfaces.

Second, investigators sometimes attach live organisms to settlement plates as an initial assay to see if their presence affects larval settlement (Hubbard 1988; Toonen 1993). A response to this treatment need not be due to chemical mediation. The physical structure of the organism, its effect on flow, the glue used to attach the organism, feeding by the organism, etc., are all realistic alternative mechanisms that could affect larval recruitment. These alternative hypotheses need to be evaluated. This may be accomplished using inert structural models or dead organisms (for effects on flow or effects of feeding), using controls for the effects of the glue, etc. (Wethy 1986, Pawlik 1989).

Third, it is important to consider the amount of time that the assay will run. It has been shown that some larvae will settle and metamorphose in response to organic solvents, with increased settlement occurring with increased assay length (Pennington & Hadfield 1989). In addition, some larvae have been shown to become "desperate" as the length of their stay in the plankton increases. Over time these desperate larvae will cease to be discriminating in their substrate selectivity and will begin to settle randomly (Knight-Jones 1953; Wilson 1953; but see also Toonen & Pawlik 1994). Appropriate timing of assays will vary depending on the natural history of the species being investigated, but assays should be performed over the least amount of time possible. Many successful assays have been performed over the course of 24 h. This will require that larvae be grown to competence before using them in assays or that they be collected from brooding species that release competent larvae.

Fourth, the volume of water that will be used for assays needs to be considered. A balance must be achieved between maintaining enough water for ecological realism and avoiding providing an overabundance of water so that larvae are prevented from encountering the experimental substrata, or so that larvae are lost during the attempted recovery process. Appropriate experimental volumes will vary with the species being investigated.

Finally, larval density should be considered. It is generally not possible to use natural larval densities during experimentation, either because natural larval concentrations are not known, or because field densities of larvae are too dilute to be used for laboratory experimentation. Because of this, it is necessary to select a larval density that will provide for sufficient statistical power, while at the same time preventing larvae from becoming overcrowded to the extent that they influence each others' behavior. As with the previous parameters, the ideal larval density will vary from species to species; as a rule, however, it is probably better to assay larvae in a dilute rather than concentrated environment. In many studies, larval densities of 1–10 larvae/ml have been used.

2.10. Allelopathy and Antifouling

2.10.1. Invertebrates

Numerous studies explore allelopathy targeted against marine invertebrates. Early studies were strictly observational. Existence of a competitive interaction and a chemical mechanism were both assumed, and uninvestigated patterns were often interpreted as allelopathy. Goodbody (1961) observed inhibited development of a sessile community when settling plates were placed in the presence of a mature invertebrate community. The effect was not seen on settling plates 20 ft away, where a sessile community developed normally. Contact between the organisms was not required, suggesting the presence of a waterborne allelopathic chemical; however, alternative explanations such as localized effects on hydrodynamics or the mature community depleting larvae from the water column due to their feeding activities were not considered. Although field observations provide a sound foundation on which to base future study, strictly observational studies cannot unambiguously document allelopathy.

More rigorous investigations attempt to document allelopathy more directly. A number of studies (Sammarco et al. 1983, 1985; La Barre et al. 1986; Leone et al. 1995) use manipulative experiments with appropriate controls, and successfully demonstrate the existence of competition. La Barre et al. (1986) investigated allelopathic interactions between alcyonacean corals. The manipulation involved establishing independent pairs of soft coral species by cementing them into flower pots. Colonies were cemented into either contact or noncontact situations (in which expanded colonies were separated by 2.5 cm). Individuals cemented into flower pots alone served as controls, and colonies of each species occurring naturally isolated from other soft corals in the field served as transplant controls. All colonies were monitored at 1 day and 4, 7, 12, and 20 weeks for visible effects of allelopathic interactions including tissue death and growth modification (bending). Manipulative experiments of this type are valuable in that they can demonstrate direct inhibition of one species by another. However, these studies can not unequivocally show that the competitive effect has a chemical basis. They suggest, but do not document, allelopathy.

At the opposite end of the spectrum, investigators demonstrate a chemical basis for a competitive interaction in the lab but it remains unclear whether the results are meaningful under ecologically relevant conditions (i.e., natural flow regimes). Sullivan et al. (1983) isolated siphonodictidine as the major secondary metabolite from the Indo-Pacific sponge *Siphonodictyon* sp. This species burrows into corals, maintaining a zone of dead coral polyps around the exposed osculum. Laboratory bioassays in standing water demonstrated that siphonodictidine, at concentrations of 10^{-2} to 10^{-1} ppm, stimulates respiration in the hard coral *Acropora formosa* and causes cell lysis at concentrations above 100 ppm, suggesting siphonodictidine may be responsible for maintaining the zone of dead coral around the osculum. Studies of this type are more desirable from a chemical standpoint than those based on observations alone, but interactions seen in these laboratory assays may or may not occur in nature, where flow would promote dilution of allelopathic compounds and where corals might be less stressed and better able to tolerate allelopathic metabolites.

The most convincing studies of allelopathy (Porter & Targett 1988; de Nys et al. 1991) use manipulative field experiments to demonstrate direct inhibition of one species by another and also assay extracts in the field to show a chemical mechanism for the inhibition. de Nys et al. (1991) investigated chemically mediated interactions between the red alga *Plocamium hamatum* and the soft coral *Sinularia cruciata*. The study was grounded with field observations of sessile organisms that experienced tissue necrosis when in physical contact with the alga. Manipulative field experiments were done in which the alga and coral were transplanted onto galvanized wire grids anchored to the reef, and placed in direct contact or in close proximity (<5 cm) but having no direct contact. Controls were also relocated, but corals and algae were placed at least 30 cm apart. Additionally, soft corals that were in natural contact and noncontact situations with the alga were marked and served as transplant controls. Observations of coral tissue damage were made at 24 and 48 h and 14 days. Finally, a chemical mechanism for the inhibition was established by coating the lipophilic extract of the alga onto artificial aquarium plants and placing these in contact with soft corals in the field. Other artificial plants, coated with solvent only, served as controls for physical abrasion and observations of tissue damage were again made at 24 and 48 h and 14 days. Deleterious effects were observed only among corals in contact with extract-treated plants, making a strong case for allelopathy.

Future studies of allelopathy might be improved by putting more effort into documenting that the concentrations of metabolites or extracts being used in assays are ecologically relevant. In many previous studies, metabolite concentrations were tested at unnaturally high concentrations because the entire volume of a three-dimensional organism was extracted and applied to a planar surface. Using these methods, compounds from the interior of the organism will be extracted and tested although they may never be present at the surface of the organism in nature, and therefore can not be involved in allelopathy. Since allelopathy involves surface chemistry, future studies should take the difficult step of attempting to target the surface of the organism in their extraction procedures. Schmitt et al. (1995) made a first attempt at this by swabbing the surface of an apparently deterrent organism (a brown alga) in an attempt to extract only the surface chemistry (see section 2.9.3). There are potential drawbacks to this technique as the swab may be inefficient in absorbing the active compounds (e.g., consider cleaning an oily covering off your desk surface with a cotton swab) or may break open surface cells, releasing internal chemistry that is not really present on the organism's surface.

Coll et al. (1982) developed a submersible sampling apparatus capable of *in situ* isolation of exuded allelopathic chemicals. Utilization of this sort of apparatus to isolate compounds will provide more reliable data on ecologically relevant concentrations in chemical assays; however, even in these instances, it will be difficult to determine the concentration of a metabolite that a settling larvae will experience when contacting the surface of an organism releasing compounds. Mucous coverings on some organisms could retain released metabolites near the surface or facilitate entrance of the metabolite into tissues of an organism contacting this mucus; mimicking these characteristics of the potential host would be useful, but has not been tried. In some of our ongoing investigations (M.E. Hay, M. Deal, N. Lindquist unpublished data), we seem to be finding more bioactive compounds that are water soluble in organisms that have mucous exteriors, suggesting that the mucous may serve to retain these type of metabolites.

Henrikson and Pawlik (1995) recently demonstrated that organic extracts from invertebrates or seaweeds could be placed into Phytagel™ and these Phytagel™ disks could be placed in the field for up to 6 weeks without being degraded. Fouling organisms colonized untreated Phytagel™ surfaces at about the same rate as they colonized Plexiglas™ plates, but Phytagel™ disks with certain extracts either facilitated or deterred colonization. They demonstrated that 56% of a sponge extract placed in Phytagel™ was still present after 21 days in flowing seawater, suggesting that the extract was slowly released from this gel. For organisms that can be shown to naturally release metabolites (see methods of Coll et al. 1982), this method appears to offer an efficient and ecologically realistic method of conducting field assays on potential allelopathic interactions.

2.10.2. *Seaweeds*

A few investigations have looked at allelopathy targeted against macroalgae. De Ruyter van Steveninck et al. (1988) studied the brown alga *Lobophora variegata* where it bordered stands of scleractinian corals. They measured reduced blade growth rates and increased blade damage in *Lobophora* that contacted corals or was in close proximity, but did not contact corals. These observations are suggestive of allelopathy, but there are also numerous alternative explanations (e.g., physical abrasion or grazing by herbivorous symbionts that shelter in the corals; see Stachowicz & Hay 1996). As with invertebrates, direct demonstrations of chemically mediated competition are needed.

Coll et al. (1987) found that colonies of the soft coral *Lobophytum pauciflorum* that were fouled by algae contained a single major diterpene, while clean colonies contained two other diterpenes in approximately equal concentrations. Culturing the fouling seaweeds in media containing 0 (control), 12.5, 25, 50, or 100 mg/l of these compounds demonstrated that the diterpene from overgrown colonies suppressed algal growth at lower concentrations than the diterpenes from clean colonies. Because the bioassays were done in standing water, it is possible that theses differences would not be seen in the field under natural flow regimes. Growth is only one aspect governing epiphyte cover; effects of these compounds on settlement of spores, density of mesograzers, etc., were not measured. Additionally, due to the lack of manipulation, the presence of allelopathy is unclear as there is only a correlative relationship between the chemical differences and the variation of epiphyte cover. However, these types of relationships are amenable to rigorous investigation in the field. If clean corals were split, one half of each infected with epiphytic algae, and the chemistry of infected and uninfected halves followed over time in the field (see Cronin and Hay 1996a for a similar methodology with seaweeds), then chemical changes in direct responses to fouling could be determined. The chemical extracts from fouled versus clean halves could then be extracted, mixed into Phytagel™ discs (see Henrikson & Pawlik 1995), and assayed for seaweed antifoulant properties in the field.

Finally, Bak & Borsboom (1984) investigated allelopathic interactions between *Condylactis gigantea* (a reef anemone) and filamentous algae. To quantify the influence of anemone exudate on settlement and growth of filamentous algae, algal biomass was monitored on control and experimental microscope slides placed in the field. Both experimental and control slides were placed inside perforated Perspex™ tubes (10 cm × 2.2 cm diameter, perforated with 400 holes of 0.2 cm diameter) that permitted water flow and presumably exchange of algal propagules, but prevented anemone tentacles or macrograzers from entering the tube. The tubes were placed in the field for 2 weeks; the treatment inside the extended tentacles of *C. gigantea,* and the control in an identical orientation but 30–40 cm away from the anemone. Similar numbers of spores were observed to settle on treatment and control slides, but development of the propagules was inhibited on treatment slides. Since the anemone tentacles could not directly contact the slides, a waterborne allelopathic agent was suspected. This study was innovative and used an excellent experimental design, but there was no direct demonstration of anemone chemistry affecting algal growth (a number of physical conditions also may have differed between treatments and controls (oxygen levels, pH, nutrients, etc.) thus weakening the case for allelopathy. The reader interested in exploring allelopathy targeted against macroalgae should consult de Nys et al. (1991). This study, although concerned with allelopathy against invertebrates, contains methods useful for exploring allelopathic interactions in general.

2.11. Chemical Ecology within a Broader Environmental Context

Although there is still much work to be accomplished within the traditional confines of marine and aquatic chemical ecology, we feel that some of the most significant future advances in marine chemical ecology will be achieved by better integrating chemical ecology with physiology; neurobiology; molecular biology; and population, community, and ecosystem ecology (see Hay 1992, 1996b; Hay

& Fenical 1996 for examples). Determining how various physical parameters (UV and visible light, flow, nutrients, etc.) and biological rigors (competition, predation, microbial infection) combine to affect an organism's chemical defenses and how changes in these defenses affect that organism's interaction with its physical and biological environment is a challenge that, if successfully met, will allow a much deeper understanding of the mechanisms affecting biological change in modern communities. Improved understanding of these mechanisms will not only provide important insights into fundamental aspects of ecology and evolution, but also may allow advances in pest control, disease resistance, environmental restoration, and conservation.

Because methodologies for these more complex multivariate studies mandate a deep appreciation for the chemistry, natural history, ecology, and evolution of the organisms being studied, most of these studies may need to be uniquely designed for the organisms, questions, and environmental contexts being addressed. Given the unique problems that many of these studies will address, trying to outline or recommend specific approaches would be premature and potentially misleading. However, several recent studies approach these types of questions and can be viewed as potential templates for future methodological development. Investigators interested in these multilevel and more context dependent alterations in chemical defenses should consult the papers listed below and the earlier investigations cited therein.

Chemically defended organisms often provide indirect defense to other associated organisms (sometimes even to their own consumers and competitors) by creating spatial refuges from consumers or pathogens that are affected by these chemical defenses. The direct and indirect role of secondary metabolites in creating these associational refuges needs additional investigation, but examples of these studies can be found in Hay (1986, 1992), Littler et al. (1986), Hay et al. (1987b, 1988b, 1988d, 1989, 1990a, 1990b), Pfister and Hay (1988), Gil-Turnes et al. (1989), McClintock and Janssen (1990), and Kerr and Paul (1995). Studies of how various physical and biological stresses interact to affect concentrations of chemical defenses, the palatability of chemically defended organisms, and the interactions of those organisms with their surroundings include Harvell (1990), Renaud et al. (1990), Meyer and Paul (1992), Yates and Peckol (1993), Harvell et al. (1993), Steinberg (1994, 1995), Cronin and Hay (1996a, 1996c, 1996d), and Peckol et al. (1996).

2.12. Conclusions

Aquatic chemical ecology is a young science, and both its methodologies and general concepts are subject to rapid change. During the last decade, bioassays have progressed from ones designed to follow chemical bioactivity without much concern for its ecological relevance, to assays that are well designed, ecologically relevant, and often conducted under field conditions. The major questions that need to be addressed before performing most bioassays in aquatic chemical ecology involve:

1. Is the compound or extract being tested really the natural metabolite, or is it a degraded product that was created by the storage or extraction process?
2. What is the natural concentration of the metabolite or extract being assayed?
3. How can this be presented to the assay organism in a way that is ecologically relevant?
4. What are the appropriate assay organisms to use for this particular ecological question?
5. Will the conditions under which the assay is run affect the outcome and, if so, what does this context-dependent response mean ecologically?

2.13. Acknowledgments

The U.S. National Science Foundation provided support via grants OCE 95-29784 and DEB 94-10336 to M.E. Hay and via NSF Pre-Doctoral Fellowships to S. Bullard and J. Stachowicz.

2.14. References

Ache, B.W. 1982. Chemoreception and thermoreception. *In:* The Biology of Crustacea, Vol. 3, eds. L. Atwood & D.C. Sandeman, pp. 369–398, Academic Press, New York.

Alcaraz, M., G.-A. Paffenhöfer & J.R. Strickler. 1980. Catching the algae: a first account of visual observations on filter feeding calanoids. *In:* Evolution and Ecology of Zooplankton Communities, ed. W.C. Kerfoot, pp. 241–248, University Press of New England, Hanover, NH.

Allmon, R.A. & K.P. Sebens. 1988. Feeding biology and ecological impact of an introduced nudibranch, *Tritonia plebia,* New England, USA. Mar. Biol. **99**:375–385.

Ammerman, J.W. & F. Azam. 1985. Bacterial 5′ nucleotidase in aquatic ecosystems: a novel mechanism of phosphorous regeneration. Science **227**:1338–1340.

Anderson, G. & W.E. Dale. 1989. *Probopyrus pandalicola* (Packard) (Isopoda: Epicaridea): swimming responses of cryptoniscus larvae in water conditioned by hosts *Palaemonetes pugio* (Holthius) (Decapoda; Palaemonidae). J. Exp. Mar. Biol. Ecol. **130**:9–18.

Atema, J., K. Holland & W. Ikehara. 1980. Olfactory responses of yellowfin tuna (*Thunnus albacares*) to prey odors: chemical search image. J. Chem. Ecol. **6**:457–465.

Atema, J., R.R. Fay, A.N. Popper & W.N. Tavolga, eds. 1988. Sensory Biology of Aquatic Animals. Springer-Verlag, New York.

Bak, R.P.M. & J.L.A. Borsboom. 1984. Allelopathic interaction between a reef coelenterate and benthic algae. Oecologia **63**:194–198.

Bärlocher, F. & S.Y. Newell. 1994. Growth of the salt marsh periwinkle *Littoraria irrorata* on fungal and cordgrass diets. Mar. Biol. **118**:109–114.

Barnes, D.J., R.N. Braver & M.R. Jordan. 1970. Locomotory response of *Acanathaster planci* to various species of coral. Nature **228**:342–344.

Bernays, E.A. 1991. Relationship between deterrence and toxicity of plant secondary compounds for the grasshopper *Schistocerca americana.* J. Chem. Ecol. **17**:2519–2526.

Bernays, E.A. & R. Chapman. 1987. The evolution of deterrent responses in plant-feeding insects. *In:* Perspectives in Chemoreception and Behavior, eds. R.F. Chapman, E.A. Bernays & J.G. Stoffolano, Jr., pp. 159–173, Springer-Verlag, New York.

Bernays, E.A. & M. Cornelius. 1992. Relationship between deterrence and toxicity of plant secondary compounds for the alfalfa weevil *Hypera brunneipennis.* Entomol. Exp. App. **64**:289–292.

Bernays, E.A., G.C. Driver & M. Bilgener. 1989. Herbivores and plant tannins. Adv. Ecol. Res. **19**:263–302.

Boettcher, A.A. & N.M. Targett. 1993. Role of polyphenolic molecular size in reduction of assimilation efficiency in *Xiphister mucosus.* Ecology **74**:891–903.

Bolser, R.C. & M.E. Hay. 1996. Are tropical plants better defended? Palatability and defenses of temperate versus tropical seaweeds. Ecology **77**:2269–2286.

Borowsky, B. 1984. Effects of receptive females' secretions on some male reproductive behavior in the amphipod crustacean *Microdeutopus gryllotalpa.* Mar. Biol. **84**:183–187.

Boudreau, B., E. Bourget & Y. Simard. 1993. Behavioral responses of competent lobster postlarvae to odor plumes. Mar. Biol. **117**:63–69.

Brancato, M.S. & R.M. Woollacott. 1982. Effects of microbial films on settlement of bryozoan larvae (*Bugula simplex, B. stolonifera,* and *B. turrita*). Mar. Biol. **71**:51–56.

Branch, G.M. 1984. Limpets: their role in littoral and sublittoral community dynamics. *In:* The Ecology of Rocky Coasts, eds. P.G. Moore & R. Seed, pp. 97–116, Columbia University, New York.

Brown, G.E. & R.J.F. Smith. 1994. Fathead minnows use chemical cues to discriminate natural shoalmates from unfamiliar conspecifics. J. Chem. Ecol. **20**:3051–3061.

Bryan, P.J., D. Rittschof & J.B. McClintock. 1996. Bioactivity of echinoderm ethanolic body wall extracts: an assessment of marine bacterial attachment and macroinvertebrate larval settlement. J. Exp. Mar. Biol. Ecol. **196**:79–96.

Bryant, B.P. & J. Atema. 1987. Diet manipulation affects social behavior of catfish: importance of body odor. J. Chem. Ecol. **12**:989–1011.

Burke, R.D. 1986. Pheromones and the gregarious settlement of marine invertebrate larvae. Bull. Mar. Sci. **39**:323–331.

Burritt, E.A. & F.D. Provenza. 1989. Food aversion learning: ability of lambs to distinguish safe from harmful foods. J. Anim. Sci. **67**:1732–1739.

Buschbaum, R., I. Valiela & T. Swain. 1984. The role of phenolic compounds and other plant constituents in feeding by Canada geese in the coastal marsh. Oecologia **63**:343–349.

Buskey, E.J. 1984. Swimming pattern as an indicator of the roles of copepod sensory systems in the recognition of food. Mar. Biol. **79**:165–175.

Butman, C.A. 1987. Larval settlement of soft-sediment invertebrates: the spatial scales of pattern explained by active habitat selection and the emerging role of hydrodynamical processes. Oceanogr. Mar. Biol. Annu. Rev. **25**:113–165.

Butman, C.A. & J.P. Grassle. 1992. Active habitat selection by *Capitella* sp. larvae. I. Two-choice experiments in still water and flume flows. J. Mar. Res. **50**:669–715.

Butman, C.A., J.P. Grassle & C.M. Webb. 1988. Substrate choices made by marine larvae settling in still water and in a flume flow. Nature **333**:771–773.

Caldwell, R.L. 1979. Cavity occupation and defensive behavior in the stomatopod *Gonodactylus festai:* evidence for chemically mediated individual recognition. Anim. Behav. **27**:194–201.

Carpenter, R.C. 1984. Predator and population density control of homing behavior in the Caribbean echinoid *Diadema antillarum.* Mar. Biol. **82**:101–108.

Carr, W.E.S. & C.D. Derby. 1986a. Chemically stimulated feeding behavior in marine animals: implications of chemical mixtures and involvement of mixture interactions. J. Chem. Ecol. **12**:989–1011.

Carr, W.E.S. & C.D. Derby. 1986b. Chemoreception in the shrimp, *Palaemonetes pugio,* identification of active compounds in food extracts and evidence of synergistic mixture interactions. Chem. Senses **11**:49–64.

Carte, B. & D.J. Faulkner. 1986. Role of secondary metabolites in feeding associations between a predatory nudibranch, two grazing nudibranchs, and a bryozoan. J. Chem. Ecol. **12**:795–803.

Chanas, B. & J.R. Pawlik. 1995. Defense of Caribbean sponges against predatory reef fish. II. spicules, tissue toughness, and nutritional quality. Mar. Ecol. Prog. Ser. **127**:195–211.

Chanas, B. & J.R. Pawlik. 1996. Does the skeleton of a sponge provide a defense against predatory reef fish? Oecologia **107**:225–231.

Cherry, D.S. & J. Cairns. 1982. Biological monitoring. Part V. Preference and avoidance studies. Wat. Res. **16**:263–301.

Chivers, D.P., G.E. Brown & R.J.F. Smith. 1995. Familiarity and shoal cohesion in fathead minnows (*Pimephales promelas*): implications for antipredator behaviour. Can. J. Zool. **73**:955–960.

Chausen, T.P., F.D. Provenza, E.A. Burritt, P.B. Reichardt & J.P. Bryant. 1990. Ecological implications of condensed tannin structure: a case study. J. Chem. Ecol. **16**:2381–2392.

Coll, J.C., B.F. Bowden, D.M. Tapiolas, & W.C. Dunlap. 1982. In situ isolation of allelochemicals released from soft corals (Coelenterata: Octocorrallia): A totally submersible sampling apparatus. J. Exp. Mar. Biol. Ecol. **60**:293–299.

Coll, J.C., I.R. Price, G.M. König & B.F. Bowden. 1987. Algal overgrowth of alcyonacean soft corals. Mar. Biol. **96**:129–135.

Collins, A.R.S. 1974. Biochemical investigation of two responses involved in the feeding behaviour of *Acanthaster planci* (L.) I. Assay methods and preliminary results. J. Exp. Mar. Biol. Ecol. **15**:173–184.

Cooper, J.C. & P.J. Hirsch. 1982. The role of chemoreception in salmonid homing. *In:* Chemoreception in Fishes, ed. T.J. Hara, pp. 343–362, Elsevier, Amsterdam.

Costero, M. & S.P. Meyers. 1993. Evaluation of chemoreception by *Penaeus vannamei* under experimental conditions. Prog. Fish Culturist **55**:157–162.

Cronin, G. & M.E. Hay. 1996a. Induction of seaweed chemical defenses by amphipod grazing. Ecology **77**:2287–2301.

Cronin, G. & M.E. Hay. 1996b. Within-plant variation in seaweed palatability and chemical defenses: optimal defense theory versus the growth-differentiation balance hypothesis. Oecologia **105**:361–368.

Cronin, G. & M.E. Hay. 1996c. Susceptibility to herbivores depends on recent history of both plant and animal. Ecology **77**:1531–1543.

Cronin, G. & M.E. Hay. 1996d. Effects of light and nutrient availability on the growth, secondary chemistry, and resistance to herbivory of two brown seaweeds. Oikos. **77**:93–106.

Cronin, G., M.E. Hay, W. Fenical & N. Lindquist. 1995a. Distribution, density, and sequestration of host chemical defenses by the specialist nudibranch *Tritonia hamnerorum* found at high densities on the sea fan *Gorgonia ventalina.* Mar. Ecol. Prog. Ser. **119**:177–189.

Cronin, G., N. Lindquist, M.E. Hay & W. Fenical. 1995b. Effects of storage and extraction procedures on yields of lipophilic metabolites from the brown seaweeds *Dictyota ciliolata* and *D. menstrualis.* Mar. Ecol. Prog. Ser. **119**:265–273.

Daldorph, P.W.G. & J.D. Thomas. 1988. The chemical ecology of some British freshwater gastropod molluscs: behavioural responses to short chain carboxylic acids and maltose. Freshwater Biol. **19**:167–178.

Daniel, P.C. & R.C. Bayer. 1987a. Attraction of predatorily naive postlarval lobsters to extracts of metabolites of common prey: *Mytilus edulis, Mya arenaria, Cancer irroratus,* and *Asterias vulgaris.* J. Chem. Ecol. **13**:1201–1215.

Daniel, P.C. & R.C. Bayer. 1987b. Development of chemically mediated prey-search response in postlarval lobsters (*Homarus americanus*) through feeding experience. J. Chem. Ecol. **13**:1217–1233.

Davin, W.T., Jr., C.C. Kohler & D.R. Tindall. 1986. Effects of ciguatera toxins on the bluehead. Trans. Am. Fish. Soc. **115**:908–912.

Davin, W.T., Jr., C.C. Kohler & D.R. Tindall. 1988. Ciguatera toxins adversely affect piscivorous fishes. Trans. Am. Fish. Soc. **117**:374–384.

Davis, A.R., A.J. Butler & I. Van Altena. 1991. Settlement behaviour of ascidian larvae: preliminary evidence for inhibition by sponge allelochemicals. Mar. Ecol. Prog. Ser. **72**:117–123.

Decho, A.W. & J.W. Moriarty. 1990. Bacterial exopolymer utilization by a harpacticoid copepod: a methodology and results. Limnol. Oceanogr. **35**:1039–1049.

Deibel, D. & G.-A. Paffenhöfer. 1988. Cinematographic analysis of the feeding mechanism of the pelagic tunicate *Doliolum nationalis.* Bull. Mar. Sci. **43**:404–412.

DeMott, W.R. & M.D. Watson. 1991. Remote detection of algae by copepods: responses to algal size, odors and motility. J. Plankton Res. **13**:1203–1222.

de Nys, R., J.C. Coll & I.R. Price. 1991. Chemically mediated interactions between the red alga *Plocamium hamatum* (Rhodophyta) and the octocoral *Sinularia cruciata* (Alcyonacea). Mar. Biol. **108**:315–320.

Derby, C.D. & J. Atema. 1988. Chemoreceptor cells in aquatic invertebrates: peripheral mechanisms of chemical signal processing in decapod crustaceans. *In:* Sensory Biology of Aquatic Animals, eds. J. Atema, R.R. Fay, A.N. Popper & W.N. Tavolga, pp. 365–386, Springer-Verlag, New York.

De Ruyter van Steveninck, E.D., L.L. Van Mulekom & A.M. Breeman. 1988. Growth and inhibition of *Lobophora variegata* (Lamouroux) Womersley by scleractinian corals. J. Exp. Mar. Biol. Ecol. **115**:169–178.

Distel, R.A. & F.D. Provenza. 1991. Experience early in life affects voluntary intake of blackbrush by goats. J. Chem. Ecol. **17**:431–450.

Dittman, A.H. & T.P. Quinn. 1996. Homing in pacific salmon: mechanisms and ecological basis. J. Exp. Biol. **199**:83–91.

Dittman, A.H., H. Quinn & G.A. Nevitt. 1996. Timing of imprinting to natural and artificial odors by coho salmon, *Oncorhynchus kisutch.* Can. J. Fish. Aquat. Sci. **53**:434–447.

Dodson, J.J. & M.E. Bitterman. 1989. Compound uniqueness and the interactive role of morpholine in fish chemoreception. Biol. Behav. **14**:13–27.

Dodson, S.I. 1989. The ecological role of chemical stimuli for the zooplankton: predator induced morphology in *Daphnia.* Oecologia **78**:361–367.

Duffy, J.E. 1990. Amphipods on seaweeds: partners or pests? Oecologia **83**:267–276.

Duffy, J.E. & M.E. Hay. 1991. Food and shelter as determinants of food choice by an herbivorous marine amphipod. Ecology **72**:1286–1298.

Duffy, J.E. & M.E. Hay. 1994. Herbivore resistance to seaweed chemical defense: the role of mobility and predation risk. Ecology **75**:1304–1319.

Duffy, J.E. & V.J. Paul. 1992. Prey nutritional quality and the effectiveness of chemical defenses against tropical reef fishes. Oecologia **90**:333–339.

Dunham, P.J. 1978. Sex pheromones in Crustacea. Biol. Rev. **53**:555–583.

Durante, K.M. 1991. Larval behavior, settlement preference, and induction of metamorphosis in the temperate solitary ascidian *Molgula citrina* Alder & Hancock. J. Exp. Mar. Biol. Ecol. **145**:175–187.

Eckman, J.E. 1983. Hydrodynamic processes affecting benthic recruitment. Limnol. Oceanogr. **28**:241–257.

Fagerlund, U.H.M., J.R. McBride, M. Smith & N. Tomlinson. 1963. Olfactory perception in migrating salmon. III. Stimulants for adult sockeye salmon (*Oncorhynchus nerka*) in home stream waters. J. Fish. Res. Board Can. **20**:1457–1463.

Faulkner, D.J. 1996. Marine natural products. Nat. Prod. Rept. **13**:75–125.

Folin, O. & W. Denis. 1915. A colorimetric method for the determination of phenols (and phenol derivatives) in urine. J. Biol. Chem. **22**:305–308.

Forward, Jr., R.B. & K.J. Lohmann. 1983. Control of egg hatching in the crab *Rhithropanopeus harrisii* (Gould). Biol. Bull. **165**:154–166.

Gallager, S.M. 1988. Visual observation of particle manipulation during feeding in larvae of a bivalve mollusc. Bull. Mar. Sci. **43**:344–365.

Gates, J.E. & W.B. Wilson. 1960. The toxicity of *Gonyaulax monilata* Howell to *Mugil cephalus.* Limnol. Oceanogr. **5**:171–174.

Gerhart, D.J. 1984. Prostaglandin A_2: an agent of chemical defense in the Caribbean gorgonian *Plexaura homomalla.* Mar. Ecol. Prog. Ser. **19**:181–187.

Gerhart, D.J., D. Rittschof & S.W. Mayo. 1988. Chemical ecology and the search for marine antifoulants. Studies of a predator-prey symbiosis. J. Chem. Ecol. **14**:1905–1917.

Giattina, J.D. & R.R. Garton. 1983. A review of the preference-avoidance responses of fishes to aquatic contaminants. Residue Rev. **87**:43–90.

Giattina, J.D., R.R. Garton & D.G. Stevens. 1982. Avoidance of copper and nickel by rainbow trout as monitored by a computer-based data acquisition system. Trans. Am. Fish. Soc. **111**:491–504.

Gil-Turnes, M.S., M.E. Hay & W. Fenical. 1989. Symbiotic marine bacteria chemically defend crustacean embryos from a pathogenic fungus. Science **246**:116–118.

Gilbert, J.J. 1966. Rotifer ecology and embryological induction. Science **151**:1234–1237.

Goodbody, I. 1961. Inhibition of the development of a marine sessile community. Nature **190**:282–283.

Gosselin, S., L. Fortier & J.A. Gagné. 1989. Vulnerability of marine fish larvae to the toxic dinoflagellate *Protogonyaulax tamarensis.* Mar. Ecol. Prog. Ser. **57**:1–10.

Grant, J.W.G. & I.A.E. Bayly. 1981. Predator induction of crests in morphs of the *Daphnia carinata* King complex. Limnol. Oceanogr. **26**:201–218.

Hadfield, M.G. & D. Scheuer. 1985. Evidence for a soluble metamorphic inducer in *Phestilla:* ecological, chemical, and biological data. Bull. Mar. Sci. **37**:556–566.

Halstead, B.W. 1978. Poisonous and Venomous Marine Animals of the World. Darwin Press, Princeton, NJ.

Hanlon, R.T. & J.B. Messenger. 1996. Cephalopod Behaviour. Cambridge University Press, Cambridge.

Hara, T.J. 1982. Structure-activity relationships of amino acids as olfactory stimuli. *In:* Chemoreception in Fishes, ed. T.J. Hara, pp. 29–57, Elsevier, Amsterdam.

Harpaz, S., D. Kahan, R. Galun & I. Moore. 1987. Responses of freshwater prawn, *Macrobrachium rosenbergii,* to chemical attractants. J. Chem. Ecol. **13**:1957–1965.

Harvell, C.D. 1986. The ecology and evolution of inducible defenses in a marine bryozoan: cues, costs and consequences. Am. Nat. **128**:810–823.

Harvell, C.D. 1990. The ecology and evolution of inducible defenses. Quart. Rev. Biol. **65**:323–337.

Harvell, C.D. 1992. Inducible defenses and allocation shifts in a marine bryozoan. Ecology **73**:1567–1576.

Harvell, C.D., W. Fenical & C.H. Green. 1988. Chemical and structural defenses of Caribbean gorgonians (*Pseudopterogorgia* spp.). I. development of an in situ feeding assay. Mar. Ecol. Prog. Ser. **49**:287–294.

Harvell, C.D., W. Fenical, V. Roussis, J.L. Ruesink, C.C. Griggs & C.H. Greene. 1993. Local and geographic variation in the defensive chemistry of a West Indian gorgonian coral (*Briareum asbestinum*). Mar. Ecol. Prog. Ser. **93**:165–173.

Hasler, A.D. & A.T. Scholz. 1983. Olfactory Imprinting and Homing in Salmon. Springer-Verlag. Berlin.

Hay, M.E. 1981. The functional morphology of turf forming seaweeds: persistence in stressful marine habitats. Ecology **62**:739–750.

Hay, M.E. 1984. Predictable spatial escapes from herbivory: how do these affect the evolution of herbivore resistance in tropical marine communities? Oecologia **64**:396–407.

Hay, M.E. 1985. Spatial patterns of herbivore impact and their importance in maintaining algal species richness. Proc. Fifth Int. Coral Reef Congr. **4**:29–34.

Hay, M.E. 1986. Associational plant defenses and the maintenance of species diversity: turning competitors into accomplices. Am. Nat. **128**:617–641.

Hay, M.E. 1991. Fish-seaweed interactions on coral reefs: effects of herbivorous fishes and adaptations of their prey. *In:* The Ecology of Fishes on Coral Reefs, ed. P. F. Sale, pp. 96–119, Academic Press, San Diego, California, USA.

Hay, M.E. 1992. Seaweed chemical defenses: their role in the evolution of feeding specialization and in mediating complex interactions. *In:* Ecological Roles for Marine Secondary Metabolites; Explorations in Chemical Ecology Series, ed. V.J. Paul, pp. 93–118, Comstock, Ithaca.

Hay, M.E. 1996a. Defensive synergisms? A reply to Pennings. Ecology **77**:1950–1952.

Hay, M.E. 1996b. Marine chemical ecology: what's known and what's next? J. Exp. Mar. Biol. Ecol. **200**:103–134.

Hay, M.E. & W. Fenical. 1988. Marine plant-herbivore interactions: the ecology of chemical defense. Ann. Rev. Ecol. Syst. **19**:111–145.

Hay, M.E. & W. Fenical. 1996. Chemical ecology and marine biodiversity: insights and products from the sea. Oceanography **9**:10–20.

Hay, M.E. & P.D. Steinberg. 1992. The chemical ecology of plant-herbivore interactions in marine versus terrestrial communities. *In:* Herbivores: Their Interaction with Secondary Plant Metabolites, Vol. II, Evolutionary and Ecological Processes, 2nd ed., eds. G. Rosenthal & M. Berenbaum, pp. 371–413, Academic Press, San Diego.

Hay, M.E., T. Colburn & D. Downing. 1983. Spatial and temporal patterns in herbivory on a Caribbean fringing reef: the effects on plant distribution. Oecologia **58**:299–308.

Hay, M.E., W. Fenical & K. Gustafson. 1987a. Chemical defense against diverse coral reef herbivores. Ecology **68**:1581–1591.

Hay, M.E., J.E. Duffy & C.A. Pfister. 1987b. Chemical defense against different marine herbivores: are amphipods insect equivalents? Ecology **68**:1567–1580.

Hay, M.E., J.E. Duffy & W. Fenical. 1988a. Seaweed chemical defenses: among-compound and among-herbivore variance. Proc. Sixth Int. Coral Reef Symp. **3**:43–48.

Hay, M.E., J.E. Duffy, W. Fenical & K. Gustafson. 1988b. Chemical defense in the seaweed *Dictyopteris delicatula:* differential effects against reef fishes and amphipods. Mar. Ecol. Prog. Ser. **48**:185–192.

Hay, M.E., V.J. Paul, S.M. Lewis, K. Gustafson, J. Tucker & R.N. Trindell. 1988c. Can tropical seaweeds reduce herbivory by growing at night? diel patterns of growth, nitrogen content, herbivory, and chemical versus morphological defenses. Oecologia **75**:233–245.

Hay, M.E., P.E. Renaud & W. Fenical. 1988d. Large mobile versus small sedentary herbivores and their resistance to seaweed chemical defenses. Oecologia **75**:246–252.

Hay, M.E., J.R. Pawlik, J.E. Duffy & W. Fenical. 1989. Seaweed-herbivore-predator interactions: host-plant specialization reduces predation on small herbivores. Oecologia **81**:418–427.

Hay, M.E., J.E. Duffy & W. Fenical. 1990a. Host-plant specialization decreases predation on a marine amphipod: an herbivore in plant's clothing. Ecology **71**:733–743.

Hay, M.E., J.E. Duffy, V.J. Paul, P.E. Renaud & W. Fenical. 1990b. Specialist herbivores reduce their susceptibility to predation by feeding on the chemically-defended seaweed *Avrainvillea longicaulis.* Limnol. Oceanogr. **35**:1734–1743.

Hay, M.E., Q. Kappel & W. Fenical. 1994. Synergisms in plant defenses against herbivores: interactions of chemical defenses, calcification, and plant quality. Ecology **75**:1714–1726.

Henrikson, A.A. & J.R. Pawlik. 1995. A new antifouling assay method: results from field experiments using extracts of four marine organisms. J. Exp. Mar. Biol. Ecol. **194**:157–165.

Hessinger, D.A. & H.M. Lenhoff. eds. The Biology of Nematocysts. Academic Press, San Diego.

Hews, D.K. 1988. Alarm response in larval western toads, *Bufo boreas:* release of larval chemicals by a natural predator and its effect on predator capture efficiency. Anim. Behav. **36**:125–133.

Himmelman, J.H. 1988. Movement of whelks (*Buccinum undulatum*) toward a baited trap. Mar. Biol. **97**:521–531.

Hollibaugh, J.T. 1979. Metabolic adaptation in natural bacterial populations supplemented with selected amino acids. Estuarine Coastal Mar. Sci. **9**:215–230.

Horn, M.H. 1989. Biology of marine herbivorous fishes. Oceanogr. Mar. Biol. Ann. Rev. **27**:167–272.

Howe, N.R. & Y.M. Sheikh. 1975. Anthopleurine: a sea anemone alarm pheromone. Science **189**:386–389.

Hubbard, E.J.A. 1988. Larval growth and the induction of metamorphosis of a tropical sponge-eating nudibranch. J. Molluscan Stud. **54**:259–269.

Huntley, M., P. Sykes, S. Rohan & V. Marin. 1986. Chemically-mediated rejection of dinoflagellate prey by the copepods *Calanus pacificus* and *Paracalanus parvus*: mechanism, occurrence and significance. Mar. Ecol. Prog. Ser. **28**:105–120.

Huntley, M.E., P. Ciminello & M.D.G. Lopez. 1987. Importance of food quality in determining development and survival of *Calanus pacificus* (Copepoda: Calanoida). Mar. Biol. **95**:103–113.

Hurlbert, S.H. 1984. Pseudoreplication and the design of ecological field experiments. Ecol. Monogr. **54**:187–211.

Idler, D.R., J.R. McBride, R.E.E. Jonas & N. Tomlinson. 1961. Olfactory perception in migrating salmon. II. Studies on a laboratory bio-assay for home-stream water and mammalian repellent. Can. J. Biochem. Physiol. **29**:1575–1584.

Ireland, C.D. & M.H. Horn. 1991. Effects of macrophyte secondary chemicals on food choice and digestive efficiency of *Cebidichthys violaceus* (Girard), an herbivorous fish of temperate marine waters. J. Exp. Mar. Biol. Ecol. **153**:179–194.

Ives, J.D. 1985. The relationship between *Gonyaulax tamarensis* cell toxin levels and copepod ingestion rates. *In:* Toxic Dinoflagellates. Proc. 3rd Int. Conf., eds. D.M. Anderson, A.W. White & D.G. Baden, pp. 413–418, Elsevier, New York.

Ives, J.D. 1987. Possible mechanisms underlying copepod grazing responses to levels of toxicity in red tide dinoflagellates. J. Exp. Mar. Biol. Ecol. **112**:131–145.

Jensen, R.A. & D.E. Morse. 1984. Intraspecific facilitation of larval recruitment: gregarious settlement of the polychaete *Phragmatopoma californica* (Fewkes). J. Exp. Mar. Biol. Ecol. **83**:107–126.

Jones, D.A., J.G. Munford & P.A. Gabbott. 1974. Microcapsules as artificial food particles for aquatic filter feeders. Nature **247**:233–235.

Jones, D.A., A. Kanazawa & K. Ono. 1979. Studies on the nutritional requirements of the larval stages of *Penaeus japonicus* using microencapsulated diets. Mar. Biol. **54**:261–267.

Jones, K.A. & T.J. Hara. 1985. Behavioural responses of fishes to chemical cues: results from a new bioassay. J. Fish Biol. **27**:495–504.

Kats, L.B., J.W. Petranka & A. Sih. 1988. Antipredator defenses and the persistence of amphibian larvae with fishes. Ecology **69**:1865–1970.

Kem, W.R. 1988. Worm toxins. *In:* Handbook of Natural Toxins: Marine Toxins and Venoms, Vol. 3, ed. A.T. Tu, pp. 353–378, Marcel Dekker, Inc., New York.

Kerr, J.N.Q. & V.J. Paul. 1995. Animal-plant defense association: the soft coral *Sinularia* sp. (Cnidaria, Alcyonacea) protects *Halimeda* species from herbivory. J. Exp. Mar. Biol. Ecol. **186**:183–205.

Knight-Jones, E.W. 1951. Gregariousness and some other aspects of settling behavior of *Spirorbis*. J. Mar. Biol. Assoc. UK **30**:201–222.

Knight-Jones, E.W. 1953. Laboratory experiments on gregariousness during settling in *Balanus balanoides* and other barnacles. J. Exp. Biol. **30**:584–599.

Knutsen, J.A. 1992. Feeding behavior of North Sea turbot (*Scopthalmus maximus*) and Dover sole (*Solea solea*) larvae elicited by chemical stimuli. Mar. Biol. **113**:543–548.

Kobayashi, M., J. Kobayashi & Y. Ohizumi. 1989. Cone shell toxins and the mechanisms of their pharmacological action. Bioorg. Mar. Chem. **3**:71–84.

Krueger, D.A. & S.I. Dodson. 1981. Embryological induction and predation ecology in *Daphnia pulex*. Limnol. Oceanogr. **26**:219–223.

Kvitek, R.G. 1991. Sequestered paralytic shellfish poisoning toxins mediate glaucous-winged gull predation on bivalve prey. Auk **108**:381–392.

Kvitek, R.G., A.R. DeGange & M.K. Beitler. 1991. Paralytic shellfish poisoning toxins mediate feeding behavior of sea otters. Limnol. Oceanogr. **36**:393–404.

La Barre, S.C., J.C. Coll & P.W. Sammarco. 1986. Competitive strategies of soft corals (Coelenterata: Octocorallia): III. Spacing and aggressive interactions between alcyonaceans. Mar. Ecol. Prog. Ser. **28**:147–156.

Lawrence, J.M. 1991. A chemical alarm response in *Pycnopodia helianthoides* (Echinodermata: Asteroidea). Mar. Behav. Physiol. **19**:39–44.

Lenhoff, H.M. 1983. Hydra: Research Methods. Plenum Press, New York.

Leone, P.A., B.F. Bowden, A.R. Carroll, & J.C. Coll. 1995. Chemical consequences of relocation of the soft coral *Lobophytum compactum* and its placement in contact with the red alga *Plocamium hamatum*. Mar. Biol. **122**:675–679.

Lewis, S.M., J.N. Norris & R.B. Searles. 1987. The regulation of morphological plasticity in tropical reef algae by herbivory. Ecology **68**:636–641.

Lindquist, N. 1996. Palatability of invertebrate larvae to corals and sea anemones. Mar. Biol. **126**:745–755.

Lindquist, N. & M.E. Hay. 1995. Can small rare prey be chemically defended? The case for marine larvae. Ecology **76**:1347–1358.

Lindquist, N. & M.E. Hay. 1996. Palatability and chemical defenses of marine invertebrate larvae. Ecol. Monogr. **66**:431–450.

Lindquist, N., M.E. Hay & W. Fenical. 1992. Defense of ascidians and their conspicuous larvae: adult vs. larval chemical defenses. Ecol. Monogr. **62**:547–568.

Littler, M.M., P.R. Taylor & D.S. Littler. 1986. Plant defense associations in the marine environment. Coral Reefs **5**:63–71.

Lubchenco, J. & J. Cubit. 1980. Heteromorphic life histories of certain marine algae as adaptations to variations in herbivory. Ecology **61**:676–687.

Luczkovich, J.J. 1988. The role of prey detection in the selection of prey by pinfish *Lagodon rhomboides* (Linnaeus). J. Exp. Mar. Biol. Ecol. **123**:15–30.

Lumbang, W.A. & V.J. Paul. 1996. Chemical defense of the tropical green seaweed *Neomeris annulata* Dickie: effects of multiple compounds on feeding by herbivores. J. Exp. Mar. Biol. Ecol. **201**:185–195.

Magnelia, S.J., C.C. Kohler & D.R. Tindall. 1992. Acanthurids do not avoid consuming cultured toxic dinoflagellates yet do not become ciguatoxic. Trans. Am. Fish. Soc. **121**:737–745.

Manahan, D.T., S.H. Wright & G.C. Stephens. 1983. Simultaneous determination of net uptake of 16 amino acids by a marine bivalve. Am. J. Physiol. **244**:R832–R838.

Marsden, J.R. 1987. Coral preference behaviour by planktonic larvae of *Spirobranchus giganteus corniculatus* (Serpulidae: Polychaeta). Coral Reefs **6**:71–74.

Marsden, J.R., B.E. Conlin & W. Hunte. 1990. Habitat selection in the tropical polychaete *Spirobranchus giganteus.* II. Larval preferences for corals. Mar. Biol. **104**:93–99.

Martel, G. & L.M. Dill. 1993. Feeding and aggressive behaviors in juvenile coho salmon (*Oncorynchus kisutch*) under chemically-mediated risk of predation. Behav. Ecol. Sociobiol. **32**:365–370.

Mathis, A. & R.J.F. Smith. 1993a. Chemical alarm signals increase the survival time of fathead minnows (*Pimephales promelas*) during encounters with northern pike (*Esox lucius*). Behav. Ecol. **4**:260–265.

Mathis, A. & R.J.F. Smith. 1993b. Intraspecific and cross-superorder responses to chemical alarm signals by brook stickleback. Ecology **74**:2395–2404.

Mathis, A., D.P. Chivers & R.J.F. Smith. 1995. Chemical alarm signals: predator deterrents or predator attractants. Am. Nat. **145**:994–1005.

McClintock, J.B. & J. Janssen. 1990. Pteropod abduction as a chemical defense in a pelagic antarctic amphipod. Nature **346**:462–464.

McClintock, J.B., M. Slattery, J. Heine & J. Weston. 1992. Chemical defense, biochemical composition and energy content of three shallow-water Antarctic gastropods. Polar Biol. **11**:623–629.

McClintock, J.B., B.J. Baker, M. Slattery, M. Hamann, R. Kopitkze & J. Heine. 1994. Chemotactic tube-foot responses of a spongivorous sea star *Perknaster fuscus* to organic extracts from Antarctic sponges. J. Chem. Ecol. **20**:859–870.

McConnell, O.J., P.A. Hughes, N.M. Targett & J. Daley. 1982. Effects of secondary metabolites on feeding by the sea urchin, *Lytechinus variegatus.* J. Chem. Ecol. **8**:1437–1453.

McCosker, J.E. 1975. Feeding behavior of Indo-Australian Hydrophyiidae. *In:* The biology of sea snakes, ed. W.A. Dunson, pp. 217–232, University Park Press, Baltimore.

Meier, J. & J. White. 1995. Handbook of clinical toxicology of animal venoms and poisons. CRC Press, Boca Raton, FL.

Meyer, K.D. & V.J. Paul. 1992. Intraplant variation in secondary metabolite concentration in three species of *Caulerpa* (Chlorophyta: Caulerpales) and its effects on herbivorous fishes. Mar. Ecol. Prog. Ser. **82**:249–257.

Meyer, K.D. & V.J. Paul. 1995. Variation in secondary metabolite and aragonite concentrations in the tropical green seaweed *Neomeris annulata:* effects on herbivory by fishes. Mar. Biol. **122**:537–545.

Meyer, K.D., V.J. Paul, H.R. Sanger & S.G. Nelson. 1994. Effects of seaweed extracts and secondary metabolites on feeding by the herbivorous surgeonfish *Naso lituratus.* Coral Reefs 13:105–112.

Moore, P.A., & J. Atema. 1991. Spatial information in the three-dimensional fine structure of an aquatic odor plume. Biol. Bull. **181**:408–418.

Moore, P.A., M.J. Weissburg, J.M. Parrish, R.K. Zimmer-Faust & G.A. Gerhardt. 1994.

Spatial distribution of odors in simulated benthic boundary layer flows. J. Chem. Ecol. **20**:255–279.

Newman, R.M., W.C. Kerfoot & Z. Hanscom, III. 1990. Watercress and amphipods: potential chemical defense in a spring stream macrophyte. J. Chem. Ecol. **16**:245–259.

Norris, J.N. & W. Fenical. 1985. Natural products chemistry: uses in ecology and systematics. *In:* Handbook of Phycological Methods, Ecological Field Methods: Macroalgae, eds. M.M. Littler & D.S. Littler, pp. 121–145, Cambridge University Press, Cambridge.

Nowell, A.R.M. & P.A. Jumars. 1987. Flumes: theoretical and experimental considerations for simulation of benthic environments. Oceanogr. Mar. Biol. Ann. Rev. **25**:91–112.

Ogura, N. 1975. Further studies on decomposition of dissolved organic matter in coastal seawater. Mar. Biol. **31**:101–111.

Olivera, B.M., W.R. Gray & L.J. Cruz. 1988. Marine snail venoms. *In:* Handbook of Natural Toxins, Marine Toxins and Venoms, Vol. 3, ed. A.T. Tu, pp. 327–352, Marcel Dekker, Inc., New York.

Paerl, H.W. 1988. Nuisance phytoplankton blooms in coatal, estuarine, and inland waters. Limnol. Oceanogr. **33**:823–847.

Paul, V.J. 1987. Feeding deterrent effects of algal natural products. Bull. Mar. Sci. **41**:514–522.

Paul, V.J., ed. 1992. Ecological Roles of Marine Natural Products. Comstock, Ithaca, NY.

Paul, V.J. & W. Fenical. 1986. Chemical defense in tropical green algae, order Caulerpales. Mar. Ecol. Prog. Ser. **34**:157–169.

Paul, V.J. & M.E. Hay. 1986. Seaweed susceptibility to herbivory: chemical and morphological correlates. Mar. Ecol. Prog. Ser. **33**:255–264.

Paul, V.J. & K.L. Van Alstyne. 1992. Activation of chemical defenses in the tropical green algae *Halimeda* spp. J. Exp. Mar. Biol. Ecol. **160**:191–203.

Paul, V.J., M.E. Hay, J.E. Duffy, W. Fenical & K. Gustafson. 1987. Chemical defense in the seaweed *Ochtodes secundiramea* (Montagne) Howe (Rhodophyta): effects of its monoterpenoid components upon diverse coral-reef herbivores. J. Exp. Mar. Biol. Ecol. **114**:249–260.

Paul, V.J., C.R. Wylie & H.R. Sanger. 1988. Effects of algal chemical defenses toward different coral-reef herbivorous fishes. Proc. Sixth Int. Coral Reef Symp. **3**:73–78.

Pawlik, J.R. 1989. Larvae of the sea hare *Apylsia californica* settle and metamorphose on an assortment of macroalgal species. Mar. Ecol. Prog. Ser. **51**:195–199.

Pawlik, J.R. 1990. Natural and artificial induction of metamorphosis of *Phragmatopoma lapidosa californica* (Polychaeta: Sabellariidae), with a critical look at the effects of bioactive compounds on marine invertebrate larvae. Bull. Mar. Sci. **46**:512–536.

Pawlik, J.R. 1992. Chemical ecology of the settlement of benthic marine invertebrates. Oceanogr. Mar. Biol. Annu. Rev. **30**:273–335.

Pawlik, J.R. 1993. Marine invertebrate chemical defenses. Chem. Rev. **93**:1911–1922.

Pawlik, J.R. & W. Fenical. 1989. A re-evaluation of the ichthyodeterrent role of prostaglandins in the Caribbean gorgonian coral *Plexaura homomalla.* Mar. Ecol. Prog. Ser. **52**:95–98.

Pawlik, J.R., M.T. Burch & W. Fenical 1987. Patterns of chemical defenses among Caribbean gorgonian corals: a preliminary survey. J. Exp. Mar. Biol. Ecol. **108**:55–66.

Pawlik, J.R., C.A. Butman & V.R. Starczak. 1991. Hydrodynamic facilitation of gregarious settlement of a reef-building tube worm. Science **251**:421–424.

Pawlik, J.R., B. Chanas, R.J. Toonen & W. Fenical. 1995. Defenses of Caribbean sponges against predatory reef fish. I. Chemical deterrency. Mar. Ecol. Prog. Ser. **127**:183–194.

Peckarsky, B.L. 1980. Predator-prey interaction between stoneflies and mayflies: behavioral observations. Ecology **61**:932–943.

Peckol, P., J.M. Krane & J.L. Yates. 1996. Interactive effects of inducible defenses and resource availability on phlorotannins in the north Atlantic brown algae *Fucus vesiculosus.* Mar. Ecol. Prog. Ser. **138**:209–217.

Pennings, S.C. 1990. Multiple factors promoting narrow host range in the sea hare, *Aplysia californica.* Oecologia **82**:192–200.

Pennings, S.C. 1996. Testing for synergisms between chemical and mineral defenses—a comment. Ecology **77**:1948–1949.

Pennings, S.C. & V.J. Paul. 1992. Effects of plant toughness, calcification and chemistry on herbivory by *Dolabella auricularia.* Ecology **73**:1606–1619.

Pennings, S.C., S.R. Pablo, V.J. Paul & J.E. Duffy. 1994. Effects of sponge secondary metabolites in different diets on feeding by three groups of consumers. J. Exp. Mar. Biol. Ecol. **180**:137–149.

Pennings, S.C., M.P. Puglisi, T.J. Pitlik, A.C. Himaya & V.J. Paul. 1996. Effects of secondary metabolites and $CaCO_3$ on feeding by surgeonfishes and parrotfishes: within-plant comparisons. Mar. Ecol. Prog. Ser. **134**:49–58.

Pennington, J.T. & M.G. Hadfield. 1989. Larvae of a nudibranch mollusc (*Phestilla sibogae*) metamorphose when exposed to common organic solvents. Biol. Bull. **177**:350–355.

Peterson, C.H. & P.E. Renaud. 1989. Analysis of feeding preference experiments. Oecologia **80**:82–86.

Petranka, J.W., L.B. Kats & A. Sih. 1987. Predator-prey interactions among fish and larval amphibians: use of chemical cues to detect predatory fish. Anim. Behav. **35**:420–425.

Pfeiffer, W. 1977. The distribution of fright reaction and alarm substance cells in fishes. Copeia **1977**:653–655.

Pfeiffer, W., G. Riegelbauer, G. Meier & B. Scheibler. 1985. Effect of hypoxanthine-3(N)-oxide on central nervous excitation of the black tetra *Gymnocorymbus ternetzi* (Characidae, Ostariophysi, Pices) indicated by dorsal light response. J. Chem. Ecol. **11**:507–523.

Pfister, C.A. & M.E. Hay. 1988. Associational plant refuges: convergent patterns in marine and terrestrial communities result from differing mechanisms. Oecologia **77**:118–129.

Porter, J.W. & N.M. Targett. 1988. Allelochemical interactions between sponges and corals. Biol. Bull. **175**:230–239.

Poulet, S.A. & P. Marsot. 1978. Chemosensory grazing by marine calanoid copepods (Arthropoda: Crustacea). Science **200**:1403–1405.

Poulet, A.A. & P. Marsot. 1980. Chemosensory feeding and food-gathering by omnivorous marine copepods. *In:* The Evolution and Ecology of Zooplankton Communities, ed. W.C. Kerfoot, pp. 198–218, University Press of New England, Hanover, NH.

Price, H.J., G. Paffenhöfer & J.R. Strickler. 1983. Modes of cell capture in calanoid copepods. Limnol. Oceanogr. **28**:116–123.

Provenza, F.D. & D.F. Balph. 1987. Diet learning by domestic ruminants: theory, evidence and practical implications. Appl. Anim. Behav. Sci. **18**:211–232.

Provenza, F.D. & J.C. Malechek. 1986. A comparison of food selection and foraging behavior in juvenile and adult goats. Appl. Anim. Behav. Sci. **16**:49–61.

Provenza, F.D., E.A. Burritt, T.P. Clausen, J.P. Bryant, P.B. Reichardt & R.A. Distel. 1990. Conditioned flavor aversion: a mechanism for goats to avoid condensed tannins in black brush. Am. Nat. **136**:810–828.

Ragan, M.A. & K.W. Glombitza. 1986. Phlorotannins, brown algal polyphenols. *In:* Progress in Phycological Research, Vol. 4, eds. F.E. Round & D.J. Chapman, pp. 129–241, Biopress Limited, Bristol, England.

Raimondi, P.T. 1988. Settlement cues and determination of the vertical limit of an intertidal barnacle. Ecology **69**:400–407.

Ray, S.M. & D.V. Aldrich. 1967. Ecological interactions of toxic dinoflagellates and mollusks in the Gulf of Mexico. *In:* Animal Toxins, eds. F.E. Russell & P.R. Saunders, pp. 75–83, Pergamon, Oxford.

Regnault, M. 1981. Respiration and ammonia excretion of the shrimp *Crangon crangon* L: metabolic response to prolonged starvation. J. Comp. Physiol. **141B**:549–555.

Rehnberg, B.G. & C.B. Schreck. 1987. Chemosensory detection of predators by coho salmon (*Oncorhynchus kisutch*): behavioral reaction and the physiological stress response. Can. J. Zool. **65**:481–485.

Reiswig, H.M. 1970. Porifera: sudden sperm release by tropical Demospongiae. Science **170**:538–539.

Renaud, P.E., M.E. Hay & T.M. Schmitt. 1990. Interactions of plant stress and herbivory: intraspecific variation in the susceptibility of a palatable versus an unpalatable seaweed to sea urchin grazing. Oecologia **82**:217–226.

Ricker, W.E. 1972. Hereditary and environmental factors affecting certain salmonid populations. *In:* The Stock Concept in Pacific Salmon, eds., R.C. Simon & P.A. Larkin, pp. 27–160, Columbia Press, Vancouver, BC.

Rittschof, D., R. Shepherd & L.G. Williams. 1984. Concentration and preliminary characterization of a chemical attractant of the oyster drill, *Urosalpinx cinerea.* J. Chem. Ecol. **10**:63–75.

Rittschof, D., D.W. Tsai, P.G. Massey, L. Blanco, G.L. Keuber, Jr. & R.J. Haas, Jr. 1992. Chemical mediation of behavior in hermit crabs: alarm and aggregation cues. J. Chem. Ecol. **18**:959–984.

Robbins, C.T., S. Mole, A.E. Haggerman & T.A. Hanley. 1987. Role of tannins in defending plants against ruminants: reduction in dry matter digestion? Ecology **68**:1606–1615.

Rosenthal, G.A. & M.R. Berenbaum, eds. 1992. Herbivores Their Interaction with Secondary Plant Metabolites, Vol. II: Ecological and Evolutionary Processes, 2nd ed. Academic Press, San Diego.

Russell, F.E. 1984. Marine toxins and venomous and poisonous marine plants and animals (invertebrates). Adv. Mar. Biol. **21**:59–217.

Russell, F.E. & P.R. Saunders. 1967. Animal Toxins. Pergamon Press, Oxford.

Sakata, K. 1989. Feeding attractants and stimulant for marine gastropods. *In:* Bioorganic Marine Chemistry, Vol. 3, ed. P.J. Scheuer, pp. 115–129, Springer-Verlag, New York.

Sakata, K., T. Itoh & K. Ina. 1984. A new bioassay method for phagostimulants for a young abalone, *Halitosis discus* Reeve. Agric. Biol. Chem. **48**:425–429.

Sakata, K., M. Tsuge & K. Ina. 1986. A simple bioassay for feeding stimulants for the young sea hare *Aplysia juliana.* Mar. Biol. **91**:509–511.

Sale, P.F. 1991. The Ecology of Fishes on Coral Reefs. Academic Press, San Diego, CA.

Sammarco, P.W. & J.C. Coll. 1992. Chemical adaptations in the Octocorallia: evolutionary considerations. Mar. Ecol. Prog. Ser. **88**:93–104.

Sammarco, P.W., J.C. Coll, S. La Barre & B. Willis. 1983. Competitive strategies of soft corals (Coelenterata: Octocorallia): allelopathic effects on selected scleractinian corals. Coral reefs. **1**:173–178.

Sammarco, P.W., J.C. Coll & S. La Barre. 1985. Competitive strategies of soft corals (Coelenterata: Octocorallia). II. Variable defensive responses and susceptibility to scleractinian corals. J. Exp. Mar. Biol. Ecol. **91**:199–215.

Scheltema, R.S., I.P. Williams, M.A. Shaw & C. London. 1981. Gregarious settlement by larvae of *Hydroides dianthus* (Polychaeta: Serpulidae). Mar. Ecol. Prog. Ser. **5**:69–74.

Schmitt, B.C. & B.W. Ache. 1979. Olfaction: responses of a decapod crustacean are enhanced by flicking. Science **205**:204–206.

Schmitt, T.M., M.E. Hay & N. Lindquist. 1995. Constraints on chemically mediated coevolution: multiple functions for seaweed secondary metabolites. Ecology **76**:107–123.

Scholz, A.T. 1980. Hormonal Regulation of Smolt Transformation and Olfactory Imprinting in Coho Salmon. Ph.D. Dissertation, University of Wisconsin, Madison.

Scholz, N. & J. Atema. 1991. Effect of flow velocity on chemical signal dispersal and hermit crab orientation. Chem. Senses **16**:577–578.

Scholz, A.T., R.M. Horrall, J.C. Cooper & A.D. Hasler. 1976. Imprinting to chemical cues: the basis for home stream selection in salmon. Science **192**:1247–1249.

Schupp, P.J. & V.J. Paul. 1994. Calcification and secondary metabolites in tropical seaweeds: variable effects on herbivorous fishes. Ecology **75**:1172–1185.

Shaw, B.A., P.J. Harrison & R.J. Andersen. 1994. Evaluation of the copepod *Tigriopus californicus* as a bioassay organism for the detection of chemical feeding deterrents by marine phytoplankton. Mar. Biol. **121**:89–95.

Shaw, B.A., P.J. Harrison & R.J. Andersen. 1995b. Feeding deterrence properties of apo fucoxanthinoids from marine diatoms. II. Physiology of production of apo fucoxanthinoids by the marine diatoms *Phaeodactylum tricornutum* and *Thalassiosirea pseudo-*

nana, and their feeding deterrent effects on the copepod *Tigriopus californicus.* Mar. Biol. **124**:473–481.

Shaw, B.A., R.J. Andersen & P.J. Harrison. 1995a. Feeding deterrence properties of apo-fucoxanthinoids from marine diatoms. I. Chemical structures of apo-fucoxanthinoids produced by *Phaeodactylum tricornutum.* Mar. Biol. **124**:467–472.

Sievers, A.M. 1969. Comparative toxicity of *Gonyaulax monilata* and *Gymnodinium breve* to annelids, crustaceans, molluscs, and a fish. J. Protozool. **16**:401–404.

Sih, A. 1986. Antipredator responses and the perception of danger by the mosquito larvae. Ecology **67**:434–441.

Sih, A., L.B. Kats & R.D. Moore. 1992. Effects of predatory sunfish on the density, drift, and refuge use of stream salamander larvae. Ecology **73**:1418–1430.

Slansky, F., Jr. 1992. Allelochemical-nutrient interactions in herbivore nutritional ecology. *In:* Herbivores Their Interaction with Secondary Plant Metabolites, Vol. II: Ecological and Evolutionary Processes, 2nd eds., eds. G.A. Rosenthal & M.R. Berenbaum, pp. 135–174, Academic Press, San Diego.

Slattery, M., J.B. McClintock & J.N. Heine. 1995. Chemical defenses in Antarctic soft corals: evidence for antifouling compounds. J. Exp. Mar. Biol. Ecol. **190**:61–77.

Sleeper, H.L., V.J. Paul & W. Fenical. 1980. Alarm pheromones from the marine opisthobranch *Navanax inermis.* J. Chem. Ecol. **6**:57–70.

Sloan, N.A. 1980. The arm curling and terminal tube-foot response of the asteriod *Crossaster papposus* (l.). J. Nat. Hist. **14**:469–482.

Smith, R.J.F. 1992. Alarm signals in fishes. Rev. Fish Biol. Fishes **2**:33–63.

Snyder, N. & H. Snyder. 1970. Alarm response of *Diadema antillarum.* Science **168**:276–278.

Sola, C. 1995. Chemoattraction of upstream migrating eels *Anguilla anguilla* to earthy and green odorants. Env. Biol. Fishes **43**:179–185.

Stachowicz, J.J. & M.E. Hay. 1996. Facultative mutualism between an herbivorous crab and a coralline alga: advantages of eating noxious seaweeds. Oecologia **105**:377–387.

Stachowicz, J.J. & N. Lindquist. 1997. Chemical defense among hydroids on pelagic *Sargassum:* predator deterrence and absorption of UV radiation by secondary metabolites. Mar. Ecol. Prog. Ser. **155**:115–126.

Standing, J.D., I.R. Hooper & J.D. Costlow. 1984. Inhibition and induction of barnacle settlement by natural products present in octocorals. J. Chem. Ecol. **6**:823–834.

Stanhope, M.J., M.M. Connelly & B. Hartwick. 1992. Evolution of a crustacean chemical communication channel: behavioral and ecological genetic evidence for a habitat-modified, race-specific pheromone. J. Chem. Ecol. **18**:1871–1887.

Stasko, A.B. 1971. Review of field studies on fish orientation. Ann. N.Y. Acad. Sci. **188**:12–29.

Steele, C.W., A.D. Scarfe & D.W. Owens. 1991. Effects of group size on the responsiveness of zebrafish, *Brachydanio verio* (Hamilton Buchanan), to alanine, a chemical attractant. J. Fish. Biol. **38**:553–564.

Steinberg, P.D. 1984. Algal chemical defense against herbivores: allocation of phenolic compounds in the kelp *Alaria marganita.* Science **223**:405–407.

Steinberg, P.D. 1988. Effects of quantitative and qualitative variation in phenolic compounds on feeding in three species of marine invertebrate herbivores. J. Exp. Mar. Biol. Ecol. **120**:221–237.

Steinberg, P.D. 1994. Lack of short-term induction of phlorotannins in the Australian brown algae *Ecklonia radiata* and *Sargassum vestitum.* Mar. Ecol. Prog. Ser. **112**:129–133.

Steinberg, P.D. 1995. Interaction between the canopy dwelling echinoid *Holopneustes purpurescens* and its host kelp *Ecklonia radiata.* Mar. Ecol. Prog. Ser. **127**:169–181.

Steinberg, P.D. & I. van Altena. 1992. Tolerance of marine invertebrate herbivores to brown algal phlorotannins in temperate Australasia. Ecol. Monogr. **62**:189–222.

Steneck, R.S. & W.H. Adey. 1976. The role of environment in control of morphology in *Lithophyllum congestum,* a Caribbean algal ridge builder. Bot. Mar. **19**:197–215.

Sullivan, B., D.J. Faulkner & L. Webb. 1983. Siphonodictidine, a metabolite of the burrowing sponge *Siphonodictyon* sp. that inhibits coral growth. Science **221**:1175–1176.

Sutterlin, A.M. & R. Gray. 1973. Chemical basis for homing in Atlantic Salmon (*Salmo salar*). J. Fish Res. Board Can. **30**:985–989.

Sweatman, H.P.A. 1988. Field evidence that settling coral reef fish larvae detect resident fishes using dissolved chemical cues. J. Exp. Mar. Biol. Ecol. **124**:163–174.

Sykes, P.F. & M.E. Huntley. 1987. Acute physiological reactions of *Calanus pacificus* to selected dinoflagellates: direct observations. Mar. Biol. **94**:19–24.

Tachibana, K. 1988. Chemical defense in fishes. BioOrg. Mar. Chem. **2**:117–138.

Tamburri, M.N. & R.K. Zimmer-Faust. 1996. Suspension feeding: basic mechanisms controlling recognition and ingestion of larvae. Limnol. Oceanogr. **41**:1188–1197.

Tamiya, N. 1975. Sea snake venoms and toxins. *In:* The Biology of Sea Snakes, ed. W.A. Dunson, pp. 386–415, University Park Press, Baltimore.

Targett, N.M. & J.E. Ward. 1991. Bioactive microalgal metabolites: mediation of subtle ecological interactions in phytophagous suspension-feeding marine invertebrates. *In:* Bioorganic Marine Chemistry, Vol. 4, ed. P.J. Scheuer, pp. 91–118, Springer-Verlag, Berlin.

Targett, N.M., T.E. Targett, N.H. Vrolijk & J.C. Ogden. 1986. Effect of macrophyte secondary metabolites on feeding preferences of the herbivorous parrotfish *Sparisoma radians.* Mar. Biol. **92**:141–148.

Targett, N.M., A.A. Boettcher, T.E. Targett & N.H. Vrolijk. 1995. Tropical marine herbivore assimilation of phenolic-rich plants. Oecologia **103**:170–179.

Targett, T.E. & N.M. Targett. 1990. Energetics of food selection by the herbivorous parrotfish *Sparisoma radians:* roles of assimilation efficiency, gut evacuation rate, and algal secondary metabolites. Mar. Ecol. Prog. Ser. **66**:13–21.

Thompson, J.E. 1985. Exudation of biologically active metabolites in the sponge *Aplysina fistularis* I. Biological evidence. Mar. Biol. **88**:23–26.

Thompson, J.E., R.P. Walker & D.J. Faulkner. 1985. Screening and bioassays for biologically-active substances from forty marine sponge species from San Diego, California, USA. Mar. Biol. **88**:11–21.

Thuesen, E.V., K. Kogure, K. Hashimoto & T. Nemoto. 1988. Poison arrowworms: a tetrodotoxin venom in the marine phylum Chaetognatha. J. Exp. Mar. Biol. Ecol. **116**:249–256.

Toonen, R.J. 1993. Environmental and Heritable Components of Settlement Behavior of *Hydroides dianthus* (Serpulidae: Polychaeta). Master's Thesis. University of North Carolina at Wilmington.

Toonen, R.J. & J.R. Pawlik. 1994. Foundations of gregariousness. Nature **370**:511–512.

Turner, E.J., R.K. Zimmer-Faust, M.A. Palmer, M. Luckenbach & N.D. Pentcheff. 1994. Settlement of oyster (*Crassostrea virginica*) larvae: effects of water flow and a water-soluble chemical cue. Limnol. Oceanogr. **39**:1579–1593.

Uye, S. & K. Takamatsu. 1990. Feeding interactions between planktonic copepods and red-tide dinoflagellates from Japanese coastal waters. Mar. Ecol. Prog. Ser. **59**:97–107.

Vadas, R.L., M.T. Burrows & R.N. Hughes. 1994. Foraging strategies of dogwhelks, *Nucella lapillus* (L.): interacting effects of age, diet and chemical cues to the threat of predation. Oecologia **100**:439–450.

Van Alstyne, K.L. 1986. Effects of phytoplankton taste and smell on feeding behavior of the copepod *Centropages hamatus.* Mar. Ecol. Prog. Ser. **34**:187–190.

Van Alstyne, K.L. 1995. Comparison of three methods for quantifying brown algal polyphenolic compounds. J. Chem. Ecol. **21**:45–58.

Van Alstyne, K.L. & V.J. Paul. 1992. Chemical and structural defenses in the sea fan *Gorgonia ventalina:* effects against generalist and specialist predators. Coral Reefs **11**:155–159.

Van Alstyne, K.L., C.R. Wylie, V.J. Paul & K. Meyer. 1992. Antipredator defenses in tropical Pacific soft corals (Coelenterata: Alycyonacea). I. Sclerites as defenses against generalist carnivorous fishes. Biol. Bull. **182**:231–240.

Van Alstyne, K.L., C.R. Wylie & V.J. Paul. 1994. Antipredator defenses in tropical Pacific soft corals (Colenterata: Alcyonacea) II. The relative importance of chemical and structural defenses in three species of *Sinularia.* J. Exp. Mar. Biol. Ecol. **178**:17–34.

Van-Praët, M. 1985. Nutrition of sea anemones. Adv. Mar. Biol. **22**:65–99.

Vannini, M. & S. Cannicci. 1995. Homing behavior and possible cognitive maps in crustacean decapods. J. Exp. Mar. Biol. Ecol. **193**:67–91.

Wahl, M. 1989. Marine epibiosis. I. Fouling and antifouling: some basic aspects. Mar. Ecol. Prog. Ser. **58**:175–189.

Walls, M. & M. Ketola. 1989. Effects of predator-induced spines on individual fitness in *Daphnia pulex.* Limnol. Oceanogr. **34**:390–396.

Wahl, M., P.R. Jensen & W. Fenical. 1994. Chemical control of bacterial epibiosis on ascidians. Mar. Ecol. Prog. Ser. **110**:45–57.

Walton, I. 1653. The Compleat Angler or the Contemplative Man's Recreation. Marriot, London.

Ward, J.E. & N.M. Targett. 1989. Influence of marine microbial metabolites on the feeding behavior of the blue mussel *Mytilus edulis.* Mar. Biol. **101**:313–321.

Waterman, P.G. & S. Mole. 1994. Analysis of Phenolic Plant Metabolites. Blackwell Scientific, Oxford.

Weissburg, M.J. & R.K. Zimmer-Faust. 1993. Life and death in moving fluids: hydrodynamic effects on chemosensory-mediated predation. Ecology **74**:1428–1443.

Weissburg, M.J. & R.K. Zimmer-Faust. 1994. Odor plumes and how blue crabs use them in finding prey. J. Exp. Biol. **197**:349–375.

Wells, M.J. 1978. Octopus: Physiology and Behaviour of an Advanced Invertebrate. Chapman and Hall, London.

Wethey, D.S. 1986. Ranking of settlement cues by barnacle larvae: influence of surface contour. Bull. Mar. Sci. **39**:393–400.

White, A.W. 1981. Sensitivity of marine fishes to toxins from the red-tide dinoflagellate *Gonyaulax excavata* and implications for fish kills. Mar. Biol. **65**:255–260.

Wilson, D.P. 1953. The settlement of *Ophelia bicornis* Savigny larvae. J. Mar. Biol. Assoc. UK **32**:209–233.

Winter, F.C. & J.A. Estes. 1992. Experimental evidence for the effects of polyphenolic compounds from *Dictyoneurum californicum* Ruprecht (Phaeophyta: Laminariales) on feeding rate and growth in the red abaolone *Haliotus rufescens* Swainson. J. Exp. Mar. Biol. Ecol. **155**:263–277.

Wisby, W.J. & A.D. Hasler. 1954. The effect of olfactory occlusion on migrating silver salmon (*O. kisutch*). J. Fish. Res. Board Can. **11**:472–478.

Wylie, C.R. & V.J. Paul. 1989. Chemical defenses in three species of *Sinularia* (Coelenterata, Alcyonaceae): effects against generalist predators and the butterfly fish *Chaetodon unimaculatus* Bloch. J. Exp. Mar. Biol. Ecol. **129**:141–160.

Yang, C.S., J.F. Brady & J. Hong. 1992. Dietary effects on cytochromes P450, xenobiotic metabolism, and toxicity. FASEB J. **6**:737–744.

Yates, J.L. & P. Peckol. 1993. Effects of nutrient availability and herbivory on polyphenolics in the seaweed *Fucus vesiculosus.* Ecology **74**:1757–1766.

Young, C.M., P.A. Tyler, R.H. Emson & J.D. Gage. 1993. Perception and selection of macrophyte detrital falls by the bathyal echinoid *Stylocidaris lineata.* Deep-sea Res. I **40**:1475–1486.

Zimmer-Faust, R.K. 1987. Towards a theory on optimal chemoreception. Biol. Bull **172**:10–29.

Zimmer-Faust, R.K. 1989. The relationship between chemoreception and foraging behavior in crustaceans. Limnol. Oceanogr. **34**:1367–1374.

Zimmer-Faust, R.K. 1991. Chemical signal-to-noise detection by spiny lobsters. Biol. Bull. **181**:419–426.

Zimmer-Faust, R.K. 1993. ATP: A potent prey attractant evoking carnivory. Limnol. Oceanogr. **38**:1271–1275.

Zimmer-Faust, R.K. & J.F. Case. 1983. A proposed dual role of odor in foraging by the California spiny lobster, *Panulirus interruptus* (Randall). Biol. Bull. **164**:341–353.

Zimmer-Faust, R.K. & M.N. Tamburri. 1994. Chemical identity and ecological implications of a waterborne larval settlement cue. Limnol. Oceanogr. **39**:1075–1087.

Zimmer-Faust, R.K., J.E. Tyre, W.C. Michel & J.F. Case. 1984. Chemical mediation of adaptive feeding in a marine decapod crustacean: the importance of suppression and synergism. Biol. Bull. **167**:339–353.

Zimmer-Faust, R.K., J.M. Stanfill & S.B. Collard III. 1988. A fast, multi-channel fluorometer for investigating aquatic chemoreception and odor trails. Limnol. Oceanogr. **33**:1586–1595.

Zimmer-Faust, R.K., C.M. Finelli, N.D. Pentcheff & D.S. Wethey. 1995. Odor plumes and animal navigation in turbulent water flow: a field study. Biol. Bull. **188**:111–116.

Zimmer-Faust, R.K., J.E. Commins, D.W. Schar, K.A. Browne, D.S. Wethey, N.D. Pentcheff & C.M. Finelli. 1996a. Mechanisms regulating predation: the role of free amino acids as prey attractants. Eos **76**(3):OS70.

Zimmer-Faust, R.K., P.B. O'Neill & D.W. Schar. 1996b. The relationship between predator activity state and sensitivity to prey odor. Biol. Bull. **190**:82–87.

3

Bioassay Methods for Fungi and Oomycetes

James L. Kerwin and Melinda J. Semon

3.1. Introduction 142
3.2. Intraspecific interactions—reproduction 143
  3.2.1. Chytridiomycetes 144
  3.2.2. Oomycetes 145
  3.2.3. Yeast 147
  3.2.4. Basidiomycetes and ascomycetes 148
3.3. Intraspecific population interactions 148
3.4. Interspecific interactions 151
  3.4.1. Exploitation of resources—saprophytes 151
  3.4.2. Exploitation of resources—parasites and pathogens 151
    3.4.2.1. Plant parasites and pathogens 151
    3.4.2.2. Animal parasites and pathogens 157
    3.4.2.3. Fungal and lichen parasites and pathogens 159
  3.4.3. Mutualistic and symbiotic interactions 160
    3.4.3.1. Vesicular-arbuscular mycorrhizal fungi 160
    3.4.3.2. Ectomycorrhizal fungi 162
    3.4.3.3. Fungal-insect associations 162
  3.4.4. Competition—allelopathy 163
  3.4.5. Defense against herbivores and pathogens 163
3.5. Conclusions 165
3.6. Acknowledgments 165
3.7. References 165

3.1. Introduction

Fungi are an integral part of the ecosystem, with roles ranging from primary decomposers to virulent pathogens. Recycling of plant and other organic matter is dependent in part on saprophytic fungi. Mycorrhizal interactions with plants regulate survival, growth, and reproduction of economically important crops.

Mutualistic symbioses of lichens may involve as many as 20% of all known fungal species. Control of pathogenic and parasitic fungi requires millions of dollars annually. These fungal interactions, ranging from lignicolous saprophytes to pulmonary pathogens and their animal hosts, all involve signal transduction. Whether a fungal propagule remains dormant, germinates, dies, or proceeds along another developmental path will be determined largely by chemical signals it receives from its environment. This chapter describes a number of these interactions, underlying mechanisms, and approaches taken to describe and quantify responses to these signals.

Oomycetes, commonly referred to as water molds, are a group of organisms characterized by laterally biflagellate zoospores and production of a thick-walled, dormant sexual spore known as an oospore (Sparrow 1960). Although classically treated as fungi, this group is now recognized as being derived from heterokont chrysophyte algae on the basis of a variety of ultrastructural, biochemical, and molecular biological features (Beakes 1987; Cavalier-Smith 1986, 1987, 1989; Barr 1992). Oomycetes include some intensively studied species such as *Achlya* spp., fish pathogenic species of *Saprolegnia,* plant pathogenic species of *Pythium* and *Phytophthora,* and the mosquito parasite *Lagenidium giganteum.* This group, therefore, which has a fungal life style even if it is not closely related to true fungi, is also discussed in this chapter.

Compendia and reviews cover aspects of some of the topics discussed in this chapter in greater detail. These include secondary metabolism and fungal differentiation (Bennett & Ciegler 1983); fungal ecophysiology, including interaction with other heterotrophs (Cooke & Whipps 1993); coevolution of parasitic and symbiotic fungi with their hosts (Batra 1979; Pirozynski & Hawksworth 1988); fungal differentiation and sporulation (Turian & Hohl 1981; Smith 1983); fungal-plant interactions (Arora et al. 1991; Isaac 1992); plant-microbe symbiosis (Werner 1992); microbial effects on plant herbivory (Barbosa et al. 1991); mycotoxins (Lee et al. 1992); microbiology of soil epiphytes (Fokkema & van den Heuvel 1986); host specificity of plant pathogens (Kohmoto et al. 1995); lichens (Longton 1988; Nash & Wirth 1988); mycorrhizae (Allen 1991); and medical and veterinary mycology (Kwon-Chung & Bennett 1992; Crissey et al. 1995).

3.2. Intraspecific Interactions—Reproduction

Fungi are used extensively as model developmental systems in biochemistry and genetics. An early classical study of steroid-mediated morphogenesis involved the Oomycete *Achlya ambisexualis* (Raper 1939, 1940). The ascomycetous yeast *Saccharomyces cerevisiae* is the most extensively studied eukaryote in many areas of signal transduction (Kurjan 1993; Ferguson et al. 1994; Grishin et al. 1994; Schultz et al. 1995), especially as it relates to cell division and sexual reproduction.

3.2.1. Chytridiomycetes

An early example of pheromone-regulated reproduction by a fungus involved the chytrid *Allomyces macrogynus* (Machlis 1958, 1969; Fig. 3.1). Gametes and the female reproductive structures (gametangia) of this species produce and excrete a terpenoid compound, sirenin, that attracts male gametes (Pommerville 1987a, 1987b). This will serve as a useful example of handling zoosporic fungi for a variety of bioassays. First, sporulation with these fungi, as with many other species, is triggered by starvation. Occasionally, as with *A. macrogynus,*

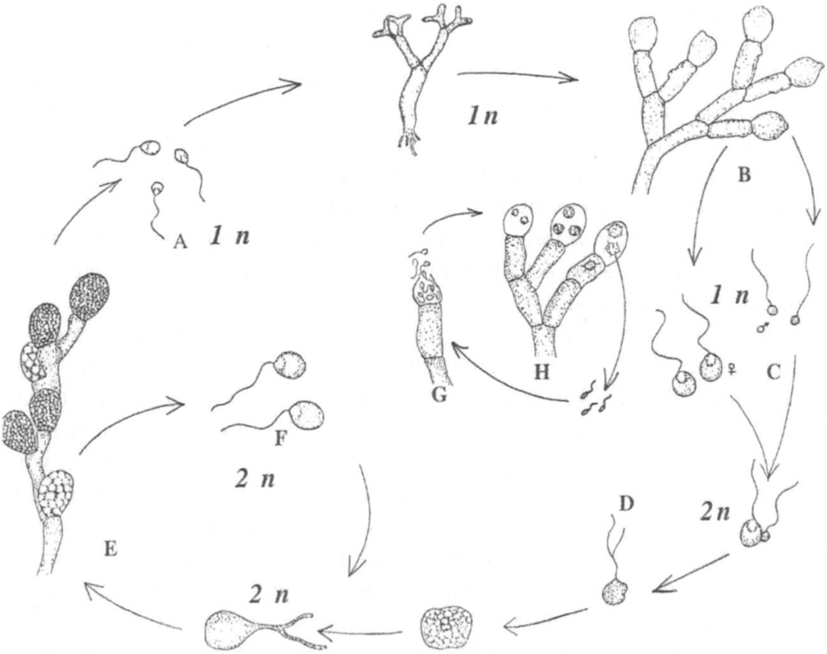

Figure 3.1. Life cycles of *Allomyces macrogynus* and a mycoparasite, *Rozella allomycis.* Haploid zoospores of *A. macrogynus* (*A*) germinate to form a multicellular thallus (*B*), at the tips of the thallus terminal orange male gametangia and subterminal female gametangia are delimited. Posteriorly uniflagellate male and female gametes (*C*) are released, with the female gametes and gametangia producing a pheromone known as sirenin to which male gametes are attracted. Gametes fuse and germinate (*D*), ultimately forming a diploid sporothallus (*E*) that forms thin-walled mitosporangia (MS) and thick-walled resistant sporangia (RS). Diploid zoospores that differentiate in MS (*F*) germinate to produce thalli, and haploid zoospores form following meiosis in the RS, completing the life cycle. Zoospores of the obligate parasite *Rozella allomycis* (*G*) are attracted to gametothalli and sporothalli of *A. macrogynus* by an unidentified chemotactic substance. They attach to, encyst on, and then invade their host, developing into zoospores or resistant sporangia (*H*). Adapted from Sparrow (1960) and Alexopoulos (1979).

sporulation (but not gamete/zoospore release) will occur on agar media, although often not for 2–6 weeks after inoculation. The dark-colored resistant sporangia (RS) can be collected aseptically using a drawn out capillary pipette. The RS are dried on filter paper for 48 h at 30°C, and stored at 5–15°C indefinitely. Approximately 2 weeks before the material is needed, RS are flooded in dilute salts (DS) medium, and the haploid spores (meiospores) that release from the RS are used to inoculate agar media. Vegetative gametophytes should be produced in 12–14 days. Water agar plates are prepared, and a #1 cork borer is used to cut out an agar plug in the center and at one or more places a few millimeters from the center. Gametangia are dislodged from mature cultures, and several hundred male (smaller, orange) and female (larger, colorless) structures are collected. Ten to twenty female gametangia are placed in the central well and 100–200 male thalli are placed in the peripheral well, with additional DS added to fill each well. Within 10–20 min after spore discharge, sirenin will have diffused from the central well, and the male gametes will congregate toward the central well, an effect that can persist for hours.

A similar approach can be taken to assay for the production of sexual pheromones by other chytrids exhibiting this type of sexual reproduction, for instance species of *Coelomomyces*, which are obligate parasites that alternate between mosquito larvae and microcrustacean hosts, usually copepods or ostracods (Couch & Bland 1985). A complication that is encountered for many species is that the gametangia of male and female (or + and –) thalli are morphologically indistinguishable. For instance the mating type of gametes and gametangia of *Coelomomyces dodgei* and *C. punctatus* can be distinguished, with those designated as males being bright orange in color, and those designated females pale amber (Federici 1977; Federici & Thompson 1979); however, this is not the case for most species of this or other genera. For parasites such as *Coelomomyces* a common strategy is to use a very small amount of inoculum to infect individual hosts. This will optimize the chances that only one mating type will be present in a given host. Production of a mating pheromone by *Coelomomyces* can then be monitored by adding copepods containing the same mating type to the central well and the complementary mating type to the peripheral wells using the agar plate assay described above. For this species and other chytrids it is relatively easy to microscopically observe gamete fusion, and therefore monitor whether only one mating type is present in a given well (Federici 1977).

3.2.2. *Oomycetes*

There is abundant evidence for the presence of pheromones regulating sexual reproduction by Oomycetes, the classical example being the steroid pheromones produced by *Achlya* spp. (reviewed by Barksdale 1969; Horgen 1977, 1981; McMorris 1989). Sexual reproduction (oosporogenesis) is initiated by secretion of a steroid pheromone, antheridiol, and a variety of less biologically active derivatives of this lipid (Barksdale et al. 1974; McMorris et al., 1993) by the female mating type. Antheridiol is bound by specific receptor proteins (Riehl et al. 1984; Brunt & Silver 1986; Brunt et al. 1990) in male strains of the fungus, inducing formation of antheridial hyphae (Gow & Gooday 1987). The antheridia subsequently secrete a related steroid, oogoniol, that induces the induction of oogonial (female sex organs) cells. These or similar compounds are usually produced by heterothallic species, but homothallic species can also synthesize them. Stock cultures of male and female strains are commonly grown on agar, or liquid cultures are initiated using dilute suspensions of encysted spores. Agar plugs from near the colony edge are removed from agar cultures, or spherical mycelial mats from liquid cultures, and suspended in dilute pond water or a defined mating medium (Barksdale & Lasure 1974) on opposite sides of a large petri dish. Antheridial branches usually form within 8–12 h, and oogonial initials form within 2–4 days.

Oomycete sexual reproduction requires communication between the progenitors of male and female sex cells. Even in homothallic species, such as the mosquito larval parasite *Lagenidium giganteum*, communication must take place between two morphologically identical hyphal cells such that one is induced to form an antheridium and the second initiates the oogonium development. This initial communication must continue until fusion of the gametangia is completed (Domnas & Warner 1991). The importance of continuous communication can be demonstrated by highly enriching culture media of *L. giganteum* with saturated fatty acids. When the fungus incorporates high levels of palmitic (16:0) or stearic (18:0) acid, many of the antheridia attempt to fuse with the same cell from which they are emanating (Kerwin & Washino, 1986), resulting in abortion of the sexual process.

A problem that will be encountered with many fungi and Oomycetes is breaking the dormancy of resting spores, either to provide experimental material or to maintain fungal cultures in the laboratory. Methods used to break dormancy are as diverse as the organisms, and include temperature shock; repeated desiccation and rehydration; altered light regimens; long-term storage at reduced temperatures in water or soil; reduced oxygen tension; exposure to a wide variety of defined or undefined organic chemicals, natural product extracts, or exudates; and combinations of these treatments. Reviews of protocols successfully used to break dormancy are available (Sussman & Halvorson 1966; Hutchinson 1973; Cochrane 1974, Sussman 1976; Tommerup 1983; van Etten et al. 1983; French 1985, French 1992; French et al. 1994).

In addition to *Achlya,* there is some evidence supporting the existence of the production of sexual pheromones in other Oomycetes, especially among the plant pathogenic members of the Pythiaceae (Sherwood 1966; Hendrix 1970; Pratt et al. 1972; Brasier 1975; Ko 1978; Zentmeyer 1979; Shen et al. 1983). Several other methods have been developed for indirectly following the production of these compounds. One involves the use of polycarbonate membranes, which are too fine to permit penetration by hyphae (Ko 1978, 1980; Shen et al. 1983). Agar blocks containing opposite mating types are placed in a petri dish on either side of the membrane (0.2 μM pore size). Dehydration of the agar blocks is minimized by adding moistened filter paper on a thin layer of water agar on the top plate of the petri dish. Sealing the plates with parafilm is advisable if longer incubation periods are used. After 2–24 days oogonial primordia or oospores will be observed if A^1 and A^2 mating types have been paired.

A second approach used cheese cloth impregnated with various biological extracts (Zaki et al. 1983). Extracts (or commercially available standards) are dissolved in an appropriate volatile solvent and applied to sterilized cheese cloth. The solvent is allowed to evaporate, and agar blocks with the experimental organism placed between two sheets of the cloth. The blocks are incubated in sterile distilled water in a petri plate for 2–24 days with periodic microscopic examination for sexual or asexual propagules.

3.2.3. Yeast

Ascomycetous and basidiomycetous yeasts are among the most intensively studied organisms at the molecular level (Sprague & Thorner 1983; Sprague et al. 1983; Dyer et al. 1992; Raju 1992; Springer 1993; Bardwell et al. 1994; Cid et al. 1995). Elucidation of the structure and function of the **a** and α mating factors in *Saccharomyces cerevisiae* and subsequent biochemical events have been instrumental in dissecting eukaryotic meiotic and mitotic processes. The usual way to monitor the production of pheromones by yeasts that produce peptide sex hormones is to microscopically observe initial events in conjugation, e.g., shmoo formation by *S. cerevisiae* (Sprague et al. 1983) or crozier formation by *Neurospora crassa* (Springer 1993). Another option for rapid assay of pheromone production by heterothallic yeasts (i.e. those with more than one mating type) is to mix two cultures and observe whether agglutination occurs (Wickerham 1958). Rapid agglutination indicates the presence of opposite mating types and complementary binding sites on cell walls (and probably pheromone production) of the two strains. This phenomenon does not occur for all species, so except for previously characterized strains, it is of questionable reliability. Some species such as *Ustilago maydis* respond to mating pheromones by switching from yeastlike to filamentous growth, and this allows rapid macroscopic screening for production of these compounds (Spellig et al. 1994).

Egel et al. (1994) have developed a rapid, semiquantitative assay for yeast pheromone production. Strains of *Schizosaccharomyces pombe* that are heterozygous for mating type are grown in liquid media, and suspensions of haploid source and responder strains are mixed 1:100. After placing small aliquots on nutrient agar, the plates are incubated for 2–3 days, and then exposed to iodine vapor. This reveals sporulating areas due to a positive reaction with starch in the spores, with the diameter of the halo roughly proportional to the amount of pheromone produced by the source strain. This has obvious advantages over assays based on other morphological changes associated with pheromone production such as cell elongation (Fukui et al. 1986), especially when screening large numbers of uncharacterized isolates for pheromone production.

3.2.4. Basidiomycetes and Ascomycetes

Mating type, especially among Basidiomycetes, can become very complicated at the genetic level (Koltin et al. 1967; Casselton et al. 1995), where, for example, there are probably more than 20,000 mating types in the Hymenomycete *Schizophyllum commune* (Raper 1966). Despite this underlying complexity, assays for compatible or incompatible interactions can rely solely on microscopic examination. Despite a paucity of evidence, it is evident that communication between hyphae must occur to initiate fusion by uninucleate, haploid homokaryotic mycelia (monokaryons) with another compatible monokaryon to produce binucleate dikaryotic mycelia. These mycelia often have characteristic hooked appendages at each septum, commonly referred to as clamp connections (Fig. 3.2A).

The Ascomycetes *Neurospora crassa* and *Aspergillus nidulans* are model systems for study of developmental regulation on a genetic and biochemical level (Marshall & Timberlake 1991; Springer 1993). An example of a typical life cycle for Ascomycetes is presented in Figure 3.2B for *N. crassa,* showing developmental options induced by environmental signals including desiccation and carbon or nitrogen deprivation.

A simple but useful assay developed for sclerotial formation in basidiomycetes can be applied to fungi that require spatial separation and/or starvation to induce reproductive propagules. The split-plate method (Amir et al. 1992) uses square petri plates divided in half by a 4-mm polycarbonate barrier. A nutrient medium is placed in half of the plate, and a minimal medium in the other. Hyphae are seeded in the nutrient agar and allowed to grow to the far edge of the minimal medium side, allowing partition of a culture into a nutrient source and a nutrient sink side, which can induce vigorous translocation in some species. This is often accompanied by differentiation, e.g., sclerotium (thick-walled resting spores) formation by *Morchella esculenta.* Addition of extracts or standard compounds such as polyphenols, alkaloids, or sugars to either side of the agar plate can be followed by monitoring their effects on fungal growth and differentiation. For fungi that will grow primarily on the agar surface, the two sides of the plate can be covered with autoclaved cellophane, facilitating semiquantitative removal of mycelia to quantitate changes in growth, differentiation, and dry weight.

3.3. Intraspecific Population Interactions

Population biology studies, especially those done under field conditions or using simulated field conditions in the laboratory, are difficult due to the large number

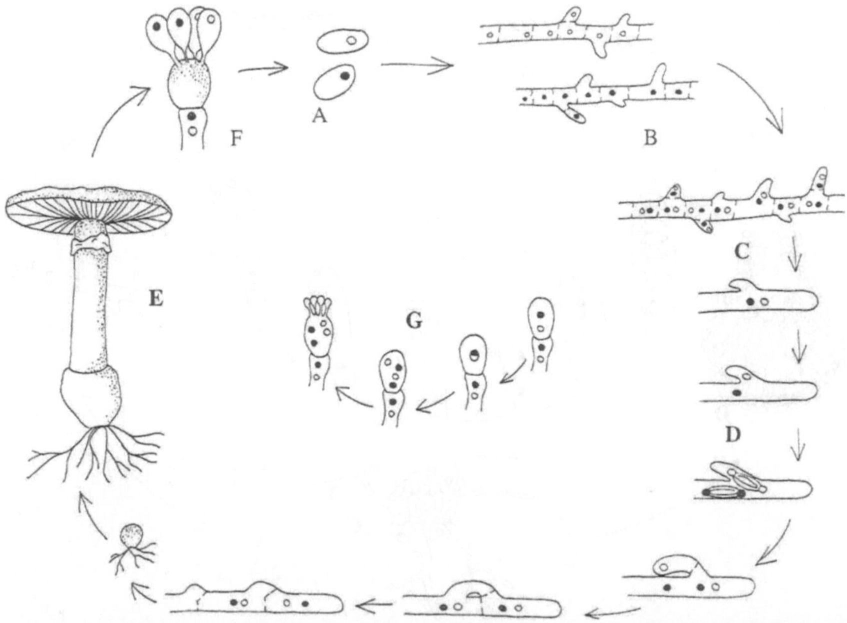

Figure 3.2. (**A**) Life cycle of a typical Basidiomycete. Basidiospores (*A*) germinate to form monokaryotic hyphae (*B*). Most Basidiomycetes are heterothallic, and fusion of compatible hyphae (*C*) usually involves reciprocal exchange of nuclei. Clamp connections (*D*) form on some but not all dikaryotic hyphae during nuclear division at the apex. Formation of these structures can be used as a bioassay for mating type and sexual compatibility. Dikaryotic hyphae are the dominant growth form for this class of fungi. Following nutrient assimilation and maturation, chemical and/or physical signals trigger the dikaryon to sporulate. Induction of basidiocarps (*E*) lead to the formation of basidia and basidiospores (*F* and more detailed insert *G*). Adapted from Ross (1979a) and Gow and Gadd (1995).

of variables that must be considered (reviewed in Kranz & Rotem 1988; Levin et al. 1992). The discussion that follows is pertinent to all intra- and interspecific interactions (especially involving field studies), including mutualistic and parasitic interactions, that are discussed in following sections. One approach is to simplify the experimental system so that one or a few parameters are varied; an alternative is to observe a single response such as growth or plant colonization as a function of a minimum number of variables assumed to be of primary importance. Such assumptions can be misleading, but may be the only approach feasible when dealing with complex biological systems. In any study involving chemical interactions within microbial communities, whether these are intraspecific or interspecific, involving epiphytic, mutualistic, antagonistic, or parasitic species, the first series of questions to be answered is the identity, number, and relative proportions

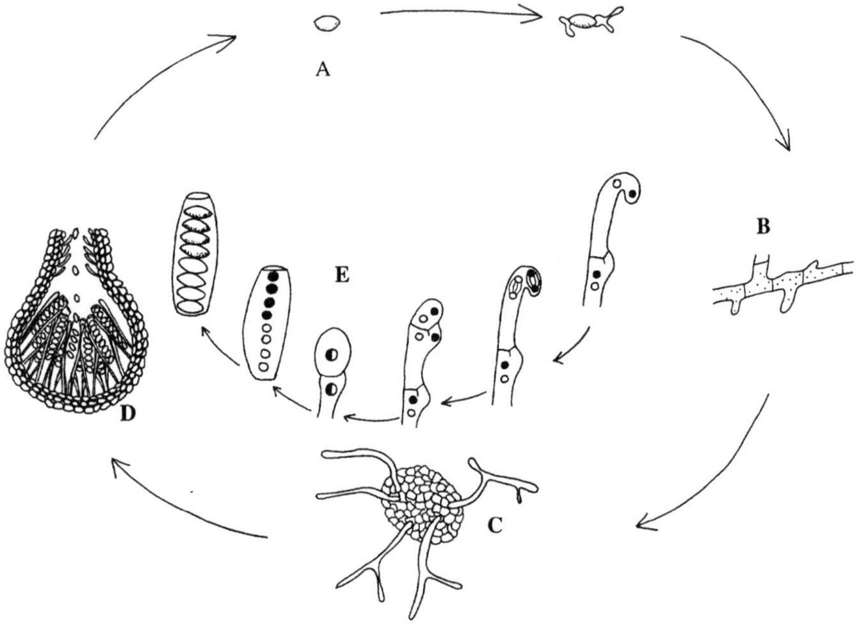

Figure 3.2. (**B**) Life cycle of a typical perithecial Ascomycete. Ascospores (*A*), usually haploid and uninucleate, germinate to form multinucleate, septate vegetative hyphae (*B*). Asexual conidiogenesis is common but not depicted here. After passing through an intermediate stage involving densely interwoven mycelia (*C*), one of five major types of ascocarps or fruiting bodies are produced. Only the perithecium (*D*) type is shown here. The insert (*E*) shows details of crozier formation that occurs within the perithecium, one of many variations of Ascomycete sexual reproduction. This culminates in the production of eight ascospores within a single ascus. Chemical signals control in part spore germination and sexual (in)compatibility. There is strong evidence for hormonal control of plasmogamy initiation in some Ascomycetes. Adapted from Ross (1979b) and Smith (1983).

of different species or biologically diverse populations of the same species (Fokkema & van den Heuval 1986). Reviews providing detailed discussions of sampling, measurement, and data manipulation for quantification of microbial community structure are available (Pielou 1975; Kranz & Rotem 1988; Magurran 1988; Barbosa et al. 1991).

A variety of techniques have been used to estimate microbial biomass in natural habitats. Visual measurements of cell volume are laborious, and subject to a variety of errors when converted to dry-weight equivalents (Jones & Mollison 1948; Carroll et al. 1980). Hydrolysis of fluorescein diacetate has been shown to be correlated to microbial biomass, especially for epiphytic fungi (Swisher & Carroll 1980). Other methods of sampling include dilution plating of tissue homogenates or surface wash fluids, incubation of leaves or other biological samples in moist chambers to promote fungal growth, leaf printing and impression films, or spore fall and spore trapping (Kranz & Rotem 1988; Magurran 1988; Barbosa et al. 1991). The spore-fall method as modified by Dickinson (1971) involves cutting disks of tissue and sticking them on a petri dish lid using petroleum jelly or cellulose paste. The lid is inverted over the base of a petri dish containing nutrient agar and incubated for 1–7 days. Sporulation is optimized by exposure to near 100% humidity.

It is often necessary to track species or subspecies over space and time (Andrews 1986, Dahlberg & Stenlid 1990). Tracking can be accomplished for small-scale studies by direct counting for morphologically distinctive organisms, e.g., those with unique morphologies or color variants. Other options include radio-labeling, use of drug-resistance markers, fluorescent antibodies, growth (or failure to grow) on selective nutrient media, or mating type evaluations (Dahlberg & Stenlid 1990).

Other intraspecific interactions such as production of self-inhibitors or promoters of spore germination are easier to monitor. Some germination assays can be done in water or dilute salt solutions, and quantitative assays can be completed using a hemocytometer; however, conidia or other spores of many fungal species must adhere to a solid substrate before they proceed through their germination cycle (Macko 1981; Nicholson & Epstein 1991; Leite & Nicholson 1992). For many species adherence to glass slides, polystyrene or cellophane is sufficient to initiate the germination process (Bartnicki-Garcia & Sing 1987), the only requirement being that the surface is hydrophilic, although even this does not appear to be an absolute requisite for some species.

3.4 Interspecific Interactions

3.4.1. Exploitation of Resources—Saprophytes

The distinction between saprophytic, commensal, mutualistic, and parasitic interactions is not always clear. For instance, the distinction between sooty molds and black mildews is based in part on taxonomy, the latter group consisting primarily of members of the Meliolaceae and the former of a variety of pleomorphic dematiaceous hyphomycetes. The other major distinction is the saprophytic nature of the sooty molds and the parasitic life style of the black mildews; however, there is very limited evidence of deleterious effects of black mildews on their host plants (Parbery & Brown 1986), and the distinction between this group and, e.g., endophytes, or between endophytes that transition into parasitic forms, is blurred. Fortunately, many bioassays developed for one type of interaction can be adapted for use in other systems.

3.4.2. Exploitation of Resources—Parasites and Pathogens

3.4.2.1. Plant Parasites and Pathogens

Because of their economic importance, plant pathogenic fungi have been studied extensively (Tuite 1969), often with emphasis on factors affecting spore adhesion, germination, and formation of appressoria (flattened hyphal organs from which infection pegs are initiated). It is at this point in the infection process that the fungus is most susceptible to environmental factors, especially temperature and humidity, that can inhibit spore germination or kill the propagules. The approach to be taken to investigate chemical interactions between fungi and plants is determined by the preferred route of pathogen/parasite entry, the presence or absence of toxins, and whether the parasitism is opportunistic, facultative or obligate. Oomycetes such as the pathogenic members of *Pythium* and *Phytophthora* primarily infect plants via roots, while most Deuteromycete (asexual stage of a fungus, also referred to as Fungi Imperfecti) plant pathogens such as *Colletotrichum* and *Helminthosporium* as well as rusts and smuts initiate infection through the leaf (Nicolson & Epstein 1991; Deising et al. 1992, Nicolson 1996).

Detection and identification of plant pathogens is necessary when unknown species are isolated from field material, and to confirm species identity (completion of Koch's postulates) during or on completion of a given bioassay using plant material. Until recently this has been accomplished using a combination of microscopy and, in the case of nonobligate pathogens, plating out plant homogenates on nutrient-based agars. A number of molecular techniques are now available to identify many species, including those based on monoclonal antibodies, random amplified polymorphic DNA, enzyme-linked immuunoassays, restriction fragment length polymorphism (RFLP) techniques, and polymerase chain reaction (PCR). Very often aggressive and nonaggressive strains of the same species cannot be differentiated morphologically, and these newer techniques can often separate pathotypes, even during latent phases of infection. The reader is referred to a recent review of these techniques (Schots et al. 1994).

Many assays involving plant pathogenic fungi monitor tropic and tactic responses. Although presented anthropomorphically as "mistakes" made by these fungi during pre- and post-infection, a review by Wynn (1981) of common causes of failure to complete host plant infection is a useful compendium of variables that can confound design and interpretation of these kinds of experiments.

In many cases of qualitative assays for the presence or absence of unidentified compounds that inhibit spore germination, the method used for application of spores to plant surfaces may not be important. Commonly, aqueous suspensions of spores, often supplemented with 0.1–2% of a wetting agent such as members of the Tween® series, are sprayed using a fine mist, or applied in discrete drops on specific areas (Robinson & Hodges 1976; Mercure et al. 1994). A disadvantage of this technique is that surfactants can alter sporulation and enzymatic activities (Reese & Maguire 1969), confounding data interpretation. Settling towers are widely used for foliage inoculation (Bell et al 1952; Petersen 1959), and Schein (1964) has developed a quantitative inoculator that permits precise, highly reproducible delivery of spores for many types of plant pathogens. These are fairly complex devices, and the reader is referred to the primary references for detailed

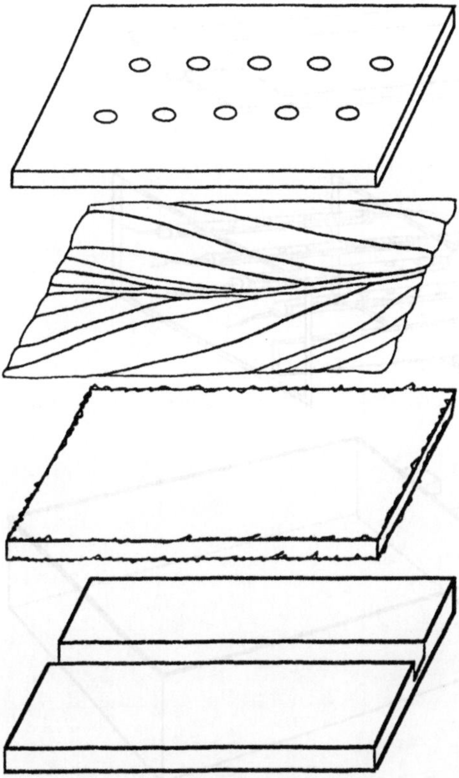

Figure 3.3. A microhumidity chamber for quantitative inoculation of plant leaves with fungal spores. Adapted from Bergstrom and Nicholson (1983).

descriptions of design and operation of these instruments (Bell et al. 1952; Petersen 1959; Schein 1964).

A microhumidity chamber is a simpler device that allows uniform inoculation rates over a defined area (Bergstrom & Nicholson 1983). Upper and lower plates are made of clear acrylic plastic (Fig. 3.3), with a 3-mm-deep by 5-mm-wide groove cut in the center of the bottom plate. A sponge backing is placed on the bottom plate as a cushion, and the leaf placed in the chamber with the midrib in the bottom groove. Holes are bored in the upper plate for spore inoculation, at a size and distribution optimum for a given series of investigations. Spore suspensions can then be pipetted into each well, adding test compounds if desired, the wells sealed with transparent tape, and germination monitored. For some plant species, monitoring of fungal disease progression can be followed quantitatively using the inoculation apparatus depicted in Figure 3.4 (Robinson & Hodges 1976). The bottom of a refrigerator crisper is covered with damp cheese cloth to maintain humidity. Leaves are mounted in Pyrex™ tubes (8 cm long, 8 mm

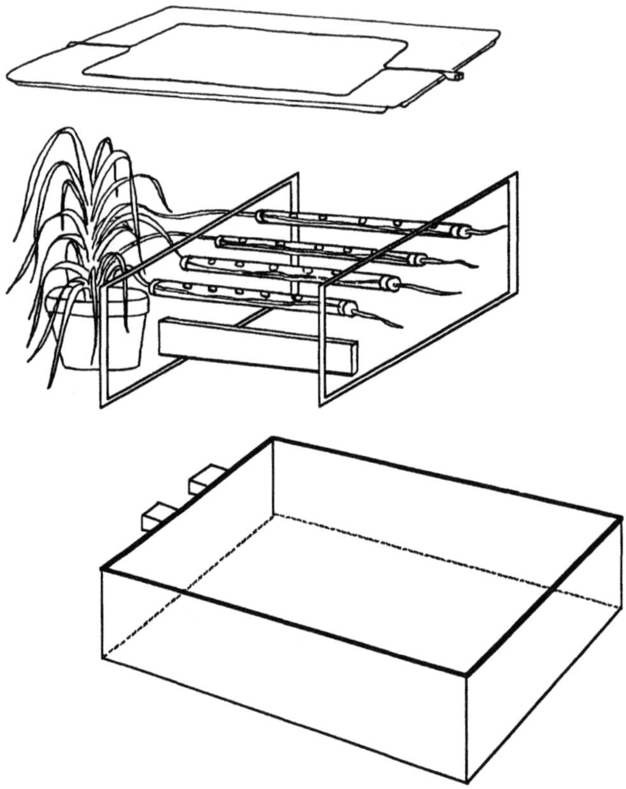

Figure 3.4. A plant leaf inoculation apparatus devised for monitoring fungal development in leaves from the same plant of different ages. See text for details. Adapted from Robinson and Hodges (1976).

i.d.) with evenly spaced inoculation ports. Tygon tubing can be placed at either end of the glass tube to prevent slipping, and Styrofoam™ end plugs hold the leaves in position. The top is held just above the leaf surface by a wood or metal spacer. Both of these are relatively simple to construct and will allow reproducible spore inoculation without the need for wetting or dispersing agents.

Spore development of many plant pathogens that enter through leaf tissue is regulated at least initially by waxes that cover the leaf surface (Podila et al. 1993; Kolattukudy et al. 1995). Spore behavior can be examined using leaf wax extracts, commercially available lipids or inhibitors of specific enzymes layered on glass slides. These compounds can also be incorporated into agar media, adsorbed onto glass beads or microcarriers such as SPEX® beads (JRH Biosciences) placed in microcentrifuge tubes (Flaishman et al. 1995), or layered onto isolated leaf or fruit cuticle (Doss et al. 1993).

Spore germination and host invasion by some plant pathogenic fungi, especially rusts including species of *Uromyces*, *Puccinia*, and *Melampsora*, are triggered by the physical structure of leaf stomata (Hoch et al. 1987; Allen et al. 1991). This response to physical features (thigmotropic response) can be mimicked in some cases by scratching polystyrene (Allen et al. 1991) to form ridges ~ 0.4—1 μm in height. An alternative is to make polystyrene replicas of an appropriate leaf surface as described by Wynn (1976). This is done by polymerizing silicon rubber with stannous octanoate, spreading the mixture over a leaf surface, and then forming a positive impression of the rubber mold by dissolving polystyrene (20% polystyrene in ethyl acetate, wt/vol) in the mold, drying, and peeling it out. The effect of chemicals on spore development can then be assessed by layering them on appropriately microfabricated topographies. There are a wide variety of physical and chemical treatments that stimulate spore germination by rust fungi (Mendgen et al. 1988), and care should be taken to choose among the many possibilities best suited for a particular species and experiment.

One strategy to obtain rapid quantitative data on fungal production of elicitors that induce phytoalexin production and other plant defense responses, which in turn affect fungal growth and morphogenesis, is to use plant cell culture or protoplast microassays (DiCosmo & Misawa 1985; Ebel 1986; Hohl 1991; Parker et al. 1991). The advantages of this technique include the ability to use any preparation of the fungus, from mycelial homogenates to partly or completely purified compounds for the assay, very little material is necessary, and dose-response data is easily generated. For these assays 50–1000 μl of cell or protoplast suspensions are added to multiwell tissue culture plates, and the elicitor preparations are added at the desired dosage to each well. After an appropriate incubation period, the plant cells are examined for elicitor recognition. Responses can range from autofluorescence under ultraviolet (UV) light due to production of phytoalexins, to triggering a tanning reaction indicative of a hypersensitive response to parasite invasion. A variation of this is to embed plant cells or protoplasts in agar, and add mycelia or spores of the pathogen to observe the full range of possible host-parasite interactions. This type of assay is limited, of course, to plant species for which protoplasts or cell cultures are available, and it should be first confirmed that the cultured cells act similarly to the intact plant. The reader is referred to reviews of plant cell culture and protoplast generation (Hegelson & Deverall 1983; Reinert & Binding 1986; Dixon & Gonzales 1994; Terzi et al. 1995). A variation of this assay in which fungal spores adhering to microinfection chambers on microscope slides are bathed in plant cell or protoplast suspensions (Gross et al. 1993) is particularly useful when the assay involves high-resolution light (or even electron) microscopy or video microscopy.

Another technique that avoids use of the entire plant is to place stem, leaf or root segments on damp filter paper or agar in sterile petri dishes. Fungal spore suspensions or extracts are then either placed in discrete drops or, especially in the case of pathogens that only enter their host through wounds, injected into the plant tissue using a narrow bore syringe or inoculated following a pin prick

(Kolattukudy & Gamble 1995). In the case of fungal infection of leaves of some plant species, infection is also promoted by removal of leaf surface waxes by immersion of chloroform or another suitable organic solvent prior to the infection assay (Carver et al. 1990). Inoculation of the fungus is then followed by assaying for a necrotic or other hypersensitive response or phytoalexin production, and the effect of these responses on fungal proliferation.

Discussion in the next few paragraphs is applicable to screening for toxic metabolites in many experimental systems, including those produced by plant and animal parasites and pathogens. Mutagenicity assays are not commonly employed in many routine investigations, but can reveal subtle effects. The Ames *Salmonella*/mammalian-microsome assay is a fairly standard assay of this type (Ames et al. 1975; Anke et al. 1989). Using frameshift and base-pair mutant strains of bacteria or fungi, LD_{50} (dose causing 50% mortality) values are first determined for the test compounds using plating of the test organism on nutrient agar. The number of spontaneous revertants is determined, and a standard mutagenic compound, e.g., 2-aminoanthracene, is used as a positive control. The LD_{50} concentration of the test compound is then used in the assay and the number of revertants per plate determined. Details are provided in Ames et al. (1975).

Antimicrobial activity is usually evaluated using agar plate diffusion assays, growth inhibition in microtiter plates, direct application of test compounds to mycelia, broth serial dilution assays, or variations of these techniques (Homans & Fuchs 1970; Anke et al. 1989). Microspectrophotometric monitoring using a microplate reader allows rapid, quantitative determination of antifungal activity using very small (microliter) amounts of test compounds and test organisms (Broekaert et al. 1990). A useful and rapid method to expose populations to a range of compound concentrations is the gradient plate technique (Bryson & Brown 1952). A bottom layer of nutrient agar is poured and allowed to harden with the plate held at an angle such that the entire bottom is barely covered. An overlaying layer of medium containing the test compound is then poured with the plate now in a horizontal position, allowing the agar and compound to diffuse downward, becoming diluted in proportion to the thickness of the agar layer. Double-gradient plates and other variations are reviewed by Hartman (1968).

A relatively new class of fungicides, increasingly investigated using plant pathogenic fungi, acts by inhibiting melanin biosynthesis (Yamaguchi & Kubo 1992). An assay developed for rapid screening of melanin biosynthesis inhibition (Thines et al. 1995) suggests an approach that can be taken to screening a variety of inhibitory compounds. In Ascomycetes and their related imperfect forms dark brown/black melanins are produced via a pathway that has 1,8-dihydroxynaphthalene (DHN) as the immediate precursor of these polymers. The assay involves growing a fungal species, *Lachnellula* sp., that synthesizes large amounts of melanin, in liquid culture, and incorporating mycelial homogenates into warm, unsolidified nutrient agar media. Paper disks containing the test compounds are placed onto the plates, and after an appropriate incubation period (48–72 h), the plates were examined for white zones signaling the inhibition of melanin synthesis.

An often ignored aspect of designing bioassays for screening of antimicrobial and other toxic substances is the wide variety of culture conditions than can affect toxin production. These include fungal strain, culture medium and culture conditions, age and developmental stage of the fungus, maintenance media, and even the stage of the fungus used to inoculate the culture from which putative toxins are ultimately extracted (Zapf et al. 1995). Although it is not possible to eliminate all of these variables, awareness of potential complications can minimize these problems.

Light can profoundly affect *in vivo* and *in vitro* toxin bioassays, as has been documented by photosensitization experiments using livestock and poultry (Towers 1980; Ivie 1982). Photoactivation of quinones, certain alkaloids, and other compounds can be assessed by, for example, exposure of agar plates used in diffusion assays to UV-A light (320–400 nm, 5 W/m^2) for 2 h while keeping control plates in the dark (Taylor et al. 1995).

Epidemiological studies, as outlined previously, are difficult, and when bioassays are based on measuring spore survival, infection versus colonization rates, disease intensity, inoculum potential, or any other characteristic, careful experimental design is vital to generating meaningful results (Kranz & Rotem 1988).

3.4.2.2. Animal Parasites and Pathogens

Many species of fungi are obligate or facultative parasites of arthropods (McCoy et al. 1988), humans, and other animals (Rippon 1988; Kwon-Chung & Bennett 1992; Crissey et al. 1995). Many of these are virulent pathogens, but do not produce toxins as part of the infection process. Differentiating between fungi that kill their host by starvation, proliferation in vital organs, or disruption of host metabolism rather than by toxin production often is not trivial. Even in instances where toxin production had been verified *in vivo* and *in vitro,* e.g., the production of peptide toxins of the beauvericin group by the entomopathogenic fungus *Beauveria bassiana,* although suspected as being involved in pathogenesis, these compounds have yet to be directly related to insect mortality (Gillespie & Claydon 1989; Khachatourians 1991). A further complication is that many animal parasitic fungi are infective only during certain stages of their life cycle, or when a potential host is exposed to the fungus in the proper form. For instance, under natural conditions the infective stage of the virulent human pulmonary pathogen *Coccidioides immitis* are arthroconidia, which must be inhaled into the lungs of its host (Huppert & Sun 1980). This pathogen also undergoes extreme morphological changes during its life cycle, and is difficult to recognize following dissemination into the blood stream. Another example is *Lagenidium giganteum,* a facultative parasite of mosquito larvae, which must produce motile zoospores by either a sexual or asexual process to infect a host (Domnas 1981). A final complication when dealing with complex organisms is that many parasites will only affect specific organs, parts of organs, or organelles.

Entomopathogenic fungi are unique in that, unlike bacteria and viruses, they initiate host infection by spore attachment to the outer cuticle, germination, and piercing of the cuticle using a combination of mechanical and enzymatic forces. Although spore adhesion of many deuteromycete pathogens requires only a hydrophobic surface (Boucias et al. 1988), subsequent events in the infection process involve constant communication between the host and fungus (St. Leger 1993). Virulence and host specificity of deuteromycetes are largely determined by postgermination events. Examination of spore germination and subsequent development of entomopathogenic fungi is facilitated by the fact that most of the initial steps are regulated by physical and chemical characteristics of the insect cuticle. Cuticular preparations can be obtained either from exuviae shed by insects during ecdysis (Kerwin 1983), or for larger insects such as grasshoppers or the beet armyworm (*Spodoptera exigua*), by mechanical removal and washing. This provides a much simpler experimental system than the intact insect for spore development studies. Another convenient feature of terrestrial insect cuticle is that, like many leaves, a well-defined wax layer makes up the outermost layer. Since fatty acids and other lipids affect spore development (Kerwin 1982), removal of these by organic solvents and their subsequent incorporation into agar media or inert substrates (Kerwin 1984), as previously outlined for plant parasites, provides defined and quantitative bioassays.

Unlike terrestrial deuteromycete entomopathogens, host specificity exhibited by fungi and Oomycetes that infect their hosts via motile spores is usually regulated by parasite recognition of chemical and physical signals on the cuticular surface (Kerwin 1983; Maniara et al. 1984; Kerwin et al. 1991; Ricker & Bostock 1992). As with terrestrial insects, use of isolated cuticle preparations simplifies examining the chemical interactions of host and parasite. Harsh organic extractions or enzymatic hydrolyses can be carried out without the complication of hemolymph and other contents leaking from the insect. Another strategy is to isolate carbohydrates, lipids, polyphenols, or proteins from insect cuticle and incorporate the extracts into an inert matrix followed by assays of zoospore attachment, encystment, and appressorium formation. One convenient matrix is films of chitosan, which unlike chitin, a ubiquitous component of insect cuticle, is soluble in dilute acid. Chitosan is dissolved in 1–2% aqueous HCl or acetic acid, and then dried overnight at 50°C on glass slides or petri dishes (Muzzarelli et al. 1994). Proteins or other cuticular extracts can be incorporated into the solubilized chitosan prior to drying. After neutralization with dilute base and washing with water, these transparent films are very useful for dissecting the chemical nature of cuticular components that regulate zoospore behavior (Petersen et al. 1997).

Research on mammalian pathogens, including those infecting humans, is obviously centered on the discovery of new and efficacious antifungal agents. *In vitro* bioassays of compound toxicity are largely centered on determining the minimal inhibitory concentration (MIC), despite the acknowledged advantages of using animal models (Hare & Loebenberg 1988). These tests, which can be performed by incorporating compounds into agar or liquid culture media, are straightforward and not further discussed.

Adhesion of spores and cells of some pathogens such as species of *Candida* is a pivotal event in infection, and can be investigated using either natural substrates such as fibrin-platelet matrices or neutrophils, or inert materials including Teflon® and denture acrylic (Douglas 1987). For instance, proteins or other compounds suspected to act as ligands for fungal recognition can be immobilized on microtitration plates, incubated with spore or cell suspensions with gentle shaking, followed by microscopic observation of stained cells after washing. A variation of this assay uses flow cytometric analysis of fluorescently labeled proteins in which fluorescence provides a dynamic estimate of spore binding to a given substrate (Henwick et al. 1993; Coulet et al. 1994).

A novel method for examining the binding of *Candida albicans* with blood platelets can be used to define ligand-receptor interactions for a number of systems. Fungal spores, germ tubes, or cells are immobilized on a Sepharose® column onto which platelet suspensions are added. The number of cells added and eluted from the column can be determined using a hemocytometer (Calderone et al. 1994). By altering the suspension buffer, the developmental stage of the fungal propagules, or pretreatment of the column with antibodies or chelators, the nature and specificity of these interactions can be investigated.

The mycelium-yeast transition for dimorphic pathogens often coincides with changing from a saprobic to a pathogenic phase in the fungal life cycle (Szaniszlo 1985). This transition can be induced by changes in temperature, pH, or nutrition. Chemical signals involved in dimorphism have been identified, and bioassays usually involve simple microscopic examination of treated cultures; however, if other factors are being evaluated, e.g., stage-specific protein production or gene expression, care must be taken to differentiate between morphology-specific and inducer-specific effects (Cannon et al. 1994).

3.4.2.3. Fungal and Lichen Parasites and Pathogens

Despite few commercial successes, interest remains in developing biological control of pest organisms including fungi, due to continual development of resistance to existing chemicals and an increasingly restrictive regulatory climate. Among the many biotic and abiotic factors that affect successful implementation of biological control (Cook 1993), understanding the chemical basis of host recognition, fungal attachment and invasion is vital. Direct parasitism is not necessarily a prerequisite for successful control of a plant pathogen. For instance, endo- and ectomycorrhizae can suppress root rot and other soil-borne fungi by improving plant host vigor, displacement of the pathogen from the rhizosphere by precolonization, establishment of a protective mantle on the root sheath, or induction of cross-protection by triggering phytoalexin production (Deacon & Fleming 1992; Linderman 1994; Traquair 1995). Another possible complication is alteration of the rhizosphere by secreted chemicals such that the growth of antibiotic-producing actinomycetes or hyperparasitic fungi is promoted. Bioassay of root disease control, therefore, can be complicated, especially under field or simulated field conditions.

Over 1000 species of lichenicolous fungi have been described (Egan 1987; Hawksworth 1988). There are parasitic, commensal, and mutualistic species, all of which must be able to detoxify or otherwise inactivate the plethora of secondary metabolites produced by lichens (Lawrey 1993; Lawrey et al. 1994). Lichens are parasitized less than most groups of fungi and algae, and secondary metabolites apparently play a role in limiting parasitism (Lawrey 1995). The role of these compounds in these interactions can be determined by removing phenols, which can constitute up to 20% of the dry weight of some lichens, using acetone extraction of dry, powdered lichens. This material is then used as a growth substrate for selected fungal species, inoculated using water suspensions of comminuted mycelia, and weight loss of the substrate measured after 30 days (Lawrey et al. 1994). There are instances where growth of hyperparasites is actually promoted by secondary metabolites secreted by the lichen host (Lawrey 1993), and comparable laboratory assays can be used to demonstrate this.

3.4.3. Mutualistic and Symbiotic Interactions

3.4.3.1 Vesicular-Arbuscular Mycorrhizal Fungi

Vesicular-arbuscular mycorrhizal (VAM) fungi are found in most terrestrial ecosystems, where they have significant effects on host plant growth and development (Gerdemann 1968; Gianinnazi 1991). Plants supporting VAM colonization, referred to as mycotrophic plants, make up the majority of plant species (Schreiner & Koide 1993a, 1993b). This association apparently originated early in terrestrial plant evolution, and may have contributed to adaptation to the terrestrial environment (Pirozinsky & Dalpé 1989). Since non-mycotrophic species are rare, there has been strong selective pressure to maintain plant associations with VAM. Significant effects on crop plant growth by VAM has lead to renewed interest in understanding the bases of host recognition, the role of VAM in regulating plant growth, the absence of significant host defense response to endocellular fungal invasion, with the paradoxical limited extent of VAM proliferation in the host plant. Such a close interaction of host and symbiont must involve the exchange of a series of chemical and physical signals. A related problem is the basis of incompatability between VAM and nonmycotrophic species. Progress in understanding these processes has been limited due to the obligate biotrophy of VAM symbionts, the complex nature of these interactions and the absence of reliable experimental systems and techniques.

Spore germination and early hyphal growth can occur in the absence of a host plant, but subsequent development requires the presence of a suitable host. Carbon dioxide stimulates, and in some cases appears to be required, for spore germination of these fungi (R.P. Schreiner, personal communication 1995). Exposure of spores to CO_2 may aid in synchronizing their development and increasing the amount of fungus available for experimentation. Since VAM spores invariably have to be isolated from soil, methods devised to separate them from the substrate and from other fungal species are helpful. A novel method recently developed by Okuda et al. (1995) uses Percoll® (Pharmacia) gradients. Sieved soil samples (1–10 g) were added to 10-ml Percoll®, sonicated for 2 min, and let stand for 10 min to allow soil particles to sink. Five milliliters of the supernatant was decanted, diluted with 45 ml distilled water, and centrifuged for 20 min at 2150 × *g*. The pellet was resuspended in 0.5 ml distilled water and gently layered on the top of a Percoll® pregradient (7.0 ml Percoll®, 1.0 ml 1.5-M NaCl, 2.0 ml distilled water, 16- × 76-mm polyallomer tube [Beckman], centrifuged at 60,000 × *g* for 20 min). This was centrifuged for 15 min at 3000 × *g*, and 100–500 μl fractions serially collected from the bottom. This method works well for separating the spores of many common soil fungi (Okuda et al. 1995), and is much more efficient than simple dilution plate methods commonly employed.

An innovation that provides a more defined experimental system for VAM is the use of nontransformed or transformed roots to determine critical components involved in establishing and maintaining mycorrhizal associations (Bécard & Fortin 1988; Chabot et al. 1992). Using either surface-sterilized excised roots (Mosse & Hepper 1975) or Ri T DNA-transformed roots as plant partners, fungal spores are first isolated using a combination of wet sieving and density gradient centrifugation (Gerdemann & Nicolson 1963; Gerdemann 1968; Furlan et al. 1980). Surface-sterilized spores (1% gentamycin + 2% streptomycin) can be stored on water agar at reduced temperature (4–10°C), and germinated in the dark in a petri dish at 25–30°C. Root explants are subsequently added in close proximity to the spores. Compounds or other factors such as CO_2 that are suspected as being involved in mycorrhizal interactions can then be added or adjusted to the culture system.

A less-defined system that still avoids the use of soil involves hydroponic culture of host plants, an assay that is also useful for monitoring attachment of zoospores or conidia of parasitic fungi to plant roots (Schuerger et al. 1993). Behavior of zoospores and germ tubes can be more easily monitored using this setup, provided the hydroponic solution does not adversely affect fungal behavior. Replacement of the hydroponic solution prior to initiation of the assay, or transfer of the growing plant to another container are alternatives. Stir bars can be used to obtain homogeneous mixtures of both spores and added chemicals, allowing quantitation of behavior such as spore attachment or appressorium formation in response to chemical signals.

3.4.3.2. Ectomycorrhizal Fungi

First described by Frank in 1885, ectomycorrhizae are primarily Basidiomycetes (more than 5000 species) and Ascomycetes (at least several hundred species) associated with secondary and tertiary roots of trees (Harley 1989; Werner 1992). They can be beneficial to plants by increasing soil nutrient and mineral uptake, reducing heavy metal toxicity, and minimizing below-ground attack by producing antimicrobial and insect deterrent chemicals; however, the fungus also alters the carbon balance of the host plant, which can alter growth rates and affect how the plant reacts to herbivory and microbial attack. Besides the three systems used for self and non-self recognition characteristic of Basidiomycetes—somatic compatibility, mating system compatibility, and prevention of interspecific mating—communication with the host plant regulates formation of the fungal sheath (also referred to as the hartig net) and growth of the fungus between cortical root cells. Most of the techniques described previously for VAM can be used for ectomycorrhizae, and the reader is referred to that section.

3.4.3.3. Fungal-Insect Associations

Beginning with the pioneering work of Buchner (Buchner 1998, 1965) on intracellular insect symbionts, research on microorganisms and insects has produced fascinating examples of the mutual interactions of these two groups of organisms (Martin 1987, 1992). The best known of these interactions is the nearly axenic culture by attine ants (leafcutter ants) of one or a few fungal species as a food source (Weber 1955, 1966; Martin 1992), but comparable interactions involving fungi and beetles, woodwasps and termites have been documented (Batra 1979). A complication with examining many of these interactions is that many insects, including most termites, survive for very short periods when removed from their natural habitat. Acrylic plastic rearing chambers have been constructed to simulate termitaria (Batra & Batra 1979), with some queens surviving up to 3 weeks under these conditions. Until survival of laboratory colonies is improved, it will be difficult to elucidate the nature of signals involved in choosing the proper fungal species to maintain feeding habits of the different castes, production of fungitoxic or fungistatic compounds that promote monoxenic fungal culture, and how fungal growth is controlled by the termites.

Another major (but little studied) group of fungi are the Laboulbeniales (Thaxter 1931). These obligate ectoparasites of beetles and other insects are tiny, obscure Ascomycetes that often exhibit extreme host specificity (Benjamin 1971; Tavares 1985). Specificity can be limited to a single species, a defined location on the host integument, or even to a single sex of one species (Benjamin 1965).

Host range is determined by a combination of chemical and physical signals and possibly host behavior. Since they are ectoparasites, ascospore germination can be investigated in many species by using insect exuviae or dissected parts of the cuticle, e.g., wings or antennae. In those instances, chemical and enzymatic modification of the isolated cuticle followed by monitoring spore behavior can provide clues to the bases of host specificity. This assay can be done on isolated cuticle, or more often by placing the samples on damp filter paper, water agar, or dilute nutrient media to maintain the high relative humidity required for optimum spore germination.

3.4.4. Competition—Allelopathy

Isolation of the plant growth regulator indole-3-acetic acid from *Rhizopus suinis* served as the foundation for development of the entire field of plant growth regulators (Thimann 1935). Since then thousands of compounds have been identified from microbial sources, many of which display phytotoxic, plant growth regulatory, fungistatic, fungitoxic, or other anti-microbial activities. For fungi the classic example involves the "killer" phenomenon (Woods & Bevan 1968), in which proteins secreted by one strain are toxic to other strains of the same species. Many of these are screened by streak-plate assays (Stumm et al. 1977) in which the target species is seeded in cooled but unsolidified nutrient agar. After the plates solidify, the killer strain is streaked on top of the agar, the cultures incubated for several days, and the test yeast grows as a background lawn except when there is a toxic level of excreted protein. A variation of this method involves incorporating methylene blue into the agar, where instead of clear zones of growth inhibition, blue-stained (dead) cells mark areas of toxin secretion (Walker et al. 1995). These methods can be applied to ecological studies of natural communities (Starmer et al. 1987) as long as care is taken in choosing appropriate test species.

3.4.5. Defense Against Herbivores and Pathogens

Some endophytic fungi produce toxins in their plant host and/or in culture (Carroll 1988, 1991) where they can deter feeding by or kill foraging insects and other animals (Funk et al. 1983, Clay 1988; Clark et al. 1989; Barbosa et al. 1991; Siegel et al. 1991). The (presumed) mutualistic relationship between endophytic fungi and grasses and other plant taxa is very widespread, involving thousands of grasses and other plant species. Disadvantages of these relationships are centered on poisoning of livestock by, for instance, clavicipitaceous fungal endophytes colonizing pasture grasses (Bacon et al. 1986; Siegel et al. 1987), or deleterious effects on humans following ingestion of cereal grains colonized by *Fusarium*, *Claviceps*, and other toxin-producing species (Marasas & Nelson 1987). The large animal toxicity assays are usually inadvertent, and not useful to most researchers; however, avian and especially murine models are widely used to evaluate the effects of putative fungal toxins on mammals, the host defense responses, and subsequent effects on fungal development.

Fungal-mediated defense of plants from herbivory is often assumed to be toxin-mediated, but this assumption is not always warranted. If it is not a goal of the study to elucidate the mechanisms of deterrence, this aspect can be ignored; however, besides mycotoxin production by the endophyte, other possible mechanisms include altered physical and chemical properties of plant cell walls, changes in nutrient status, or fungal-induced phytoalexin production. There are dozens of physiological and morphological changes associated with endophyte colonization of host plants, so interpretation of experiments monitoring herbivory on plants with and without the fungus, especially when performed over several days or weeks, will be difficult. As summarized by Clay (1991), some of the costs often associated with endophytic colonization, including decreased growth under suboptimum conditions, reduced seed set and/or maturation, and various morphological changes, can mimic conditions induced by herbivory. Even if toxins are involved, reduced oviposition rather than feeding deterrence may be the primary cause of reduced plant damage.

To evaluate the fate of endophytic fungi ingested by foraging insects, it is possible to use pure cultures of the fungus, infected versus uninfected plants, crude extracts of any of the former, or compounds purified from these sources. The choice of insect (and stage of insect) can be limited by plant host preferences, availability or ease of colonization, with favorites including the fall armyworm (*Spodoptera frugiperda*), the flour beetle (*Tribolium castaneum*), the house cricket (*Acheta domesticus*), or the milkweed bug (*Oncopeltus fasciatus*). Bioassay techniques can be as variable as the species chosen for investigation. These methods can include observations on weight loss, mortality, altered behavior, deterrence given a choice of plant species (or infected versus noninfected plants of the same species), larval survival, days to pupation, oviposition preferences, or insect density (Clay 1988; Dahlman et al. 1991; chapter 5). A common technique to evaluate oviposition preference involves placing a fungal colony or extract in a wire mesh ball (5–10 cm diameter) and wrapping this and a control ball with cheese cloth (Honda et al. 1988), on the surface of which many insect species will oviposit. Alternatively, data on plant damage can be collected, assessing root/leaf/stem damage, growth rates, seed set, or seed maturation.

It is difficult to justify mammalian toxicity tests, especially when morbidity or mortality has previously been observed; however, there are well-established protocols for evaluating mammalian toxicity based on methods devised by the U.S. Environmental Protection Agency for registering chemicals and microbial pest control agents (Anonymous 1988; Siegel & Shadduck 1992). These tests involve intravenous or intraperitoneal injections of the agent, dermal and eye irritation/infectivity tests, and acute oral and pulmonary studies, using mice, rats, or rabbits, followed by reisolation of the fungus to determine its dissemination and growth. They should not be undertaken without well-designed and highly controlled protocols and sufficient justification.

3.5. Conclusions

This chapter has outlined general approaches to bioassays of chemical interactions involving fungi and Oomycetes. An appreciation for the diversity of these interactions should provide the reader with the background needed to approach many of these areas of research, while tempering interpretation of data with the realization that even the simplest experimental systems can lead to unexpected and contradictory results. With the advent of molecular biology and the development of sophisticated analytical instruments, there is a tendency to resort to increasingly complex equipment and molecular approaches in all areas of research; however, there is currently no substitute for using fungi and other (intact) organisms when assessing the biological effects of other organisms or chemicals. Understanding the subtleties of these interactions will lead to advances in many fields of basic and applied research.

3.6. Acknowledgments

Writing of this manuscript was supported in part by a grant from the National Institutes of Health (AI 34339). The following people kindly provided reprints and preprints of their work: J.W. Bennett, G.C. Carroll, T.L.W. Carver, A. Dahlberg, J. Gloer, J.D. Lawrey, P.E. Kolattukudy, U. Matern, R.L. Nicholson, R.P. Schreiner, J. Thorner, and J.A. Traquair.

3.7. References

Alexopoulos, C.J. & C.W. Mims. 1979. Introductory Mycology, 3rd edition. John Wiley & Sons, New York.

Allen, E.A., B.E. Hazen, H.C. Hoch, G.M.E. Leinhos, R.C. Staples, M.A. Stumpf & B.T. Terhune. 1991. Appressorium formation in response to topographical signals by 27 rust species. Phytopathology **81**:323–331.

Allen, M.F. 1991. The Ecology of Mycorrhizae. Cambridge University Press, Cambridge.

Ames, B.N., J. McCann & E. Yamasaki. 1975. Methods for detecting carcinogens and mutagens with the *Salmonella* mammalian-microsome mutagenicity test. Mutation Res. **31**:347–364.

Amir, R., D. Lavanon, Y. Hadar & I. Chet. 1992. Formation of sclerotia by *Morchella esculenta:* relationship between media composition and turgor potential in the mycelium. Mycol. Res. **96**:943–948.

Andrews, J.H. 1986. How to track a microbe. *In:* Microbiology of the Phyllosphere, eds. N.J. Fokkema & J. van den Heuvel, pp. 14–34, Cambridge University Press, Cambridge.

Anke, H., O. Bergendorf & O. Sterner. 1989. Assays of the biological activities of guaiane sesquiterpenoids isolated from the fruit bodies of edible *Lactarius* species. Food Chem. Toxicol. **27**:393–397.

Anonymous. 1988. Toxicology Guidelines for Microbial Pest Control Agents, Subdivision M. U.W. Environmental Protection Agency, Office of Pesticide and Toxic Substances, Washington, DC.

Arora, D.K., B. Rai, K.G. Mukerji & G.R. Knudsen. eds. 1991. Handbook of Applied Mycology, Vol. 1. Soil and Plants. Marcel Dekker, New York.

Bacon, C.W., P.C. Lyons, J.K. Porter & J.D. Robbins. 1986. Ergot toxicity from endophyte-infected grasses: a review. Agron J. **78**:106–116.

Barbosa, P., Krischik, V.A. & C.G. Jones. eds. 1991. Microbial Mediation of Plant-Herbivore Interactions. John Wiley & Sons, New York.

Bardwell, L., Cook, J.G. Inouye, C.J. & Thorner, J. 1994. Signal propagations and regulation in the mating pheromone response pathway of the yeast *Saccharomyces cerevisiae.* Develop. Biol. **166**:363–379.

Barksdale, A.W. 1969. Sexual hormones of *Achlya* and other fungi. Science **166**:831–837.

Barksdale, A.W. & L.L. Lasure. 1974. Production of hormone B by *Achlya heterosexualis.* Appl. Microbiol. **28**:544–546.

Barksdale, A.W., T.C. McMorris, R. Seshadri, T. Aranchalam, J.A. Edwards, J. Sundeen & D.M. Green. 1974. Response of *Achlya ambisexualis* E87 to the hormone antheridiol and certain other steroids. J. Gen. Microbiol. **82**:295–299.

Barr, D.J.S. 1992. Evolution and kingdoms of organisms from the perspective of a mycologist. Mycologia **84**:1–11.

Bartnicki-Garcia, S. & V.O. Sing. 1987. Adhesion of zoospores of *Phytophthora* to solid surfaces. *In:* Zoosporic Fungi in Teaching and Research, eds. M.S. Fuller & A. Jaworski, pp. 279–283, Southeastern Publishing Company, Athens, GA.

Batra, L.R., ed. 1979. Insect-Fungus Symbiosis—Nutrition, Mutualism, and Commensalism. John Wiley & Sons, New York.

Batra, L.R. & S.W.T. Batra. 1979. Termite-fungus mutualism. *In:* Insect-Fungus Symbiosis—Nutrition, Mutualism, and Commensalism, (ed.) L.R. Batra pp. 117–163, John Wiley & Sons, New York.

Beakes, G.W. 1987. Oomycete phylogeny: ultrastructural perspectives. *In:* Evolutionary Biology of the Fungi, eds., A.D.M. Rayner, C.M. Brasier & D. Moore, pp. 405–421, Cambridge University Press, Cambridge.

Bécard, G. & J.A. Fortin. 1988. Early events of vesicular-arbuscular mycorrhiza formation on Ri T-DNA transformed roots. New Phytol. **108**:211–218.

Bell, F.H., C.G. Schmidt, W.E. Miller & C.H. Kingsolver. 1952. A technique of obtaining uniform deposition of urediospores on cereal leaves. Phytopathology **42**:340.

Benjamin, R.K. 1965. Study in specificity: minute fungi parasitize living arthropods. Nat. History **74**:41–48.

Benjamin, R.K. 1971. Introduction and supplement to Roland Thaxter's contribution towards a monograph of the Laboulbeniaceae. Biblio. Mycol. **30**:1–155.

Bennett, J.W. & A. Ciegler. eds. 1983. Secondary Metabolism and Differentiation in Fungi. Mercel Dekker, New York.

Bergstrom, G.C. & R.L. Nicholson. 1983. Microhumidity chamber for quantitative inoculation of attached corn leaves with fungal pathogens. Phytopathology **73**:1040–1042.

Boucias, D.G., J.C. Pendland & J.-P. Latge. 1988. Nonspecific factors involved in the attachment of entomopathogenic deuteromycetes to host insect cuticle. Appl. Environ. Entomol. **54**:1795–1805.

Brasier, C.M. 1975. Stimulation of sex organ formation in *Phytophthora* by antagonistic species of *Trichoderma.* I. The effect *in vitro.* New Phytol. **74**:183–194.

Brodowski, I.D. & E.H. Oliw. 1993. Biosynthesis of 8*R*-hydroperoxylinoleic acid by the fungus *Laetisaria arvalis.* Biochim. Biophys. Acta **1168**:68–72.

Broekaert, W.F., F.R.G. Terras, B.P.A. Cammue & J. Vanderleyden. 1990. An automated quantitative assay for fungal growth inhibition. FEMS Microbiol. Lett. **69**:55–60.

Brunt, S.A. & J.C. Silver. 1986. Steroid hormone-induced changes in secreted proteins in the filamentous fungus *Achlya.* Exp. Cell Res. **163**:22–34.

Brunt, S.A., R. Riehl & J.C. Silver. 1990. Steroid hormone regulation of the *Achlya ambisexualis* 85-kilodalton heat shock protein, a component of the *Achlya* steroid receptor complex. Mol. Cell. Biol. **10**:273–281.

Bryson, V. & B.J. Brown. 1952. Microbial selection. Science **116**:45–51.

Buchner, P. 1928. Holzhahrung und Symbiose. Springer, Berlin.

Buchner, P. 1965. Endosymbiosis of Animals with Plant Microorganisms. Interscience, New York.

Calderone, R., R. Diamond, J.-M. Senet, J. Warmington, S. Filler & J.E. Edwards. 1994. Host cell-fungal cell interactions. J. Med. Vet. Mycol. **32** (suppl. 1):151–168.

Cannon, R.D., W.E. Timberlake, N.A.R. Gow, D. Bailey, A. Brown, G.W. Gooday, B. Hube, M. Monod, C. Nombela, F. Navarro, R. Perez, M. Sanchez & J. Pla. 1994. Molecular biological and biochemical aspects of fungal dimorphism. J. Med. Vet. Mycol. **32** (suppl 1):53–64.

Carroll, G.C. 1988. Fungal endophytes in stems and leaves: from latent pathogen to mutualistic symbiont. Ecology **69**:2–9.

Carroll, G.C. 1991. Fungal associates of woody plants as insect antagonists in leaves and stems. *In:* Microbial Mediation of Herbivore-Plant Interactions, eds. P. Barbosa, V.A. Krischik & C.G. Jones, pp. 253–271, John Wiley & Sons, New York.

Carroll, G.C., L.H. Pike, J.R. Perkins & M.A. Sherwood. 1980. Biomass and distribution patterns of conifer twig microepiphytes in a Douglas fir forest. Can. J. Bot. **58**:624–630.

Carver, T.L.W., B.J. Thomas, S.M. Ingerson-Morris & H.W. Roderick. 1990. The role of the abaxial leaf surface waxes of *Lolium* spp. in resistance to *Erysiphe graminis.* Plant Pathol. **39**:573–583.

Casselton, L.A., R.N. Asante-Owusu, A.H. Banham, C.S. Kingsnorth, U. Kues, S.F.

O'Shea & E.H. Pardo. 1995. Mating type control of sexual development in *Coprinus cinereus.* Can. J. Bot. **73** (suppl):S266–S272.

Cavalier-Smith, T. 1986. The kingdom Chromista: origin and systematics. Prog. Phycol. Rers. **4**:309–347.

Cavalier-Smith, T. 1987. The origin of fungi and pseudofungi. *In:* Evolutionary Biology of the fungi, eds. A.D.M. Rayner, C.M. Brasier & D. Moore, pp. 339–353, Cambridge University Press, Cambridge.

Cavalier-Smith, T. 1989. The kingdom Chromista. *In:* The Chromophyte Algae: Problems and Perspectives, eds. J.C. Green, B.S.C. Leadbeater & W.L. Diver, Systematics Association Special Vol. No. 38, pp. 381–407, Clarendon Press, Oxford.

Chabot, S., G. Bécard & Y. Piché. 1992. Life cycle of *Glomus intraradix* in root organ culture. Mycologia **84**:315–321.

Cid, V.J., A. Duran, F. del Ray, M.P. Snyder, C. Nombela & M. Sanchez. 1995. Molecular basis of cell integrity and morphogenesis in *Saccharomyces cerevisiae.* Microbiol. Rev. **59**:345–386.

Clark, C.L., J.D. Miller & N.J. Whitney. 1989. Toxicity of conifer needle endophytes to spruce budworm. Mycol. Res. **93**:508–512.

Clay, K. 1988. Fungal endophytes of grasses: a defensive mutualism between plants and fungi. Ecology **69**:10–16.

Clay, K. 1991. Fungal endophytes, grasses, and herbivores. *In:* Microbial Mediation of Plant-Herbivore Interactions, eds. P. Barbosa, V.A. Krischik & C.G. Jones, pp. 199–226, John Wiley & Sons, New York.

Cochrane, V.W. 1974. Dormancy in spores of fungi. Trans. Amer. Microscopical Soc. **93**:599–609.

Cook, R.C. & J.M. Whipps. 1993. Ecophysiology of Fungi. Blackwell Scientific Publications, London.

Cook, R.J. 1993. Making greater use of introduced microorganisms for biological control of plant pathogens. Ann. Rev. Phytopathol. **31**:31–53.

Couch, J.N. & C.E. Bland, eds. 1985. The genus *Coelomomyces.* Academic Press, Orlando, FL.

Coulet, P., J.-P. Bouchara, G. Renier, V. Annaix, C. Planchenault, G. Tronchin & D. Chabasse. 1994. Specific interaction of *Aspergillus fumigatus* with fibrinogen and its role in cell adhesion. Infect. Immun. **62**:2169–2177.

Crissey, J.T., H. Lang & L.C. Parish. 1995. Manual of Medical Mycology. Blackwell Scientific, Cambridge, MA.

Dahlberg, A. & J. Stenlid. 1990. Population structure and dynamics in *Suillus bovinus* as indicated by spatial distribution of fungal clones. New Phytol. **115**:487–493.

Dahlman, D.L., H. Eichenseer & M.R. Siegel. 1991. Chemical perspectives on endophyte-grass interactions and their implications to insect herbivory. *In:* Microbial Mediation of Plant-Herbivore Interactions, eds. P. Barbosa, V.A. Krischik & C.G. Jones, pp. 227–252, John Wiley & Sons, New York.

Deacon, J.W. & L.V. Fleming. 1992. Interactions of ectomycorrhizal fungi. *In:* Mycorrhizal

Functioning: An Integrative Plant-Fungal Process, ed. M.F. Allen, pp. 249–300, Chapman & Hall, New York.

Deising, H., R.L. Nicholson, M. Haug, R.J. Howard & K. Mengden. 1992. Adhesion pad formation and the involvement of cutinase and esterases in the attachment of uredospores to the host cuticle. Plant Cell **4**:1101–1111.

Dickinson, C.H. 1971. Cultural studies of leaf saprophytes. *In:* Ecology of Leaf Surface Microorganisms, eds. T.F. Preece & C.H. Dickinson, pp. 129–137, Academic Press, New York.

DiCosmo, F. & M. Misawa. 1985. Eliciting secondary metabolism in plant cell cultures. Trends Biotechnol. **3**:318–322.

Dixon, R.A. & R.A. Gonzales, eds. 1994. Plant Cell Culture: A Practical Approach. IRL Press, Oxford.

Domnas, A.J. 1981. Biochemistry of *Lagenidium giganteum* infection in mosquito larvae. *In:* Pathogenesis of Invertebrate Microbial Diseases, ed. E.W. Davidson, pp. 425–449, Allanheld, Osmun, Totowa, NJ.

Domnas, A.J. & S.A. Warner. 1991. Biochemical activities of entomophagous fungi. Crit. Rev. Microbiol. **18**:1–13.

Doss, R.P., S.W. Potter, G.A. Chastagner & J.K. Christian. 1993. Adhesion of nongerminated *Botrytis cinerea* conidia to several substrata. Appl. Environ. Microbiol. **59**:1786–1791.

Douglas, L.J. 1987. Adhesion of *Candida* species to epithelial surfaces. CRC Crit. Rev. Microbiol. **15**:27–43.

Dyer, P.S., D.S. Ingram & K. Johnstone. 1992. The control of sexual morphogenesis in the Ascomycotina. Biol. Rev. **67**:421–458.

Ebel, J. 1986. Phytoalexin synthesis: the biochemical analysis of the induction process. Ann. Rev. Phytopathol. **24**:235–264.

Egan, R.S. 1987. A fifth checklist of the lichen-forming, lichenicolous and allied fungi of the continental United States and Canada. Bryologist **90**:77–173.

Egel, R., M. Willer, S. Kjaerulff, J. Davey & O. Nielsen. 1994. Assessment of pheromone production and response in fission yeast by a halo test of induced sporulation. Yeast **10**:1347–1354.

Federici, B.A. 1977. Differential pigmentation in the sexual stage of *Coelomomyces.* Nature **267**:514–515.

Federici, B.A. & S.N. Thompson. 1979. β-Carotene in the gametophytic phase of *Coelomomyces dodgei.* Exp. Mycol. **3**:281–284.

Ferguson, B., J. Horecka, J. Printen, J. Schultz, B.J. Stevenson & G.F. Sprague, Jr. 1994. The yeast pheromone response pathway: new insights into signal transmission. Cell Mol. Biol. Res. **40**:223–228.

Flaishman, M.A., C.-S. Hwang & P.E. Kolattukudy. 1995. Involvement of protein phosphorylation in the induction of appressorium formation in *Colletotrichum gloeosporioides* by its host surface wax and ethylene. Physiol. Molec. Plant Pathol. **47**:103–117.

Fokkema, N.J. & J. van den Heuvel. eds. 1986. Microbiology of the Phyllosphere. Cambridge University Press, Cambridge.

Frank, A.B. 1885. Über die auf Wurzelsymbiose beruhende Ernährung gewisser Bäume durch unterirdische Pilze. Ber Dtsch. Bot. Ges. **3**:128–145.

French, R.C. 1985. The bioregulatory action of flavor compounds on fungal spores and other propagules. Ann. Rev. Phytopathol. **23**:173–199.

French, R.C. 1992. Volatile chemical germination stimulators of rust and other fungal spores. Mycologia **84**:277–288.

French, R.C., S.E. Nester & R.G. Binder. 1994. Volatiles from germinating Canada thistle seed and root cuttings that stimulate germination of teliospores of the Canada thistle rust fungus, *Puccinia punctiformis.* J. Agric. Food Chem. **42**:2937–2941.

Fukui, Y., Y. Kaziro & M. Yamamoto. 1986. Mating pheromone-like diffusible factor released by *Schizosaccharomyces pombe.* EMBO J. **5**:991–993.

Funk, C.R., P.M. Halisky, M.C. Johnson, M.R. Siegel, A.V. Stewart, S. Ahmad, R.H. Hurley & I.C. Harvey. 1983. An endophytic fungus and resistance to sod webworms: association in *Lolium perenne* L. Biotechnology **1**:189–191.

Furlan, V., H. Bartschi & J.A. Fortin. 1980. Media for density gradient extraction of endomycorrhizal spores. Trans. Br. Mycol. Soc. **75**:336–338.

Gerdemann, J.W. 1968. Vesicular-arbuscular mycorrhiza and plant growth. Ann. Rev. Phytopathol. **6**:397–418.

Gerdemann, J.W. & T.H. Nicolson. 1963. Spores of mycorrhizal *Endogone* species extracted from soil by wet sieving and decanting. Trans. Br. Mycol. Soc. **46**:235–244.

Gianinazzi, V. 1991. Vesicular-arbuscular mycorrhizas: cellular, biochemical and genetic aspects. Agric. Ecosyst. Environ. **35**:105–119.

Gillespie, A.T. & N. Claydon. 1989. The use of entomogenous fungi for pest control and the role of toxins in pathogenesis. Pestic. Sci. **27**:203–215.

Gow, N.A.R., and G.M. Gadd. 1995. The Growing Fungus. Chapman & Hall, London.

Gow, N.A.R., and G.W. Gooday. 1987. Effects of antheridiol on growth, branching and electrical currents of hyphae of *Achlya ambisexualis.* J. Gen. Microbiol. **133**:3531–3535.

Grishin, A.V., J.L. Weiner, J.L. & Blumer, K.J. 1994. Control of adaptation to mating pheromone by G protein beta subunits of *Saccharomyces cerevisiae.* Genetics **138**:1081–1092.

Gross, P., C. Julius, E. Schmelzer & K. Hahlbrock. 1993. Translocation of cytoplasm and nucleus to fungal penetration sites is associated with depolymerization of microtubules and defense gene activation in infected, cultured parsley cells. EMBO J. **12**:1735–1744.

Hare, R.S. & D. Loebenberg. 1988. Animal models in the search for antifungal agents. ASM News **54**:235–243.

Harley, J.L. 1989. The significance of mycorrhiza. Mycol. Res. **92**:129–139.

Harnish, W.N., L.A. Berg & B.G. Lilly. 1964. Factors in lima bean and hemp seed required for oospore formation by species of *Phytophthora.* Phytopathol. **54**:895.

Hartman, P.A. 1968. Miniaturized Microbiological Methods. Academic Press, New York.

Hawksworth, D.L. 1988. The variety of fungal-algal symbioses, their evolutionary significance, and the nature of lichens. Bot. J. Linn. Soc. **96**:3–20.

Hegelson, J.P. & B.J. Deverall. eds. 1983. Use of Tissue Culture and Protoplasts in Plant Pathology. Academic Press, New York.

Hendrix, J.W. 1970. Sterols in growth and reproduction of fungi. Ann. Rev. Phytopathol. **8**:111–130.

Henwick, S., S.V. Hetherington & C.C. Patrick. 1993. Complement binding to *Aspergillus* conidia correlates with pathogenicity. J. Lab. Clin. Med. **122**:27–35.

Hoch, H.C., R.C. Staples, B. Whitehead, J. Comeau & E.D. Wolf. 1987. Signaling for growth orientation and cell differentiation by surface topography in *Uromyces*. Science **235**:1659–1662.

Hohl, H.R. 1991. Surface-related host-pathogen interactions in *Phytophthora. In: Phytophthora,* eds. J.A. Lucas, R.C. Shattock, D.S. Shaw & L.R. Cooke, pp. 70–89, Cambridge University Press, Cambridge.

Homans, A.C. & A. Fuchs. 1970. Direct bioautography on thin-layer chromatograms as a method for detecting fungitoxic substances. J. Chromatogr. **51**:327–329.

Honda, H., T. Ishiwatari & Y. Matsumoto. 1988. Fungal volatiles as oviposition attractants for the yellow peach moth, *Congethes punctiferalis* (Guenee) (Lepidoptera: Pyralidae). J. Insect Physiol. **34**:205–211.

Horgen, P.A. 1977. Steroid induction of differentiation: *Achlya* as a model system. *In:* Eukaryotic Microbes as Model Developmental Systems, eds. D.H. O'Day & P.A. Horgen, pp. 272–292, Marcel Dekker, New York.

Horgen, P.A. 1981. The role of the steroid sex pheromone antheridiol in controlling the development of male sex organs in the water mold *Achlya ambisexualis. In:* Sexual Interactions in Eukaryotic Microbes, eds. D.H. O'Day & P.A. Horgen, pp. 155–178, Academic Press, New York.

Huppert, M. & S.H. Sun. 1980. Overview of mycology, and the mycology of *Coccidioides immitis. In:* Coccidioidomycosis, ed. D.A. Stevens, pp. 21–46, Plenum, New York.

Hutchinson, S.A. 1973. Biological activities of volatile fungal metabolites. Ann. Rev. Phytopathol. **11**:233–246.

Isaac, S. 1992. Fungal-Plant Interactions. Chapman & Hall, London.

Ivie, G.W. 1982. Chemical and biochemical aspects of photosensitization in livestock and poultry. J. Natl. Cancer Inst. **69**:259–266.

Jones, P.C.T. & J.E. Mollison. 1948. A technique for the quantitative estimation of soil microorganisms. J. Gen. Microbiol. **2**:54–68.

Kerwin, J.L. 1982. Chemical control of the germination of asexual spores of *Entomophthora culicis,* a fungus parasitic on dipterans. J. Gen. Microbiol. **128**:2179–2186.

Kerwin, J.L. 1983. Biological aspects of the interaction between *Coelomomyces psorophorae* zygotes and the larvae of *Culiseta inornata:* host-mediated factors. J. Invertebr. Pathol. **41**:233–237.

Kerwin, J.L. 1984. Fatty acid regulation of the germination of *Erynia variabilis* conidia on adults and puparia of the lesser housefly, *Fannia canicularis.* Can. J. Microbiol. **30**:158–161.

Kerwin, J.L. & R.K. Washino. 1986. Regulation of oosporogenesis by *Lagenidium giganteum:* promotion of sexual reproduction by unsaturated fatty acids and sterol availability. Can. J. Microbiol. **32**:294–300.

Kerwin, J.L., D.F. Grant & M.L. Berbee. 1991. Specific induction of encystment of *Lagenidium giganteum* zoospores by concanavalin A and derivatives of chitin and chitosan. Protoplasma **161**:43–51.

Khachatourians, G.G. 1991. Physiology and genetics of entomopathogenic fungi. *In:* Handbook of Applied Mycology, Vol. 2 Humans, Animals and Insects., eds. D.K. Arora, K.G. Mukerji & E. Drouhet, pp. 613–663, Marcel Dekker, New York.

Ko, W.-H. 1978. Heterothallic *Phytophthora:* evidence for hormonal regulation of sexual reproduction. J. Gen. Microbiol. **107**:15–18.

Ko, W.-H. 1980. Hormonal regulation of sexual reproduction in *Phytophthora.* J. Gen. Microbiol. **116**:459–463.

Kohmoto, K., U.S. Singh & R.P. Singh. eds. 1995. Pathogenesis and Host Specificity in Plant Diseases—Histological, Biochemical, Genetic and Molecular Bases, Volume II Eukaryotes. Pergamon, Elsevier, New York.

Kolattukudy, P.E. & D.L. Gamble. 1995. *Nectria haematococca. In:* Pathogenesis and Host Specificity in Plant Diseases—Histological, Biochemical, Genetic and Molecular Bases, Vol. II Eukaryotes, eds. K. Kohmoto, U.S. Singh & R.P. Singh, pp. 83–101, Pergamon, Elsevier, New York.

Kolattukudy, P.E., D. Li, C.-S. Hwang & M.A. Flaishman. 1995. Host signals in fungal gene expression involved in penetration into the host. Can. J. Bot. **73** (suppl):S1160–S1168.

Koltin, Y., J.R. Raper & G. Simchen. 1967. Genetics of the incompatibility factors of *Schizophyllum commune:* the B factor. Proc. Natl. Acad. Sci USA **57**:55–62.

Kranz, J. & J. Rotem, eds. 1988. Experimental Techniques in Plant Disease Epidemiology. Springer-Verlag, Berlin.

Kuc, J. & C. Preisig. 1984. Fungal regulation of disease resistance mechanisms in plants. Mycologia **76**:767–784.

Kurjan, J. 1993. The pheromone response pathway in *Saccharomyces cerevisiae.* Annu. Rev. Genet. **27**:147–179.

Kwon-Chung, K.J. & J.E. Bennett. 1992. Medical Mycology. Lea and Febiger, Philadelphia.

Lawrey, J.D. 1993. Chemical ecology of *Hobsonia christiansenii,* a lichenicolous hyphomycete. Am. J. Bot. **80**:1109–1113.

Lawrey, J.D. 1995. The chemical ecology of lichen mycoparasites: a review. Can. J. Bot. **73**(suppl. 1):S603–S608.

Lawrey, J.D., A.Y. Rossman & R. Lowen. 1994. Inhibition of selected hypocrealean fungi by lichen secondary metabolites. Mycologia **86**:502–506.

Lee, L.S., P. Bayman & J.W. Bennett. 1992. Mycotoxins. *In:* Biotechnology of Filamentous

Fungi—Technology and Products, eds. D.B. Finkelstein & C. Ball, pp. 463–503, Butterworth-Heinemann, Boston.

Leite, B. & R.L. Nicholson. 1992. Mycosporine-alanine: a self-inhibitor of germination from the conidial mucilage of *Colletotrichum graminicola.* Exp. Mycol. **16**:76–86.

Levin, M.A., R.J. Seidler & M. Rogul. eds. 1992. Microbial Ecology—Principles, Methods and Applications, McGraw-Hill, New York.

Linderman, R.G. 1994. Role of VAM fungi in biocontrol. *In:* Mycorrhizae and Plant Health, eds. F.L. Pfleger & R.G. Linderman, pp. 1–25, APS Press, St. Paul, MN.

Longton, R.E. 1988. The Biology of Polar Bryophytes and Lichens. Cambridge University Press, Cambridge.

McCoy, C.W., R.A. Samson & D.G. Boucias. 1988. Entomogenous fungi. *In:* CRC Handbook of Natural Pesticides, Vol. V Microbial Insecticides, Part A Entomogenous Protozoa and Fungi, ed. C.M. Ignoffo, pp. 151–243, CRC Press, Boca Raton, FL.

McMorris, T.C. 1989. Analysis of brassinosteroids, antheridiol, and oogoniols. *In:* Analysis of Sterols and Other Biologically Significant Steroids, eds. W.D. Nes & E.J. Parish, pp. 61–79, Academic Press, New York.

McMorris, T.C. & R. White. 1977. Biosynthesis of the oogonials, steroidal sex hormones of *Achlya:* the role of fucosterol. Phytochemistry **16**:359–362.

McMorris, T.C., D.O. Toft, S. Moon & W. Wang. 1993. Biological response of the female strain *Achlya ambisexualis* 734 to dehydro-oogoniol and analogues. Phytochemistry **32**:833–837.

Machlis, L. 1958. Evidence for a sexual hormone in *Allomyces*. Physiol. Plant. **11**:181–192.

Machlis, L. 1969. Zoospore chemotaxis in the watermold *Allomyces.* Physiol. Plant. **22**:126–139.

Macko, V. 1981. Inhibitors and stimulants of spore germination and infection structure formation in fungi. *In:* The Fungal Spore: Morphogenetic Controls, eds. G. Turian & H.R. Hohl, pp. 565–584, Academic Press, New York.

Magurran, A.E. 1988. Ecological Diversity and its Measurement. Princeton University Press, Princeton, NJ.

Maniara, G., R. Laine & J. Kuc. 1984. Oligosaccharides from *Phytophthora infestans* enhance the elicitation of sesquiterpenoid stress metabolites by arachidonic acid in potato. Physiol. Plant Pathol. **24**:177–186.

Marasas, W.F.O. & P.E. Nelson. 1987. Mycotoxicology—Introduction to the Mycology, Plant Pathology, Chemistry, Toxicology, and Pathology of Naturally Occurring Mycotoxicoses in Animals and Man. Pennsylvania State University Press, University Park and London.

Marshall, M.A. & W.E. Timberlake. 1991. *Aspergillus nidulans wetA* activates spore-specific gene expression. Mol. Cell. Biol. **11**:55–62.

Martin, M.M. 1987. Invertebrate-Microbial Interactions. Comstock Publishing Associates, Ithaca, NY.

Martin, M.M. 1992. The evolution of insect-fungus associations: from contact to stable symbiosis. Am. Zool. **32**:593–605.

Mendgen, K., A. Schneider, M. Sterk & W. Fink. 1988. The differentiation of infection structures as a result of recognition events between some biotrophic parasites and their hosts. J. Phytopathol. **123**:259–272.

Mercure, E.W., B. Leite & R.L. Nicholson. 1994. Adhesion of ungerminated conidia of *Colletotrichum graminicola* to artificial hydrophobic surfaces. Physiol. Molec. Plant Pathol. **45**:421–440.

Mirsch, O., J. Schmidt, G. Sembdner & K. Schreiber. 1989. Jasmonic acid-like substances from the culture filtrate of *Botryodiplodia theobromae.* Phytochemistry **28**:1303–1305.

Mosse, B. & C. Hepper. 1975. Vesicular-arbuscular mycorrhizal infections in root organ cultures. Physiol. Plant Pathol. **5**:215–223.

Muzzarelli, R.A.A., P. Ilari, W. Xia, M. Pinotti & M. Tomasetti. 1994. Tyrosinase-mediated quinone tanning of chitinous materials. Carbohyd. Polym. **24**:295–300.

Nash, T.H. & V. Wirth. eds. 1988. Lichens, Bryophytes and air quality. J. Cramer, Berlin.

Nicholson, R.L. 1996. Adhesion of fungal propagules. *In:* Histology, Ultrastructure and Molecular Cytology of Plant-Microorganism Interactions, eds. M. Nicole & V. Gianinazzi-Pearson, Kluwer Academic Publishers, Dordrecht & London.

Nicolson, R.L. & L. Epstein. 1991. Adhesion of fungi to the plant surface: prerequisite for pathogenesis. *In:* The Fungal Spore and Disease Initiation in Plants and Animals, eds. G.T. Cole & H.C. Hoch, pp. 3–23, Plenum, New York.

Okuda, T., M. Yanagishawa, F. Fujimori, Y. Nishizuka, Y. Takehana & M. Sugiyama. 1995. New isolation methods and polymerase chain reaction strain discrimination techniques for natural products screening programs. Can. J. Bot. **73** (suppl. 1):S946–S954.

Parbery, I.H. & J.F. Brown. 1986. Sooty molds and black mildews in extra-tropical rainforests. *In:* Microbiology of the Phyllosphere, eds. Fokkema, N.J. & J. van den Heuvel, pp. 101–120, Cambridge University Press, Cambridge.

Parker, J.E., W. Knogge & D. Scheel. 1991. Molecular aspects of host-pathogen interactions in *Phytophthora. In: Phytophthora,* eds. J.A. Lucas, R.C. Shattock, D.S. Shaw & L.R. Cooke, pp. 90–103, Cambridge University Press, Cambridge.

Petersen, J.L. 1959. Relations between inoculum density and infection of wheat by uredospores of *Puccinia graminis* var. *tritici.* Phytopathology **49**:607–614.

Petersen, E.E., M.J. Semon, J.L. Kerwin & J.M. Brower. 1997. Regulation of attachment, germination, and appressorium formation by zoospores of *Lagenidium giganteum* and related oomycetes by chitin, chitosan, and catecholamines. Protoplasma **197**:96–110.

Pielou, E.C. 1975. Ecological Diversity. Wiley Interscience, New York.

Pirozinsky, K.A. & Y. Dalpé. 1989. Geological history of the Glomaceae with particular reference to mycorrhizal symbiosis. Symbiosis **7**:1–36.

Pirozynski, K.A. & D.L. Hawksworth. eds. 1988. Coevolution of Fungi with Plants and Animals. Academic Press, London.

Podila, G.K., L.M. Rogers & P.E. Kolattukudy. 1993. Chemical signals from avocado surface wax trigger germination and appressorium formation in *Colletotrichum gloeosporioides.* Plant Physiol. **103**:267–272.

Pommerville, J. 1987a. *Allomyces macrogynus*—sexual fusion. *In:* Zoosporic Fungi in

Teaching and Research, eds. M.S. Fuller & A. Jaworski, pp. 46–47, Southeastern Publishing Corporation, Athens, GA.

Pommerville, J. 1987b. Sirenin. *In:* Zoosporic Fungi in Teaching and Research, eds. M.S. Fuller & A. Jaworski, pp. 274–275, Southeastern Publishing Corporation, Athens, GA.

Pratt, B.H., J.H. Sedgley, W.A. Heather & C.J. Shepherd. 1972. Oospore production in *Phytophthora cinnamomi* in the presence of *Trichoderma koningii.* Austr. J. Biol. Sci. **25**:861–863.

Raju, N.B. 1992. Genetic control of the sexual cycle in *Neurospora.* Mycol. Res. **96**:241–262.

Raper, J.R. 1939. Sexual hormones in *Achlya.* I. Indicative evidence for a hormone coordinating mechanism. Am. J. Bot. **26**:639–650.

Raper, J.R. 1940. Sexual hormones in *Achlya.* II. Distance reactions, conclusive evidence for a hormonal co-ordinating mechanism. Am. J. Bot. **27**:162–173.

Raper, J.R. 1966. Genetics of Sexuality in Higher Fungi. Ronald Press Company, New York.

Reese, E.T. & A. Maguire. 1969. Surfactants as stimulants of enzyme production by microorganisms. Appl. Microbiol. **17**:242–245.

Reinert, J. & H. Binding. eds. 1986. Differentiation of Protoplasts and Transformed Plant Cells. Springer-Verlag, Berlin.

Ricker, K.E. & R.M. Bostock. 1992. Evidence for release of the elicitor arachidonic acid and its metabolites from sporangia of *Phytophthora infestans* during infection of potato. Physiol. Mol. Plant Pathol. **41**:61–72.

Riehl, R.M., D.O. Toft, M.D. Meyer, G.L. Carlson & T.C. McMorris. 1984. Detection of a pheromone-binding protein in the aquatic fungus *Achlya ambisexualis.* Exp. Cell Res. **153**:545–549.

Rippon, J.W. 1988. Medical Mycology: The Pathogenic Fungi and the Pathogenic Actinomycetes. Saunders, Philadelphia.

Robinson, P.W. & C.F. Hodges. 1976. An inoculation apparatus for evaluation of *Bipolaris sorokiniana* lesion development on progressively older leaves of *Poa pratensis.* Phytopathology **66**:360–362.

Ross, I.K. 1979a. Biology of the Fungi. McGraw-Hill, New York.

Schein, R.D. 1964. Design, performance, and use of a quantitative inoculator. Phytopathology **54**:509–513.

Schots, A., F.M. Dewey & R. Oliver. eds. 1994. Modern Assays for Plant Pathogenic Fungi. CAB International, Oxford, UK.

Schreiner, R.P. & R.T. Koide. 1993a. Antifungal compounds from the roots of mycotrophic and non-mycotrophic plant species. New Phytol. **123**:99–105.

Schreiner, R.P. & R.T. Koide. 1993b. Stimulation of vesicular-arbuscular mycorrhizal fungi by mycotrophic and nonmycotrophic plant root systems. Appl. Environ. Microbiol. **59**:2750–2752.

Schuerger, A.C., D.J. Mitchell & D.T. Kaplan. 1993. Influence of carbon source on attachment and germination of macroconidia of *Fusarium solani* f. sp. *phaseoli* on roots of *Vigna radiata* grown in hydroponic nutrient solution. Phytopathology **83**:1171–1177.

Schultz, J., Ferguson, B., and Sprague, Jr., G.F. 1995. Signal transduction and growth control in yeast. Curr. Opin. Gen. Develop. **5**:31–37.

Shen, C.-Y., L.A. Bower, D.C. Erwin & P.H. Tsao. 1983. Formation of sex organs in the A^1 mating type of *Phytophthora infestans* induced chemically by A^2 isolates of other species of *Phytophthora.* Can. J. Bot. **61**:1462–1466.

Sherwood, W.A. 1966. Evidence for a sexual hormone in the water mold *Dictyuchus.* Mycologia **58**:215–220.

Siegel, J.P. & J.A. Shadduck. 1992. Testing the effects of microbial pest control agents on mammals, *In:* Microbial Ecology—Principles, Methods, and Applications, eds. M.A. Levin, R.J. Seidler & M. Rogul, pp. 745–759, McGraw-Hill, New York.

Siegel, M.R., G.C.M. Latch & M.C. Johnson. 1987. Fungal endophytes of grasses. Ann. Rev. Phytopathol. **25**:293–315.

Siegel, M.R., G.C.M. Latch, L.P. Bush, N.F. Fannin, D.D. Rowan, B.A. Tapper, C.W. Bacon & M.C. Johnson. 1991. Alkaloids and aphid response in grasses infected with fungal endophytes. J. Chem. Ecol. **16**:3301–3315.

Smith, J.E. ed. 1983. Fungal Differentiation. Marcel Dekker, Inc., New York.

Sparrow, F.K. Jr. 1960. Aquatic Phycomycetes. University of Michigan Press, Ann Arbor.

Spellig, T., M. Bölker, F. Lottspeich, R.W. Frank & R. Kahmann. 1994. Pheromones trigger filamentous growth in *Ustilago maydis.* EMBO J. **13**:1620–1627.

Sprague, G.F. Jr. & J. Thorner. 1983. Pheromone response and signal transduction during the mating process of *Saccharomyces cerevisiae. In:* The Molecular and Cellular Biology of the Yeast *Saccharomyces:* Gene Expression, (eds.) J.N. Strathern, E.W. Jones & J.R. Broach, Vol. II, pp. 657–744, Cold Spring Harbor Laboratory Press.

Sprague, G.F. Jr., L.C. Blair & J. Thorner. 1983. Cell interactions and regulation of cell type in the yeast *Saccharomyces cerevisiae.* Ann. Rev. Microbiol. **37**:623–660.

Springer, M.L. 1993. Genetic control of fungal differentiation: the three sporulation pathways of *Neurospora crassa.* BioEssays **15**:365–374.

St. Leger, R.J. 1993. Biology and mechanisms of invasion of deuteromycete fungal pathogens. *In:* Parasites and Pathogens of Insects, Vol. 2, eds. N.C. Beckage, S.N. Thompson & B.A. Federici, pp. 211–229, Academic Press, New York.

Starmer, W.T., P.F. Ganter, V. Aberdeen, M.-A. Lachance & H.J. Pfaff. 1987. The ecological role of killer yeasts in natural communities of yeasts. Can. J. Microbiol. **33**:783–796.

Stumm, C., J.M.H. Hermans, E.J. Middlebeek, A.F. Croes & G.J.M. De Vries. 1977. Killer sensitive relationships in yeasts from natural habitats. Antonie van Leeuwenhoek **43**:125–128.

Sussman, A.S. 1976. Activators of fungal spore germination. *In:* The Fungal Spore, eds. D.J. Webster & W.M. Hess, pp. 99–110, John Murray, London.

Sussman, A.S. & H.O. Halvorson. 1966. Spores: Their Dormancy and Germination. New York, Harper Row.

Swisher, R. & G.C. Carroll. 1980. Fluorescein diacetate hydrolysis as an estimator of microbial biomass on coniferous needle surfaces. Microbial Ecol. **6**:217–226.

Szaniszlo, P.J., M.S. Kang & E. Cabib. 1985. Stimulation of β-1,3-glucan synthetase of various fungi by nucleoside triphosphates: generalized regulatory mechanism for cell wall biosynthesis. J. Bacteriol. **161**:1188–1194.

Tavares, I.I. 1985. Laboulbeniales, J. Cramer, Braunschweig, Germany.

Taylor, R.S., N.P. Manandhar & G.H.N. Towers. 1995. Screening of selected medicinal plants of Nepal for antimicrobial activities. J. Ethnopharmacol. **46**:153–159.

Terzi, M., R. Cella & A. Falavigna. eds. 1995. Current Issues in Plant Molecular and Cellular Biology. Proceedings of the VIIIth International Congress on Plant Tissue and Cell Culture. Kluwer, Dordrecht.

Thaxter, R. 1931. Contribution towards a monograph of the Laboulbeniaceae. Part V. Mem. Am. Acad. Arts Sci. **16**:1–435.

Thimann, K.V. 1935. On the plant growth hormone produced by *Rhizopus suinus*. J. Biol. Chem. **109**:279–291.

Thines, E., T. Dauβmann, M. Semar, O. Sterner & H. Anke. 1995. Fungal melanin biosynthesis inhibitors: introduction of a test system based on the production of dihydroxynaphthalene (DHN) melanin in agar cultures. Z. Naturforsch. **50c**:1–7.

Tommerup, J.C. 1983. Spore dormancy in vesicular-arbuscular mycorrhizal fungi. Trans. Br. Mycol. Soc. **81**:37–45.

Towers, G.H.N. 1980. Interactions of light with phytochemicals in some natural and novel systems (a review). Prog. Phytochem. **6**:183–202.

Traquair, J.A. 1995. Fungal biocontrol of root diseases: endomycorrhizal suppression of cylindrocarpon root rot. Can. J. Bot. **73** (suppl. 1):S89–S95.

Tuite, J. 1969. Plant Pathological Methods: Fungi and Bacteria. Burgess Publishing Company, Minneapolis.

Turian, G. & H.R. Hohl. eds. 1981. The Fungal Spore: Morphogenetic Controls. Academic Press, London.

van Etten, J.L., K.R. Dahlberg & G.M. Russo. 1983. Fungal spore germination. *In:* Fungal Differentiation: A Contemporary Synthesis, ed. J.E. Smith, pp. 235–266, Marcel Dekker, New York.

Walker, G.M., A.H. McLeod & V.J. Hodgson. 1995. Interactions between killer yeasts and pathogenic fungi. FEMS Microbiol. Lett. **127**:213–222.

Weber, N.A. 1955. Pure cultures of fungi produced by ants. *Science* **121**:109.

Weber, N.A. 1966. Fungus-growing ants. Science **153**:587–604.

Webster, J. 1970. Coprophilous fungi. Trans. Br. Mycol. Soc. **54**:161–180.

Werner, D. 1992. Symbiosis of Plants and Microbes. Chapman & Hall, London.

Wickerham, L.J. 1958. Sexual agglutination of heterothallic yeasts in diverse taxonomic areas. Science **158**:1504–1505.

Woods, D.R. & E.A. Bevan. 1968. Studies on the nature of the killer factor produced by *Saccharomyces cerevisiae*. Heredity **21**:121–130.

Wynn, W.K. 1976. Appressorium formation over stomates by the bean rust fungus: Response to a surface contact stimulus. Phytopathol. **66**:136–146.

Wynn, W.K. 1981. Tropic and taxic responses of pathogens to plants. Ann. Rev. Phytopathology, **19**:237–255.

Yamaguchi, I. & Y. Kubo. 1992. Target sites of melanin biosynthesis inhibitors. *In:* Target Sites of Fungicide Action, ed. W. Köller, pp. 101–118, CRC Press, Boca Raton, FL.

Zaki, A.I., G.A. Zentmeyer, J.J. Sims & N.T. Keen. 1983. Stimulation of sexual reproduction in the A2 mating type of *Phytophthora cinnamomi* by oleic acid and lipids from avocado roots. Phytopathology **73**:199–203.

Zapf, F., M. Hossfeld, H. Anke, R. Velten & W. Steglich. 1995. Darlucins A and B, isocyanide antibiotics from *Sphaerellopsis filum* (*Darluca filum*). J. Antibiotics **48**:36–41.

Zentmeyer, G.A. 1979. Stimulation of sexual reproduction in the A^2 mating type of *Phytophthora cinnamomi* by a substance in avocado roots. Phytopathology **69**:129–1131.

4

Bioassays for Allelopathy in Terrestrial Plants

John T. Romeo and Jeffrey D. Weidenhamer

4.1.	Introduction	179
4.2.	Case studies illustrating appropriate bioassays	184
	4.2.1. Bracken fern	184
	4.2.1.1. Vegetation pattern	184
	4.2.1.2. Environmental factors	185
	4.2.1.3. Allelopathy	186
	4.2.2. The California chaparral	188
	4.2.2.1. Role of allelopathy	188
	4.2.2.2. Physical and biotic factors	189
	4.2.2.3. Bioassays to establish mechanism of allelochemical transport	189
	4.2.3. Rye	191
	4.2.3.1. Greenhouse experiments and bioassays	192
	4.2.3.2. Chemical studies	193
	4.2.4. The Florida scrub	195
	4.2.4.1. Vegetation pattern and soil	195
	4.2.4.2. Resource competition	195
	4.2.4.3. Allelopathy	196
	4.2.4.4. Chemical studies	196
	4.2.5. Crowberry	198
	4.2.5.1. Laboratory bioassays	198
	4.2.5.2. Greenhouse and field studies	199
4.3.	Density-dependent phytotoxicity	200
4.4.	Practical considerations	202
4.5.	Acknowledgments	205
4.6.	References	205

4.1. Introduction

Using the broadest definition, allelopathy encompasses biochemical interactions among plants at all levels of complexity, including microorganisms (Molisch

1937). Both inhibitory and stimulatory reactions originally were considered part of this concept. Since the appearance of the first major compendium on allelopathy (Rice 1974), however, the term has commonly been accepted to mean any direct or indirect harmful effect by one plant (including microorganisms) on another through the production of chemical compounds released into the environment. Although Rice reverted to the Molisch definition in his second edition (Rice 1984), the term continues to be associated with negative effects among most workers. Those working in allelopathy are often involved in the search for potential herbicides and growth inhibitors, and commonly ignore observed positive effects, publishing only the inhibitory data. The narrower definition is perhaps unwise, since most compounds are both stimulatory and inhibitory depending on the concentration used.

Certainly no area of plant science has provoked as much controversy as the study of allelopathy. Despite elegant and exhaustive studies demonstrating the phenomenon in the California chaparral by C.H. Muller and his students in the 1950s, 1960s, and early 1970s (e.g. C.H. Muller 1965, 1966; Muller & Chou 1972), and those by Rice and his students working in Midwestern U.S. old fields in the 1960s, 1970s, and early 1980s (summarized in Rice 1974, 1984), much criticism has followed the field. Often, what at first blush had appeared to be classic allelopathy in the field, with further scrutiny, was attributed to other factors such as animal herbivory (Bartholomew 1970; Halligan 1973). Harper (1977) harshly criticized the pioneering work of many, including the classic studies of Wilson and Rice (1968), maintaining that alternative interpretations were possible, such as growth inhibition as the result of the depletion of available soil nitrogen by microorganisms, following the addition of plant debris to soil. A single paper by Kaminsky (1981) in which he showed that much, if not all, of the allelopathic activity of *Adenostoma fasciculatum*, one of Muller's chaparral shrubs, was due to microbial activity, had a significant negative impact on the perception of allelopathy. Connell, as recently as 1990, claimed that no published study had ever demonstrated direct chemical interference in the field.

It should be noted that similar difficulties plague research on other mechanisms of plant-plant interference. It is commonly assumed that competition for light, nutrients, water, and other resources is responsible for differential growth of plants, species distribution, and succession, without concrete data to verify that this is the case. Most studies of competitive effects do not exclude the potential for chemical interference. It is rarely recognized that sensitivity to chemical toxins may be dependent on nutrient status and other environmental factors, and that the production of chemical toxins by source plants (and hence potential for chemical interference) may be greatly influenced by nutrient levels and other environmental factors (Einhellig 1996). Allelopathy may exert subtle but significant effects on plants and communities. For example, chemical interference with mycorrhizal growth may manifest itself as a reduced competitive ability for soil nutrients and water. Effects of allelochemicals on the nitrogen cycle (and thereby on succession and community structure) have been the subject of extensive study (Rice & Pancholy 1974; White 1991). It is likely that the interaction of competitive and allelopathic effects (Nilsson 1994) is the norm rather than the exception.

Clearly, for those looking for ways to confirm the existence of allelopathy in nature, the challenge of developing appropriate field bioassays is a daunting one. Many are unprepared for the task, and the burden of establishing proof of a significant role for allelopathy in field situations has been tackled by few, due to the inherent difficulties involved in performing the studies (see, for example, Rice & Pancholy 1974; Gliessman & Muller 1978; Richardson & Williamson 1988; Weidenhamer & Romeo 1989; Fischer et al. 1994; Nilsson 1994). A collaboration between botanists and chemists is almost essential if the work is to be done properly, and studies must be long-term. In the United States, the number of workers has dramatically decreased during the past 10 years. In the developing world, interest in allelopathy has remained high because of the recognition that agroecological applications of allelopathy may provide alternatives to synthetic herbicides for weed control and may be more compatible with traditional agricultural systems (Anaya et al. 1987a, 1987b, 1990, 1995, 1996; Chou et al. 1989; Pereda-Miranda et al. 1993; Calera et al. 1995; Inderjit & Dakshini 1996; Narwal & Tauro 1996; Anaya & Pelayo-Benavides 1997).

Literally hundreds of plant extracts have been tested for bioactivity and shown to be active. In fewer cases, compounds have been isolated from these extracts and tested, and individually they often have shown activity. Scores of commercially available natural products also have been tested in a myriad of bioassays. Paper after paper has been written purporting that many of them are allelopathic. Such studies are done quickly and, unfortunately, often carelessly. The common flaw of most of these laboratory reports is that they fail to consider such obvious essentials as the need to test for activity of chemicals at concentrations in which they occur in nature; use appropriate bioassay species that occur in the field, rather than the traditional, easily obtained crop species such as lettuce or tomato; and test mixtures of compounds that simulate nature's release of chemicals into the soil, and in which synergistic effects are often commonplace. Thus, because of such neglect, there is both garbage and confusion in the literature. Many reported effects are not real, and probably many real ones have been overlooked simply because the bioassay used was faulty.

A sharp distinction must be drawn between bioassay methods used to screen for potentially herbicidal molecules and those used to study allelopathy. If one is interested only in identifying potential herbicides, standardized bioassays using selected crop and weed species under specified protocols are highly appropriate. Such assays are also useful in gauging the relative phytotoxicity of compounds. They are not adequate, however, for demonstration of allelopathy!

The complexity of allelopathic interactions enormously complicates the choice of an appropriate bioassay. The pitfalls of lab and greenhouse bioassays for allelopathy were clearly demonstrated by Stowe (1979) who used nine commonly employed bioassays to test the seven most abundant species in an Illinois old field. Numerous statistically significant cases of inhibition were found, but there was no correlation with the distribution of species in the field. Stowe concluded:

> that the types of allelopathy which were tested by these bioassays were not demonstrably effective under field conditions . . . that perhaps any species can be shown to have allelopathic properties in bioassays, and that bioassays may, for many communities, have no ecological meaning.

For those trying to conduct *ecologically* meaningful bioassays, there are several challenges. One must consider

1. The possible *transport mechanism* of the allelochemical to the target plant. Is it through root exudation, precipitation throughfall, or volatilization?
2. The *mechanism(s) of inhibition.* Does the phytotoxin inhibit or delay germination, affect seedling growth directly by disrupting nutrient uptake or water balance, or by toxic effects on essential metabolic processes such as photosynthesis? Indirect effects on mycorrhizal or *Rhizobium* symbioses may also be determining factors on plant growth in the field, but will not be tested for by most allelopathy bioassays. Such concerns emphasize the crucial importance of observations of patterns and symptoms of growth inhibition in selecting appropriate bioassays.
3. *Density-dependent effects* in laboratory, greenhouse, and field bioassays. These, proposed almost a decade ago by Weidenhamer et al. (1987, 1989), are almost never considered. Simply put, the hypothesis states that growth reductions (phytotoxicity) observed at low plant densities will diminish as the density of the test species is increased.
4. The *role of environmental factors* in allelopathic inhibition. Moisture and nutrient stress, high temperature, light intensity, plant density, and the presence of pathogens can all alter the response of target plants to allelochemicals and also the production or release rates of allelochemicals by source plants. Chemicals that are not toxic at certain concentrations in a laboratory assay may, in fact, be significantly inhibitory under stressful field conditions.
5. The *transformation of plant allelochemicals* by environmental and/or microbial action. The creation of less toxic or more toxic compounds is always a possibility.

The last two concerns are an issue for bioassay-guided fractionation of plant extracts; it is possible that such an approach may not result in identification of the allelopathic compound(s).

Many authors have identified the need for both standardization and greater rigor of field and laboratory tests for allelopathy. The plea by Fuerst and Putnam (1983), in which they attempted to put methodological studies for proving the hypothesis of allelopathy on firmer ground by proposing criteria modeled after Koch's postulates in microbiology, is already 15 years old. The basic four tenets they proposed are identify and quantify specific symptoms of interference; isolate, identify, and synthesize the toxin, characterizing its biological activity through bioassays; simulate the interference through applications mimicking natural rates; and quantify the amount of toxin released to the environment and taken up by the target plant. Many workers tend to work on one or two aspects, and then overinterpret their work with outlandish claims that they have demonstrated the phenomenon.

The Koch's postulates approach has not been without its critics. Williamson (1990) argued that the analogy to microbiology has several flaws. Unlike microorganisms, allelochemicals do not reproduce, so any application of a chemical cannot merely be a onetime inoculation. In addition, allelochemicals can degrade to more or less toxic substances. Furthermore, whereas there is a direct correspondence between a disease and infection with a microorganism, allelochemicals may interact in their effects. Williamson, like Harper (1977), is skeptical that experiments can be designed to conclusively demonstrate allelopathy.

Long-term investigations of allelopathy in the Florida scrub by many investigators, including ourselves, have borne out the wisdom of C.H. Muller (1953), who argued that a description of the factors controlling plant communities "will be at least as intricate as the situation it seeks to describe." While the unambiguous proof promised by a Koch's postulates approach may not be attainable, the challenge of developing compelling evidence for chemical interference amid other competing hypotheses, remains.

A number of important recent papers attempted to put allelopathy methodology on a more solid footing (see Leather & Einhellig 1986; Weidenhamer 1996; Blum 1998 for both philosophical and practical reviews). Having certain basic tenets at hand, in a volume such as this one, may serve a useful purpose to those now working or those contemplating work in the field. It is in this spirit that we attempt this review. We drew heavily from these recent papers, and we have singled out several case studies to examine the role of appropriate bioassays in long-term ecological studies of allelopathy. The review is not intended to be comprehensive, and many carefully done papers are omitted here. Thus, with no little trepidation and recognizing that given the complexities of this phenomenon, some of the arguments we make may later be found to be flawed, we summarize our thoughts in this regard.

4.2. Case Studies Illustrating Appropriate Bioassays

4.2.1. Bracken Fern

The major challenge involved in establishing credibility for the phenomenon of allelopathy is the sheer number and variety of experiments and bioassays that must be performed. Whenever allelopathy is suspected, it is always important to eliminate the obvious alternate explanations before embarking on months, possibly years, of laboratory chemical work and field tests. An early classic paper, entitled "The Allelopathic Mechanisms of Dominance in Bracken (*Pteridium aquilinum*) in Southern California," that remains a model for this process is Gliessman and Muller (1978). The authors systematically eliminate the various factors that could account for interference by bracken on the growth of associated plants, and then proceed to make the case for allelopathy.

Bracken fern, *Pteridium aquilinum*, exerts a strong dominance over its associated plants worldwide. The plant is morphologically heterogeneous, but its phytotoxic effect is widespread. Other plants are often severely growth inhibited or eliminated entirely from dense stands of the fern. The phenomenon extends from the northern limits of the boreal forests to the tropics. A number of factors have been implicated, including competitive interference for light, water, and nutrients. As early as 1917, however, Jeffries showed that associated plants were inhibited by the breakdown of bracken litter. Gliessman and Muller (1972) showed that inhibitory compounds are washed out of fronds in rain, fog, or dew drip, the mechanism of toxic release presumably differing under differing climate regimes. In the study discussed here, Gliessman and Muller (1978) examined the dominance of bracken in one geographic situation (coastal California) to determine what factors might be involved. The paper illustrates beautifully how the "elimination of the obvious" ought to proceed in allelopathic studies. The work is briefly summarized.

They chose two sites—one in the Santa Ynez Mountains of California (2200′) near Santa Barbara, and another on Santa Cruz Island, off Santa Barbara (2300′). Both sites have a Mediterranean climate with winter rains (November to March) followed by several dry months. Annual rainfall is 25 cm, and other environmental factors are similar. The areas studied are densely covered by oak woodlands in which meadows occur. These are vegetated by a diverse mixture of annual grass and dicotyledonous herbs. In some, bracken has established dominance and dramatically altered the species composition.

4.2.1.1. Vegetation Pattern

The pattern of bracken dominance at both sites was quantified by counting the number of species and determining their dry weights. Ten 100-cm-diameter plots in areas of bracken dominance and in adjacent grasslands were sampled on 10 different dates. Data were pooled providing nearly 200 samples of both vegetation types. Based on species composition alone, there were no differences between fern-covered areas and open grasslands. The number of individuals of each species encountered, however, was significantly higher in the grassland, a fourfold difference being found. Dry weights of the species also showed significantly higher yields in the grasslands. The most common species were identified. The pattern of bracken communities was similar to that found worldwide. Dense mature fronds dominated the stands with a building phase on the edge. Deep fern rhizomes invaded into the grassland as much as 75 cm/year where fronds emerged in the pioneer phase.

4.2.1.2. Environmental Factors

4.2.1.2.1. Soil

Soil profile pits were dug to a depth of 1 m in four locations in both grassland and fern stands. Because of homogeneity at lower depths and lack of interaction between herbs and ferns, analyses were restricted to the upper 60 cm. Samples were taken at depths of 0–10, 10–20, 20–40, and 40–60, with eight samples analyzed from each depth. Soils were checked for five characteristics: *texture* (particle size, percentage sand, silt and clay); *pH*; total *organic matter* (by loss on ignition); presence of *essential elements* (four samples each at two sites)—total N, PO_4, K, Mg, Ca, Fe, Zn, Mn, and Cu; and *soil moisture* (gravimetric determinations of available moisture made throughout the growing season in the upper 10 cm of soil with silting coefficients computed). While some minor differences were found among these various parameters, none was significant enough to account for variation in vegetation patterns.

4.2.1.2.2. Light

The factor most often evoked to explain bracken dominance, light, was measured at ground level, inside bracken stands, and outside in open grassland, at regular intervals throughout the growing season and during the dry season when fronds were dead. During the growing season, when allelopathy would be operating, there was actually more light in bracken stands than in grassland. By the time shading could become a factor, the growing season was essentially over and the herbs had completed their life cycles. Additionally, as a check, shade frames, devices constructed to provide artificial reduction of light (3 × 3-m wooden frames with upper surfaces 0.6 m above ground and covered with a layer of ⅜-in. hardware cloth, fiberglass window screen, and aluminum window screen), were placed in open grasslands. This degree of shading approximated that which occurred inside the bracken stands at the peak of herb development. Herb species were monitored under the devices. No differences were found in shaded versus nonshaded areas.

4.2.1.2.3. Herbivorous Mammals

Bracken stands could be expected to provide cover to support large populations of herbivorous mammals that might, by selective grazing behavior, be responsible for a paucity of individual plants within the stands. To estimate the number of mammals, live traps were set in open grasslands, on the edge of the bracken stands, and within the stands. More animals were captured within the bracken, and thus a series of experiments were performed to determine if animals were having an effect on plants.

Petri dishes with 50 or 100 seeds covered with 3 mm of soil were placed in the area. Dishes were either covered completely with small mesh enclosures at ground level or left uncovered. Some of the covered dishes were elevated on stakes. The results indicated effective amounts of seed predation on a few of the species, especially *Bromus rigidus*. The most common herb species in fern stands, however, *Festuca megalura*, was not touched, nor were several other common ones. To test for grazing of young green shoots, small plastic greenhouse trays (20 × 30 × 5 cm) were half-filled with a water-retaining soil mix and planted with at least 30 seeds. Trays were kept in the greenhouse until herbs averaged 5–10 cm. Half were placed in the bracken, the others in the grassland. Two species were essentially totally removed from the bracken trays while they were little touched in the grassland location. The most common herb species, however, were untouched.

A bracken stand known for lack of rodents on Santa Cruz Island was further studied. Traps were set, and no animals were captured. Trays were set out with baited seeds and they remained untouched. Herbaceous species counts at the island site indicated that the major species excluded by mammals on the mainland, *Bromus rigidus*, was a major component of vegetation there, being by far the major species encountered. The obvious conclusion is that on the mainland, *B. rigidus* is excluded by mammals. With few exceptions, other species that were not much grazed by herbivores, occurred less frequently in bracken stands. Thus, herbivorous mammals play a role in excluding some herbs from bracken stands, but some other factor is more important in maintaining the complete vegetation pattern.

The above experiments, analyses, and bioassays give an idea of the many factors that must be excluded to establish a hypothesis of allelopathy. In the second part of their study, Gleissman and Muller proceeded to make the case.

4.2.1.3. Allelopathy

A series of eight experiments were performed that established the source of toxic compounds in the dominant plant (bracken); differential susceptibility of associated species to toxic effects; the probable route of the toxins into the soil; toxicity in the natural and toxin-modified soils; and reversal of the phenomenon by removal of the toxic source. A brief summary follows.

Extracts were made from live and dead bracken fronds soaked in distilled water. The effects of these on radicle growth of several associated herb species were tested using standard sponge bioassays developed by McPherson and Muller (1969). Toxicity was identified as coming not from live green fronds, or very much from yellow dying ones, but rather from the dead fronds. Dead fronds extracted before leaching by rainfall were more toxic than those leached. Toxicity was severely reduced in those exposed to one rainy season. Raindrip was collected beneath fronds in the field throughout the season. After filtering, the initial raindrip was found to be most effective in causing growth inhibition.

To ascertain whether or not osmotic effects of extracts were responsible for the inhibitory activity, manitol solutions were made up to osmotic concentrations slightly in excess of frond extracts. Osmotic concentrations were shown to be in part responsible for some of the bioassay inhibition observed, but a significant component was due to organic inhibition.

To test for differential effects on different species, extracts of fronds collected before rains were made by soaking them for 2 h in distilled water. Variation in effects on herb species were observed: (1) the species most abundant within bracken stands was the most tolerant, (2) the rodent-excluded species was somewhat tolerant, and (3) several species least frequently encountered inside bracken stands were the most susceptible.

Soils were collected from underneath bracken stands following the first rain of the season. The wetted upper 2 to 3-cm layer was screened and placed in storage dishes. Grassland soils similarly collected served as controls. Seeds were planted in these soils, watered with distilled water, and the dishes sealed with parafilm and stored. All species tested, except for *Bromus rigidus*, were inhibited in radicle growth.

Soils collected from grassland during the summer were placed in trays under bracken fronds or in grassland during the winter growing season. Similarly, summer soils from bracken stands were placed in both environments. Immediately following the first rains, the trays were brought into the laboratory and transferred to storage dishes in which seeds were planted. The results, as expected, showed the greatest inhibition in bracken/bracken soils and least in grassland/grassland.

Fronds from bracken stands were collected at the end of the dry season. Over wooden frames, 2.5 m square, 40 cm above the ground, these fully mature and dead fronds were woven into a grid of cotton twine. The complete grid was placed in the grassland and another grid covered with a shade fabric was placed as a control. Light intensities under both grids was equal to that under frond stands. At the end of each growing season, fronds from that season were added to the grid. Thus, over a period of two consecutive growing seasons, leachates from these frond grids were being added to the grassland environment. Fronds, however, were not in contact with the soil, so there was no smothering effect. At the end of the second season, herb growth under the frond grids had been severely altered. Species highly susceptible to the toxins were reduced in numbers compared to grasslands controls. The tolerant species, in the absence of rodent pressure in the grassland, however, had increased in number, although their dry weight was decreased.

Areas of bracken (3 × 3 m) were cleared of fronds and litter for 4 years. By the end of the first growing season, there was a significant increase in number of herb individuals. By the end of the third season, there were few differences in herbs in bracken plots compared to grassland ones. Removal of the source of toxicity resulted in restoration of normal species.

What this study did not do was isolate and characterize any of the toxic components. As a result, known compounds were not applied directly to the soil or tested in laboratory bioassays. What makes this work unique among allelopathic studies, however, is its scope and the breadth of experiments performed. In a logical stepwise series, alternative explanations that could have been expected to explain plant dominance were eliminated. A convincing series of experiments and bioassays established the source of the phytotoxins, their route of entry into the soil, and demonstrated the toxicity of the soil. Sadly, such in-depth studies are mostly lacking nearly 20 years later.

4.2.2. The California Chaparral

One of the classic examples of allelopathy seemingly exerting a dominant influence in a natural ecosystem is that of the California chaparral. The chaparral is characterized by large stands of deep-rooted sclerophyllous shrubs, and generally lacks an herbaceous layer (C.H. Muller et al. 1968), In areas where shrubs border grassland communities, there is often a zone of bare soil of 60–90 cm around shrubs, next to a zone of decreasing inhibition of growth for 2–9 m (C.H. Muller et al. 1964). Species for which such zones of inhibition are observed include *Salvia leucophylla*, *Artemisia californica*, *Adenostoma fasciculatum*, and *Arctostaphylos glandulosa* var. *zacensis*. The focus of this discussion is on the first two species.

4.2.2.1. Role of Allelopathy

On the basis of a long series of studies, Muller and coworkers postulated that bare zones are caused primarily by chemical interference (C.H. Muller & W.H. Muller 1964; Muller 1966; Muller & Chou 1972; Muller & del Moral 1966; C.H. Muller et al. 1968). Principal aspects of the hypothesis include the following.

1. Volatile terpenes are produced by *S. leucophylla* and *A. californica*, with greatest volatilization during hot periods when the soil is dry.
2. Terpenes adsorb to clay soil colloids, reaching maximum concentrations at the end of the dry season. Microbial activity dissipates the toxins once the winter rains begin.

3. Germinating seeds are exposed to terpenes. Effects are greatest on annuals, and are intensified by periodic drought stress.
4. Old stands of *S. leucophylla* deteriorate due to autotoxicity.

4.2.2.2. Physical and Biotic Factors

As in the bracken study, Muller and coworkers carefully eliminated many physical and biotic factors. Differences in soil physical factors and nutrient content were ruled out by soil analyses. Salinity stress was shown not to be a factor, and soil types were the same in all zones (Muller 1966). Soil moisture, however, was found to be slightly lower in the bare zone (W.H. Muller 1965), and field observations indicated that drought intensified the effects of phytotoxins (Muller 1966). The role of herbivorous animals remains a contentious issue to this day. Data are conflicting (Muller 1966; Bartholomew 1970; Muller & del Moral 1971). For *A. californica*, Halligan (1973, 1975) proposed that, similar to what was observed with bracken, allelopathy was important in excluding some species from the bare zone, while animal activity excluded others.

As a check to eliminate resource competition as a cause, manure was applied to various areas. No stimulation of growth occurred in the bare zones, only slight stimulation occurred in the inhibited zones, while in neighboring grassland, significant growth increase occurred. These results are opposite of what would be expected if nutrients were a limiting factor (Muller 1966). Additionally, the lateral spread of shrub roots reached only to the closest herbs of the zone of inhibition.

4.2.2.3. Bioassays to Establish the Mechanism of Allelochemical Transport

A series of bioassays established the credibility of the route of volatile terpenes produced by the plants to the soil. Toxin production from roots was eliminated in the preliminary experiments, but leaf material, in contrast, was highly toxic when assayed in direct contact with germinating seeds and also when crushed leaves were present in a sealed bioassay chamber in which only vapor transport of toxins to seeds was possible. *Avena fatua* root growth was reduced 50% by volatiles from 0.5 g of *Artemisia californica*, and 64% by 0.5 g of *S. leucophylla* (Muller et al. 1964). Volatile inhibitors were also indicated by observations that the bare and inhibition zones were similar in extent on both the uphill and downhill sides of *S. leucophylla* stands (Muller 1966).

Several monoterpenes were identified in the leaves of three *Salvia* species. They included α-pinene, 1,8-cineole, dipentene, and camphor (Muller & Muller 1964). Camphor and cineole were the most toxic in bioassays with cucumber. Analyses of the atmospheres in sealed containers of crushed *Salvia* leaves showed significant quantities of monoterpenes. It is important to note that although monoterpenes are typically thought of as volatile compounds, they are also sufficiently water soluble to be transported in aqueous solution, as will be pointed out later (Weidenhamer et al. 1993). Both cineole and camphor were present in the air around *Salvia* shrubs. This occurred both in the field and in greenhouse-grown plants. Experiments indicated that the terpenes were rapidly adsorbed by paraffin shavings in sealed flasks, suggesting that similar adsorption by the cutin of seedlings in direct contact with soil terpenes might be responsible for the production of toxic effects.

It was shown that soil colloids could adsorb volatile terpenes from the atmosphere (Muller & del Moral 1966). Fresh soil exposed to volatile terpenes and also soil that had been stored several months were both toxic when bioassayed with a grass, *Bromus rigidus*. Other work showed that cineole and camphor could be recovered from soil in bare zones next to *Salvia*, but not in adjacent grassland (C.H. Muller 1965). Growth of the grass in surface soils of the bare zone was reduced compared to controls from nearly artificial bare areas. By 4–6 weeks into the growing season, terpenes were no longer detectable in the soil (Muller & Chou 1972).

Tyson et al. (1974) took the work a step further when they conducted studies of the rate of terpene volatilization from foliage of *S. mellifera*. The calculated rate of volatilization was 13.3 $\mu g/dm^2$ per day. Total loss of photosynthate was estimated at 0.06%. On an area basis, it was calculated that 3.1 kg/km^2 were released daily. The total terpene content of the leaves was more than 1000 times the amount volatilized.

Several studies by Muller and Hauge (1967) and W.H. Muller et al. (1968, 1969) showed that *S. leucophylla* terpenes affected development of cucumber seedlings. Root and hypocotyl growth were reduced, lipid globules built up within cells, outer cell walls had deposits of excess cutin in the epidermis, fewer lateral roots were produced (Muller & Hauge 1967), respiration of roots diminished (W.H. Muller et al. 1968), and uptake of oxygen by mitochondria was reduced (W.H. Muller et al. 1969). Unfortunately, similar studies were not conducted with chaparral and grassland species.

In complementary studies with *A. californica*. Halligan (1973, 1976) observed apparent allelopathic exclusion of *Media sativa* and *Hypochoeris glabra* around these shrubs. The latter was only rarely found under *A. californica*, although it was common beneath other chaparral shrubs and in grazed grassland areas. *M. sativa* did not grow well in the inhibition zone even within animal exclosures although it grew well in grazed grassland areas. It was never found in the shrub zone. In bioassays of leaf and soil volatiles and of natural and artificial raindrip from beneath the plant, germination and growth of both species were substantially inhibited. Two other species, *Bromus rigidus* and *Festuca megalura,* were less susceptible in bioassays. The former is excluded by animal activity, and the latter grows in all three zones.

Five major terpenes were identified in the leaves of *A. californica*: 1,8-cineole, artemisia ketone, α-thujone, isothujone, and camphor. Minor amounts of β-pinene, borneol and caryophyllene were also identified (Halligan 1975). Camphor and cineole were again the most toxic in bioassays with *M. sativa*. Production of the terpenes varied seasonally, and many compounds were also isolated from soil and litter beneath *A. californica*.

The bare zone phenomenon in the chaparral emerges as a product of at least three primary factors—periodic drought, animal activity, and allelopathy. Different factors are more important for various species. Collectively, the many studies on chaparral illustrate the difficulty of demonstrating allelopathy by simply eliminating other factors. The most striking omission from these studies was the failure to measure the actual amounts of terpenes present in the soil of the shrub and bare zones throughout the year. While the plausibility of the hypothesized mechanism of volatilization of monoterpene phytotoxins, adsorption to solid colloids, and subsequent inhibition of seedling growth was demonstrated by carefully done experiments and bioassays, it remains to be shown that actual soil concentrations and flux rates of the implicated chemicals are sufficient to produce the bare zones.

4.2.3. Rye

In a series of papers, Putnam and his colleagues (Barnes & Putnam 1983, 1986, 1987; Putnam et al. 1983; Nair et al. 1990; Chase et al. 1991a, 1991b) examined the use of allelopathic cover crops in reduced tillage cropping systems for weed control. While the aim of the studies was development of an ecologically sound and environmentally safe management strategy for weed control, the experiments and bioassays employed are models of those that can be used in excluding the effects of physical factors from allelopathy in a field situation.

In the no-tillage system, a cover crop can be used to suppress weeds. Sites are covered with some plant residue, which reduces soil erosion and improves water retention. Crop seeds are drilled into the soil through the sod or previous crop residues and the cover. The cover crop may be killed by overwintering or with a contact herbicide. Potential cover crops include barley, rye, sorghum, buckwheat, sudangrass, sweetclover, and sunflower. All have been reported as allelopathic to test species (Rice 1984). Putnam's group examined several species including rye, *Secale cereale*.

They first demonstrated that rye is toxic to weeds by comparing weed biomass in open (unplanted) versus rye plots. Biomass of three common weed species was suppressed 98–42% in rye plots compared to unplanted controls. Spring-planted living rye reduced weed biomass overall by 93% after 41 days. In a second experiment, a field of fall-planted rye was killed with glyphosate (*N*-phosphonomethylglycine). Residual biomass was calculated at 4.9 metric tons/ha. Peas, inoculated with *Rhizobium*, were planted through the rye residue using a specialized no-tillage seeder. Weeds were harvested from several plots 25 days after planting peas, and pea plants were harvested after 46 days. As a control, separate plots had poplar excelsior (PE), a nontoxic mulch whose appearance is similar to rye, secured to the ground with plastic netting. PE sprayed with glyphosate was also applied. Thus, allelopathic effects could be separated from the physical effects (difference in soil water, light levels, soil temperature). The PE mulch controls showed diminished growth effects due to physical factors. The effects were magnified in rye residue plots where total weed biomass was reduced 63% compared to PE mulch controls, demonstrating that allelopathic effects as well as physical ones were operating. Pea yields were not reduced under rye residues. Wheat and barley residues produced similar results. The experimental design for both tests was a randomized complete block with four replicates (Barnes & Putnam 1983; Putnam et al. 1983).

4.2.3.1. Greenhouse Experiments and Bioassays

In greenhouse experiments, a series of tests further evaluated the influence of surface residues on germination and growth of weeds and crops. Weed seeds were added to plants in individual pots, which had soils amended with rye residues. Germination rates were reduced when residues remained on the surface, but were lost when residues were mixed throughout the soils. Apparently a zone of inhibition develops near the surface where weeds germinate. PE, in contrast, had little effect on germination. Rye residues also reduced growth (Putnam et al. 1983).

In other greenhouse studies, rye root leachates, collected by manually transferring half-strength Hoagland's solution through either a four-pot system of four indicator or 2/2 rye/indicator pots, were tested against lettuce and tomato. Different ages of rye were evaluated. Leachates reduced plant weights and these were independent of the age of the rye (Barnes & Putnam 1983). This demonstrated that, in addition to material from residual plant material in the soil, leachates were also a source of toxins.

Petri dish bioassays of residues in soil—(1) segments of ground dried shoots cut from the soil surface 35 days after planting, (2) 2-cm segments of roots rinsed free of soil, and (3) aqueous extracts of shoot segments, 50 g extracted with 1 L distilled H_2O—confirmed phytotoxicity. Petri dishes were filled with oven-dried sterile soil or nonsterile soils, and dried residues were uniformly spread and watered with distilled water, or in the case of extracts, added directly. Dishes were incubated for 72 h at 27°C in a growth chamber.

Radicle elongation was a more sensitive measure than germination. Phytotoxicity decreased as the distance from rye shoot material to assay seeds increased. Generally, cress and lettuce were more sensitive indicators than the grass species tested. Phytotoxic compounds from shoots were water extractable. To evaluate the effect of plant part and quantity of residue, roots, shoots, or both were assayed at 0.0, 0.25, 0.5, 1.0, and 2.0 g/dish and placed 5 mm away from the assay seed. Similar amounts of paper towels were used as controls. Shoot tissue was about twice as toxic as root tissue. The data suggested that under field conditions, a sufficient quantity of residues would be present to affect plant growth. Phytotoxicity decreased in nonsterile soil, indicating that the compounds were degraded by microorganisms, and that breakdown products were not the probable source of allelopathic activity by rye (Barnes & Putnam 1986).

4.2.3.2. Chemical Studies

With greenhouse experiments confirming field data, the next stage of the work involved identification of the toxic substances. Using bioassay-guided fractionation, aqueous extracts of 35-day-old greenhouse-grown rye shoot tissue and rye root tissue were separated via solvent partitioning, and all fractions tested for bioactivity. Finding the greatest activity in the diethyl ether fraction, two phytotoxic cyclic hydroxamic acids were isolated and identified from this fraction: 2,4-dihydroxy-1,4(2H)-benzoxazin-3-one (DIBOA) and its breakdown product, 2(3H)-benzoxazolinone (BOA). Both compounds were also detected in greenhouse grown root tissue and field-grown shoot tissue. Concentrations varied in tissues, with greatest amounts in shoots, and they were higher in greenhouse than field-grown shoots. Their activity was tested in laboratory bioassays at concentrations approximating those calculated as being in residue fields. They were found inhibitory to germination and seedling growth of several monocot and dicot species including crops, grasses, and weeds, with dicots being more sensitive (Barnes & Putnam 1983, 1986, 1987).

DIBOA was quantified and purified. Phytotoxicity was compared in bioassays for germination and seedling growth inhibition against weeds and tomato with β-phenyl-lactic acid (PLA) and β-hydroxybutyric acid (HBA), two previously identified and known phytotoxins from rye. Both DIBOA and BOA were consistently more toxic than PLA and HBA (Barnes & Putnam 1987).

Compounds were added to soils immediately after test species were planted in individual pots. Ten seeds were added to pots (43 × 43 mm) filled with soil (loamy sand). They were planted 3 mm deep. Pots were subirrigated to reduce soil disturbance. Chemicals were dissolved in 4 ml of 50% Me_2CO in H_2O and applied by spraying at rates of 25, 50, and 100 kg/ha. Controls were sprayed with equivalent concentrations of Me_2CO in H_2O. Pots were placed in a greenhouse (28°/18°C), with 15-h day length under fluorescent lights (168 $\mu E \cdot m^{-2} \cdot s^{-1}$). Emergence was recorded at 4 and 14 days. Experimental design was a randomized complete block with four replicates. Although concentrations were higher than those that would normally be applied in field situations, relative to those expected from fall-planted, spring-harvested rye, allelopathic activity was demonstrated against lettuce and two weed species. The activity did not take into account expected relative contributions from root tissue as well (Barnes & Putnam 1987).

2,2′-Oxo-1,1′-azobenzene (AZOB), a compound with strong herbicidal activity, was isolated from a soil to which BOA had been added. A parallel experiment with 6-methoxy-2,3-benzoxazolinone (MBOA) also yielded AZOB as well as its mono-(MAZOB) and dimethoxy-(DIMAZOB) derivatives. These compounds were produced only in the presence of microorganisms. When BOA (10 mg) was mixed with soil (25 g), rye residues (5 g fresh), and water (2.5 ml) either under field conditions or *in vitro*, and then incubated at 26°C for 10 days, AZOB was detected. Levels of free BOA were greatly reduced on incubation. AZOB was found to be more toxic to cress and barnyard grass than either DIBOA or BOA in seedling bioassays in petri dishes (Nair et al. 1990). Incorporation of DIBOA into soil also resulted in the formation of AZOB. Complementary experiments with sterile soils produced no AZOB. The soil transformation was determined to be mediated by the microorganism *Acinetobacter calcoaceticus* (Chase et al. 1991b). In summary, primary allelochemicals from rye, DIBOA and BOA, were shown to be able to leach into the soil. Once there, they can undergo microbial transformation resulting in AZOB, a more toxic compound.

The three compounds, DIBOA, BOA, and AZOB, were evaluated singly and in mixtures for activity against a number of crop species and barnyard grass. All combinations were active with significant synergistic activity observed, particularly when AZOB was present (Chase et al. 1991a). Generally, they were more inhibitory to crops than weeds. Rapid conversion to AZOB in the soil would be highly desirable.

In contrast, Gagliardo and Chilton (1992), working with different soil samples (North Carolina versus Michigan), found that the soil transformation product of BOA was not AZOB but rather 2-amino-3H-phenoxazin-3-one, a spectroscopically similar compound. This compound is also an order of magnitude more toxic to barnyard grass than BOA, and the transformation occurs only in nonsterile soils, probably through the intermediate, *o*-aminophenol. Several microorganisms can oxidize *o*-aminophenol to aminophenoxazinone, but their presence has not yet been determined in soils that convert this compound or BOA to aminophenoxazinone (Gagliardo & Chilton 1992). 2-Amino-3H-phenoxazin-3-one also has been shown to be the phytotoxin of the plant pathogen *Acrospermum viticola* (Kinjo et al. 1987). Certainly such a microbe produced compound has the potential for increasing the allelopathic effect of rye mulch. The differing North Carolina and Michigan soil results possibly illustrate how different soil microbial populations can impact experimental results, a precaution that workers should note.

The papers summarized above represent the efforts of several workers over many years. They well demonstrate the Koch's postulates approach to allelopathy. The identification of the source of the chemicals, their structural characterization and specific testing individually and in mixtures in bioassays, elimination of obvious physical factors accounting for some of the inhibition, demonstration of activity in soil in both the field as well as in complementary laboratory bioassays, and demonstration of a role for microorganisms have all been convincingly shown. While information on flux rates and soil concentrations is still needed, this is perhaps the best-characterized instance of allelopathy in an agroecological setting.

4.2.4. The Florida Scrub

Over the past 15 years, several investigators have studied the role of allelopathic interference in the Florida scrub community. This work has recently been reviewed (Williamson et al. 1992b; Fischer et al. 1994; Weidenhamer 1996).

4.2.4.1. Vegetation Pattern and Soil

The Florida scrub is conspicuous in its lack of an herbaceous understory. Younger scrub sites are characterized by widely spaced perennial shrubs, including the widespread Florida rosemary (*Ceratiola ericoides* Michx.) and several other locally abundant shrubs [*Calamintha ashei* (Weatherby) Shinners, *Chrysoma pauciflosculosa* (Michx.) Greene, *Conradina canescens* (Torrey & A. Gray ex Benth.) A. Gray, and *Polygonella myriophylla* (Small) Horton]. At some sites, extensive mats of lichens (*Cladina* and *Cladonia* spp.) occur on the soil surface.

The scrub forms distinct boundaries with roadsides consisting primarily of ruderal species, with a characteristic bare zone 1 m or more in extent between the scrub and roadsides. Similar sharp ecotones occur between the scrub and Florida sandhill community, which is dominated by longleaf pine (*Pinus palustris* Mill.), slash pine (*P. elliotti* var. densa Little & Dorman), and oaks (*Quercus laevis* Walt. and *Q. geminata* Small). Unlike the scrub, the sandhill has a dense herbaceous understory. The scrub vegetation is fire-sensitive, and Richardson and Williamson (1988) proposed that chemical interference by fire-sensitive scrub perennials prevents the invasion of the scrub by herbs and grasses, which would provide fuel for fires.

For many years, the prevailing hypothesis was that soil differences accounted for the existence of the scrub and sandhill as distinct plant communities. Sandhill communities generally occur on yellow sands, and scrub on white sands. However, no consistent differences in soil physical and chemical properties have been found. Kalisz and Stone (1984) studied soil properties along sandhill-scrub gradients in Ocala National Forest, and found no difference in soil profile morphology, particle size distribution, or available nutrients that could account for the differences in vegetation type.

4.2.4.2. Resource Competition

For *Polygonella myriophylla*, a prostrate woody perennial that has numerous primary and secondary branches that form dense mats, field observations of root mass distribution rule out belowground competition as a significant factor in the maintenance of the bare zones surrounding this plant (Weidenhamer & Romeo 1989). At a scrub site that had been subjected to disturbance several years previously, a large patch of *P. myriophylla* was found that was more than 4 × 9 m in extent. This patch bordered an area of bahiagrass (*Paspalum notatum* Flügge) and was surrounded by a bare zone 0.6–1.0 m wide. Dead or dormant rhizomes of bahiagrass were found within the *Polygonella myriophylla* stand and in the bare zone, so it appeared that over time, this stand of *P. myriophylla* had invaded the bahiagrass. Quantitative measurements of root mass showed that few *P. myriophylla* roots extended into the bare zone. Compared to the *P. myriophylla* root mass beneath the patch, root biomass was 63% lower at the edge of the patch and 97% lower in the bare zone. The prostrate nature of this shrub precludes aboveground competition from being a significant factor in the bare zone.

Richardson and Williamson (1988) conducted a field study in which plugs of wiregrass (*Aristida stricta* Michx.) were transplanted from the sandhill into PVC pipes (10 cm diameter × 20 cm deep) at both scrub and sandhill sites. Other plugs of sandhill soil were seeded with three species of pine [*Pinus clausa* (Chapm. ex Engelm) Vasey ex. Sarg., *P. palustris*, and *P. elliottii*]. Wire grass survival and pine growth were both reduced at scrub sites.

4.2.4.3. Allelopathy

The sandhill grass *Andropogon gyrans* was germinated in flats of sandhill soil, and watered with either distilled water, leaf drip collected after misting potted *Ceratiola ericoides*, or leaf drip collected after misting potted *Conradina canescens*. Flats were watered to field capacity every 3–4 days and fertilized every 18 days. The total dry mass of seedlings was reduced 52% by *Conradina canescens* leaf mist and 79% by *Ceratiola ericoides* leaf mist (Richardson & Williamson 1988).

Bioassays of soils collected biweekly for one year from beneath *Polygonella myriophylla*, the bare zone and adjacent grassed areas were conducted by Weidenhamer & Romeo (1989). All soil samples were stored frozen before assaying in 6-month sets. Compared to adjacent grassed area soil, the average germination of bahiagrass was 71% in *P. myriophylla* soil and 81% in bare zone soil, and average shoot dry weight was 48% in *P. myriophylla* soil and 81% in bare zone soil.

4.2.4.4. Chemical Studies

The chemistry of five scrub perennials have now been investigated (Macias et al. 1989; Menelaou et al. 1993; Tanrisever et al. 1987, 1988; Weidenhamer 1987; Weidenhamer et al. 1993, 1994; Williamson et al. 1989). For *P. myriophylla*, the major secondary components of the plant and leaf wash were identified, with the assumption that these compounds or their derivatives were most likely to be involved in any allelopathic interactions. For the other four species, *Calamintha ashei*, *Ceratiola ericoides*, *Chrysoma pauciflosculosa*, and *Conradina canescens*, bioassays with native species were used to guide fractionation of extracts of the leaf washes. In most cases, pronounced differences were seen in the sensitivity of native sandhill assay species (*Schizachyrium scoparium*, *Leptochloa dubia*, and *Rudbeckia hirta*) and the commonly used indicator species for many allelopathy studies, *Lactuca sativa* L.

Two species produce phytotoxic monoterpenes: *Conradina canescens* contains a mixture of monoterpenes, including 1,8-cineole, camphor, and borneol (Williamson et al. 1989). *Calamintha ashei* contains a mixture of menthofuran monoterpenes (Tanrisever et al. 1988; Macias et al. 1989; Weidenhamer et al. 1994). Both species are mints, and presumably contain the terpenes in glandular trichomes on the leaf surface. Studies of the solubility of these compounds showed that their aqueous solubility is far in excess of their toxicity (Weidenhamer et al. 1993), which is necessary if these compounds are to be transported to the soil in aqueous solution by precipitation throughfall.

For three species, environmental transformation of plant natural products into more toxic substances seems to be significant. Leaf washes of *Ceratiola ericoides* contain the inactive dihydrochalcone ceratiolin, which in aqueous solution degrades to phytotoxic hydrocinnamic acid on exposure to light (Tanrisever et al. 1987; Fischer et al. 1994). Leaf washes of *Chrysoma pauciflosculosa* contain the diterpene 17-hydroxygrindelic acid as a major component. This compound is readily oxidized to 17-oxogrindelic acid, which is more toxic (Menelaou et al. 1993). *Polygonella myriophylla* produces large quantities of hydroquinone and gallic acid (Weidenhamer 1987). Both of these phenols have been identified previously as allelopathic agents (Rice 1984), oxidize readily in aqueous solutions, and whole plant bioassays with bahiagrass showed increasing inhibition of growth with time, long after detectable levels of hydroquinone and gallic acid disappeared from the soil (Weidenhamer et al. 1989).

Quantitation of allelochemical release rates has been attempted for one species, *Calamintha ashei*. Concentrations of menthofuran monoterpenes were measured in leaf soaks and washes collected biweekly for 1 year (Weidenhamer et al. 1994). In the leaf washes, the major compounds found, (+)-evodone and desacetylcalaminthone, were also the most phytotoxic in laboratory assays. However, concentrations (as high as 0.021 mM) appear to be too low to cause inhibitory effects.

Obviously, much more quantitative work remains to establish the hypothesis of allelopathy in the Florida scrub. It is likely that sophisticated bioassays will also be required, in which low doses of allelochemicals are applied for long periods of time under environmental conditions that mimic the harsh environment of the scrub. Environmental stress factors such as temperature and nutrient limitation are known to increase the toxicity of allelochemicals (Einhellig & Eckrich 1984; Einhellig 1987, 1989). Soil surface temperatures in the scrub are high (above 50°C on sunny days; D.R. Richardson, personal communication 1985), and levels of macronutrients are very low (Kalisz & Stone 1984; Weidenhamer & Romeo 1989). In a 4-month greenhouse experiment with the native sandhill grass *Schizachyrium scoparium*, hydrocinnamic acid toxicity was increased under reduced N and reduced K treatments (Williamson et al. 1992a).

In the Florida scrub, strong evidence from field and greenhouse bioassays that have ruled out other possible factors and implicated allelopathy has been supported by chemical studies that have identified phytotoxic inhibitors in the leaf dip of five scrub perennials. In several cases, the breakdown products of these compounds are more phytotoxic than the original compound. However, measured concentrations of these compounds in the environment are low, and long-term bioassays and field studies that measure the effects of phytotoxins on plant growth under harsh environmental conditions comparable to those of the scrub will be required to confirm the hypothesis of allelopathy. The studies of allelopathy in the Florida scrub illustrate the crucial role that both bioassays and chemical studies must play in the examination of allelopathy.

4.2.5. Crowberry

Several recent papers by Nilsson, Zackrisson, and coworkers (Nilsson & Zackrisson 1992; Zackrisson & Nilsson 1992; Odén et al. 1992; Nilsson et al. 1993; Nilsson 1994) have presented evidence for the allelopathic effects of crowberry, *Empetrum hermaphroditum* Hagerup. *E. hermaphroditum* is a dwarf boreal shrub that is fire-sensitive and spreads primarily by vegetative propagation. In northern Scandinavian forests, *E. hermaphroditum* has become a dominant species as the frequency of fires has been reduced. At *E. hermaphroditum*-dominated sites, regeneration failure of conifers is common (Nilsson & Zackrisson 1992).

4.2.5.1. Laboratory Bioassays

Bioassays of aqueous leaf extracts (1.25–5.0%) demonstrated dose-dependent inhibition of Scots pine (*Pinus sylvestris* L.) germination. Activated carbon removed the toxicity of the extracts. Aqueous extracts passed through *E. hermaphroditum* soil lost toxicity, while extracts passed through sterilized soil did not (Zackrisson & Nilsson 1992). Other experiments showed that even a relatively short exposure (7 h) to aqueous leaf extracts had a persistent effect on subsequent growth of pine seeds (Nilsson & Zackrisson 1992).

Leaf rinse collected from plants grown in a greenhouse was assayed by placing petri dishes containing 1.0 g of paper pulp at the soil surface underneath *E. hermaphroditum* for several days. Subsequent germination of aspen (*Populus tremula* L.) seeds showed significant inhibition of germination and radicle elongation (Zackrisson & Nilsson 1992).

Spring snow melt was assayed by placing mesh containers with 50 Scots pine seeds in contact with snow melt in the field for 48 h. Germination of pine seeds (measured in the laboratory) was reduced more than 25% by this brief exposure. Soil containing undecomposed litter strongly inhibited Scots pine germination. Activated carbon removed the toxicity of litter (Zackrisson & Nilsson 1992).

Leaves with different densities of surface glands were assayed, and it was found that inhibition was associated with the presence of glandular trichomes on the leaf surface (Zackrisson & Nilsson 1992). In subsequent work, 5-methoxy-3,3′-dihydroxystilbene (batatasin III) was identified in aqueous leaf extracts and found to account for 28% of the extract's inhibitory effects (Odén et al. 1992).

4.2.5.2. Greenhouse and Field Studies

In studies that lasted 3 years, Scots pine was grown with *E. hermaphroditum* in the greenhouse and field in four treatments designed to separate allelopathy and belowground resource competition. In the experiments,

1. Interspecific root competition was reduced by the use of PVC pipes.
2. Allelopathy was reduced by the application of activated carbon to the soil (which had previously been shown to detoxify aqueous leaf extracts).
3. PVC pipes were supplemented by the application of activated carbon to the soil to reduce both allelopathy and belowground competition.
4. Untreated controls provided a measure of the combined effects of competition and allelopathy on pine growth.

Results of these experiments were striking, and showed that the greatest inhibition occurred when pines were subject to both allelopathy and resource competition, while the effects of allelopathy and resource competition separately caused intermediate inhibition (Nilsson 1994). Other greenhouse experiments (Nilsson et al. 1993) established that aqueous extracts of *E. hermaphroditum* reduced N uptake by both mycorrhizal and nonmycorrhizal Scots pine seedlings, and reduced the spread of mycorrhizal infection.

In a series of carefully conducted, logically constructed bioassays, Nilsson, Zackrisson and their coworkers have provided compelling evidence for the allelopathic effects of *E. hermaphroditum*. The field and greenhouse studies using PVC pipes and activated carbon to control belowground resource competition and allelopathy, respectively, are an elegant demonstration that the separation of resource competition and allelopathy is possible. The isolation and characterization of batatasin III is a beginning of the work needed to fully elucidate the chemical nature of *E. hermaphroditum* interference with conifer establishment.

4.3. Density-Dependent Phytotoxicity

One promising approach for bioassays of allelopathic effects is to vary the density of the assay species. Weidenhamer et al. (1987, 1989) found that phytotoxic effects in both petri dish radicle elongation assays and greenhouse growth studies were density-dependent. At low seedling densities, the amount of toxin available *per seedling* is greater than it is at higher densities. The result is greater toxicity at low densities, which disappears at higher densities. Such a result is contrary to the expected consequence of resource competition. For example, growth of bahiagrass seedlings in soil with gallic acid and hydroquinone (400 μg each g^{-1} soil) was unaffected after 5 weeks with 16 seedlings per pot, but was reduced 37% with 2 seedlings per pot (Weidenhamer et al. 1989).

Density-dependent phytotoxic effects also cause characteristic deviations from expected yield-density functions (Fig. 4.1). As plant density increases, the yield per area is generally expected to increase until neighboring plants are extracting a maximal amount of environmental resources. Beyond this point, plants compete with one another and individual plant growth is reduced by competition, although total yield will remain constant (Fig. 4.1A). In the range of densities where yield is constant, the relationship of log mean mass per plant and log plant density is linear, with a slope of −1 (Fig. 4.1B). This is often referred to as Kira's law or the law of constant total yield (Kira et al. 1953). However, in the presence of a phytotoxin, and the resulting density-dependent toxic effects, which will be greatest at the lowest plant densities, it may be predicted that deviations from this line will occur (Fig. 4.1.C). At high phytotoxin concentrations, plant size may be reduced at high density due to resource competition (with no difference in growth compared to an untreated control), and reduced at low density due to phytotoxicity (with a significant difference in growth compared to an untreated control).

When tomato (*Lycopersicon esculentum* Mill.) was grown at three different densities in soil collected beneath black walnut and from an adjacent field, the slope of the yield-density line was reduced at 5 weeks (Fig. 4.1D). This effect disappeared with time, presumably as toxins from the soil degraded (Weidenhamer et al. 1989). Choesin & Boerner (1991) used the lack of density-dependent deviations in growth as evidence against allelopathic interference by *Brassica napus* L.

Density-dependent phytotoxic effects have also been used to study plants grown in interspecific competition (Thijs et al. 1994). Corn (*Zea mays* L.) and soybean (*Glycine max* [L.] Merrill) were grown together in the target-neighbor design proposed by Goldberg and Werner (1983). In this design, varying densities of one species are planted around a target plant. Atrazine, an herbicide that kills broadleaf weeds, was used with corn and soybean as a model of natural situations in which two species differing in their sensitivity to a toxin grow together in the presence of a third species or leaf litter, which is the source of the phytotoxin.

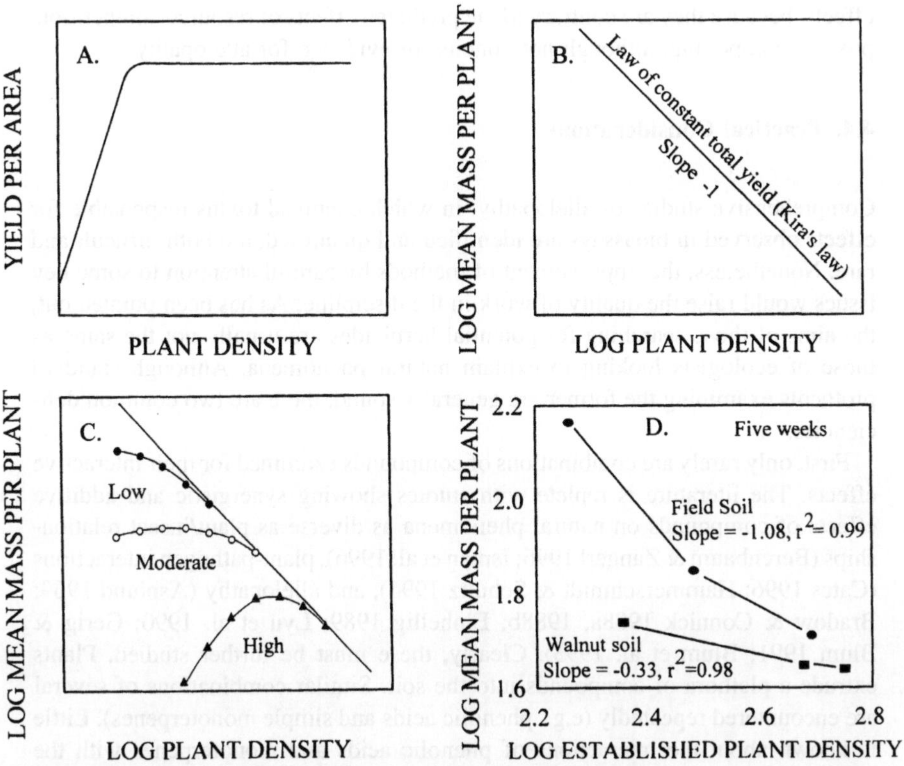

Figure 4.1. **Effects of phytoxins on yield-density relationships: (A) expected response of yield to increasing plant densities; (B) expected relationship of log mean mass per plant and log plant density in the range of plant densities where total yield is constant; (C) predicted deviations in the log mean mass-log density relationship in the presence of low, moderate, and high concentrations of phytotoxins; (D) relationship of log mean mass per plant and log density for tomatoes grown in black walnut and normal field soil for 5 weeks in the greenhouse. (Original figures appeared in Weidenhamer et al. 1989; used with permission of Blackwell Science, Ltd.)**

Such situations presumably occur in the understory beneath allelopathic trees and shrubs, or in the release of toxins from leaf litters or crop residues. With atrazine applications of 3.0 mg/L, the dry weight of soybean target plants *increased* from 0.2 g with no neighbors to 0.5 g with 9–12 neighboring corn plants (Thijs et al. 1994). The increased growth of the soybean target at higher densities of neighboring competitors is contrary to the expected effects of resource competition. In this case, "competition" for and uptake of the atrazine by the corn plants, which are insensitive to the toxin, decreased the amount available to the soybean target. Thus, the use of assays in which plant density is varied would appear to be a potentially useful tool in the study of allelopathy. Density-dependent phytotoxic effects, because they are contrary to the predicted effects of resource competition, provide compelling, although not conclusive evidence for allelopathy.

4.4. Practical Considerations

Comprehensive studies of allelopathy, in which chemical toxins responsible for effects observed in bioassays are identified and quantified, are both difficult and rare. Nonetheless, the improvement of methods by careful attention to some key issues would raise the quality of work in the discipline. As has been pointed out, the aims of those searching for potential herbicides are usually not the same as those of ecologists looking to explain natural phenomena. Although standard protocols examining the former are generally sound, there are two common deficiencies.

First, only rarely are combinations of compounds examined for their interactive effects. The literature is replete with studies showing synergistic and additive effects of compounds on natural phenomena as diverse as plant/insect relationships (Berenbaum & Zangerl 1996; Isman et al. 1996), plant-pathogen interactions (Cates 1996; Hammerschmidt & Schultz 1996), and allelopathy (Asplund 1969; Bradow & Connick 1988a, 1988b; Einhellig 1989; Lyu et al. 1990; Gerig & Blum 1991; Blum et al. 1993). Clearly, these must be further studied. Plants extrude a plethora of compounds into the soil. Similar combinations of several are encountered repeatedly (e.g., phenolic acids and simple monoterpenes). Little is known about the interactions of phenolic acids and monoterpenes with the more active substances that have recently been identified, such as sesquiterpene lactones, polyacetylenes, and thiophenes (Campbell et al. 1982; Duke 1986; Duke et al. 1987; Fischer et al. 1989, 1990; Chase et al. 1991a, 1991b; Einhellig et al. 1993). Einhellig (1996) has recently shown that the effect of compounds from different chemical classes can be cumulative. In addition, many compounds degrade into products that can intensify the activity of those already present. Studies must incorporate bioassays that employ the testing of combinations of compounds mimicking those found in the plant or litter source of the toxins (Langenheim 1994).

Second, density-dependent effects must be considered in bioassays. Plants "compete" for toxins just as they do for nutrients (Hoffman & Lavy 1978; Weidenhamer et al. 1987, 1989; Thijs et al. 1994). In bioassays, the practical consequence of this is that the greatest inhibitory effects will be observed with the lowest density of assay species. In assays of a toxic soil or litter, the greatest effects will be observed with the largest pot volumes. While the ability to use low plant densities and large pot volumes may be limited depending on the availability of experimental material, the influence of these factors needs to be remembered.

With regard to ecological and agroecological studies of allelopathy, there are a number of basic concerns. Too often the terms "allelopathy" and "allelopathic" are used in papers, when in reality they are misleading and validity in the natural system has not been established. Allelopathy should be proposed when supported by concrete evidence based on carefully done bioassays that simulate natural conditions and exclude other obvious factors. Several considerations must be kept in mind:

1. The vegetative pattern of the field situation must be known, qualitatively and quantitatively. The importance of careful field observation in suggesting appropriate experiments cannot be overstated.
2. Chemical interference must be distinguished from resource competition. The studies of Nilsson and colleagues in which PVC pipes were used to reduce belowground competition, and activated carbon was used to reduce allelopathy, illustrate one approach to this problem. Muller's application of manure to bare zones in the chaparral represents another. Demonstrating density-dependent inhibition of growth offers another means of distinguishing the operation of these interference mechanisms. The probability that allelopathy and resource competition together cause the observed inhibitory effects (Nilsson 1994) must be kept in mind.
3. An examination of obvious competing physical factors should occur. Shading, soil moisture, soil composition and texture, and temperature all affect plant growth and represent potential alternative explanations to chemical interference. The studies by Gleissman and Muller (1972, 1978) and by Putnam and colleagues (Barnes & Putnam 1983, 1986, 1987; Chase et al. 1991a, 1991b; Putnam et al. 1983) illustrate excellent ways of identifying these. Enhanced toxicity in response to environmental factors or stresses is well known and should also be considered (Einhellig 1987, 1989, 1996; Williamson et al. 1992a). It must also be kept in mind that the addition of plant materials to soil also influences soil pH, nutrient status, and microbial populations (Blum 1998).
4. The importance of biotic factors such as herbivores and microorganisms in influencing plant growth must be considered the norm rather than the exception. Traps and baiting experiments will quickly indicate whether or not herbivory is likely. Experiments in sterile soil to complement those in field site soils will assist in ascertaining the role of microorganisms.
5. Bioassays, be they in the laboratory, greenhouse, or field, must be conducted with associated plants. An effect on tomato or lettuce or other commonly used crop plants is a demonstration of phytotoxicity, not allelopathy. The confusion of these two terms has caused much undue criticism of the field.

6. Bioassays must be conducted with realistic concentrations approximating those produced by plants and which enter the environment and persist long enough to cause an effect. The assays of Zackrisson and Nilsson (1992), in which they showed that a brief exposure to snow melt was sufficient to significantly inhibit the growth of pine, are convincing because of this realism. Bioassays of concentrated plant extracts are perhaps the most susceptible to criticism that they bear no relation to the natural situation.

7. Critical evaluation of the appropriateness of bioassay procedures, including some of those used in the case studies reviewed here, and careful attention to experimental detail, are a must. The nuts and bolts of many different bioassays have been ably reviewed by others (Leather & Einhellig 1986). A few points deserve emphasis. Even today, papers appear in the literature in which osmotic effects of concentrated plant extracts (Anderson & Loucks 1966) are ignored. Where osmotic effects are possible, appropriate osmotic controls are essential. In petri dish and pot assays, test plant numbers and solution or pot volumes always should be chosen that maximize the sensitivity to chemical effects (Weidenhamer et al. 1987). For native species, an understanding of the seed dormancy stage is necessary. Bioassay-guided fractionation has been used by many investigators to search for active phytotoxins in plant extracts. One potential pitfall of this methodology is that the active compound may be a product of environmental or microbial transformation of a less toxic plant natural product. If evidence from the field or bioassays suggests such a transformation may be important, it may be necessary to fractionate extracts exposed to environmental conditions and/or soil microorganisms.

8. The choice of the appropriate growth parameter(s) to measure is a critical one. In petri dish assays, germination and radicle elongation are commonly measured. Radicle elongation is typically more sensitive to toxin concentration. Many investigators, ourselves included, have measured both germination and radicle growth in the same assay. The difficulty with this is that effects on radicle elongation may be confounded by delays in germination, and thus separate tests for germination and radicle growth (using pregerminated seeds) should be preferred. In pot or field bioassays, shoot biomass is most commonly measured, but this does not provide much information about the potential mechanism of interference. Other parameters, including (but not limited to) root biomass, degree of root mycorrhizal infection or *Rhizobium* nodulation, leaf water potential and diffusive resistance (Einhellig & Schon 1982), relative growth rate, protein synthesis (Cameron & Julian 1980), anthocyanin and chlorophyll production (Leather & Ein-

hellig 1985), and nutrient content can provide greater insight about the mode of action of the chemical toxins. Studies of root cell morphology (Lovett et al. 1989) can provide visual evidence of the effects of chemical interference.

9. Activity in a petri dish is not sufficient to demonstrate allelopathy. Ultimately, proof of activity in the soil must be sought. Toxins must be identified, their concentrations and flux rates in the soil analyzed (Williamson & Weidenhamer 1990), and their interactions with soil components and microorganisms determined. Bioassays following applications of toxins to soil in ways that mimic the natural input, and under environmental conditions similar to those in the natural system, should be done if at all possible.

It should be clear that easy explanations of complex ecological phenomena such as allelopathy will not be forthcoming. Establishing the mechanism of any allelopathic interaction will likely be a long and arduous task. Carefully done bioassays are essential in this process, and we hope that the concerns identified in this review, as well as the selected examples, will be helpful to those working in the field.

4.5. Acknowledgments

We wish to thank those individuals who helped in our preparation for writing this chapter by generously providing reprints and references to papers. Comments by two anonymous reviewers were both useful and encouraging. The numerous colleagues and students who have influenced our thinking over many years have been fondly remembered on many occasions during this writing task.

4.6. References

Anaya, A.L. & H.R. Pelayo-Benavides. 1997. Allelopathic potential of *Mirabilis jalapa* L. (Nyctaginaceae): effects on germination, growth and cell division of some plants. Allelopathy J **4**:57–68.

Anaya, A.L., L. Ramos, R. Cruz, J. Hernandez & V. Nava. 1987a. Perspectives on allelopathy in Mexican traditional agroecosystems: a case study in Tlaxcala. J. Chem. Ecol. **13**:2083–2101.

Anaya, A.L., L. Ramos, J. Hernandez & R. Cruz. 1987b. Allelopathy in Mexico. *In*: Allelochemicals: Role in Agriculture and Forestry, ed. G.R. Waller, pp. 89–101, ACS Symposium series No. 330. American Chemical Society, Washington DC.

Anaya, A.L., M.R. Calera, R. Mata & R. Pereda-Miranda. 1990. Allelopathic potential of compounds isolated from *Ipomoea tricolor* Cav. (Convolvulaceae). J. Chem. Ecol. **16**:2145–2152.

Anaya, A.L., D.J. Sabourin, B.E. Hernandez-Bautista & I. Mendez. 1995. Allelopathic potential of *Ipomoea tricolor* (Convolvulaceae) in a greenhouse experiment. J. Chem. Ecol. **21**:1085–1102.

Anaya, A.L., B.E. Hernández-Bautista, A. Torres-Barragán, León-Cantero & M. Jiménez-Estrada. 1996. Phytotoxicity of cacalol and some derivatives obtained from the roots of *Psacalium decompositum* (A. Gray) H. Rob. and Brettell (Asteraceae), Matarique or Maturin. J. Chem. Ecol. **22**:393–403.

Anderson, R.C. & O.L. Loucks. 1966. Osmotic pressure influence in germination tests for antibiosis. Science **152**:771–773.

Asplund, R.O. 1969. Some quantitative aspects of phytotoxicity of monoterpenes. Weed Sci. **17**:454–455.

Barnes, J.P. & A.R. Putnam. 1983. Rye residues contribute weed suppression in no-tillage cropping systems. J. Chem. Ecol. **13**:889–906.

Barnes, J.P. & A.R. Putnam. 1986. Evidence for allelopathy by residues and aqueous extracts of rye (*Secale cereale*). Weed Sci. **34**:384–390.

Barnes, J.P. & A.R. Putnam. 1987. Role of Benzoxazinones in allelopathy by rye (*Secale cereale* L.). J. Chem. Ecol. **13**:889–906.

Bartholomew, B. 1970. Bare zone between California shrub and grassland communities: the role of animals. Science **170**:1210–1212.

Berenbaum, M.R. & A.R. Zangerl. 1996. Phytochemical diversity: adaptation or random variation? *In*: Phytochemical Diversity and Redundancy in Ecological Interactions, eds. J.T. Romeo, J.A. Saunders & P. Barbosa, pp. 1–24, Plenum Press, New York.

Blum. U. 1998. Designing laboratory plant debris-soil bioassays: some reflections. *In*: Principles and Practices in Chemical Ecology, ed. Inderjit. in Press. CRC Press, Boca Raton, FL.

Blum, U., T.M. Gerig, A.D. Worsham & L.D. King. 1993. Modification of allelopathic effects of p-coumaric acid on morning-glory seedling biomass by glucose, methionine, and nitrate. J. Chem. Ecol. **19**:2791–2811.

Bradow, J.M. & W.J. Connick, Jr. 1988a. Volatile methyl ketone seed-germination inhibitors from *Amaranthus palmeri* S. Wats. residues. J. Chem. Ecol. **14**:1617–1631.

Bradow, J.M. & W.J. Connick, Jr. 1988b. Seed-germination inhibition by volatile alcohols and other compounds associated with *Amaranthus palmeri* residues. J. Chem. Ecol. **14**:1633–1648.

Calera, M.R., A.L. Anaya & M. Gavilanes-Ruiz. 1995. Effect of phytotoxic resin glycoside on activity of H^+-ATPase from plasma membrane. J. Chem. Ecol. **21**:289–297.

Cameron, H.J. & G.R. Julian. 1980. Inhibition of protein synthesis in lettuce (*Lactuca sativa* L.) by allelopathic compounds. J. Chem. Ecol. **6**:989–995.

Campbell, G., J.D.H. Lambert, T. Arnason & G.H.N. Towers. 1982. Allelopathic properties of α-terthienyl and phenylheptatriene, naturally occurring compounds from species of Asteraceae. J. Chem. Ecol. **8**:961–972.

Cates, R.G. 1996. The role of mixtures and variation in the production of terpenoids in conifer-insect-pathogen interactions. *In*: Phytochemical Diversity and Redundancy in

Ecological Interactions, eds. J.T. Romeo, J.A. Saunders & P. Barbosa, pp. 179–216, Plenum Press, New York.

Chase, W.R., M.G. Nair & A.R. Putnam. 1991a. 2,2′-Oxo-1,1′-azobenzene: selective toxicity of rye (*Secale cereale* L.) allelochemicals to weed and crop species: II. J. Chem. Ecol. **17**:9–19.

Chase, W.R., M.G. Nair, A.R. Putnam & S.K. Mishra. 1991b. 2.2′-Oxo-1,1′-azobenzene: microbial transformation of rye (*Secale cereale* L.) allelochemical in field soils by *Acinetobacter calcoaceticus*. III. J. Chem. Ecol. **17**:1575–1584.

Choesin, D.N. & R.E.J. Boerner. 1991. Allyl isothiocyanate release and the allelopathic potential of *Brassica napus* (Brassicaceae). Am. J. Bot. **78**:1083–1090.

Chou, C.-H, S-J. Chang, C.-M. Cheng, Y.-C Wang, F.-H. Hsu & W-H. Den. 1989. The selective allelopathic interaction of a pasture-forest intercropping in Taiwan. II. Interaction between kikuyu grass and three hardwood plants. Plant Soil **116**:207–215.

Connell, J.H. 1990. Apparent versus "real" competition in plants. *In*: Perspectives on Plant Competition, eds. J.B. Grace & D. Tilman, pp. 9–25, Academic Press, San Diego.

Duke, S.O. 1986. Naturally occurring chemical compounds as herbicides. Rev. Weed Sci. **2**:15–44.

Duke, S.O., K.C. Vaughn, E.M. Croom, Jr. & H.N. Elsohly. 1987. Artemisinin, a constituent of annual wormwood (*Artemisia annua*), is a selective phytotoxin. Weed Sci. **35**:499–505.

Einhellig, F.A. 1987. Interaction among allelochemicals and other stress factors of the plant environment. *In*: Allelochemicals: Role in Agriculture and Forestry, ed. G.R. Waller, pp. 343–357, American Chemical Society Symposium Series Vol. 330, American Chemical Society, Washington, DC.

Einhellig, F.A. 1989. Interactive effects of allelochemicals and environmental stress. *In*: Phytochemical Ecology: Allelochemicals, Mycotoxins, and Insect Pheromones and Allomones, eds. C-H. Chou & G.R. Waller, pp. 101–118, Inst. of Botany, Academia Sinica Monogr. 9., Inst of Bot., Acad. Sinica, Taipei, Taiwan.

Einhellig, F.A. 1996. Interactions involving allelopathy in cropping systems. Agron. J. **88**:886–893.

Einhellig, F.A. & P. Eckrich. 1984. Interactions of temperature and ferulic acid stress on grain sorghum and soybeans. J. Chem. Ecol. **10**:161–170.

Einhellig, F.A., J.A. Rasmussen, A.M. Hejl & I.F. Souza. 1993. Effects of root exudate sorgoleone on photosynthesis. J. Chem. Ecol. **19**:369–375.

Einhellig, F.A. & M.K. Schon. 1982. Noncompetitive effects of *Kochia scoparia* on grain sorghum and soybeans. Can. J. Bot. **60**:2923–2930.

Fischer, N.H., J.D. Weidenhamer & J.M. Bradow. 1989. Inhibition and promotion of germination by several sesquiterpenes. J. Chem. Ecol. **15**:1785–1793.

Fischer, N.H., J.D. Weidenhamer, J.L. Riopel, L. Quijano & M.A. Menelaou. 1990. Stimulation of witchweed germination by sesquiterpene lactones: a structure-activity study. Phytochemistry **29**:2479–2483.

Fischer, N.H., G.B. Williamson, J.D. Weidenhamer & D.R. Richardson. 1994. In search of allelopathy in the Florida scrub: the role of terpenoids. J. Chem. Ecol. **20**:1355–1380.

Fuerst, E.P. & A.R. Putnam. 1983. Separating the competitive and allelopathic components of interference: theoretical principles. J. Chem. Ecol. **9**:937–944.

Gagliardo, R.W. & W.S. Chilton. 1992. Soil transformation of 2(3H)-benzoxazolone of rye into phytotoxic 2-amino-3H-phenoxazin-3-one. J. Chem. Ecol. **18**:1683–1691.

Gerig, T.M. & U. Blum. 1991. Effects of mixtures of four phenolic acids on leaf area expansion of cucumber seedlings grown in Portsmouth B1 soil materials. J. Chem. Ecol. **17**:29–40.

Gliessman, S.R. & C.H. Muller. 1972. The phytotoxic potential of bracken (*Pteridum aquilinum* (L), Kuhn.). Madroño **21**:299–304.

Gliessman, S.R. & C.H. Muller. 1978. The allelopathic mechanisms of dominance in bracken (*Pteridum aquilinum*) in southern California. J. Chem. Ecol. **4**:337–362.

Goldberg, D.E. & P.A. Werner. 1983. Equivalence of competitors in plant communities: a null hypothesis and a field experimental approach. Am. J. Bot. **70**:1098–1104.

Halligan, J.P. 1973. Bare areas associated with shrub stands in grassland: the case of *Artemisia californica*. Bioscience **23**:429–432.

Halligan, J.P. 1975. Toxic terpenes from *Artemisia californica*. Ecology **56**:999–1003.

Halligan, J.P. 1976. Toxicity of *Artemisia californica* to four associated herb species. Am. Midl. Natur. **95**:406–421.

Hammerschmidt, R. & J.C. Schultz. 1996. Multiple defenses and signals in plant defense against pathogens and herbivores. *In*: Phytochemical Diversity and Redundancy in Ecological Interactions, eds. J.T. Romeo, J.A. Saunders & P. Barbosa, pp. 121–154, Plenum Press, New York.

Harper, J.L. 1977. Population Biology of Plants. Academic Press, London.

Hoffman, D.W. & T.L. Lavy. 1978. Plant competition for atrazine. Weed Sci. **26**:94–99.

Inderjit & K.M.M. Dakshini. 1996. Allelopathic potential of *Pluchea lanceolata*: comparative studies of cultivated fields. Weed Sci. **44**:393–396.

Isman, M.B., H. Matsuura, S. MacKinnon, T. Durst, G.H.N. Towers & J.T. Arnason. 1996. Phytochemistry of the Meliaceae: so many terpenoids, so few insecticides. *In*: Phytochemical Diversity and Redundancy in Ecological Interactions, eds. J.T. Romeo, J.A. Saunders & P. Barbosa, pp. 155–178, Plenum Press, New York.

Jeffries, H. 1917. On the vegetation of four Durham coal-measure fells. III. On water-supply as an ecological factor. J. Ecol. **5**:129–154.

Kalisz, P.J. & E.L. Stone. 1984. The longleaf pine islands of the Ocala National Forest, Florida: a soil study. Ecology **65**:1743–1754.

Kaminsky, R. 1981. The microbial origin of the allelopathic potential of *Adenostoma fasciculatum* H. & A. Ecol. Monog. **51**:365–382.

Kinjo, I., K. Yokomizo, Y. Awata, M. Shibata & T. Nohara. 1987. Structures of phytotoxins, AV-toxins, C,D, and E produced by zonate leaf spot fungus of mulberry. Tetrahedron Lett. **28**:3697–3698.

Kira, T., H. Ogawa & N. Sakazaki. 1953. Intraspecific competition among higher plants: I. Competition-density yield interrelationship in regularly dispersed populations. J. Inst. Polytech. Osaka City Univ. **D4**:1–16.

Langenheim, J.H. 1994. Higher plant terpenoids: a phytocentric overview of their ecological roles. J. Chem. Ecol. **20**:1223–1280.

Leather, G.R. & F.A. Einhellig. 1985. Mechanisms of allelopathic action in bioassay. *In*: The Chemistry of Allelopathy, ed. A.C. Thompson, pp. 197–205, American Chemical Society, Washington, DC.

Leather, G.R. & F.A. Einhellig. 1986. Bioassays in the study of allelopathy. *In*: The Science of Allelopathy, eds. A. Putnam & C.-S. Tang, John Wiley & Sons, New York.

Lovett, J.V., M.Y. Ryuntyu & D.L. Liu. 1989. Allelopathy, chemical communication, and plant defense. J. Chem. Ecol. **15**:1193–1202.

Lyu, S-W., U. Blum, T.M. Gerig & T.E. O'Brien. 1990. Effects of mixtures of phenolic acids on phosphorus uptake by cucumber seedlings. J. Chem. Ecol. **16**:2559–2567.

Macias, F.A., F.R. Fronczek & N.H. Fischer. 1989. Menthofurans from *Calamintha ashei* and the absolute configuration of desacetylcalaminthone. Phytochemistry **28**:79–82.

McPherson, J.K. & C.H. Muller. 1969. Allelopathic effect of *Adenostoma fasciculatum*, "chamise," in the California chaparral. Ecol. Monogr. **39**:177–198.

Menelaou, M.A., J.D. Weidenhamer, G.B. Williamson, F.R. Fronczek, H.D. Fischer, L. Quijano & N.H. Fischer. 1993. Diterpenes from *Chrysoma pauciflosculosa*: effects on Florida sandhill species. Phytochemistry **34**:97–105.

Molisch, H. 1937. Der Einfluss einer Pflanze auf die andere-Allelopathie. Fischer, Jena.

Muller, C.H. 1953. The association of desert annuals with shrubs. Am. J. Bot. **40**:53–60.

Muller, C.H. 1965. Inhibitory terpenes volatilized from *Salvia* shrubs. Bull. Torrey Bot. Club **92**:38–45.

Muller, C.H. 1966. The role of chemical inhibition (allelopathy) in vegetational composition. Bull. Torrey Bot. Club **93**:332–351.

Muller, C.H. & C.H. Chou. 1972. Phytotoxins: An ecological phase of phytochemistry. *In*: Phytochemical Ecology, ed. J.B. Harborne, pp. 201–216, Academic Press, London.

Muller, C.H. & R. del Moral. 1966. Soil toxicity induced by terpenes from *Salvia leucophylla*. Bull. Torrey Bot. Club **93**:130–137.

Muller, C.H. & R. del Moral. 1971. Role of animals in suppression of herbs by shrubs. Science **173**:462–463.

Muller, C.H. & W.H. Muller. 1964. Antibiosis as a factor in vegetation patterns. Science **144**:889–890.

Muller, C.H., W.H. Muller & B.L. Haines. 1964. Volatile growth inhibitors produced by aromatic shrubs. Science **143**:471–473.

Muller, C.H., R.B. Hanawalt & J.K. McPherson. 1968. Allelopathic control of herb growth in the fire cycle of California chaparral. Bull. Torrey Bot. Club **95**:225–231.

Muller, W.H. 1965. Volatile materials produced by *Salvia leucophylla*: effects on seedling growth and soil bacteria. Botan. Gaz. **126**:195–200.

Muller, W.H. & R. Hauge. 1967. Volatile growth inhibitors produced by *Salvia leucophylla*: effect on seedling anatomy. Bull. Torrey Bot. Club **94**:182–191.

Muller, W.H. & C.H. Muller. 1964. Volatile growth inhibitors produced by *Salvia* species. Bull. Torrey Bot. Club **91**:327–330.

Muller, W.H., P. Lorber & B. Haley. 1968. Volatile growth inhibitors produced by *Salvia leucophylla*: effect on seedling growth and respiration. Bull. Torrey Bot. Club **95**:415–422.

Muller, W.H., P. Lorber, B. Haley & K. Johnson. 1969. Volatile growth inhibitors produced by *Salvia leucophylla*: effect on oxygen uptake by mitochondrial suspensions. Bull. Torrey Bot. Club **96**:89–95.

Nair, M.G., C.J. Whitenack & A.R. Putnam. 1990. 2,2′-Oxo-1,1′-azobenzene, a microbially transformed allelochemical from 2,3-benzoxazolinone: I. J. Chem. Ecol. **16**:353–364.

Narwal, S.S., & P. Tauro. eds. 1996. Allelopathy in Pest Management for Sustainable Agriculture. Scientific Publishers, Jodhpur, India.

Nilsson, M.-C. 1994. Separation of allelopathy and resource competition by the boreal dwarf shrub *Empetrum hermaphroditum* Hagerup. Oecologia **98**:1–7.

Nilsson, M.-C. & O. Zackrisson. 1992. Inhibition of Scots pine seedling establishment by *Empetrum hermaphroditum*. J. Chem. Ecol. **18**:1857–1870.

Nilsson, M.-C., P. Högberg, O. Zackrisson & W. Fengyou. 1993. Allelopathic effects by *Empetrum hermaphroditum* on development and nitrogen uptake by roots and mycorrhizae of *Pinus sylvestris*. Can. J. Bot. **71**:620–628.

Odén, P.C., P.-O. Brandtberg, R. Andersson, R. Gref, O. Zackrisson & M.-C. Nilsson. 1992. Isolation and characterization of a germination inhibitor from leaves of *Empetrum hermaphroditum* Hagerup. Scand. J. For. Res. **7**:497–502.

Pereda-Miranda, R., R. Mata, A.L. Anaya, D.B. Mahinda Wickramaratne, J.M. Pezzuto & A.D. Kinghorn. 1993. Tricolorin A, major phytogrowth inhibitor from *Ipomoea Tricolor*. J. Nat. Prod. **56**:571–582.

Putnam, A.R., J. Defrank & J.P. Barnes. 1983. Exploitation of allelopathy for weed control in annual and perennial cropping systems. J. Chem. Ecol. **9**:1001–1010.

Rice, E.L. 1974. Allelopathy. Academic Press, New York.

Rice, E.L. 1984. Allelopathy, 2nd ed. Academic Press, New York.

Rice, E.L. & S.K. Pancholy. 1974. Inhibition of nitrification by climax ecosystems. III. Inhibitors other than tannins. Am. J. Bot. **61**:1095–1103.

Richardson, D.R. & G.B. Williamson. 1988. Allelopathic effects of shrubs of the sand pine scrub on pines and grasses of the sandhills. Forest Sci. **34**:592–605.

Stowe, L.G. 1979. Allelopathy and its influence on the distribution of plants in an Illinois old-field. J. Ecol. **67**:1065–1085.

Tanrisever, N., N.H. Fischer & G.B. Williamson. 1987. Ceratiolin and other flavonoids from *Ceratiola ericoides*. Phytochemistry **26**:175–179.

Tanrisever, N., N.H. Fischer & G.B. Williamson. 1988. Menthofurans from *Calamintha ashei*: effects on *Schizachyrium scoparium* and *Lactuca sativa*. Phytochemistry **27**:2523–2536.

Thijs, H., J.R. Shann & J.D. Weidenhamer. 1994. The effect of phytotoxins on competitive outcome in a model system. Ecology **75**:1959–1964.

Tyson, B.J., W.A. Dement & H.A. Mooney. 1974. Volatilisation of terpenes from *Salvia mellifera*. Nature **252**:119–120.

Weidenhamer, J.D. 1987. Allelopathic properties of *Polygonella myriophylla*. Ph.D. diss. Univ. of South Florida, Tampa (Diss. Abstr. AAG 8806204).

Weidenhamer, J.D. 1996. Distinguishing resource competition and chemical interference: overcoming the methodological impasse. Agron. J. **88**:866–875.

Weidenhamer, J.D. & J.T. Romeo. 1989. Allelopathic properties of *Polygonella myriophylla*: field evidence and bioassays. J. Chem. Ecol. **15**:1957–1969.

Weidenhamer, J.D., T.C. Morton & J.T. Romeo. 1987. Solution volume and seed number: overlooked factors in allelopathic bioassays. J. Chem. Ecol. **13**:1481–1491.

Weidenhamer, J.D., D.C. Hartnett & J.T. Romeo. 1989. Density-dependent phytotoxicity: distinguishing resource competition and allelopathic interference in plants. J. Appl. Ecol. **26**:613–624.

Weidenhamer, J.D., F.A. Macias, N.H. Fischer & G.B. Williamson. 1993. Just how insoluble are monoterpenes? J. Chem. Ecol. **19**:1827–1835.

Weidenhamer, J.D., M.A. Menelaou, F.A. Macias, N.H. Fischer, D.R. Richardson & G.B. Williamson. 1994. Allelopathic potential of menthofuran monoterpenes from *Calamintha ashei*. J. Chem. Ecol. **20**:3345–3359.

White, C.S. 1991. The role of monoterpenes in soil nitrogen cycling processes in ponderosa pine. Biogeochemistry **12**:43–68.

Williamson, G.B. 1990. Allelopathy, Koch's postulates, and the neck riddle. *In*: Perspectives on Plant Competition, eds. J.B. Grace & D. Tilman, pp. 143–161, Academic Press, San Diego.

Williamson, G.B. & J.D. Weidenhamer. 1990. Bacterial degradation of juglone: evidence against allelopathy? J. Chem. Ecol. **16**:1739–1742.

Williamson, G.B., N.H. Fischer, D.R. Richardson & A. de la PeTa. 1989. Chemical inhibition of fire-prone grasses by fire-sensitive shrub, *Conradina canescens*. J. Chem. Ecol. **15**:1567–1577.

Williamson, G.B., E.M. Obee & J.D. Weidenhamer. 1992a. Inhibition of *Schizachyrium scoparium* (Poaceae) by the allelochemical hydrocinnamic acid. J. Chem. Ecol. **18**:2095–2105.

Williamson, G.B., D.R. Richardson & N.H. Fischer. 1992b. Allelopathic mechanisms in fire-prone communities. In: Allelopathy: Basic and Applied Aspects, eds. S.J.H. Rizvi & V. Rizvi, pp. 58–75, Chapman and Hall, London.

Wilson, R.E. & E.L. Rice. 1968. Allelopathy as expressed by *Helianthus annuus* and its role in old-field succession. Bull. Torrey Bot. Club **95**:432–448.

Zackrisson, O. & M-C. Nilsson. 1992. Allelopathic effects of *Empetrum hermaphroditum* on seed germination of two boreal tree species. Can. J. For. Res. **22**:1310–1319.

5

Bioassay Methods with Terrestrial Invertebrates

J. Daniel Hare

5.1. Introduction 213
- 5.1.1. Scope of the chapter 214
- 5.1.2. General considerations for laboratory bioassays 214
- 5.1.3. Statistical analysis of choices 215

5.2. Behavioral bioassays for odors, pheromones, and other volatile compounds 216
- 5.2.1. General considerations 216
- 5.2.2. Moving air bioassays (olfactometers and wind tunnels) 217
- 5.2.3. Still-air bioassays 223
- 5.2.4. Modifications for specific purposes 225

5.3. Bioassays for contact oviposition stimulants—two case studies 228
- 5.3.1. Butterflies 229
- 5.3.2. Parasitic Wasps 233
- 5.3.3. Data Analysis 237

5.4. Measurement of preference 237
- 5.4.1. Host-plant selection for feeding 237
  - 5.4.1.1. Neutral substrates 238
  - 5.4.1.2. Artificial diets 240
  - 5.4.1.3. Plant tissue substrates 241
  - 5.4.1.4. Leaf disks 243
  - 5.4.1.5. Other natural substrates 243
- 5.4.2. Details of bioassay design 243
- 5.4.3. Data collection 244
- 5.4.4. Data analysis 247

5.5. Postingestive bioassays 247

5.6. Measurements in diet studies: growth rate, consumption rate, and efficiency of conversion of food to biomass 248
- 5.6.1. Data acquisition 252
  - 5.6.1.1. Consumption 252
  - 5.6.1.2. Weight gain 253

5.6.1.3. Measurements of excreta 253
5.6.2. Alternatives to gravimetry 254
5.6.2.1. Uric acid determination 254
5.6.2.2. Calorimetry and respirometry 255
5.6.3. Summary 256
5.7. Alternative methods to separate preingestive and postingestive effects 257
5.8. Contact and volatile toxicity 259
5.9. Conclusions 260
5.10. Acknowledgments 260
5.11. References 261

5.1. Introduction

Terrestrial invertebrates include a large and diverse group of organisms. These organisms live in a great variety of habitats and obtain their resources for survival and reproduction in uncountable ways. Because of the range of species, the range of chemicals, and the diversity of responses to those chemicals, it is not possible to describe all of the potential means to bioassay all species for all interactions. Rather, this review concentrates on presenting a selected group of bioassays that illustrate how some of the general problems in bioassays design have been solved. This review focuses mostly on insects, reflecting both my greater familiarity with that body of literature and the relatively greater quantity of work done with insects and other arthropods compared to that done with other groups of terrestrial invertebrates.

This chapter is organized largely on the ways in which organisms are exposed to chemicals, not on the kinds of interactions those chemicals mediate. Bioassays for behavioral responses to volatile compounds that are detected at some distance from their site of production or release are discussed first, followed by bioassay procedures for behavioral responses to chemicals that are detected only after contact.

Bioassays for chemicals that may have postingestive metabolic effects in addition to preingestive behavioral effects are reviewed next, as well as additional procedures to isolate the behavioral from the metabolic effects. Bioassays for both inter- and intraspecific interactions are discussed within each section, as appropriate, because the bioassay procedures for chemicals mediating intraspecific interactions do not necessarily differ from those mediating interspecific interactions. Nevertheless, it must be kept in mind that because intraspecific interactions are necessarily more focused than interspecific interactions, one may be more likely to observe a clear response intraspecifically than interspecifically. For example, a plant-feeding female insect must mate with a male of the same species to produce viable progeny. In contrast, the same female may deposit eggs on any of several host-plant species that are all suitable for the growth and survival of her progeny. Thus, while males and females may show clear, qualitative responses when exposed to each other's sex pheromones in a wind tunnel, females may show a range of more subtle olfactory responses to host plant chemicals in the same wind tunnel.

There have been several recent reviews of the methodology for bioassays of volatile chemicals mediating intraspecific interactions (i.e., pheromones), including those by Baker and Cardé (1984) and Baker and Linn (1984). Experimental techniques to bioassay chemicals mediating one important class of interactions—host-plant location and identification by phytophagous insects—were also compiled and reviewed by Finch (1986) and by Singer (1986). Bioassays for host-plant selection and ingestion by plant-feeding insects have been reviewed by Lewis and van Emden (1986), Berenbaum (1986), and Kogan (1986). The goals of this chapter are to draw together and update these methodological areas and to review bioassay applications for other interspecific interactions. Both the interactions involving potential hosts and natural enemies are included.

5.1.1. Scope of the Chapter

"An assay in which the detector is a living organism or part of its sensory system is referred to as a bioassay" (Finch 1986). Bioassays are the basic tools required to assess the biological activity of behavioral stimuli used during host or mate finding and the physiological responses to compounds that determine the quality of hosts for utilization. The main types of bioassay are based on either behavioral or physiological responses of whole organisms, or electrophysiological responses of isolated organs or structures (e.g., isolated insect antennae). If an organism, or part of an organism, responds immediately to a crude chemical extract, it may be possible to incorporate this response into a suitable, relatively simple bioassay. The results of a series of such bioassays may not be interpreted easily if the crude extracts are poorly defined and/or reproduced. Bioassay results can be critically evaluated only when both the stimuli and the background in which the stimuli are presented are tightly controlled. Other problems of interpretation may occur if the isolated behavioral response component that is evaluated does not reflect the overall behavioral response. Most problems in designing bioassays focus on controlling the stimulus and maintaining a uniform background to evaluate responses.

Aspects of electrophysiological measurement has been reviewed elsewhere (volume 1). In this chapter, the focus is on the behavioral and physiological responses of whole organisms, with an emphasis on terrestrial insects and other arthropods.

5.1.2. General Considerations for Laboratory Bioassays

The first step in bioassay design is to determine exactly what behavior is to be measured. The next step is to standardize the experimental design and evaluate the significance of potential endogenous and exogenous sources of variation.

Appropriate methods of statistical design should be identified and considered *before* experiments are undertaken to ensure that the data collected will conform with the assumptions of the chosen statistical test. Finally, laboratory results must be evaluated with respect to both previous findings and field observations.

Laboratory studies offer several advantages over the field for bioassay. In the laboratory, temperature, humidity, the physiological state of the organism, and most external stimuli can be controlled to a greater or lesser extent. Laboratory-reared organisms, however, may exhibit responses different from those of wild organisms, so laboratory results are best interpreted with reference to field observations.

It often takes substantial ingenuity and patience to develop a laboratory bioassay that "works," i.e., to develop a method of presentation that elicits a behavioral response that can be scored and quantified. New investigators should not be discouraged by false starts when designing a laboratory protocol. Many endogenous factors, including age, physiological state, nutrition, and sex of the individual can affect behavioral responses (e.g., Hardee et al. 1966; Hawkes & Coaker 1976). Exogenous factors including level of illumination (Sparks & Cheatham 1970), humidity (e.g., Traynier 1967), time of day (Hardee et al. 1966), and even barometric pressure (Steinberg et al. 1992) also may influence responses. Ultimately, the careful use of extensive controls, both positive and negative, is the best approach to identify and account for effects of such variables. It may take a number of preliminary experiments and trial runs to determine the most important factors to control to optimize the difference between experimental and control treatments while minimizing the variance within treatments.

An important point to consider in the design of bioassays is the complexity of the behavior being observed. In a complex process such as the identification and acceptance of a host for feeding or oviposition, several different behaviors must be released in sequence by unique stimuli. Visual or volatile chemical cues may be used to bring the insect in contact with the potential host, after which tactile or nonvolatile chemical cues may be used to actually identify the host and assess its quality for utilization. In such cases, it may be necessary to dissect the overall response into its components and develop specific bioassays for each stimulus separately, rather than to attempt to develop a consistent, discriminating bioassay that utilizes the whole behavioral process.

5.1.3. Statistical Analysis of Choices

The method of statistical analysis in many bioassays focuses on analyzing the number and pattern of choices made by subjects. In general, these assays will not involve truly continuous variables, but will involve counts, e.g., the number of times that each branch of an olfactometer was chosen, the number of times that upwind flight was observed, the number of eggs deposited on test or control substrates, or the number of times that test or control feeding substrates were selected. Such data often are distributed following a Poisson distribution and can be analyzed by statistics for continuous variables, such as Student's *t*-test or analysis of variance (ANOVA), after employing the square root transformation. Where there are two independent treatments, *t*-tests or the nonparametric analogue, the Wilcoxon two-sample test, can be used. In many cases, the experiment is designed so that paired *t*-tests would be most appropriate (Horton 1995). For more than two treatments, a single classification ANOVA (parametric) or the Kruskal-Wallis test (nonparametric) is used; the multiple treatment analog of the paired *t*-test is the two-factor ANOVA. For multiple treatments, the overall analysis is followed by tests to distinguish nonsignificant subsets of treatments through the use *a posteriori* simultaneous test procedures, or carefully designed *a priori* tests of significance (Sokal & Rohlf 1995).

If the two choices are not independent (i.e., subjects must accept one of two choices and reject the other), then data should be analyzed as frequencies. For simple choice tests, the *G*-test or Fisher's exact test (Sokal & Rohlf 1995) are often used to test for deviations of the observed pattern of choices from a random pattern. The choice of test depends on sample size and calculated expected values. Sometimes, the proportion of subjects on the test substrate is calculated as $T/(T + C)$, where T is the number of subjects on the "test" treatment and C is the number of subjects on the "control" treatment. These proportions often follow a binomial distribution and can be analyzed as continuous variables after employing the arcsine square root transformation. However, analysis of frequencies is usually preferred to analysis of proportions (Sokal & Rohlf 1995).

Another statistic often calculated is an overall index of acceptability of the form $(T - C)/(T + C)$, where T and C are defined as above. This statistic has an expected value of 0 when there is no difference in preference of the test and control treatments and a range from -1 (when the test treatment is never chosen) to $+1$ where the test treatment is completely preferred over the control. While this statistic allows the expression of the relative attraction and deterrence of a range of compounds, its statistical properties are not well understood. The statistic is not normally distributed, and the occurrence of negative values precludes the use of common transformations (e.g., log, arcsine, square root) to remove some deviations from normality. Thus, if this statistic is used, it should be analyzed only by nonparametric, distribution-free statistical tests based on ranks. Further discussion of procedures and criteria for selecting statistical tests can be found in many standard texts and manuals for a number of statistic analysis packages for use with computers (e.g., Steel & Torrie 1980; SAS Institute 1989; Sokal & Rohlf 1995).

5.2. Behavioral Bioassays for Odors, Pheromones, and Other Volatile Compounds

5.2.1. General Considerations

Bioassays for volatile chemicals can be grouped into two general classes: those that utilize moving air, and those that do not (Baker & Cardé 1984). In bioassays not requiring a moving air stream, the volatile chemical evaporates from a dispenser and a decreasing gradient of concentration is established from the emitter by simple diffusion. Organisms then are released into a suitable arena, and particular behavioral events, such as movement toward or away from the emitter, are noted and recorded. The equipment for such bioassays often is relatively simple and may require test arenas constructed from only petri dishes or small cages. Bioassays utilizing moving air require additional equipment to consistently produce a controlled airstream of uniform strength and direction from trial to trial. In these bioassays, the volatile chemical is carried downwind in an odor plume, and the behavioral responses of the organism, are observed as they enter and/or leave the plume.

Test chemicals or chemical fractions can be dispensed from a variety of materials, depending on their volatility and solubility. Chemicals may be evaporated from such materials as glass rods or beads, metal surfaces, filter paper disks, cotton balls or wicks, rubber septa, polyethylene vials, glass capillary tubes, etc. It is important to ensure that the dispenser emits the compound at a fairly constant rate over the course of the assay. This may be more difficult for very volatile chemicals, thus, fresh dispensers uniformly prepared ahead of time may be required. This may be most important in the early stages of investigation where fractions of unknown composition and concentration are being used in a bioassay-driven fractionation scheme to isolate and, ultimately, identify specific chemical compounds eliciting particular behaviors.

Another decision is the quantity, or dose, of the material to be used. For fractions of a natural extract, this is usually expressed in terms of equivalent concentration based upon a known mass of starting material or known number of emitting individuals. This concentration need not be the one that elicits the optimal response, so, for initial studies, it is desirable to carry out bioassays over a range of concentrations over log series (0.1 ×, 1 ×, 10 ×, etc.), then through a series of finer ranges (0.5 ×, 1 ×, 2 ×, etc.) until the form of the dose-response curve is determined. Finally, it is necessary to relate a specific dose eliciting maximal response to those likely to be encountered by the organism in its natural context.

5.2.2. *Moving Air Bioassays (Olfactometers and Wind Tunnels)*

Olfactometers and wind tunnels are similar in that both involve observations of freely moving organisms in a stream of air. Olfactometers may be manufactured with one or multiple arms, each with a separate odor source (or odor sources and a "control" source), while wind tunnels have only one source of air. Within the airstream, however, wind tunnels may have multiple chemical emitters that can be charged with different test compounds (e.g., Fein et al. 1982; Knols et al. 1994).

Olfactometers are most often used to bioassay the rate of movement and choice of direction during walking, while wind tunnels are most often used to assess the rate of movement and choice of direction during flight. However, wind tunnels also have been used to assess the orientation behavior of walking insects (e.g., Visser & Ave 1978), and multiarm olfactometers also can be used to assess flight responses of small organisms, such as parasitic Hymenoptera, if the insects are sufficiently small.

All olfactometers and wind tunnels must be custom-made. No commercial products are available to my knowledge. Although this makes construction of wind tunnels and olfactometers time-consuming and expensive, it also permits the design and size of each to be optimized for the organism to be tested. Wind tunnels are usually made of wood, aluminum, plastic, or glass. Simple olfactometers are often made from standard laboratory glass stock with ground glass fittings for easy disassembly for cleaning. More complex olfactometers often are made from aluminum or Plexiglas™. The choice of materials must consider both the characteristics of the chemical (e.g., many chemicals are absorbed into plastics) and the characteristics of the test subject (e.g., the surface of pure glass is often too slippery for insects to walk on).

In the use of both wind tunnels and olfactometers, a comparative approach is used. In the two-arm olfactometer, for example, the simplest analysis would be to compare the number of times that the arm with the semiochemical is chosen relative to the "control" arm and test these numbers statistically with a chi-square or *G*-test (Sokal & Rohlf 1995) against the null hypothesis that the two arms will be chosen equally. In wind tunnels, an equally simple analysis would be to test for more or faster movement in the "upwind" direction when the test chemical is released into the airstream than when it is not. In both cases, however, care must be taken in the design and operation of the apparatus to avoid biases in the data, as follows.

First, the airstream should be purged of potentially interfering organic compounds by passing the airstream through a filter and/or an adsorbent such as glass wool and/or activated charcoal. After filtering, it may be necessary to rehumidify the air by bubbling it through water before it enters the olfactometer. False positive responses may be obtained if the comparison is made between living and nonliving substrates. This often occurs when the organism exhibits nonspecific responses to common substances such as water vapor, carbon dioxide, or other ubiquitous compounds (e.g., general "green-leaf" volatiles from plants) that are emitted by living organisms whether they are suitable hosts or not.

For multiarm olfactometers, it is advisable to have a flowmeter and valve on each of the arms to ensure even airflow throughout. If only the exhaust airflow is monitored, then there may be unequal airflow through the different arms due to small constrictions, or flow restrictions if the delivery tubes are of unequal length. The bioassay chamber should be isolated from any potential visual stimuli (e.g., light sources) that may induce a preference for one arm over another. Finally, the olfactometer must be cleaned between trials, and it is strongly advisable to rotate the "test" and "control" treatments between arms for each replication to control for any other nonspecific directional biases. Cleaning is important not only to remove traces of test chemicals that may have been deposited on the walls, but also to remove sources of other potential cues deposited by the previous test organisms (scales, trail pheromones, fecal material) that may affect the choice of olfactometer arm by subsequent organisms independently of the test stimulus.

Most researchers have employed two-arm olfactometers constructed in the form of a "Y" or a "T" (Fig. 5.1). Control and test airstreams enter each of the arms and exit into a common collector, where the test organism is released. The organism then walks upwind and chooses one of the branches. If constructed of glass, the glass tubes may need to be lined with materials, such as strips of filter paper or screen, on which the test organism can walk. These walking surfaces also must be either disposed of and replaced, or cleaned between trials.

A serious limitation of the two-arm olfactometers is that the design has low resolving power (i.e., it may require a large number of trials to demonstrate nonrandom choice when only two choices are available). The fact that only two choices are available also makes it difficult to interpret results involving two odor sources. First, one cannot discriminate between attraction of one versus repellence of the other without additional tests comparing each alone to a blank air stream. Second, the combination of odors from two arms after they mix may be unattractive, causing only random movement in that portion of the olfactometer where the odors are mixed. Additionally, the design actually may limit some organisms to only one choice per trial, if the organisms are unable or reluctant to switch directions in an arm, traverse back to the junction, and resample the airstreams.

Many of these problems may be overcome by using a multiple olfactometer, with the arms opening into a relatively large arena with a central exhaust port (e.g., Pettersson 1970; Vet et al. 1983, Fig. 5.2). The first potential advantage includes the more powerful statistical design provided by multiple choices. In addition, multiple airstreams may enter the arena with relatively little mixing, thereby creating several zones that the test insect may be able to sample by walking among zones prior to selecting one of the arms. Multiple choices can be made and observed by a single test subject (Vet et al. 1983). This may require that the flow in each arm be monitored to ensure the lack of mixing. Multiple-arm olfactometers have a substantial disadvantage in that additional care may be needed in construction to ensure that all arms are of equal size and spacing throughout.

If flight is essential for behavioral choices to be made, then there may be no alternative to the wind tunnel. These are larger and more expensive than olfactometers, and few allow the diversity of choices provided by multiarm olfactometers. Nevertheless, a number of variables can be measured in wind tunnels relating to the time and intensity of various behavior responses of test organisms to odor (e.g., Baker & Cardé 1984) (Table 5.1).

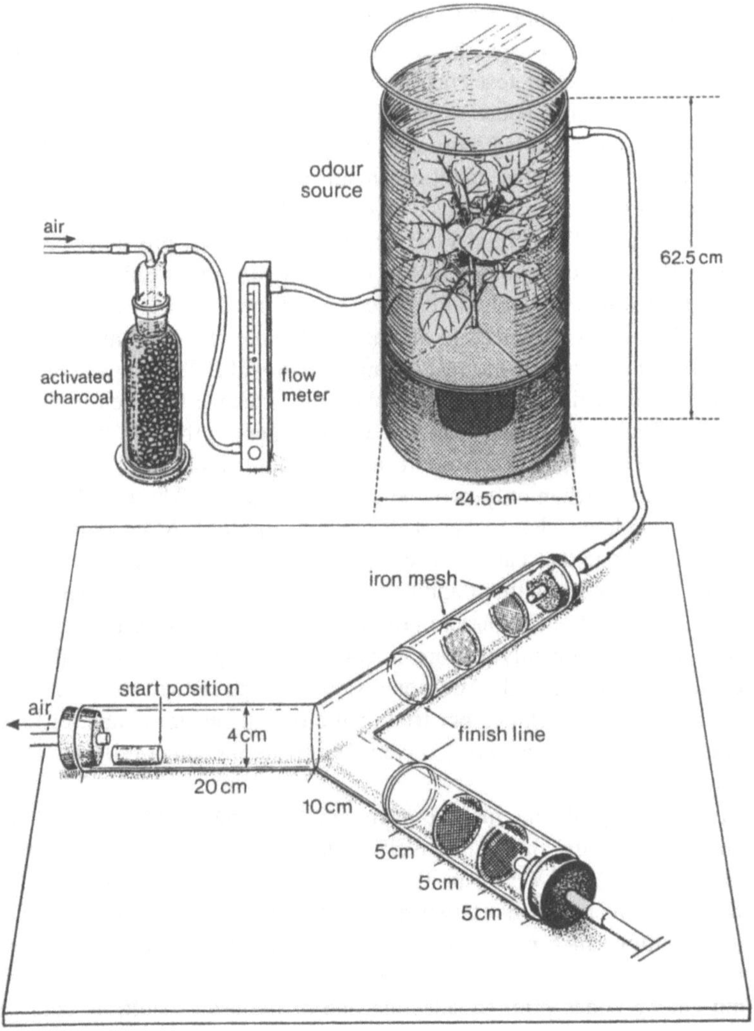

Figure 5.1. Schematic representation of a Y-tube olfactometer. From Steinberg et al. (1992), reprinted with kind permission from the authors and Kluwer Academic Publishers.

A typical wind tunnel used to assay insect behavior to pheromones is shown in Figure 5.3. A more complete description of its design and construction can be found in Miller and Roelofs (1978). Fundamental components are a variable-speed fan to generate an airstream, an expansion duct to lower the velocity of the stream to that appropriate for the test subject, a screen or diffuser to eliminate the circular turbulence generated by the fan, and the tunnel itself with windows for access and observation. The total length usually is 2–3 m. For most insects,

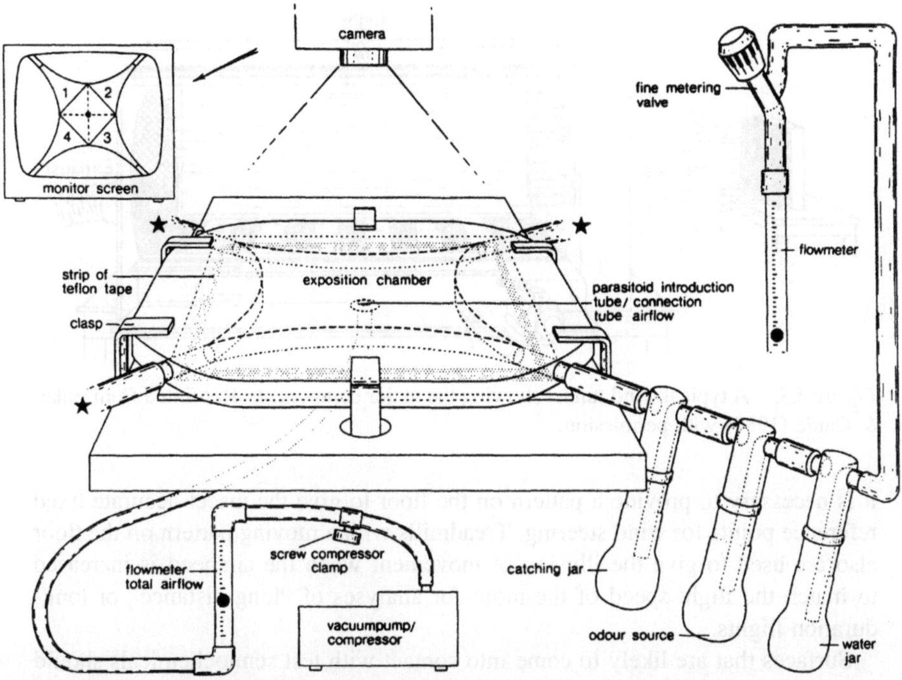

Figure 5.2. **Schematic representation of a four-arm olfactometer. Reprinted from Vet et al. (1983) with permission.**

Table 5.1. Types of Odor-Mediated Responses of Flying Insect that Can Be Monitored in a Flight Tunnel

Time from source introduction until first visible response
Time walking or running before flight
Time wing-fanning before flight
Time from source introduction until flight initiation
Time hovering in plume before making upwind progress
Flight speed in the downwind half of the tunnel
Flight speed in the upwind half of the tunnel
Number of approaches to the source before landing
Number of times dropping back and reorienting before landing
Contact with source (Yes/No)
Location where insect lands: directly on chemical source versus on structure supporting source
Elapsed time from flight initiation until landing on source
Time walking or running on source
Time wing-fanning on source
Time quiescent on source
Total time on source
Number of copulatory or oviposition attempts

Modified from Baker & Cardé (1984).

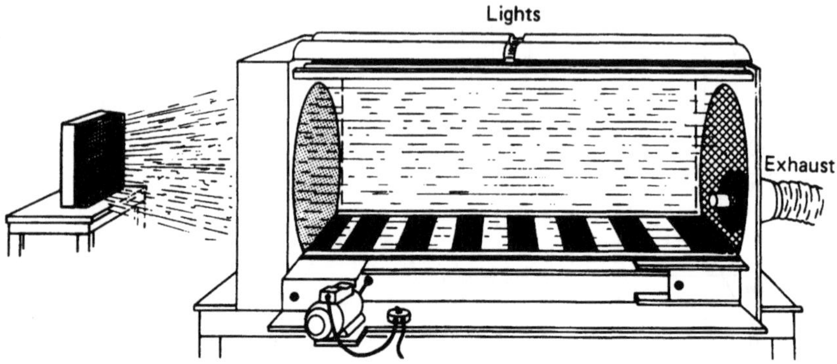

Figure 5.3. A typical wind tunnel showing the major components. Reprinted from Baker & Cardé (1984) with permission.

it is necessary to provide a pattern on the floor to give the insect accurate fixed reference points for wind steering. Treadmills with a moving pattern on the floor also are used to give the illusion of movement when the airspeed is increased to match the flight speed of the moth for analyses of "long-distance" or long-duration flights.

Surfaces that are likely to come into contact with test semiochemicals should be made of material that is either disposable or easily cleaned, such as metal, glass, or solvent-resistant, nonporous plastics, such as Plexiglas™. The use of porous materials, such as wood, should be confined to parts that do not come into contact with the test semiochemicals (see Baker & Linn 1984 for a more detailed discussion of wind tunnel design and construction for specific applications).

After the wind tunnel is built, it is necessary to evaluate the airflow for speed and uniformity using chemical smokes. To measure velocity, the passage of a smoke filament down a known length of the tunnel is timed. One way of generating smoke is by allowing the vapors from concentrated hydrochloric acid and ammonium hydroxide to mix and form particulate NH_4Cl (handle these chemicals with care and ensure that the wind tunnel is safely vented). Although smoke-tests are most often used to evaluate airflow prior to testing insects, some smokes are used in conjunction with test semiochemicals. In such cases, the chemical smoke components should be evaluated in the absence of the semiochemical to ensure that they have no behavioral activity themselves. Kellogg and Wright (1962) found that ammonium acetate smoke, generated by mixing the vapors from NH_4OH and acetic acid, was the smoke least irritating to flying insects that they tested, so they used it combined with a host attractant to visualize the passage of flying insects through and up the odor plume.

Smoke from titanium tetrachloride (corrosive!) also has been used. This smoke is generated by placing a small quantity onto a cotton wick or rubber septum.

Apparently it is "mildly irritating" to flying moths, as Baker and Linn (1984) report that "oriental fruit moths will fly in apparently normal fashion upwind in the plume of pheromone and titanium tetrachloride smoke until they reach a critical point near the source and drop to the ground, apparently dead."

In addition to the relatively simple measures discussed earlier and summarized in Table 5.1, some of the more complex behaviors and behavioral transitions of organisms in both olfactometers and wind tunnels now can be analyzed through the use of video recordings. A "permanent" record of each subject's activity (or lack thereof) and motion track is made, and several measures of that activity, such as turning rate, velocity, time at rest, etc., can be extracted from the recordings relatively quickly (e.g., Bakchine et al. 1990). Movement in three dimensions can be calculated from the simultaneous movement recorded by two cameras placed orthogonal to each other. Several new software/hardware packages (e.g., Noldus Information Technology 1995) have removed much of the tedium in analyzing motion tracks (Baker & Linn 1984). On the other hand a large quantity of such data may be difficult to sort through if only a key behavior in a long sequence is of interest.

Historically, it has been far more difficult to monitor insect responses to host odors than to their sex pheromones (e.g., Finch 1986). This may be due to the fact that responses to food may be less specific than responses to pheromones released by potential mates. Nevertheless, the wind tunnel shown in Fig. 5.3 that was originally designed for pheromone orientation studies was successfully modified to assess the responses of apple maggot flies, *Rhagoletis pomonella*, to synthetic apple volatiles. The major modification was to include models of apple trees constructed from wooden dowels that supported twigs of leaves with apple foliage. Flies were released into the wind tunnel, and their pattern of movement among plant models was monitored in the presence and absence of released apple volatiles (Aluja et al. 1993). The inclusion of plant models allowed a more natural pattern of short flights from leaf to leaf that is characteristic of this species rather than flights of longer duration. An important point in this study was that the introduction of tree models completely disrupted laminar airflow within the tunnel. As a result, it is not possible to quantify movements as precisely as in systems where laminar flow is maintained.

5.2.3. Still-Air Bioassays

Assays based on simple diffusion of volatile compounds in air are limited to behaviors expected to occur over a relatively short distance from the source. In most cases, they are limited to demonstrating attraction (i.e., oriented movements toward a chemical source) or repellence (i.e., oriented movement away from a chemical source; Dethier et al. 1960). A serious limitation of still-air bioassays is that the processes of oriented movement to a chemical and arrestment after contact with a chemical after random movement may provide the same result.

Thus, the interpretation of the mechanism for the results from still-air bioassays may be ambiguous. Still-air bioassays also are not appropriate where organisms must orient themselves in a moving airstream (e.g. exhibit anemotaxis) before responding to a behavioral stimulus. Moreover, for very volatile compounds, the arenas could become saturated with vapors relatively quickly, thus eliminating the gradient to which the test organism could respond.

Small arenas constructed from petri dishes are commonly used to screen chemicals as attractants (Matsumoto 1968; Matsumoto & Thorsteinson 1970; Jones & Coaker 1977) because they require no expensive equipment and yet are capable of showing presence or absence of biological activity clearly. Bioassays with larvae of anthomyiid flies can involve no more than placing a 5-mm-diameter absorbent disk containing the test chemical, and another "control" disk treated with a similar quantity of solvent, at opposite sides of a 9-cm petri dish and releasing freshly emerged larvae at the center of the dish (Matsumoto 1968). The percentage of larvae responding to the test chemical, within as little as 5 min, can provide a measure of its attractiveness or repellence. A modification of the still-air bioassay is to trap the subjects that respond to the test and control dispensers where they can be counted at leisure. One way to do this is to prepare pitfall traps, where the emitters are at the bottom of a shallow well. As test subjects approach the emitters, they fall into the wells where they can be counted. The wells should be made from something smooth or slippery (e.g., glass or polyethylene vials) so that once trapped, the test organisms cannot escape. For data of this type, the frequency of test subjects trapped in each well can be analyzed by chi-square or *G*-tests against the null hypothesis of equal frequencies of trapping.

More detailed information on turning rates and movements can be obtained by providing a surface on the petri dish that will record and retain the larval tracks. Jones and Coaker (1977) utilized a thin layer of agar coated with talc over which larvae of the carrot fly, *Psila rosae*, crawled. Larval tracks were evident in the talc field, and permanent records of larval movement were made by photographing the plates. Time-lapse photography also has been used to provide a permanent track of movement in relatively small arenas (Bell & Tobin 1981), and more modern video equipment provides greater flexibility for such recording.

Still-air bioassays of chemicals in traps also can be performed with flying insects, although the size of the arena (i.e., the flight cage) must be considerably larger to accommodate flight. Care must be taken to avoid biases in cage design. For example, straight-sided cages have corners that may bias the numbers of subjects caught in traps near corners, whereas cylindrical cages avoid such a bias (Weston & Miller 1985). Even in cages without corners, one or more positions within the cage may be preferred regardless of treatment. Such "position effects" may be difficult to remove. Fortunately, they are easy to control for by alternating and rotating traps in a regular sequence. The process of rotating traps to different positions can be automated by placing traps on a slowly rotating floor powered by an electric motor (Weston & Miller 1985). Random assignments of traps to positions are less satisfactory because serendipitous associations between treatment and position may persist, especially if the number of trials and subjects is small (e.g., Singer 1986).

5.2.4. Modifications for Specific Purposes

While previous sections may summarize general procedures for the development of bioassays for volatile chemicals, there are a number of other, more specialized bioassays, particularly for various kinds of pheromones. The following summarizes some of the more commonly used methods for different classes of pheromones.

Sex pheromones are produced by one sex and affect the other, while *aggregation pheromones* are produced by members of either or both sexes and induce both sexes to aggregate (Borden 1985). In many cases, the bioassay for both of these pheromones are similar. Laboratory bioassays usually employ various combinations of wind tunnels described previously. The difference is largely in determining which of the two sexes produce the pheromone, and which ones respond. Mayer and McLaughlin (1991) and Arn et al. (1992, with updates) provide a comprehensive compilation and bibliography of known sex or aggregation pheromones for several hundred species of insects. These compilations should prove useful in developing bioassays for sex or aggregation pheromones for related species.

Bioassays for less volatile *trail pheromones* used by ants, termites, and other gregarious species are simple conceptually to develop. Trails are laid down on a suitable substrate, test subjects are released, and the observer notes if the trail is found and is followed. One method of providing artificial trails is to apply a circle of the test chemical in a suitable solvent to filter paper using a fine-point drawing pen (e.g., Rapidograph™ or similar) to ensure uniform application. After the trail is found, a relative measure of activity can be determined by measuring the length of the trail that the test subject follows before leaving it (Pasteels & Verhaeghe 1974).

Choice tests for trail pheromones are similar in concept to the bioassays using Y-tube or multiple-arm olfactometers, and eventually the subject is confronted with a choice between two or more substances. Choices are recorded, and the frequency of choices is analyzed by G-test against the null hypothesis of choices of equal frequency.

One method of providing such a choice is to apply intersecting circular trails of test chemicals to filter paper (Van Vorhis Key & Baker 1982; Fig. 5.4). When the subjects reach the point of intersection, they are expected to follow the trail of one component to the left or the other component to the right.

Another approach is to use short dowels treated with test chemicals in the

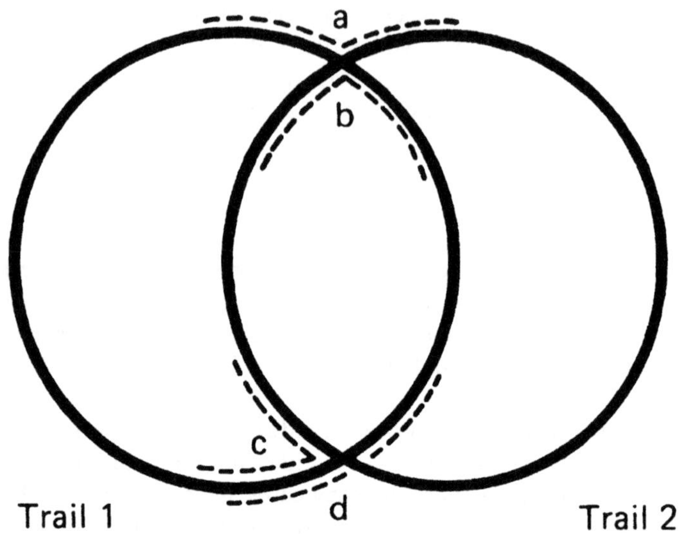

Figure 5.4. Schematic diagram of intersecting circular trails for choice tests using trail pheromones. Dashed lines indicate possible paths of trail-following by ants. Reprinted form Baker & Cardé (1984) with permission.

shape of a "Y" or "T" (Peterson & Fitzgerald 1991). The two branches can be treated with different substances, or a "test" substance and a "control," then the branch chosen by the subject can be recorded. Thus, as with olfactometers, subjects are required to make choices among semiochemicals. The major difference is that the semiochemical is not present in an airstream but on the surface that the subject traverses to the point where a choice needs to be made (also reviewed by Baker & Cardé 1984).

In contrast to sex and pheromones, where attraction is expected, the response to *alarm pheromones* depends on the social organization of the test species (Blum 1985). In noneusocial species, such as most aphids, alarm pheromones cause repellence, resulting in dispersal of the subjects. Thus, in principle, the major difference in bioassays between sex pheromones and alarm pheromones for these species is the direction of expected displacement relative to the source. While complex wind tunnels have been used in bioassays for alarm pheromones (e.g., Phelan & Miller 1982), the presence and identity of alarm pheromones often can be determined with relatively simple equipment.

For example, candidate alarm pheromones for the leaf-footed bug, *Leptoglossus zonatus*, were applied to narrow strips of filter paper and the solvent was allowed to evaporate. Then a treated strip of filter paper was inserted into a disposable glass pipette. The pipette was inserted into a rubber bulb of known volume (e.g., 3 ml), and the pipette was placed a fixed distance from test insects in a test arena. The bulb was squeezed to cause a puff of pheromone-laden air to pass over the test insect, and subsequent behavior (i.e., motion toward or away from the pipette) was noted. Test insects also were exposed to a puff of air from a similarly constructed "control" pipette, except that the filter paper strip was treated only with solvent (Leal et al. 1994). In such a test, it is imperative to carefully control and rotate the order of presentation of test and control stimuli to avoid any potential problems with sensitization or habituation to the airstream regardless of the presence of a candidate alarm pheromone.

For species such as the eusocial Hymenoptera and termites, however, alarm pheromones elicit aggressive behavior to drive off the intruder. This may include increased activity and attraction to the source of the pheromone for ants and termites (e.g., Roisin et al. 1990) to attack by bees and wasps (Blum 1985).

A widely used bioassay to assess the abilities of test chemicals to elicit attack by bees and wasps is to expose the test subjects to a mock "attack" by a threatening object. Such objects are constructed from a series of white and black balls or cubes placed near the colony. Test chemicals are usually applied to filter paper and placed in a natural airstream that carries the test chemical to the colony entrance. The ability of a test chemical to elicit attack is quantified by counting the number of test insects that attack the cubes or balls. Usually black objects receive more attacks than objects of other colors (Jeanne 1981; Kojima 1994).

To increase both the accuracy of counting the number of attacks and the safety of the investigator, Visscher and Vetter (1995) developed an automated apparatus. Their target was a hollow ping-pong ball painted black, into which a small microphone had been inserted. As various hymenopterous species attacked and attempted to sting the ball, the sound of their contact with the ball was recorded, and the number of contacts was electronically counted. From this information, the attack rate could be calculated. In their studies, Visscher and Vetter (1995) were primarily interested in quantifying the effect of smoke on reducing aggressive behavior; there seems no reason, however, not to adapt their electronic attack counter to assay the activity of potential alarm pheromone components in a series of sequential tests.

Social insects utilize various *surface pheromones* (i.e., compounds that are found on the surfaces of individuals) to elicit several behaviors, including nestmate and caste recognition, grooming, and food exchange (reviewed by Ali & Morgan 1990). Recognition pheromones in ants are produced by individual queens. These pheromones then are adsorbed on to the surfaces of other members of the colony (Holldobler & Wilson 1990). While it is relatively simple to demonstrate nestmate versus nonnestmate recognition by transferring individuals between nests and determining if the intruder is attacked (e.g., Breed 1981, 1983), it is more difficult to develop bioassays to explore the chemical basis of recognition.

In one series of bioassays, however, individual chemical compounds were applied to filter paper and placed in 0.45-L paper containers. Then 10 worker bees ("intruders") were introduced into the container and left, with appropriate food and water, for 5 days to adsorb the compound. At the same time, other groups of 10 bees ("receptors") were established in separate containers and exposed either to identical, or different, chemicals as the intruders. After 5 days of absorption, each "intruder" worker bee then was placed in a separate container with the group of 10 "receptors." Then "attack" (i.e., biting or stinging of the intruder by the receptors) or "acceptance" (i.e., the absence of attack) of the intruder was recorded. Under the hypothesis of recognition, intruders should be accepted if they had adsorbed the same chemical as the receptors but should be attacked of they had adsorbed a different chemical. Appropriate controls would include (1) utilizing closely related bees (e.g., half- or supersister colony mates) and (2) utilizing both intruders and receptors, both jointly and independently, that were exposed only to solvent-treated filter paper (Breed et al. 1992). By using esters of long-chain acids (10–16 carbons) and alcohols (6–14 carbons) differing in chain lengths, the pattern of chemical discrimination were determined. Interestingly, the limits of discrimination were not simply a function of the differences in chain length. In some tests, compounds of similar length elicited different behavior, while in other tests, compounds of differing chain lengths elicited similar behavior (Breed et al. 1992).

5.3. Bioassays for Contact Oviposition Stimulants—Two Case Studies

A choice by adult organisms for a site or host for reproduction will differ from choices for adult feeding when adults feed on different food sources than their progeny. Such a situation occurs in several groups of insects, such as most Lepidoptera and some Hymenoptera, where the adults feed, if at all, on nectar, while their progeny feed mostly on plant tissue or other insects. For other groups of insects, where adults feed on the same material as their larvae, such as most Homoptera and chrysomelid Coleoptera, then there may be no need for adults to evaluate potential hosts for their suitability for survival of progeny separately from their suitability for their own nutrition. Even among Lepidoptera, elaborate host recognition mechanisms for oviposition are unlikely in species where progeny have high dispersal abilities and can choose their hosts directly. In the gypsy moth, *Lymantria dispar*, eggs are deposited on the bark of trees, then the newly hatched larvae spin silk threads and then disperse by wind until they alight on the leaves of potential hosts. A particular oviposition stimulant for female gypsy moths therefore is not expected.

All oviposition bioassays have the same problem of inducing test organisms to exhibit oviposition behavior under tightly controlled and unnatural conditions. The solution to this problem may vary greatly from species to species, as illustrated by the examples below.

Oviposition assays for bean weevils are among the simplest. For these insects, candidate semiochemicals are often evaporated onto glass beads roughly the size of the natural host, and the number of eggs laid on treated and untreated beads is recorded. Control beads are treated with solvent only (Jermy & Szentesi 1978; Matsumoto et al. 1994).

Oviposition bioassays for flying insects often are more complicated. Rather than try to catalog all of the oviposition bioassays for flying insects that have been attempted, I will try to illustrate how various problems were overcome for two species. One involves a plant-feeding butterfly, while the other is a parasitic wasp.

5.3.1. Butterflies

Some of the problems in understanding the behavioral processes that are involved in discriminating host from nonhost are illustrated by the design of bioassays to identify specific oviposition stimulants for swallowtail butterflies. At one time, it was generally believed that insects utilized specific secondary plant compounds as "token" or "sign" stimuli to identify their hosts (Verschaffelt 1911; Dethier 1941; Fraenkel 1959, reviewed by Feeny et al. 1983). A great deal of effort was expended to identify such token stimuli, and most of it was unsuccessful; the behavior of host selection was rarely linked to a single plant chemical. Instead, host selection came to be appreciated as the end result of several behavioral responses to a number of physical, chemical, and visual cues (reviewed by Miller & Strickler 1984).

From the standpoint of developing discriminating oviposition bioassays to a single test chemical, this implies that it is often necessary to supply one or more background cues before any behavioral response is seen. For example, in the presence of citrus odor, adults of the butterfly *Papilio demoleus* fly to artificial, paper leaves and drum on the surface with their foretarsi, especially if the paper leaves were green or blue-green (Vaidya 1969). In the absence of citrus odor, or if the odor was too strong, then no discrimination among colors was observed. Saxena and Goyal (1978) investigated the chemical mediation of host selection by the same species using plant extracts applied to muslin disks. The response to disks treated with ether extracts of citrus leaves was greatly enhanced over ether-treated "control" disks when disks were moistened with water. Through a number of subsequent experiments, Saxena and Goyal (1978) concluded that the optimum oviposition response for *P. demoleus* was observed when extracts were applied to moistened yellow-green disks. Similarly, Ichinosé and Honda (1978), investigating the response of *Papilio protenor* to host plant chemicals, found most discriminating responses occurred using extracts applied to moistened filter paper, although, in contrast to *P. demoleus*, color was apparently unimportant.

Later, Feeny and collaborators (Feeny et al. 1983, 1988) compared several bioassays to determine the chemical(s) in wild carrot, *Daucus carota*, eliciting oviposition by the black swallowtail butterfly, *Papilio polyxenes*. A number of artificial substrates were tried, including artificial "leaves" made of (1) green construction paper; (2) agar, cellulose powder, and green food coloring (following

Hovanitz & Chang 1964); and (3) cellulose sponges for household use. The two latter substrates were investigated in an attempt to provide moisture for a longer time period than could paper. Eventually, an artificial "plant" (Fig. 5.5) was developed using cellulose sponges colored with green food coloring and pretreated with mineral oil. Such plants were suitable for unattended, free-flight bioassays for up to 4 h. Greatest responses were obtained in bioassaying nonvolatile, contact cues when small, intact carrot plants, covered with gauze, were placed in the cages with the artificial plant to saturate the air space with carrot odor (Feeny et al. 1988). The gauze was used to eliminate the possibility that the carrot plants were providing any visual cues. When treated with either ethanolic or aqueous extracts of carrot leaves, female butterflies confined in ~ 1 m^3 cages with artificial plants would oviposit up to 40 eggs per female per plant in 4 h (Table 4 in Feeny et al. 1983). To obtain such high oviposition rates, it is important to note that sufficient space and light (e.g., natural daylight in a greenhouse) needed to be provided for flight, as well as uniform background levels of color, odor, and moisture.

Free-flight bioassays can also be used to test chemical compounds for deterrent

Figure 5.5. A female *Papilio polyxenes* preparing to oviposit on an artificial "leaf" during a free-flight bioassay experiment. Reprinted from Feeny et al. (1983) with permission.

as well as stimulatory activity. This approach has been used, for example, by Renwick and collaborators (below) to assess the biological activity of plant chemicals governing host plant use by pierid butterflies. The procedure to assay plant chemicals for deterrence simply involves starting with a host plant of known preference, then adding test chemicals in a suitable solvent to the surface to the "experimental" plants, and pure solvent to the "control" plants. A measure of deterrent activity then can be calculated as the difference between ovipositions (e.g., the number of eggs) on the experimental and control plants (Renwick et al. 1989). The same approach also has been used to detect stimulants. Here, the bioassay starts with a plant species known not to be a host for the test insect, then candidate stimulants in solvent are applied to the foliage of test plants, and pure solvent to the control plants (Huang et al. 1993).

It is not clear how generally useful the technique of adding test chemicals to intact plants may be. First, the chemicals to be added must be soluble in relatively polar (ideally, aqueous) solvents. Surfactants (i.e., Tween® or Triton® compounds) may be needed on leaf surfaces are very waxy or otherwise hydrophobic. Nonpolar solvents rapidly dissolve the protective waxes off of the leaf surface and quickly lead to desiccation of the leaf. Second, the use of known nonhost plants raises the potential problem of overriding differences between hosts and nonhosts that cannot be overcome by an stimulant. Such differences may involve leaf shape, color, or physical and chemical characteristics of the leaf surface. Thus, care needs to be taken to demonstrate the attractiveness of nonhosts after treatment with known stimulants before results using potential stimulants can be evaluated.

Free-flight bioassays, although providing a high level of behavioral realism, are also subject to a number of potential problems. One is that the presence of eggs from prior ovipositions may inhibit females from ovipositing subsequently. Biologically, this makes sense to minimize competition among offspring. Oviposition-deterring pheromones have been demonstrated for several phytophagous (e.g., Prokopy 1981; Messina & Renwick 1985) and parasitic (e.g., van Lenteren & DeBach 1981) insect species. Thus, at high overall levels of oviposition, the presence of oviposition-deterring pheromones would cause the less preferred or "control" treatments to be selected for oviposition relatively more frequently than at lower levels. The presence of such pheromones should be suspected if the relative preference of treatments changes with the number of ovipositions that have occurred.

Another problem with free-flight bioassays is that captive insects may be less discriminating than unconfined insects. This may result from a number of mechanisms, but two of primary importance in bioassay design are (1) the absence of normal continuity and transitions between multiple behavioral events (e.g., transition from flight to postalighting behaviors), and (2) unnaturally high ovipositional motivation in captive insects that are not allowed to oviposit between trials. Both mechanisms would tend to reduce the ability of laboratory insects to discriminate compared to insects under natural conditions. The first mechanism would reduce overall oviposition, while the second may cause high levels of indiscriminate oviposition. The only way to deal with these problems is to (1) conduct preliminary experiments to determine which combinations of insect age and experimental conditions provide the greatest discrimination and (2) standardize the physiological conditions under which subsequent bioassays are conducted.

Free-flight bioassays are subject to the same sort of position effects discussed earlier (see section 5.2.3 on still-air bioassays). Such effects can be minimized, as discussed previously, by eliminating edges and corners of flight cages where possible, and by alternating and rotating the position of test oviposition substrates either manually or automatically (Weston & Miller 1985).

Instead of requiring the full behavioral sequence from flight to egg production to be the bioassay criterion, it maybe expedient to bypass, or "short circuit," some of the steps in the process and focus upon critical steps, or "indicator" behaviors. This may result in little loss in resolution and great reductions in time and effort. Whenever such indicator behaviors are used, it is important to establish that the behavior is consistently displayed in the overall behavioral investigated, and not displayed in alternative behaviors. Additionally, results using indicator behaviors should be confirmed with bioassays using the complete behavioral sequence.

In the case of the swallowtail butterfly bioassays, the important behavioral observations were simplified by noting if a female curled her ovipositor in preparation for oviposition in response to a test chemical applied to a strip of moistened filter paper (Feeny et al. 1983, 1988; Fig. 5.6). Using the "abdomen curling" bioassay, a number of oviposition stimulants and synergists were identi-

Figure 5.6. A female *Papilio polyxenes* curling her abdomen in preparation for oviposition during a filter-paper bioassay. Reprinted from Feeny et al. (1983) with permission.

fied for several papilionid species (Feeny et al. 1988; Honda 1990; Papaj et al. 1992; Nishida 1994; Honda & Hayashi 1995). An interesting modification of this bioassay is to give greater weight to the more "complete" oviposition responses, e.g., tarsal drumming followed by abdomen curling with egg deposition is given more weight than drumming and abdomen curling only, which itself is given more weight than drumming only (Honda & Hayashi 1995). Such weighting would be most appropriate when the behavioral homogeneity of the test subjects could be assured. Conversely, it may introduce biases in the results when the propensity to complete oviposition varies naturally among subjects and the number of subjects is small.

If the response to short-range contact cues are of most interest, then another way to simplify an oviposition bioassay would be to place the adult insect directly on control and test substrates in a regular sequence. This is similar to sequential assays of host-plant preference of adult butterflies, which has been described in detail by Singer (1986). An important consideration in this type of bioassay is the order of presentation. Forced prior experience with one host may influence subsequent host choice, both positively or negatively. Tephritid fruit flies become less inclined to accept plants on which oviposition has been attempted and has failed (Prokopy et al. 1982). These results suggest that "negative" experiences with relatively unattractive semiochemicals may enhance the apparent attractiveness of other such chemicals. The mechanisms underlying behavioral changes resulting from positive experiences are more difficult to identify. The seed weevil, *Callsobruchus maculatus*, is less likely to accept a less-preferred host for oviposition after having encountered a more preferred host (Mark 1982). Here, the reduction in the response to less attractive hosts may result either from learning or from reduced motivation after having recently oviposited.

5.3.2. Parasitic Wasps

Host recognition by parasitic Hymenoptera also can be a complex process involving a series of behaviors after initial host contact that must be completed in sequence before an egg is laid. One such sequence is illustrated in Figure 5.7, which diagrams the processes by which *Aphytis melinus*, a parasitoid used for biological control of California red scale, *Aonidiella aurantii*, first identifies a potential scale insect species as a potentially suitable host, then assesses the individual quality of that host for utilization (Luck et al. 1982). Egg deposition is the final event of a chain of other behaviors that must be completed in sequence. The series can be broken in several placed and restarted, but no steps can be skipped.

Unlike most other host-parasitoid systems, the system involving *Aphytis* wasps and their scale host permit complete isolation of external (i.e., surface) cues from internal cues for the assessment of host quality by the wasp. Diaspidid scale insects are protected by a covering of hardened, nonliving material secreted by

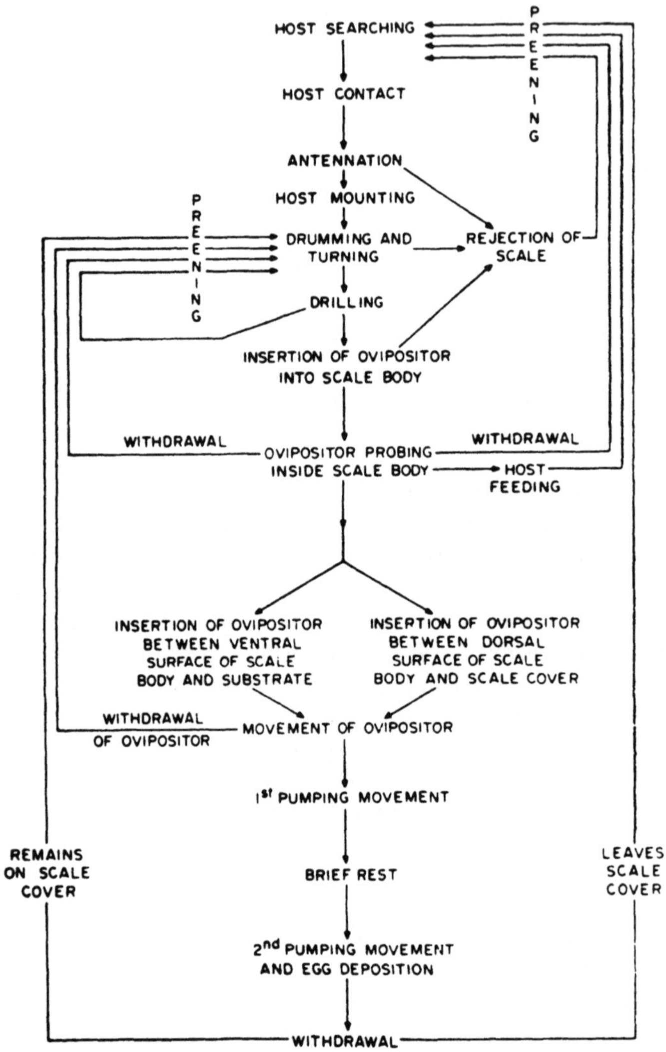

Figure 5.7. The behavioral sequence exhibited by *Aphytis melinus* during oviposition. Reprinted from Luck et al. (1982) with permission.

the scale insect. Parasitic wasps must select scales for utilization first on the basis of characteristics of the cover before they probe through the cover and penetrate the scale body. The fact that scale covers are not attached to scale bodies during the scales' growing stages provides the opportunity to assess the relative importance of surface (i.e., cover) characteristics from internal physiological characteristics in host selection and identification. One early approach was simply to switch covers among host and nonhost scale species (Baker 1976). Alternatively, if only the cues involved in selection of hosts for probing are of interest, rather than actual oviposition, then covers can be assayed in the absence of scale bodies.

To determine the identity and relative importance of chemical cues from scale covers on host identification and acceptance by *Aphytis* species, a suitable bioassay was required. The first such bioassay simply involved covers of California red scale washed with various solvents and placed in observations chambers made of glass microscope slides and Plexiglas™ (Quednau & Hübsch 1964) (Fig. 5.8). This was sufficient to demonstrate the existence of at least one chemical cue (i.e., a relatively polar, water-soluble kairomone) for oviposition. The importance of the chemical cue relative to such physical cues as scale cover size, shape, and color could not be evaluated. Also, in the absence of precise chemical measurements, it was uncertain if simple washing treatments removed all of the native chemical cues.

Both of these problems were solved in later studies. First, the washed scale covers were successfully replaced as the bioassay substrate by small (2.0-mm-diameter) disks of filter paper (Luck & Uygun 1986). In initial studies, disks

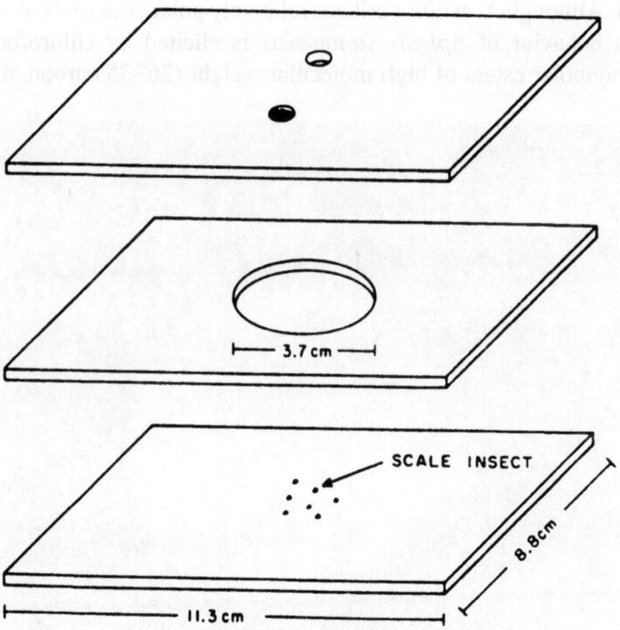

Figure 5.8. Schematic diagram of a bioassay arena made of Plexiglas™ and microscope slides for *Aphytis melinus*. The parasitoid is introduced into the hole in the top slide, which is then stoppered for the duration of the bioassay. Reprinted from Luck et al. (1982) with permission.

were treated with scale cover extracts then placed on a detached lemon fruit inside a plastic observation chamber constructed from clear plastic tubing. Adult female wasps were introduced into the chamber, and their behavior was observed under a dissecting microscope. Although kairomone concentration could be controlled only on a relative basis (i.e., the units were scalecover equivalents per disk), adult wasps exhibited normal investigative behavior and even successfully inserted their ovipositors into the filter paper disks (Fig. 5.9).

Although lemon fruit is an appropriate host for California red scale and may provide appropriate visual and olfactory background cues for *Aphytis* wasps, the quality and intensity of such cues are likely to vary among lemons as a function of lemon age and storage conditions. The bioassay was simplified and standardized still further by eliminating these background cues and simply placing treated disks in either glass petri dishes (Hare et al. 1993) or the glass and Plexiglas™ observations chambers used originally (Quednau & Hübsch 1964). Using this bioassay, the biological activity of different chemical fractions of the crude-scale cover extract were compared, and the kairomone was identified as an ester of the phenolic acid, caffeic acid and the amino acid, tyrosine (Hare et al. 1993; Millar & Hare 1993).

The potential range of chemical cues used by related *Aphytis* species may be quite broad. Although *A. melinus* utilizes relatively polar cues for host recognition, oviposition behavior of *Aphytis yanonensis* is elicited by chloroform-soluble, relatively nonpolar esters of high molecular weight (26–35 carbon atoms) acids

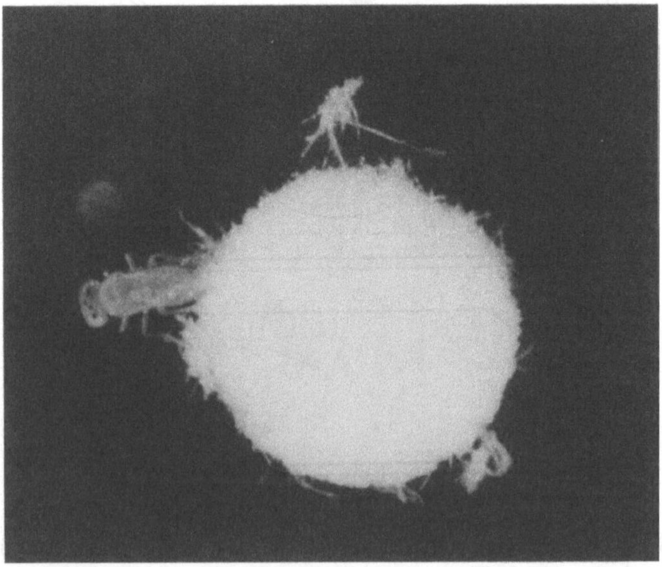

Figure 5.9. ***Aphytis melinus*** probing a filter paper disk treated with *O*-caffeoyltyrosine extracted from covers of California red scale.

and alcohols isolated from the waxy cover of its scale host, the arrowhead scale, *Unaspis yanonensis* (Takahashi et al. 1990). These compounds were bioassayed simply by applying the melted waxes to the glass surface of the observation chambers described previously. No physical substrate was required. Not only do the general chemical classes and polarities of the two oviposition stimulants differ, but *A. melinus* apparently requires a three-dimensional substrate for probing behavior to be released (J.D.H., personal observation). The differences in responses of the two *Aphytis* species suggests caution in adopting a "one-size-fits-all" approach toward developing bioassays or expectations for related species.

Bioassays for *oviposition-deterring* or *host-marking* pheromones are conceptually indistinguishable from bioassays for other oviposition deterrents. For example in studies to identify and an oviposition-deterring pheromone for the cherry fruit fly, *Rhagoletis cerasi*, test chemicals were applied to artificial fruit (i.e., made from wax). Treated and control artificial fruit then were placed in a flight cage with five adult female fruit flies. The number of ovipositions on treated and control fruit were recorded and then compared statistically against the null hypothesis of no difference and the working hypothesis that the application of candidate oviposition-deterring pheromone would reduce the number of ovipositions relative to the control (Boller & Horter 1985).

5.3.3. Data Analysis

The method of statistical analysis of an oviposition bioassay depends on the specific bioassay and the parameters measured. In general, these assays will involve counts or frequencies (e.g., the number of eggs produced, or the number of ovipositions attempted), and analysis by goodness-of-fit *G*-tests will be most appropriate. General guidelines for the analysis of these data are presented in section 5.1.3.

5.4. Measurement of Preference

5.4.1. Host-Plant Selection for Feeding

This section summarizes methods of testing nonvolatile chemicals that either stimulate or deter feeding. Chemicals that elicit feeding are called feeding stimulants or phagostimulants; those that inhibit feeding are called deterrents, feeding inhibitors, gustatory repellents, or antifeedants (Dethier et al. 1960, reviewed by Lewis & van Emden 1986; Bernays & Chapman 1994). These terms simply describe the chemical's role in a particular behavioral bioassay without regard to any potential physiological role. Compounds that stimulate feeding need not be nutrients (although many are), nor must compounds that deter feeding be toxic.

Bioassays therefore must be designed carefully so that the measured variable accurately represents a behavioral response. Because the behavioral effect of a chemical may or may not be independent of its nutritional value, these two properties must be separated experimentally. One way to accomplish is to use only short-term assays to avoid postingestional effects on food choice. Another approach is to bypass the sensory system entirely. Methods to bypass behavioral responses are discussed in more detail below (see sections 5.6 and 5.7). Moreover, the same chemical may play different roles in different contexts. Compounds that deter feeding by one species may stimulate feeding by others. Even within one test species, the same chemical may vary in its behavioral effect depending on factors such as concentration and physiological state of the insect.

Another general consideration in bioassay design concerns the type and intensity of background stimuli that the substrate may provide. Such stimuli may be required in some bioassays (e.g., those testing for deterrents) but may be an unavoidable complication in others. Investigators must be aware that the strength of behavioral responses to one chemical compound may depend on the presence and concentration of others. When working with plant tissues, for example, the measured responses to a potential phagostimulant may depend on the presence and concentration of deterrents. If the concentration of these deterrents varies, due to variation in growing conditions (leaf age, exposure to light, fertilization level), the variation in deterrents may obscure the response to controlled doses of stimulants and vice versa (Lewis & van Emden 1986; Hare 1987; Johnson & Bentley 1988). The choice of a substrate and its baseline activity of preference often depends on whether stimulants or deterrents are being assayed.

Bioassays are easier to set up for stimulants because stimulants increase feeding over the control treatment. The control treatment for a bioassay of stimulants is a substrate lacking the phagostimulants to be tested (or other phagostimulants). It may contain deterrents. In either case, the control treatment is not expected to be ingested. Deterrents will decrease feeding relative to the control. Thus, for bioassays of deterrents, "control" treatment must be palatable and provide some level of stimulus against which the reduction due to the presence of a deterrent can be measured. In contrast, the control treatment for bioassays of stimulants need only be neutral. With a substrate that provides a background, baseline level of acceptability, then both stimulants and deterrents can be assayed simultaneously.

5.4.1.1. Neutral Substrates

For short-term studies of behavior, an ideal test substrate would be a neutral, nonliving substrate with no nutrient value to which test chemicals can be added. Filter paper has been used in the past (e.g., Dethier 1941), but cellulose acetate filters (Bristow et al. 1979; Navon et al. 1993) and disks made from thin glass fibers (Städler & Hanson 1976; Adams & Bernays 1978) also have been used for their ease of chewing compared to filter paper, in addition to their chemical purity, absorbency, and uniformity.

Occasionally, it is possible to use test subjects in a bioassay directly after test chemicals have been separated on two-dimensional media. For example, Metcalf et al. (1980) bypassed transferring separated semiochemicals to neutral substrates simply by allowing cucumber beetles to feed directly on thin-layer chromatography (TLC) plates containing cucurbitacin phagostimulants. Beetles consumed the spot of silica gel containing the feeding stimulants (Fig. 5.10). Such techniques may work well when studying species-specific phagostimulants to which the insects are very sensitive, but investigators should not expect similar results from all species.

First, one should not always expect that the optimal concentration of semiochemicals applied to the plates for good separation and visualization also will

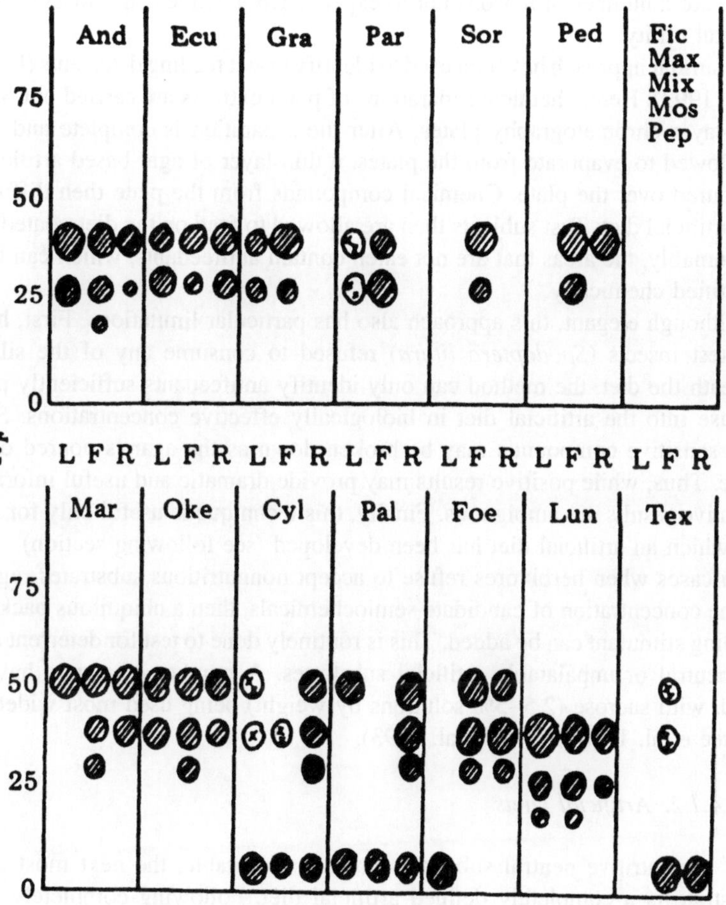

Figure 5.10. Diagrams of TLC plates from the chloroform extracts of the leaves (L), fruits (F), and roots (R) of 18 *Cucurbita* spp. showing areas eaten from the silica gel plates by *Diabrotica* spp. Reprinted from Metcalf et al. (1980) with permission.

be the optimal concentrations for feeding stimulation. Preliminary experiments using different concentrations of crude extract applied to thin layer chromatography plates prior to separation should be undertaken first. Investigators also should not be surprised that subjects of some species refuse to feed on silica gel regardless of the concentration of feeding stimulants. For example, silica gel is a sufficient irritant and desiccant for some insect species to be used in pest control. Small insects like the parasitoid, *Aphytis melinus*, will immediately began grooming themselves when placed on silica gel plates. Eventually, they cover themselves with silica gel powder, then die (J.D.H., personal observation). In summary, while it is a very powerful bioassay when insects identify their feeding stimulants on two-dimensional chromatography plates or sheets of chromatography paper, there are a number of reasons not to expect such a convenient bioassay to be of general utility.

A similar approach has been used to identify insect feeding deterrents (Escoubas et al. 1992). Here, chemical separations of plant extracts are carried out on silica thin-layer chromatography plates. After the separation is complete and solvent is allowed to evaporate from the plates, a thin layer of agar-based artificial diet is poured over the plate. Chemical compounds from the plate then diffuse into the artificial diet. Test subjects then are allowed to feed on the diet-coated plates. Presumably, the areas that are not eaten contain antifeedants, which can later be identified chemically.

Although elegant, this approach also has particular limitations. First, because the test insects (*Spodoptera litura*) refused to consume any of the silica gel beneath the diet, the method can only identify antifeedants sufficiently polar to diffuse into the artificial diet in biologically effective concentrations. Second, heat-sensitive compounds may be broken down as the agar is poured over the plate. Thus, while positive results may provide dramatic and useful information, negative results are ambiguous. Finally, this technique is useful only for species for which an artificial diet has been developed (see following section).

In cases when herbivores refuse to accept nonnutritious substrates regardless of the concentration of candidate semiochemicals, then a ubiquitous background feeding stimulant can be added. This is routinely done to test for deterrent activity on neutral or unpalatable artificial substrates. A number of sugars have been used, with sucrose (2.5–5% solutions by weight) being used most widely (e.g., Cottee et al. 1988; Navon et al. 1993).

5.4.1.2. Artificial Diets

If nonnutritive neutral substrates are not acceptable, the next most uniform substrate is a completely defined artificial diet. Following completely defined diets in uniformity are semidefined diets (e.g., those that include some plant material) and plant tissues to which test chemicals are added.

Artificial feeding substrates offer uniformity, evenness of application, and ease of preparation and measurement. For short-term preference studies, agar has been widely used. For materials to be applied to the agar surface, a simple agar solution (2.5 g per 100 ml of water) is prepared and left to cool. Disks of an appropriate size are punched from the agar using an appropriate tool such as a cork borer. Test chemicals are then applied to the surface, and treated and control disks are presented to subjects in petri dishes or other small containers. Alternatively, test chemicals can also be incorporated into the agar before it cools. This can be done directly for heat-stable, water-soluble chemicals. For water-insoluble chemicals, the test chemicals in solvent are often applied to cellulose powder, then the powder is incorporated into the liquid agar at a concentration of 5% by weight (Chan et al. 1978), before the agar cools. Control treatments must include an equivalent quantity of cellulose power treated only with solvent. Care must be taken to ensure that the powder is incorporated uniformly into the diets, and that the powder does not settle out while the agar cools.

More complex artificial diets are often used routinely to study preferences of caterpillars. For example, diet, in cups or disks, has been used to test chemicals against a number of lepidopteran species (Trial & Dimond 1979; Doskotch et al. 1980; Kaethler et al. 1982; Klocke & Chan 1982; Meisner et al. 1982; Berdegue & Trumble 1996). Unfortunately, it is remarkable how few suitable diets have been developed for nonlepidopteran insect species (e.g., Singh & Moore 1985). A few diets have been developed for some plant-feeding beetles, and there are also artificial feeding systems for some homopterans that are comprised of various chemically defined solutions enclosed in membranes (e.g., Mittler & Dadd 1962). With the exception of lepidopteran diets, most only are suitable for short-term (e.g., 24–72 h) feeding studies and do not support normal growth and development (e.g., Hsiao & Fraenkel 1968; Hare 1987). This general lack of suitability for normal growth must be considered in the design and interpretation of the bioassay, particularly when test organism must be reared on natural hosts prior to testing.

5.4.1.3. Plant Tissue Substrates

If plant tissues are to be used, then the investigator must make a number of carefully considered decisions about bioassay design. Obviously, leaves for testing should be carefully selected. Differences in leaf age, damage, disease, or water content may all be sources of differential feeding. Such differences need to be minimized if consistent behavioral results to test chemicals are to be observed. Finding a test plant that lacks the phytochemical to be tested and that is acceptable to the test subject is particularly difficult for monophagous or oligophagous insects. The most realistic bioassays would employ whole plants, but these also are the least convenient. While it may be more convenient to utilize excised leaves or disks punched from leaves, the process of excision or wounding necessary to prepare these substrates may introduce changes in leaf physiology substantial enough to alter the outcome of the test.

Ideally, the test chemical should be selectively removed from the plant without disturbing the remaining plant tissue. Sometimes this can be accomplished for chemicals on leaf surfaces (e.g., Dimock & Kennedy 1983), but selective removal of internal leaf constituents usually is impossible.

More commonly, test chemicals in solvent are applied to a plant surface by either dipping, spraying, or applications with a small paint brush (Bernays 1978; Berenbaum & Feeny 1981; Wagner et al. 1983). This technique is not particularly suitable for chemicals that degrade in the presence of oxygen or light, nor is it appropriate for the bioassay of highly volatile substances. In some cases, it may be difficult to estimate the quantity applied, so it is advisable to conduct analyses of a sample of treated leaves to determine the actual quantity applied. After the investigator determines the quantity that was applied, then replicate samples can be prepared using a standardized treatment methodology. Surfactants such as the Tween® or Triton® compounds (0.1%) are useful to ensure complete and even application. Surface applications can be used only for compounds that are soluble in moderately polar solvents (e.g., acetone, water, or aqueous mixtures of alcohol). The use of less-polar solvents (e.g., methylene chloride, hexane, ether) usually removes leaf surface waxes, causing rapid leaf desiccation and necrosis. Repeated applications can be used when large amounts must be added; each application should be allowed to dry thoroughly before the next is made. A gentle stream of air or nitrogen aimed at the leaf hastens drying. For repeated applications, solutions should be sprayed on rather than brushed on for highest uniformity of treatment.

As an alternative to solvent, chemicals may be dissolved or suspended in warm (32°C) gelatin, which is then painted uniformly on the leaves (Wolfson & Murdock 1987). The optimal gelatin concentration and other methodological factors may vary with both the insect to be tested and leaf surface characteristics of the particular plant species (Wolfson & Murdock 1987).

Presentation on leaf surfaces also may be inappropriate for chemical compounds that are normally found in more complex matrices within the leaf, where they either may be protected from herbivore's sensory apparatus, or where their activity may be subject to modification by other internal phytochemicals. For such chemicals a more realistic application procedure would be through infiltration under pressure (Schoonhoven & Jermy 1977), or through uptake through the plant's vascular tissue (e.g., Erickson & Feeny 1974; Blau et al. 1978). Again, this technique is limited to compounds that have some solubility in water. Such procedures often result in variable quantities of chemicals taken up, thus chemical analyses of infiltrated leaves must be performed to determine the quantity of chemical to which the test organisms actually were subjected. Additional potential limitations are the possibilities that novel chemicals may damage plant cells when infused into plant tissue at high concentration, or that semiochemicals may undergo chemical change (i.e., oxidation) after removal from the source plant and before uptake into the test plant.

5.4.1.4. Leaf Disks

Leaf disks are more convenient than whole leaves because they are of uniform size and shape and can be easily arranged in behavioral arenas, such as petri dishes. While such care in controlling for size, shape, and position may be necessary to develop highly discriminating bioassays, preliminary experiments should be made to ensure that results using leaf disks are consistent with those using whole leaves.

For example, the observed preference of the grasshopper, *Abracis flafolineata*, depended on the method of presentation of two *Piper* species (Marquis & Braker 1987). When presented as whole leaves with their petioles in water, the grasshopper preferred *P. sancti-felicid* over *P. arieianum*. When presented as whole leaves without water, no preference was observed. When presented as leaf disks, however, the observed preference was opposite that when intact leaves were presented in water. The difference in relative preference presumably was a function of different amounts of water loss by the leaf tissue in each treatment and variable changes in disks from wounding (Marquis & Braker 1987).

Even the size of leaf disks may affect the outcome of preference tests, as smaller disks have relatively more "wounded" edge tissue that may affect the suitability of center tissue more than for larger disks (Jones & Coleman 1988). Leaf disks are not recommended for work with sucking insects. Physiological changes seem to affect the phloem contents very rapidly (e.g., Muller 1966; van Emden & Bashford 1976) and may well induce a generally unsuitable condition.

5.4.1.5. Other Natural Substrates

Specialized substrates include twigs for bioassays of sawflies (All & Benjamin 1976; Ohigashi et al. 1981) and bark beetles (Ascher et al. 1975). Blocks of wood (Lenz & Williams 1980), board, logs, and paper towel disks (Rust & Reierson 1977) have been used for bioassays involving termites. The original references should be consulted for detailed descriptions of methods.

5.4.2. Details of Bioassay Design

For studying a chemical normally occurring in a potential host, the concentration found in or on the appropriate tissue can be tested to establish an appropriate starting point. However, because the test substrate may be substantially different from the natural substrate, activity, or lack thereof, at normal concentrations may not be meaningful. Test materials normally found within leaves may be far more active when applied to leaf surfaces. Conversely, surface chemicals may be highly diluted when incorporated into artificial diets or even when applied to filter paper. Thus, a wide range of concentrations (e.g., log series) should be employed to discern an appropriate range of activity, followed by a narrower range (e.g., linear or geometric series) to refine the pattern of activity.

Food deprivation prior to the bioassay may be desirable to increase overall activity, but hunger also may lower discrimination. Again, pilot tests can be run to determine if deprivation affects behavior. Tests are often terminated when 50% of either substrate in a choice trial is consumed. This ensures that apparent discrimination will not be lost if insects entirely consume the preferred substrate, followed by the less-preferred substrate simply due to hunger (Jermy et al. 1968). Water may need to be provided during the test but may not be needed with artificial diets or with turgid leaves. For studies with leaf disks or excised leaves without additional water, it is usually necessary to maintain high humidity, for example, with moistened filter paper lining the test arena to avoid desiccation of leaf material.

The length of the trial also may be an important consideration if insect preferences change over time. Lewis and van Emden (1986) review earlier studies indicating that prior experience may affect insect diet choice either through induction (Jermy et al. 1968), habituation (Jermy et al. 1982), food aversion learning (Dethier 1980), and associative learning (Blaney & Winstanley 1982). In experiments where such effects are to be minimized, subjects could be reared on a pretest diet not containing any of the test chemicals. This may be difficult, however, for species where no artificial diet has yet been devised. On the other hand, the process of switching insects from a rearing diet to a test diet also may be problematic for insects such as *Manduca sexta*, where specific preferences are induced by prior feeding (Städler & Hanson 1978; de Boer & Hanson 1984). In such cases, the outcome of a preference test may greatly depend on the diet prior to testing. One way to avoid problems with behavioral induction is to use neonate larvae for testing. The drawbacks to using neonates include difficulties in obtaining accurate estimates of consumption because they eat so little. Again, care must be given to the question actually to be addressed, and preliminary studies undertaken to develop the most meaningful bioassay protocol.

5.4.3. Data Collection

If the test chemical was evenly applied, the amount of substrate ingested can be used to calculate the amount of chemical ingested by chewing insects. For substrates of even and uniform consistency, such as filter paper disks, membranes, and leaf disks, ingestion is often estimated from the area consumed (but see Waller & Jones 1989). Area can be measured by tracings or photocopying eaten substrates onto graph paper then estimating area consumed, with the use of planimeter ocular grids, electronic area meters, electronic digitizers attached to a computer (e.g., Meade & Hare 1993), or through more sophisticated image processing (Hagerup et al. 1990). Implicit in the process of estimating consumption from area are (1) that the test chemical was evenly applied to the substrate and (2) for leaf disks, that there is a constant relationship between dry and fresh weight and that the insects feed completely through the disk. For small insects photographic enlargements of the substrate prior to measurement may be necessary (Bristow et al. 1979), although this increases the expense.

Where substrates are irregularly shaped (e.g., intact leaves) or where the volume consumed cannot be accurately estimated from surface area consumption (e.g., artificial diets), then consumption must be measured by weight loss of the substrate. Weight changes should be on a dry-weight basis, because diets are likely to lose water due to dehydration whether consumed or not during the experiment.

The use of measurements of weight gain of the test organism as an index of behavioral preference is less desirable, because weight gain will be determined not only by consumption, but also the efficiency with which consumed food is assimilated (see section 5.6). Fecal production is sometimes used, but this technique is limited to organisms that produce defined fecal pellets (no liquid feces), and the data may be subject to experimental error if care is not taken to eliminate variation in gut contents at the beginning of the experiment. This technique probably is appropriate only where relatively large differences in consumption are expected.

Regular, direct observation of behavior may reveal more about the behavioral effect of test chemicals than measurements of total feeding activity at the end of the experiment. In some experiments, it may be more important to document differences in the pattern and duration of feeding behavior and the process by which organisms "choose" among treatments, rather than total consumption. The simplest, but perhaps most tedious approach is to observe directly the test subjects for fixed amounts of time at regular intervals and record such information as their location (e.g., on which substrate, or on neither); whether they are actively feeding, resting, or in motion; how long each subject spent in each defined behavior; and the frequency and pattern of changes from one behavior to another. Such information can help determine how a test chemical may exert its effect on total consumption.

Indirect methods of observation also have been used, but these often are tailored to specific purposes (for a review see Lewis & van Emden 1986). One such approach involves using a digitizer and microscope to obtain and record an *X–Y* position of a test organism, while the operator enters one of a few codes for different behaviors. Elapsed time also is monitored and recorded into the data file, and from these data, a behavioral profile can be obtained (Eigenbrode et al. 1989, 1991). From such raw data, differences in such behaviors as wandering, feeding, swallowing, etc. can be extracted and compared in both choice and no-choice experiments (Berdegue & Trumble 1996). Additionally, feeding behavior can be videotaped, and much of the tedium of observations can be reduced by "fast-forward" playback of the tape if elapsed time is also recorded on the tape. The use of relatively small video cameras also may minimize the intrusion of the observer on the expression of normal, uninterrupted behavior.

Food uptake of sucking insects (e.g., aphids and leafhoppers) also is governed by a number of process that occur in several distinct phases. First, the insect must successfully probe through the surface of a plant, then insert its stylets into a plant vascular bundle before uptake can occur. Semiochemicals may modify this process primarily at two places: (1) the leaf surface or (2) the phloem. A relatively old approach to investigate the role of chemicals on ingestion after probing utilizes artificial diets in sachets made of parafilm or other similar membrane (see Mittler 1988 for a review). Consumption is determined by weighing sachets before and after groups of aphids feed on the sachets. Groups are used to increase the total consumption and minimize the effect of random errors in weighing. For more accurate measurements of uptake, radioisotopes are incorporated into the diet, and the quantity of radioactive isotopes in the honeydew produced is taken as an index of the quantity of diet ingested (Arn & Cleeve 1971). Most accurate measurements are obtained when the radioactive label is incorporated into nonnutritive substances, such as the nonmetabolizable carbohydrate, inulin (e.g., Wright et al. 1985). Major disadvantages of the use of radioactive substances are their cost and the additional care and expense necessary to maintain and dispose of radioactive substances after use.

With sucking insects that produce honeydew, the amount of honeydew excreted is assumed to be proportional to feeding. Honeydew production can be measured as number of droplets (using bromocresol blue or ninhydrin as a stain; Auclair 1958), or the stained honeydew can be collected, extracted from the collection chamber, and measured colorimetrically (Nisbet et al. 1994). Several problems must be addressed with the use of honeydew as an index of consumption. First, some aphids will produce nymphs when feeding on artificial diet, and the production of honeydew by nymphs may inflate the quantity recorded for the test insect. The simplest solution here is to frequently remove any nymphs that are produced. Second, because aphids may scatter their honeydew, the droplets will not always fall directly below the aphid. Thus, collection dishes must be sufficiently large to collect all of the honeydew that might be produced. Alternatively, the entire chamber must be lined with material used to collect the honeydew, from which it is extracted subsequently (e.g., Nisbet et al. 1994).

While the above techniques have been used to determine the role of a number of chemicals as phagostimulants or deterrents that might be found in plant vascular tissues, such techniques may provide little information to determine how semiochemicals may affect homopteran probing behavior prior to ingestion. Over the last several years, the use of electronic feeding monitors has become widespread to quantify probing behaviors of aphids, leafhoppers, and whiteflies. The methods was developed originally by McLean and Kinsey (1964), and continues to be modified extensively (e.g., Backus & Bennett 1992; Walker & Perring 1994).

The electronic feeding monitor can be used to monitor feeding behavior of small insects on plants as well as on artificial diets. A fine gold wire is attached to the insect with silver paint that conducts electricity. This wire is attached to an amplifier and oscilloscope, and a second wire is attached from the amplifier to the plant or artificial diet to complete an electric circuit. Changes in electric resistance as the insects stylet canals fill and empty of liquid are recorded continuously. Different patterns of activity are associated with different feeding behaviors by immediate killing and fixing of the insect as patterns change, followed by histological investigation to locate the position of the stylet sheath when behaviors changed. After examination of a number of specimens, general associations can be made between specific wave forms and defined feeding-related behaviors, such as probing, salivation, and continuous feeding. The effect of specific semiochemicals in altering the duration of probing and preprobing behavior, for example, then are relatively easily demonstrated (e.g., Reuter et al. 1993). This technology is developing rapidly, and specific recommendations for the setup and design of such a monitor for general use cannot be made at this time.

5.4.4. Data Analysis

General guidelines for statistical analysis were presented previously (see section 5.1.3). Where continuous variables, such as consumption, are measured, means are compared using parametric tests (e.g., *t*-test, ANOVA) or nonparametric tests (e.g., Wilcoxon two-sample test or the Kruskal-Wallis test) as appropriate. For simple choice tests, the *G*-test or Fisher's exact test (Sokal & Rohlf 1995) are often used to test for deviations of the observed pattern of choices from a random pattern.

For dose-response studies, probit analyses (Robertson & Preisler 1992) can be used analogously to their use for pesticides, although relatively large sample sizes may be needed for the fiducial limits of the probit line to exclude zero. Sample sizes should be based on the variability of the data (i.e., the coefficient of variation or the magnitude of the treatment effect relative to the magnitude of within-treatment variation; Sokal & Rohlf 1995, box 9.14 and box 10.8). However, logistical constraints on subject uniformity or availability of test chemicals from natural sources may make such optimal sample sizes unobtainable.

5.5. Postingestive Bioassays

After the behavioral role of a semiochemical has been ascertained then it may be necessary to determine if the semiochemical has any additional physiological role. This can be accomplished through a number of postingestive assays. Most of these assays measure the effect of the chemical on some aspect of the survival, growth, or reproduction of the subject (Table 5.2). Depending on the semiochemical, it may not be obvious when the bioassay should be terminated. For acutely toxic materials, observations may be taken until all exposed subjects have died. If such mortality can be observed in a reasonably short amount of time, and if multiple doses are used, then the data may be analyzed by the use of probit analysis, and the LD_{50} or LC_{50} of different compounds can be compared using appropriate procedures (Robertson & Preisler 1992).

Table 5.2. Variables Used to Measure Postingestive Effects

Parameter	Test duration
Mortality	Time period, or through a particular developmental stage
Growth rate	Specific time, or through developmental stage, time to a specific stage or event
Amount of growth	Weight change over specified time or developmental stage[a]
Fecundity	Eggs per female, total fecundity, total viability
Percent parasitization	Development time of parasitoid, number of parasitoids produced
Disease susceptibility	Time for disease expression or death of subject

[a]The time for measurements of growth usually is standardized by allowing the subjects to reach a defined developmental stage rather than by using an invariant time interval.

For less potent chemicals, their effect on such correlates of fitness as growth rate, ultimate size, or fecundity are monitored. The design of bioassays for these experiments are similar to those used to identify stimulants or deterrents through differences in consumption. One exception is that nonnutritive substrates usually are not used except when the nutritional value of the test chemical is being tested explicitly. Artificial diets are widely used for Lepidoptera because of the relatively wide availability of artificial diets for this group. The technique of applying chemicals to known hosts is used for other species where no suitable artificial diet is known (see sections 5.4.1.2–5.4.1.5).

The major difference between preingestive and postingestive bioassays is that the assays are continued for periods of time well in excess of those designed simply to demonstrate preference or deterrence. The time period may vary from the full developmental period for subjects with relatively short developmental times, to the duration of particular developmental stages, to relatively arbitrary time periods (i.e., a portion of a developmental stage) for long-lived species.

In addition, because the bioassays are continued over a longer time period, it is imperative that the quality of food be maintained during the bioassay period. Artificial diets tend to dry out over time, or support the growth of mold if no mold inhibitors are included (note that some mold inhibitors also may be toxic to test subjects!). Fresh plant tissue also will dry out and decline in quality after excision from the intact plant even if supplemental water is provided. Also, some test chemicals incorporated into artificial diets may break down when exposed to ambient temperatures and light. Therefore, test food may need to be replaced regularly with new, fresh test food during the bioassay.

5.6. Measurements in Diet Studies: Growth Rate, Consumption Rate, and Efficiency of Conversion of Food to Biomass

Perhaps the aspect of insect bioassays that has changed the most over the last decade is the measurement of insect food consumption, growth, and growth efficiency. As Slansky and Scriber (1985) pointed out, the amount, rate, and quality of food consumed by an insect influences that insect's probability of survival, growth performance, and fitness; the evolution of different lifestyles by herbivores is driven largely by constraints placed on the consumption and utilization of food.

Food consumption and utilization can be constrained at two points in time: before ingestion and after ingestion. Preingestive factors influence the rate that food will be ingested or consumed, while postingestive factors influence digestion and govern the proportion of ingested biomass that is converted into consumer biomass. Because the behavioral acceptability of potential food is often unrelated to its nutritional quality, it is important to distinguish between preingestive and postingestive mechanisms to understand why organisms utilize some potential hosts but not others.

A series of nutritional indices (Waldbauer 1968; Scriber & Slansky 1981; Table 5.3) have been used widely in the study of insect growth, consumption, and utilization efficiency. The basic method of data acquisition is gravimetric. Measurements are usually made on a fresh-weight basis, although most analyses are converted to dry weights because dry-weight data are usually less variable. If the conversion factors for fresh weight to dry weight cannot be obtained from the test subjects directly, then extra test subjects must be reared under test conditions, then sacrificed and dried to obtain conversion factors.

The age of the subject can affect the quality of the data. For very young subjects tested on plant foliage, the amount of foliage consumed may be so small that plant parts may grow more than the subject consumes. The apparent result is negative consumption. There are two solutions to this problem. One involves using cohorts of very young larvae instead of individuals (provided that rearing

Table 5.3. Nutritional Indices and Their Relationships

Parameter	Name	Formula[a]	Description
CR	Consumption rate	I/Dt	mass eaten/unit time
RCR	Relative consumption rate	$(I/w)/Dt$	(mass eaten/mass of subject)/unit time
GR	Growth rate	B/Dt	mass gained/unit time
RGR	Relative growth rate	$(B/w)/Dt$	(mass gained/mass of subject)/unit time
AD	Approximate digestibility	$(I - F)/I$	(mass eaten − mass excreted)/mass eaten[a]
ECD	Efficiency of conversion of digested food	$B/(I - F)$	mass gained/(mass eaten − mass excreted)[a]
ECI	Efficiency of conversion of ingested food	B/I	mass gained/mass eaten[b]

[a]I = mass ingested; Dt = time interval; w = mass of subject; B = mass gained; F = mass excreted; see text.

[b]Usually multiplied by 100 and expressed as a percentage.

in groups does not induce abnormal behavior or cannibalism) to increase total consumption. The other involves conducting the experiments with older subjects. For insects, these often are the ultimate or penultimate instar.

If relatively old subjects are used, then an important decision needs to be made about what the test subjects are to be reared on prior to testing. For some experiments, it is appropriate to rear subjects on a common host prior to being tested on different hosts to minimize pretreatment variation. On the other hand, for insects where preference and dietary efficiency increase after prior experience with a particular host, then switching from a rearing host to a test host may unnaturally reduce growth and feeding. Usually, the more realistic alternative is to rear subjects from hatching on the host or diet on which they are to be tested. However, if the hosts differ substantially in quality, then this may introduce substantial pretreatment variation that may affect subsequent results and interpretation.

The formulation of nutritional indices starts with the fundamental premise that the total mass of food ingested (I) can be partitioned into three quantities: the mass excreted in feces (F), the mass metabolized during respiration (M), and the mass incorporated into herbivore biomass (B). From the basic equation, $B = I - F - M$, several rates and efficiencies can be calculated (Table 5.3), and the terminology has been reviewed extensively (e.g., Slansky & Scriber 1985; Kogan 1986; Farrar et al. 1989; Raubenheimer & Simpson 1992).

As defined above, these indices have a number of important mathematical relationships. The most important of which is RGR = RCR $*$ ECI, i.e., the relative growth rate is the product of the relative consumption rate and the efficiency of conversion of ingested food. Preingestive factors will affect the consumption rate, while postingestive factors will affect the efficiency of conversion. Thus, independent measurements of RCR and ECI permit the separation and measurement of preingestive (i.e., behavioral) factors form postingestive (i.e., physiological) factors on herbivore growth rate.

These indices have been used widely in the study of insect nutritional ecology over the past 20 years or so, but recently, a number of potential problems have been identified. Among these are the fact that not all material excreted by insects is undigested food; it also includes waste products. Thus, the overall approximate digestibility, AD, can never reach 100%, even for completely digestible food. In addition, the actual logistics of these experiments are huge, as discussed in detail below, and any errors in measurement may propagate and amplify through the calculations (e.g., Schmidt & Reese 1986). Probably the most important criticism of these measurements to date, however, focuses on the calculation of *relative* consumption and growth rates.

Absolute consumption and growth rates are difficult to compare among species, or even among instars within species. The basic problem is that consumption is likely to be a function of size, with larger subjects consuming more than smaller ones simply because of the size difference. Thus, it is necessary in some way to adjust for the differences in size. The simplest adjustment probably is to divide growth or consumption by the arithmetic mean weight (e.g., Waldbauer 1968). Others have suggested that the geometric mean weight is a more appropriate divisor because insect growth tends to be exponential (e.g., Kogan 1986). On the other hand, because the weight of a test subject at any time during the feeding trial would be some combination of both the weight of the subject's biomass as well as the weight of the food that it contains, variation in food content could introduce variation in mean weight. The contribution of the weight of food to the total weight of the subject is minimized (but perhaps not entirely eliminated; Bowers et al. 1991; Barbehenn & Keddie 1992) when the insects are molting. Larvae stop feeding just prior to a molt and gradually lose weight until they resume feeding again after ecdysis and sclerotization of the new cuticle. Measurements made with larvae at molting stages minimize errors due to potential variation in amounts of food in the gut. Therefore, a minimally biased estimate of insect biomass gain is obtained if initial weights shortly after molting are used to calculate relative consumption rate and relative growth rate, rather than any mean weight during the growing period (Farrar et al. 1989).

The use of initial or mean weight to calculate ratios to adjust for the effect of variation in size on consumption has a number of statistical assumptions that may not be apparent to all investigators and are probably not realistic. Foremost among these is the implicit assumption that growth rate and consumption rate are a linear function of size, and the regression line between weight and consumption rate or growth rate has an intercept of zero (Raubenheimer & Simpson 1992). As an alternative to calculating ratios and then performing standard statistical analyses (e.g., analysis of variance) on them, Raubenheimer and Simpson (1992) suggested that the variation in initial size be addressed through analysis of covariance (ANCOVA), where the variate is the total mass consumed for analyses of consumption or final weight for analyses of growth, and the covariate is initial weight. ANCOVA has a number of advantages over the calculation of ratios, including that it does not assume an intercept of zero.

Pre- and postingestive effects on performance (e.g., weight gain) can be separated in ANCOVA by examining the adjusted treatment means of groups of subjects offered different diets (Horton & Redak 1993). In these analyses, the variate is biomass gained, and the covariate is the amount of diet consumed. Treatment means (i.e., mean biomass gain) then are compared in ANCOVA after adjusting for any variation in consumption. If treatment effects remain, then the variation in performance is most likely due to postingestive effects on assimilation rate. However, if the adjustment for variation in consumption removes any treatment effect, then the variation in weight gain was probably due to preingestive factors affecting only the amount of food consumed (Horton & Redak 1993).

At this time, relatively few studies have been completed to directly compare the results from analyses using ratio-based statistics and ANCOVA. Berenbaum and Zangerl (1994), however, analyzed their data both by conventional ratio analysis and by ANCOVA and found no significant differences in the conclusions drawn and only slight variation in probability values. Nonlinearity may still be a problem in ANCOVA as well as ratio analysis, unless there is a simple power transformation to make the final weight a linear function of the initial weight. Finally, ANCOVA also includes the assumption of homogeneity of slopes for different treatments. Until some of these controversies become better resolved, investigators would be wise to consider multiple methods of statistical analysis and ensure that they understand the assumptions that underlie each method.

5.6.1. Data Acquisition

Data for calculation of nutritional analyses are much the same for all methods of statistical analysis described above. Basic precautionary steps recommended for selection of individuals, pretest conditions, and standardization of both test animals and test diets have been discussed elsewhere (Berenbaum 1986). Nutritional indices are extremely sensitive to small errors in computation (Schmidt & Reese 1986; Stamp 1991), thus, efforts to reduce such sources of variation are of utmost importance in such studies. All consumption experiments must provide food *ad libitum* unless effects of starvation are being tested, and careful planning and some preliminary experiments may be required to determine how much food must be prepared or obtained in advance to complete the entire experiment. This is especially important when insects are to be tested on plant food to avoid introducing variation in food quality due to difference in plant or leaf age into the experimental design.

These considerations make it difficult to give firm recommendations for the number of test subjects per treatment. Such a number is invariably a compromise between the number dictated by the statistical goals of maintaining acceptably low within-treatment variation and the ability of the investigator to rear simultaneously several subjects all with unlimited and uniform-quality food. A sample size of 10 to 15 subjects per treatment is a reasonable target for preliminary studies to determine how many subjects actually will be needed in later experiments.

5.6.1.1. Consumption

The amount of food consumed by a subject over a specified time interval is obtained by measuring the initial amount of food (fresh-weight basis) and the food remaining at the end of the feeding period. The amount of food consumed is then determined by subtraction. Conversion to dry weights is made by obtaining the fresh and dry weights of an aliquot as similar to the test food as possible. When plant leaves are used as food, commonly leaves are split along the midrib, and one half-leaf is used in the feeding test and the other half is used as the aliquot for dry weight estimation. The dry weight/fresh weight ratio of the aliquot multiplied by the fresh weight of the test food provides the dry weight estimate of the food given.

When artificial diets are used, additional aliquots of diet are held under identical conditions as those consumed and used to adjust the calculated quantity of food ingested for systematic errors due to evaporation of water. For example, if a is the initial weight of the aliquot and b is the final weight, then two ratios are computed as follows: $A = (a - b)/a$ and $B = (a - b)/b$. These ratios are then used to obtain the corrected weight of food eaten as follows (Waldbauer 1968):

$$\text{amount eaten} = [1 - (A/2)] \cdot [\text{initial weight} - (\text{final weight} + B \cdot \text{final weight})].$$

There is some controversy as to how much food must be presented. Excessive food residue is a major source of error in measurement (Schmidt & Reese 1986), however, because some insects may reduce their consumption rates as available food diminishes, the quantity of food remaining should be in excess of 5% (Stamp 1991). Of course, if all food is consumed between observations, then food was limiting over the time interval, and all calculations will be in error.

Experiments with excised plant parts are usually done in closed petri dishes or coated paper containers with clear plastic lids. The containers are lined with moist filter paper to keep plant tissue from dehydrating. When possible, leaves should be placed in water reservoirs. Leaves of many herbaceous plant species can be maintained in convenient capped water reservoirs used by florists to maintain corsages (e.g., "Aqua-Pics®"), but leaf terminals of woody trees may require water reservoirs constructed from large test tubes and sealed with cork or rubber stoppers, cotton batting and parafilm, or similar materials.

5.6.1.2. *Weight Gain*

Weights are taken at the beginning and end of the growth period. Usually the growth period begins as soon as the insects have molted to the instar to be tested and ends after they have molted again or pupated, depending on which instar is to be examined. The investigator should expect variation in development times, and molting may occur at any time of the day or night for some insect species. Thus, to obtain accurate estimates of development time and the duration of the larval growth period, larvae may need to be checked frequently around the clock, and not necessarily at the convenience of the investigator. Initial fresh weights are recorded and then converted to dry weights by ratios obtained with additional subjects maintained under identical experimental conditions and sacrificed to determine their dry weights. The final dry weight can be obtained directly if there is no need to keep the test subjects alive after the growth period has completed.

5.6.1.3. *Measurements of Excreta*

With subjects that produce discrete fecal pellets, collection of excreta is simple and straightforward. When possible, it is advantageous to keep insects in containers designed so that fecal pellets collect in the bottom of the container free of food. For example, when insects are fed on intact foliage, containers may have a screen support that allows fecal material to fall through but not larvae or large pieces of foliage. When artificial media are used, diet cups can be inverted and hung from the top so that feces fall away from the diet for easy collection.

It is more difficult, if not impossible, to collect excreta of insects with highly dilute or watery feces. The Mexican bean beetle, *Epilachna varivestris*, usually excretes droplets of feces that adhere to the foliage on which they feed. The Colorado potato beetle, *Leptinotarsa decemlineata*, also excretes watery fecal material that tightly adheres to leaf tissue when dry. For such species, estimates of excreta as a function of uric acid concentration (Bhattacharya & Waldbauer 1970, see below) has been suggested (Kogan 1986) but has not been widely used.

5.6.2. Alternatives to Gravimetry

As an alternative to determining consumption by the differences in weight of food at the beginning and end of the feeding period, various markers have been used to determine consumption. After the marker is incorporated into the food, its concentration (or that of a representative aliquot) is determined. Food is offered to the test subjects as before, and fecal material is collected and weighed. The concentration of the marker in the feces then is determined. The amount of food consumed then is calculated following the procedure of Waldbauer (1968):

milligrams eaten = (marker concentration in feces/marker concentration in food) · milligrams feces.

After consumption is determined in this way, then all other nutritional indices can be calculated as described above.

In theory, if the marker is completely inert, the use of markers instead of gravimetry should provide a more accurate estimate of consumption, because plant biomass will not be lost during the feeding trial due to respiration. In practice, it appears that determining consumption using markers is not significantly more accurate than using gravimetry (e.g., Barbehenn 1993). Detailed procedures for the use of various markers, including dyes, elemental markers, and radioactive isotopes, have been presented by Kogan (1986); however, these methods have been rarely used.

5.6.2.1. Uric Acid Determination

To measure food consumption when food and feces are inseparable, Bhattacharya and Waldbauer (1969a, 1969b, 1970) developed a method based on measurement of uric acid in mixtures of food and feces. Uric acid is absent in the food but present in the feces as a product of protein digestion. The first step in the uric acid method requires separation of a sample of the feces from the mixture of feces and leftover food. This can be achieved simply by isolating well-fed insects from their food until defecation. The concentration of uric acid in the mixture is estimated both in an aliquot of the feces and in a sample of the mixture of food and feces. The amount of feces in the mixture is calculated from the concentration of the uric acid in the feces and the concentration of uric acid in the mixture, and the weight of the mixture. The calculated amount of feces in the mixture is then subtracted from the weight of the mixture to yield the amount of food remaining, from which the amount of food consumed can be determined by subtraction.

The enzymatic method for uric acid determination of Bhattacharya and Waldbauer (1969a) is outlined below. Kits for uric acid determination using these procedures are available commercially. Uric acid is extracted from a weighed feces sample with 3–3.5 ml of a 0.6% aqueous lithium carbonate solution in a tissue grinder. After centrifugation, the supernatant is decanted and brought to a standard volume with 0.1–M (pH 9.4) glycine buffer. Ten successive extractions are recommended (Bhattacharya & Waldbauer 1969a, 1969b).

A standard curve for uric acid is determined at 292 nm. Uric acid concentration is determined by measuring initial absorbance in 1–ml samples of the diluted extracts plus 2.0 ml of glycine buffer. Subsequently, 10 μg of urease is added to the cuvette and absorbance is measured again after oxidation of uric acid to allantoin is completed. The difference in absorbance is proportional to uric acid concentration. By knowing the volume in the cuvette, the quantity of uric acid in the cuvette than can be determined, and from that, the quantity of uric acid in the whole sample is easily calculated.

5.6.2.2. *Calorimetry and Respirometry*

Although most research has emphasized nutritional indices calculated on the basis of weight, occasionally it is necessary to calculate such efficiencies on the basis of energy content. This requires that the energy content of food, insect, and feces be determined by a "bomb" calorimeter. In this method, the material is burned in the presence of oxygen to produce water vapor and carbon dioxide. The heat released is determined with a sensitive thermocouple. The temperature change is proportional to the heat released, or the number of calories contained in the sample. A number of commercial bomb calorimeters are available from scientific supply houses. The parameters measured by calorimetry are similar to those measured by gravimetric analyses, as are most general patterns. However, for insect-based systems, efficiencies are generally higher when determined on a caloric than a dry weight basis (Slansky 1985). General guidelines for making calorimetric measurements are presented in Kogan (1986) and references therein.

An approach related to calorimetry is respirometry, where the volumes of oxygen consumed and carbon dioxide liberated during growth is measured to determine efficiency of growth. While such an approach may be more precise than gravimetric methods, this approach also is more technically demanding and requires far more specialized equipment. Advantages and disadvantages are reviewed by van Loon (1989). For some insects or life stages, however, respirometry may be the only useful method.

5.6.3. Summary

The process of determining nutritional indices by gravimetric methods is tedious, time-consuming, and fraught with technical difficulties and uncertain statistical assumptions. Unfortunately, none of the alternative methods are demonstrably any better. For example, in a detailed comparison of several methods to determine consumption, Kogan and Parra (1981) measured ECI and ECD for soybean loopers using gravimetric methods as well as several marker-based methods and gained little if any precision over gravimetry. Moreover, most of the alternatives were substantially more expensive. Thus, because gravimetry requires the least specialized equipment, gravimetric methods probably are justified except where food consumed cannot be measured directly (Kogan 1986).

Recommendations for statistical analyses of nutritional indices are more difficult to make because some of the questions that these indices could be used to address may no longer be of general importance. For example, much of the early research in nutritional ecology was motivated to test the hypothesis that specialist herbivores were more efficient in assimilating their hosts than generalists were in assimilating the same host. The bulk of the literature today strongly suggests that generalists and specialists do not differ markedly in their efficiencies of conversion of the same food. Rather, variation in efficiency of conversion appears to be more of a general property of the food itself, and not a property associated with the level of dietary specialization (reviewed by Scriber & Slansky 1981; Slansky & Scriber 1985).

Another early hypothesized use for analyses of ECD was to detect and quantify the metabolic cost associated with detoxifying plant defensive chemicals. According to this hypothesis, as the energy costs for detoxification increase, a greater proportion of the energy in assimilated food must be allocated to detoxification at the expense of growth, i.e., ECD should decrease. This hypothesis has been criticized on methodological grounds in that ECD is only an indirect measure metabolic rate and may not reflect those costs. The hypothesis has been criticized as well on physiological grounds that suggest that the energy needs for detoxification are a relatively small portion of the energy budgets of herbivorous insects (Appel & Martin 1992). Both of these findings suggest that additional measurements of interspecific variation in ECI or ECD are not likely provide new insights in the our understanding of the ecological processes promoting specialization of herbivorous insects.

On the other hand, the question of whether differences in growth rate (or other measures of herbivore performance) are due to preingestive or postingestive factors will continue to be central to both basic and applied studies. The relatively narrow host range of most herbivorous insects still needs to be explained, as does the need to identify the mode of action of potential mechanisms of host plant resistance for use in agricultural crops.

5.7. Alternative Methods to Separate Preingestive and Postingestive Effects

In addition to separate analyses of consumption and growth efficiency, several other techniques have been used to quantify the relative importance of semiochemicals on feeding behavior and postingestive physiology with varying levels of success. Some of these methods are summarized below. Because many of these methods are quite "invasive" and manipulative (i.e., direct injection of semiochemicals, removal of taste receptors), they may alter the behavior or physiology of subjects substantially in their own right.

One noninvasive procedure was developed by Blau et al. (1978), who were interested in separating the interacting effects of allylglucosinolate as a behavioral and physiological barrier to the utilization of plant species in the family Cruciferae by *Papilio polyxenes*, a specialist on plant species in the family Umbelliferae. Initially, a "calibration curve" was developed in which consumption rate was controlled by varying the amount of food available, and the subsequent effects on larval growth were determined by linear regression. This curve (actually a straight line) described the growth rate of insects under conditions of partial starvation and limited food supply. Food then was offered with allylglucosinolate, and consumption and growth rate were recorded as per standard procedures for the determination of nutritional indices described above. Regression lines for relative growth rate on relative consumption rate were plotted for all treatments. If allylglucosinolate was simply a feeding deterrent, then the regression line of consumption rate on growth rate should fall on the calibration curve, indicating that reduced growth was simply a function of reduced consumption. Alternatively if the compound was partially toxic, then the slope of the regression line would be significantly less (flatter) than the calibration curve. These determinations can be made using standard statistical methods to compare the slopes of regression lines (Sokal & Rohlf 1995).

Allylglucosinolate was toxic to *P. polyxenes*, as shown by a regression coefficient less than that of the calibration curve. The same compound was innocuous to the crucifer specialist, *Pieris rapae*, as shown by growth rates greater than predicted by the calibration curve. For the generalist, *Spodotera eridania*, however, allylglucosinolate was innocuous at low concentrations but moderately toxic at high concentrations (Blau et al. 1978).

This method has been used several times to identify deleterious effects that do not result in acute toxicity (e.g., Berenbaum & Feeny 1981; Rauscher 1981;

Miller & Feeny 1983; Usher & Feeny 1983). The fact that the application of the method utilizes standard ratio-based variables for relative consumption rate and relative growth rate, however, introduces the problems of the use of ratios, as described earlier.

A more manipulative method for separating the effects of deterrence and toxicity has played an important role in the development of our understanding of the importance of behavioral limitations on insect host range. Waldbauer and Fraenkel (1961) showed that the amputation of the maxillae of *Manduca sexta* permitted larvae to feed on plant species that they normally reject. Maxillae were amputated by pinching them off with a pair of sterilized watchmaker's forceps after anesthetizing larvae by chilling or exposure to CO_2 (details in Waldbauer & Fraenkel 1961). Unfortunately, this method has not been widely used on other species (perhaps because *M. sexta*, because of its large size, is unusually easy to work with), so it is not yet clear if such a pattern of host range limitation is typical.

Another approach toward bypassing the sensory apparatus is to introduce candidate chemicals directly into the digestive tube of the subject by cannulation. A small tube is introduced into the esophagus, and chemicals are then introduced with a syringe (e.g., Cottee et al. 1988). This technique may be damaging to the subject (Usher et al. 1989), and the quantities of material introduced and their rates of ingestion may be abnormal. Thus, caution is required in interpreting results of bioassays using semiochemicals administered to subjects by cannulation or direct injection into the hemolymph.

One possibly less intrusive solution to the toxicity/deterrence problem is the use of microencapsulation to incorporate test chemicals into artificial media. This method tests the physiological effect of the chemical directly by delivering it to the gut in capsules that have no taste (Szentesi & Bernays 1984; Usher et al. 1989; Clancy et al. 1992; Navon et al. 1993). After ingestion, the capsules must break down and release the test chemical in the gut lumen where the chemicals can exert their physiological effect. Incorporation of semiochemicals in microcapsules involves several steps. Methods for the preparation and use of liposomes, albumin, gelatin, or ethylcellulose microcapsules to determine the physiological activity of a number of feeding deterrents are presented by Usher et al. (1989). Clancy et al. (1992) describe procedures to encapsulate plant-derived terpenes for feeding studies, and Navon et al. (1993) present a method of encapsulating tannins in sodium alginate microcapsules.

After the microcapsules are made, they can be administered to subjects after incorporation into an artificial diet. Alternatively, the subjects can ingest the microcapsules after the microcapsules have been spread on the surface of an acceptable substrate for ingestion by normal feeding (Usher et al. 1989). After ingestion, the capsules break down, releasing the chemical, and the effect of the chemical on aspects of performance (Table 5.2) can be measured as for most other postingestive bioassays.

5.8. Contact and Volatile Toxicity

Although the potential role of semiochemicals as contact or fumigant poisons cannot be minimized, the bioassay procedures for such compounds are sufficiently similar to those for synthetic pesticides that a detailed review may not be necessary here. The simplest contact bioassay is identical to that for pesticides. The candidate toxin is dissolved in an appropriate solvent, usually acetone, and fixed quantities are applied to the subject with a microsyringe. A range of concentrations are used, and, with appropriate sample sizes, the resulting pattern of mortality is analyzed by probit analysis (Robertson & Preisler 1992) to determine LD_{50}'s and fiducial limits.

Dimock et al. (1982) developed a contact bioassay that mimics the natural means of exposure to plant-produced toxins. These researchers were interested in the contact toxicity of exudates from glandular trichomes in wild tomato (*Lycopersicon hirsutum* f. *glabratum*) toward the tomato fruitworm, *Helicoverpa zea*. For bioassay of exudates of glandular trichomes (i.e., tridecanone and analogues), solutions of extracts and test chemicals were applied in chloroform to filter paper disks 5.5 cm in diameter with a pipette. The treated disks then were placed in glass petri dishes. Solvent was evaporated off and 20 neonate *H. zea* larvae placed on the disks. The petri dishes were sealed with parafilm to keep caterpillars and vapors from escaping. The dishes were then placed in the dark for 6 h, after which time the larvae were moved to artificial diet for an additional 18 h when mortality was assessed. Those fractions and pure compounds exhibiting biological activity were reassayed at a series of dosages to determine their LD_{50} and 95% confidence limits.

In this bioassay, activity due to contact cannot be distinguished from activity through fumigant action because the containers were sealed. There are two ways to address this problem. The first is to ventilate the chamber to draw off all vapors as is done with a number of pesticide bioassays (e.g., Morse et al. 1986), thus restricting exposure to direct contact. The second simply is to isolate the test semiochemical from physical contact the subjects to that the *only* means of exposure is by fumigant action. This is illustrated by bioassays of chromenes and benzofurans from *Encelia* spp. developed by Isman and Proksch (1985) as follows.

In two bioassays, semiochemicals were applied first to the bottom and sides of glass scintillation vials, while in the second, an equivalent quantity was applied to the inside of the top of the vial. In both cases, the semiochemicals were applied in solvent, then the solvent was allowed to evaporate. Surfaces or lids of control vials were treated with solvent only. After the solvent had evaporated, five neonate larvae of the variegated cutworm, *Peridroma saucia*, were introduced into each vial for a 4-h exposure period. After exposure to the test chemicals, larvae were transferred to artificial diet, and their survival was assessed 24 h later. Because larvae could not climb the slippery surfaces of the glass vials and contact the semiochemical applied to the inside surfaces of the lids, mortality due to contact was easily separated from mortality due to fumigant action. For these compounds (encelin and related compounds), Isman and Proksch (1985) concluded that all mortality was due to exposure by contact.

5.9. Conclusions

All investigators derive great satisfaction from developing a bioassay that is simple, discriminating, and replicable. Students who read about studies using such bioassays must not forget that negative evidence (i.e., the bioassays that did not work) are seldom mentioned. Thus, it might appear that the design of elegant bioassays is simpler than is truly the case. Nevertheless, several general guidelines are worth reiterating.

Among the most important is the need to ensure that the attribute to be measured under confined experimental conditions accurately reflects the full sequence of behavior or performance under natural conditions. Equally important is a necessity to replicate all test conditions, even the most obscure.

Most behavioral bioassays reviewed here offer the subjects a small number (usually two) choices, then the frequency of choices is recorded and analyzed with relatively simple statistical techniques. The purpose of such assays usually is to determine biological activity compared to a "blank" or solvent control; quantitative rankings of activity are better developed using bioassays with multiple choices.

For feeding bioassays, where survival, growth, reproduction, or some other aspect of performance is measured, homogeneity and consistency of the diet is an absolute requirement. This is particularly difficult to achieve when natural food sources are used, but for many, perhaps most terrestrial invertebrates, there is no suitable alternative. Even when artificial diets can be used, care must be taken to ensure uniformity from batch to batch.

Despite the difficulties in developing them, bioassays are essential for the design of critical experiments. Such experiments are required to test empirically competing hypotheses about the mechanisms that generate observed patterns of association and behavior. Finally, it is exciting for an investigator to see a test organism accept a novel substrate treated with a test chemical, for such acceptance demonstrates that the investigator has developed a good understanding of the biological principles underlying the subject's behavior.

5.10. Acknowledgments

I thank the editors for the opportunity to prepare this chapter and Drs. D.J.W. Morgan, N.M. van Dam, and P.A. Weston for their comments on a previous draft. Drs. L.E.M. Vet, P. Feeny, and R.F. Luck were especially helpful in providing artwork for reproduction. Partial support for preparing this chapter was provided by USDA NRICGP award no. 94-37302-0544 and NSF DEB award no. 96-15134.

5.11. References

Adams, C.M. & E.A. Bernays. 1978. The effect of combinations of deterrents on the feeding behaviour of *Locusta migratoria*. Entomol. Exp. Appl. **23**:101–109.

Ali, M.F. & E.D. Morgan. 1990. Chemical communication in insect communities: a guide to insect pheromones with special emphasis on social insects. Biol. Rev. **65**:227–247.

All, J.N. & D.M. Benjamin. 1976. Potential of antifeedants to control larval feeding of selected *Neodiprion* sawflies (Hymenoptera: Diprionidae). Can. Entomol. **108**:1137–1143.

Aluja, M., R.J. Prokopy, J.P. Buonaccorsi & R.T. Cardé. 1993. Wind tunnel assays of olfactory responses of female *Rhagoletis pomonella* flies to apple volatiles: effect of wind speed and odour release rate. Entomol. Exp. Appl. **68**:99–108.

Appel, H.M. & M.M. Martin. 1992. Significance of metabolic load in the evolution of host specificity of *Manduca sexta*. Ecology **73**:216–228.

Arn, H. & J.S. Cleeve. 1971. A double-label choice test for the simultaneous determination of diet preference and ingestion by the aphid *Amphorophora agathonica*. Entomol. Exp. Appl. **14**:377–387.

Arn, H., M. Tóth & E. Priesner. 1992. List of Sex Pheromones of Lepidoptera and Related Attractants. International Organization for Biological Control. Montfavet, France (See http://www.nysaes.cornell.edu/fst/faculty/acree/pheronet/pherolist.html for 1997 version.)

Ascher, K.R.S., E. Gurevitz, S. Renneh & N.E. Nemny. 1975. The penetration of females of the fruit bark beetle *Scolytus mediterraneus* Eggers into antifeedant-treated twigs in laboratory tests. Z. Pflkrankh. Pflschutz. **82**:378–383.

Auclair, J.L. 1958. Honeydew excretion in the pea aphid, *Acyrthosiphon pisum* (Harr.) (Homoptera: Aphididae). J. Insect Physiol. **2**:330–337.

Backus, E.A. & W.H. Bennett. 1992. New AC electronic insect feeding monitor for fine-structure analysis of wave forms. Ann. Entomol. Soc. Am. **85**:437–444.

Bakchine, E., M.H. Pham-Delegue, L. Kaiser & C. Masson. 1990. Computer analysis of the exploratory behavior of insects and mites in an olfactometer. Physiol. Behav. **48**:183–187.

Baker, J.L. 1976. Determinants of host selection for species of *Aphytis* (Hymenoptera: Aphelinidae), parasites of Diaspine scales. Hilgardia **44**:1–25.

Baker, T.C. & R.T. Cardé. 1984. Techniques for behavioral bioassays. *In*: Techniques in Pheromone Research, eds. H.E. Hummel & T.A. Miller, pp. 45–73, Springer-Verlag, New York.

Baker, T.C. & C.E. Linn. 1984. Wind tunnels in pheromone research. *In*: Techniques in Pheromone Research, eds. H.E. Hummel & T.A. Miller, pp. 75–110, Springer-Verlag, New York.

Barbehenn, R.V. 1993. Silicon—an indigestible marker for measuring food consumption and utilization by insects. Entomol. Exp. Appl. **67**:247–251.

Barbehenn, R.V. & A. Keddie. 1992. Gut contents in molting lepidopteran larvae—a source of error in nutritional studies. Entomol. Exp. Appl. **62**:87–91.

Bell, W.J. & T. Tobin. 1981. Orientation to sex pheromone in the American cockroach: analysis of chemo-orientation mechanisms. J. Insect Physiol. **27**:501–508.

Berdegue, M. & J.T. Trumble. 1996. Effect of plant chemical extracts and physical characteristics of *Apium graveolens* and *Chenopodium murale* on host choice by *Spodoptera exigua* larvae. Entomol. Exp. Appl. **78**:253–262.

Berenbaum, M. 1986. Postingestive effects of phytochemicals on insects: on Paraclesus and plant products. *In*: Insect-Plant Interactions, eds. J.R. Miller & T.A. Miller, pp. 121–153, Springer-Verlag, New York.

Berenbaum, M. & P. Feeny. 1981. Toxicity of angular furanocoumarins to swallowtails: escalation in the coevolutionary arms race. Science **212**:927–929.

Berenbaum, M.R. & A.R. Zangerl. 1994. Costs of inducible defense: protein limitation, growth, and detoxification in parsnip webworms. Ecology **75**:2311–2317.

Bernays, E.A. 1978. Tannins: an alternative viewpoint. Entomol. Exp. Appl. **24**:44–53.

Bernays, E.A. & R.F. Chapman. 1994. Host-Plant Selection by Phytophagous Insects. Chapman & Hall, New York.

Bhattacharya, A.K. & G.P. Waldbauer. 1969a. Quantitative determination of uric acid in insect feces by lithium carbonate extraction and the enzymatic-spectrophotometric method. Ann. Entomol. Soc. Am. **62**:925–927.

Bhattacharya, A.K. & G.P. Waldbauer. 1969b. Faecal uric acid as an indicator in the determination of food utilization. J. Insect Physiol. **15**:1129–1135.

Bhattacharya, A.K. & G.P. Waldbauer. 1970. Use of the faecal uric acid method in measuring the utilization of food by *Tribobium confusum*. J. Insect Physiol. **16**:1983–1990.

Blaney, W.M. & C. Winstanley. 1982. Food selection behavior in *Locusta migratoria*. *In*: Proceedings of the 5th International Symposium on Insect-Plant Relationships, Wageningen, eds. J.H. Visser & A.K. Minks, pp. 365–366, Pudoc, Wageningen.

Blau, P.A., P. Feeny, L. Contardo & D.S. Robson. 1978. Allylglucosinolate and herbivorous caterpillars: a contrast in toxicity and tolerance. Science **200**:1296–1298.

Blum, M.S. 1985. Alarm pheromones. *In*: Comprehensive Insect Physiology, Biochemistry and Pharmacology, Vol. 9, eds. G.A. Kerkut & L.I. Gilbert, pp. 193–224, Pergamon Press, Oxford.

Boller, E.F. & J. Horter. 1985. Oviposition deterring pheromone in *Rhagoletis cerasi*: behavioral laboratory test to measure pheromone activity. Entomol. Exp. Appl. **39**:163–169.

Borden, J.H. 1985. Aggregation pheromones. *In*: Comprehensive Insect Physiology, Biochemistry and Pharmacology, Vol. 9, eds. G.A. Kerkut & L.I. Gilbert, pp. 257–285, Pergamon Press, Oxford.

Bowers, M.D., N.E. Stamp & E.D. Fajer. 1991. Factors affecting calculation of nutritional induces for foliage-fed insects—an experimental approach. Entomol. Exp. Appl. **61**:101–116.

Breed, M.D. 1981. Individual recognition and learning of queen odors by worker honeybees (*Apis mellifera*). Proc. Natl. Acad. Sci. USA **78**:2635–2637.

Breed, M.D. 1983. Nestmate recognition in honey bees. Anim. Behav. **31**:86–91.

Breed, M.D., T.M. Stiller, M.S. Blum & R.E. Page, Jr. 1992. Honeybee nestmate recognition: effects of queen fecal pheromones. J. Chem. Ecol. **18**:1633–1640.

Bristow, P.R., R.P. Doss & R.L. Campbell. 1979. A membrane filter bioassay for studying phagostimulatory materials in leaf extracts. Ann. Entomol. Soc. Am. **72**:16–18.

Chan, B.G., A.C. Waiss, W.L. Stanley & A.E. Goodban. 1978. A rapid diet preparation method for antibiotic phytochemical bioassay. J. Econ. Entomol. **71**:366–368.

Clancy, K.M., R.D. Foust, T.G. Huntsberger, J.G. Whitaker & D.M. Whitaker. 1992. Techniques for using microencapsulated terpenes in lepidopteran artificial diets. J. Chem. Ecol. **18**:543–560.

Cottee, P.K., E.A. Bernays & J.A. Mordue. 1988. Comparisons of deterrency and toxicity of selected secondary plant compounds to an oligophagous and a polyphagous acridid. Entomol. Exp. Appl. **46**:241–247.

de Boer, G. & F.E. Hanson. 1984. Foodplant selection and induction of feeding preference among host and non-host plants in larvae of the tobacco hornworm, *Manduca sexta*. Entomol. Exp. App. **35**:177–193.

Dethier, V.G. 1941. Chemical factors determining the choice of food plants by *Papilio* larvae. Am. Nat. **75**:61–73.

Dethier, V.G. 1980. Food-aversion learning in two polyphagous caterpillars, *Diacrisia virginica* and *Estigmene congrua*. Physiol. Entomol. **5**:321–325.

Dethier, V.G., L. Barton Browne & C.N. Smith. 1960. The designation of chemicals in terms of the responses they elicit from insects. J. Econ. Entomol. **53**:134–136.

Dimock, M.B. & G.G. Kennedy. 1983. The role of glandular tichomes in the resistance of *Lycopersicon hirsutum* f. *glabratum* to *Heliothis zea*. Entomol. Exp. Appl. **33**:263–268.

Dimock, M.B., G.G. Kennedy & W.G. Williams. 1982. Toxicity studies of analogs of 2-tridecanone, a naturally occurring toxicant from a wild tomato. J. Chem. Ecol. **8**:837–842.

Doskotch, R.W., Y.-Y. Cheng, T.M. O'dell & L. Girard. 1980. Nerolidol: an antifeeding sesquiterpene alcohol for gypsy moth larvae from *Malalenca leucadendron*. J. Chem. Ecol. **6**:845–851.

Eigenbrode, S.D., J. Barnard & A.M. Shelton. 1989. A system for quantifying behavior of neonate caterpillars and other small, slow-moving animals. Can. Entomol. **121**:1125–1126.

Eigenbrode, S.D., K.E. Espelie & A.M. Shelton. 1991. Behavior of neonate diamondback moth larvae [*Plutella xylostella* (L.)] on leaves and on extracted leaf waxes of resistant and susceptible cabbages. J. Chem. Ecol. **17**:1691–1704.

Erickson, J.M. & P. Feeny. 1974. Sinigrin: a chemical barrier to the black swallowtail butterfly, *Papilio polyxenes*. Ecology **55**:103–111.

Escoubas P., Y. Fukushi, L. Lajide & J. Mizutani. 1992. A new method for fast isolation of insect antifeedant compounds from complex mixtures. J. Chem. Ecol. **18**:1819–1832.

Farrar, R.R., J.D. Barbour & G.G. Kennedy. 1989. Quantifying food consumption and growth in insects. Ann. Entomol. Soc. Am. **82**:593–598.

Feeny, P., L. Rosenberry & M. Carter. 1983. Chemical aspects of oviposition behavior in butterflies. *In*: Herbivorous Insects: Host-Seeking Behavior and Mechanisms, ed. S. Ahmad, pp. 27–76, Academic Press, New York.

Feeny, P., K. Sachdev, L. Rosenberry & M. Carter. 1988. Luteolin 7-*O*-(6′-*O*-malonyl)*B*-D-glucoside and *trans*-chlorogenic acid: oviposition stimulants for the black swallowtail butterfly. Phytochemistry **27**:3439–3448.

Fein, B.L., W.H. Reissig & W.L. Roelofs. 1982. Identification of apple volatile attractive to the apple maggot. J. Chem. Ecol. **8**:1473–1487.

Finch, S. 1986. Assessing host-plant finding by insects. *In*: Insect-Plant Interactions, eds., J.R. Miller & T.A. Miller, pp. 23–63, Springer-Verlag, New York.

Fraenkel, G.S. 1959. The raison d'etre of secondary plant substances. Science **129**:1466–1470.

Hagerup, M., I. Søndergaard & J.K. Nielsen. 1990. Measurements of areas consumed from leaf disks by chewing phytophagous insects: description of a new method involving image processing. Entomol. Exp. Appl. **57**:105–113.

Hardee, D.D., E.B. Mitchell & P.M. Huddleston. 1966. Effect of age, nutrition, sex, and time of day on response of boll weevils to an attractant from cotton. Ann. Entomol. Soc. Am. **59**:1024–1025.

Hare, J.D. 1987. Growth of *Leptinotarsa decemlineata* larvae in response to simultaneous variation in protein and glycoalkaloid concentration. J. Chem. Ecol. **13**:39–46.

Hare, J.D., J.G. Millar & R.F. Luck. 1993. A caffeic acid ester mediates host recognition by a parasitic wasp. Naturwissenschaften **80**:92–94.

Hawkes, C. & T.H. Coaker. 1976. Behavioural responses to host-plant odours in adult cabbage root fly (*Erioischia brassicae* [Bouché]). Symp. Biol. Hung. **16**:85–89.

Holldobler, B. & E.O. Wilson. 1990. The Ants. Belknap Press, Cambridge, MA.

Honda, K. 1990. Identification of host-plant chemicals stimulating ovipositions by swallowtail butterfly, *Papilio protenor*, J. Chem. Ecol. **16**:325–337.

Honda, K. & N. Hayashi. 1995. A flavanoid glucoside, Phellamurin, regulates differential oviposition on a rutaceous plant, *Phellodentron amurense*, by two sympatric swallowtail butterflies, *Papilio protenor* and *P. xuthus*. J. Chem. Ecol. **21**:1531–1539.

Horton, D.R. 1995. Statistical considerations in the design and analysis of paired-choice assays. Environ. Entomol. **24**:179–192.

Horton, D.R. & R.A. Redak. 1993. Further comments on analysis of covariance in insect dietary studies. Entomol. Exp. Appl. **69**:263–275.

Hovantiz, W. & V.C.S. Chang. 1964. Adult oviposition responses in *Pieris rapae*. J. Res. Lepid. **3**:159–172.

Hsiao, T.H. & G. Fraenkel. 1968. The role of secondary plant substances in the food specificity of the Colorado potato beetle. Ann. Entomol. Soc. Am. **61**:485–493.

Huang, P., J.A.A. Renwick & K. Sachdevgupta. 1993. Oviposition stimulants and deterrents regulating differential acceptance of *Iberis amara* by *Pieris rapae* and *P. napi oleraceae*. J. Chem. Ecol. **19**:1645–1663.

Ichinosé, T. & H. Honda. 1978. Ovipositional behavior of *Papilio protenor demetrius* Cramer and the factors involved in its host plants. App. Entomol. Zool. **13**:103–114.

Isman, M. & P. Proksch. 1985. Deterrent and insecticidal chromenes from *Encelia* (Asteraceae). Phytochemistry **24**:1949–1951.

Jeanne, R.L. 1981. Evidence for an alarm substance in *Polistes canadensis*. Experientia **38**:329–30.

Jermy, T. & A. Szentesi. 1978. The role of inhibitory stimuli in the choice of oviposition site by phytophagous insects. Entomol. Exp. Appl. **24**:458–471.

Jermy, T., F.E. Hanson & V.G. Dethier. 1968. Induction of specific food preference in lepidopterous larvae. Entomol. Exp. Appl. **11**:211–230.

Jermy, T., E.A. Bernays & A. Szentesi. 1982. The effect of repeated exposure to feeding deterrents on their acceptability to phytophagous insects. *In*: Proceedings of the 5th International Symposium on Insect-Plant Relationships, Wageningen 1982, eds. J.H. Visser & A.K. Minks, pp. 25–32 Pudoc, Wageningen.

Johnson, N.D. & B.L. Bentley. 1988. Effects of dietary protein and lupine alkaloids on growth and survivorship of *Spodoptera eridania*. J. Chem. Ecol. **14**:1391–1403.

Jones, C.G. & J.S. Coleman. 1988. Leaf disk size and insect feeding preference: implications for assays and studies on induction of plant defense. Entomol. Exp. Appl. **47**:167–172.

Jones, O.T. & T.H. Coaker. 1977. Oriented responses of carrot fly larvae, *Psila rosae*, to plant odours, carbon dioxide and carrot root volatiles. Physiol. Entomol. **2**:189–197.

Kaethler, F., D.J. Pree & A.W. Brown. 1982. HCN: a feeding deterrent in peach to the oblique-banded leafroller, *Choristoneura rosaceana* (Lepidoptera: Tortricidae). Ann. Entomol. Soc. Am. **75**:568–573.

Kellogg, F.E. & R.H. Wright. 1962. The olfactory guidance of flying insects. III. A technique for observing and recording flight paths. Can. Entomol. **94**:486–493.

Klocke, J.A. & B.G. Chan. 1982. Effects of cotton condensed tannin on feeding and digestion in the cotton pest, *Heliothis zea*. J. Insect Physiol. **28**:911–915.

Knols, B.G.J., R. de Jong & W. Takken. 1994. Trapping system for testing olfactory responses of the malaria mosquito *Anopheles gambiae* in a wind tunnel. Med. Vet. Entomol. **8**:386–388.

Kogan, M. 1986. Bioassays for measuring quality of insect food. *In*: Insect-Plant Interactions, eds., J.R. Miller & T.A. Miller, pp. 155–189, Springer-Verlag, New York.

Kogan, M. & J.R.P. Parra. 1981. Techniques and application of measurements of consumption and utilization of food by phytophagous insects. *In*: Current Topics in Insect Endocrinology and Nutrition, eds., G. Bhaskaran, S. Friedman & J.G. Rodriguez, pp. 337–352, Plenum, New York.

Kojima, J. 1994. Evidence for an alarm pheromone in *Ropalidia romandi* (Le Guillou) (Hymenoptera: Vespidae). Aust. Entomol. Soc. **33**:45–47.

Leal, W.S., A.R. Panizzi & C.C. Niva. 1994. Alarm pheromone system of leaf-footed bug *Leptoglossus zonatus* (Heteroptera, Coreidae). J. Chem. Ecol. **20**:1209–1216.

Lenz, M. & E.R. Williams. 1980. Influence of container, matrix volume and group size on survival and feeding activity in species of *Coptotermes* and *Nasutitermes* (Isoptera: Rhinotermitidae, Termitidae). Mater. Org. (Berl.) **15**:25–46.

Lewis, A.C. & H.F. van Emden. 1986. Assays for insect feeding. *In*: Insect-Plant Interactions, eds., J.R. Miller & T.A. Miller, pp. 95–119, Springer-Verlag, New York.

Luck, R.F. & N. Uygun. 1986. Host recognition and selection by *Aphytis* species: response to California red, oleander, and cactus scale cover extracts. Entomol. Exp. Appl. **40**:129–136.

Luck, R.F., H. Podoler & R. Kfir. 1982. Host selection and egg allocation behaviour by *A. melinus* and *A. lingnanensis*: comparison of two facultatively gregarious parasitoids. Ecol. Entomol. **7**:397–408.

Mark, G.A. 1982. Induced oviposition preference, periodic environments and demographic cycles in the bruchid beetle *Callosobruchus maculatus*. Entomol. Exp. Appl. **32**:155–160.

Marquis, R.J. & H.E. Braker. 1987. Influence of method of presentation on results of plant-host preference tests with two species of grasshopper. Entomol. Exp. Appl. **44**:59–63.

Matsumoto, H., S. Tebayashi, Y. Kuwahara, S. Matsuyama, T. Suzuki & K Fujii. 1994. Identification of taxifolin present in the Azuki bean as an oviposition stimulant of the Azuki bean weevil. J. Pesticide Sci. **19**:181–186.

Matsumoto, Y. 1968. Volatile organic sulphur compounds as insect attractants with special reference to host selection. *In*: Control of Insect Behaviour by Natural Products, eds. D.L. Wood, R.M. Silverstein & M. Nakajima, pp. 133–160, Academic Press, New York and London.

Matsumoto, Y. & A.J. Thorsteinson. 1970. Olfactory response of larvae of the onion maggot, *Hylemya antiqua* Meigen (Diptera: Anthomyiidae) to organic sulphur compounds. Appl. Entomol. Zool. **3**:107–111.

Mayer, M.S. & J.R. McLaughlin. 1991. Handbook of Insect Pheromones and Sex Attractants. CRC Press, Boca Raton, FL.

McLean, D.L. & M.G. Kinsey. 1964. A technique for electronically recording aphid feeding and salivation. Nature **202**:1358–1359.

Meade, T. & J.D. Hare. 1993. Effects of differential host plant consumption by *Spodoptera exigua* (Lepidoptera, Noctuidae) on *Bacillus thuringiensis* efficacy. Environ. Entomol. **22**:432–437.

Meisner, J., A. Fleischer & C. Eizick. 1982. Phagodeterrency induced by (–)-carvone in the larva of *Spodoptera littoralis* (Lepidoptera: Noctuidae). J. Econ. Entomol. **75**:462–466.

Messinia, F.J. & J.A.A. Renwick. 1985. Ability of ovipositing seed beetles to discriminate between seeds with differing egg loads. Ecol. Entomol. **10**:225–230.

Metcalf, R.L., R.A. Metcalf & A.M. Rhodes. 1980. Cucurbitacins as kairomones for diabroticite beetles. Proc. Natl. Acad. Sci. USA **77**:3769–3772.

Millar, J.G. & J.D. Hare. 1993. Identification and synthesis of a kairomone inducing oviposition by parasitoid *Aphytis melinus* from California red scale covers. J. Chem. Ecol. **19**:1721–1736.

Miller, J.R. & W.L. Roelofs. 1978. Sustained-flight tunnel for mesuring insect responses to wind-borne sex pheromones. J. Chem. Ecol. **4**:187–198.

Miller, J.R. & K.L. Strickler. 1984. Finding and accepting host plants. *In*: Chemical Ecology of Insects, eds. W.J. Bell & R.T. Cardé, pp.127–157, Chapman and Hall, New York.

Miller, J.S. & P. Feeny. 1983. Effects of benzylisoquinoline alkaloids on the larvae of polyphagous Lepidoptera. Oceologia (Berl.) **58**:332–339.

Mittler, T.E. 1988. Applications of artificial feeding techniques for aphids. *In*: Aphids: Their Biology, Natural Enemies and Control, Vol. 2B, eds., A.K. Minks & P. Harrewijn, pp. 145–170, Elsevier, Amsterdam.

Mittler, T.E. & R.H. Dadd. 1962. Artificial feeding and rearing of the aphid, *Myzus persicae* (Sulzer), on a completely defined synthetic diet. Nature **195**:404.

Morse, J.G., T.S. Bellows & Y. Iwata. 1986. Technique for evaluating residual toxicity of pesticides to motile insects. J. Econ. Entomol. **79**:311–314.

Muller, H.J. 1966. Uber die Ursachen der unterschiedichen Resistenz von *Vicia faba* L. gegenuber der Bohenblattlaus, *Aphis* (*Doralis*) *fabae* Scop. IX. Der Einfluss okologischer Faktoren auf das Wachstum von *Aphis fabae* Scop. Entomol. Exp. Appl. **9**:42–66.

Navon, A., J.D. Hare & B.A. Federici. 1993. Interactions among *Heliothis virescens* larvae, cotton condensed tannin and the CryIA(c) *delta*-endotoxin of *Bacillus thuringiensis*. J. Chem. Ecol. **19**:2485–2499.

Nisbet, A.J., J.A.T. Woodford & R.H.C. Strang. 1994. Quantifying aphid feeding on non-radioactive food sources. Entomol. Exp. Appl. **72**:85–89.

Nishida, R. 1994. Oviposition stimulant of a zeryntiine swallowtail butterfly, *Luehdorfia japonica*. Phytochemistry **36**:873–877.

Noldus Information Technology. 1995. The Observer, Base Package for Windows. Reference Manual, ver. 3.0, Wageningen, the Netherlands.

Ohigashi, H., M.R. Wagner, F. Matsumura & D.M. Benjamin. 1981. Chemical basis of differential feeding behavior of the larch sawfly, *Pristiphora erichsonii* (Hartig). J. Chem. Ecol. **7**:599–614.

Papaj, D.R., P. Feeny, K. Sachdevgupta & L. Rosenberry. 1992. D-(+)-pinitol, an oviposition stimulant for the pipevine swallowtail butterfly, *Battus philenor*. J. Chem. Ecol. **18**:799–815.

Pasteels, J.M. & J.-C. Verhaeghe. 1974. Dosage biologique de la phéromone de peste ches les fourrageuses et les reines de *Myrmeca rubra*. Insects Soc. **21**:167–180.

Peterson, S.C. & T.D. Fitzgerald. 1991. Chemoorientation of eastern tent caterpillars to trail pheromone, 5β-cholestane-3,24dione. J. Chem. Ecol. **17**:1963–1972.

Pettersson, J. 1970. An aphid sex attractant. Entomol. Scand **1**:63–73.

Phelan, P.L. & J.R. Miller. 1982. Post-landing behavior of alate *Myzus persicae* as altered by *E*-β-farnesene and three carboxylic acids. Entomol. Exp. Appl. **32**:46–53.

Prokopy, R.J. 1981. Epideictic pheromones that influence spacing patterns of phytophagous insects. *In*: Semiochemicals: Their Role in Pest Control, eds. D.A. Nordlund, R.L. Jones & W.J. Lewis, pp. 181–213, Wiley & Sons, New York.

Prokopy, R.J., A.L. Averill, S.S. Cooley & B.A. Roitberg. 1982. Associative learning in egglaying site by apple maggot flies. Science **218**:76–77.

Quednau, F.W. & H.M. Hübsch. 1964. Factors influencing the host-finding and host-acceptance pattern in some *Aphytis* species (Hymenoptera: Aphelinidae). S. Afr. Agric. Sci. **7**:543–554.

Raubenheimer, D. & S.J. Simpson. 1992. Analysis of covariance—an alternative to nutritional indices. Entomol. Exp. Appl. **62**:221–231.

Rausher, M.D. 1981. Host plant selection by *Battus philenor* butterflies: the roles of predation, nutrition and plant chemistry. Ecol. Monogr. **51**:1–20.

Renwick, J.A.A., C.D. Radke, & K. Sachdevgupta. 1989. Chemical constituents of *Erysimum cheiranthoides* deterring oviposition by the cabbage butterfly, *Pieris rapae*. J. Chem. Ecol. **15**:2161–2169.

Reuter, L.L., N.C. Toscano & T.M. Perring. 1993. Modification of feeding behavior of *Myzus persicae* (Homoptera, Aphididae) by selected compounds. Environ. Entomol. **22**:915–919.

Robertson, J.L. & H.K. Preisler. 1992. Pesticide Bioassays with Arthropods. CRC Press. Boca Raton, FL.

Roisin, Y., C. Everaerts, J.M. Pasteels & O. Bonnard. 1990. Caste-dependent reactions to soldier defensive secretion and chiral alarm / recruitment pheromone in *Nasutitermes princeps*. J. Chem. Ecol. **16**:2865–2875.

Rust, M.K. & D.A. Reierson. 1977. Using wood extracts to determine the feeding preferences of the western drywood termite, *Incisitermes minor* (Hagen). J. Chem. Ecol. **3**:391–399.

SAS Institute Inc. 1989. SAS User's Guide: Statistics, Version 6.06 ed. SAS Institute, Cary, NC.

Saxena, K.N. & S. Goyal. 1978. Host-plant relations of the citrus butterfly *Papilio demoleus* L.: orientational and ovipositional responses. Entomol. Exp. Appl. **24**:1–10.

Schmidt, D.J. & J.C. Reese. 1986. Sources of error in nutritional index studies of insects on artificial diet. J. Insect Physiol. **32**:193–198.

Schoonhoven, L.M. & T. Jermy. 1977. A behavioral and electrophysiological analysis of insect feeding deterrents. *In*: Crop Protection Agents—Their Biological Evaluation, ed. N.R. MacFarlane, pp. 133–146, Academic Press, New York.

Scriber, J.N. & F. Slansky. 1981. The nutritional ecology of immature insects. Ann. Rev. Entomol. **26**:183–212.

Singer, M.C. 1986. The definition and measurement of oviposition preference in plant-feeding insects. *In*: Insect-Plant Interactions, eds., J.R. Miller & T.A. Miller, pp. 65–94, Springer-Verlag, New York.

Singh, P. & R.F. Moore. 1985. Handbook of Insect Rearing. 2 vols. Elsevier, Amsterdam.

Slansky, F. 1985. Food utilization by insects: Interpretation of observed differences between dry weight and energy efficiencies. Entomol. Exp. Appl. **39**:47–60.

Slansky, F. & J.M. Scriber. 1985. Food consumption and utilization. *In*: Comprehensive Insect Physiology, Biochemistry and Pharmacology, Vol. 4, eds. G.A. Kerkut & L.I. Gilbert, pp. 87–163, Pergamon Press, Oxford.

Sokal, R.R. & F.J. Rohlf. 1995. Biometry, 3rd. ed. Freeman, New York.

Sparks, M.R. & J.S. Cheatham. 1970. Responses of a laboratory strain of the tobacco hornworm, *Manduca sexta*, to artificial oviposition sites. Ann. Entomol. Soc. Am. **63**:428–431.

Städler, E. & F.E. Hanson. 1976. Influence of induction of host preference on chemoreception of *Manduca sexta*: behavioral and electrophysiological studies. Symp. Biol. Hung. **16**:267–273.

Städler, E. & F.E. Hanson. 1978. Food discrimination and induction of preference for artificial diets in the tobacco hornworm, *Manduca sexta*. Physiol. Entomol. **16**:121–133.

Stamp, N.E. 1991. Stability of growth and consumption rates and food utilization efficiencies when insects are given an excess of food. Ann. Entomol. Soc. Am. **84**:58–60.

Steel, R.G.D. & J.H. Torrie. 1980. Principles and Procedures of Statistics, A Biometric Approach. McGraw-Hill, New York.

Steinberg, S., M. Dicke, L.E.M. Vet & R. Wannigen. 1992. Response of the braconid parasitoid *Cotesia* (= *Apanteles*) *glomerata* to volatile infochemicals: effects of bioassay set-up, parasitoid age and experience and barometric flux. Entomol. Exp. Appl. **63**:163–175.

Szentesi, A. & E.A. Bernays. 1984. A study of behavioral habituation to a feeding deterrent in nymphs of *Schistocerca gregaria*. Physiol. Entomol. **9**:329–340.

Takahashi, S., M. Hajika, J. Takabayashi & M. Fukui. 1990. Oviposition stimulants in the coccoid cuticular waxes of *Aphytis yanonensis* DeBach and Rosen. J. Chem. Ecol. **16**:1657–1665.

Traynier, R.M.M. 1967. Effect of host plant odour on the behaviour of the adult cabbage root fly, *Erioischia brassicae*. Entomol. Exp. Appl. **10**:321–328.

Trial, H. & J.B. Dimond. 1979. Emodin in buckthorn: a feeding deterrent to phytophagous insects. Can. Entomol. **111**:207–212.

Usher, B.F. & P. Feeny. 1983. Atypical secondary compounds in the family Cruciferae: tests for toxicity to *Pieris rapae*, an adapted crucifer-feeding insect. Entomol. Exp. Appl. **34**:257–265.

Usher, B.F., E.A. Bernays, R.B. Barbehenn & R.P. Wrubel. 1989. Oral dosing of insects with feeding deterrent compounds. Entomol. Exp. Appl. **52**:119–133.

Vaidya, V.C. 1969. Investigations on the role of visual stimuli in the egg-laying and resting behaviour of *Papilio demoleus* L. (Papilionidae: Lepidoptera). Anim. Behav. **17**:350–355.

van Emden, H.F. & M.A. Bashford. 1976. The effect of leaf excision on the performance of *Myzus persicae* and *Brevicoryne brassicae* in relation to the nutrient treatment of the plants. Physiol. Entomol. **1**:67–71.

van Lenteren, J.C. & P. DeBach. 1981. Host discrimination in three ectoparasites (*Aphytis coheni, A. lingnanansis*, and *A. melinus*) of the oleander scale (*Aspidiotus nerii*). Neth. J. Zool. **31**:504–532.

van Loon, J.J.A. 1989. Measuring food utilization in plant-feeding insects: toward a metabolic and dynamic approach. *In*: Insect-Plant Interactions, Vol. 3, ed. E.A. Bernays, pp. 79–124, CRC Press, Boca Raton, FL.

Van Vorhis Key, S.E. & T.C. Baker. 1982. Specificity of laboratory trail following by the Argentine ant, *Iridomyrmex humilis* (Mayr), to Z-9-hexadecenal, analogs, and gaster extracts. J. Chem. Ecol. **8**:1057–1063.

Verschaffelt, E. 1911. The cause determining the selection of food in some herbivorous insects. Proc. Acad. Sci. Amsterdam **13**:536–542.

Vet, L.E.M., J.C. van Lenteren, M. Heymans & E. Meelis. 1983. An airflow olfactometer for measuring olfactory responses of hymenopterous parasitoids and other small insects. Physiol. Entomol. **8**:97–106.

Visscher, P.K. & R.S. Vetter. 1995. Smoke and target color effects on defensive behavior in yellowjacket wasps and bumble bees (Hymenoptera: Vespidae, Apidae) with a description of an electronic attack monitor. J. Econ. Entomol. **88**:579–583.

Visser, J.H. & D.A. Ave. 1978. General green leaf volatiles in the olfactory orientation of the Colorado beetle, *Leptinotarsa decemlineata*. Entomol. Exp. Appl. **24**:538–549.

Wagner, M.R., D.M. Benjamin, K.M. Clancy & B.A. Schu. 1983. Influence of diterpene resin acids on feeding and growth of larch sawfly, *Pristiphora erichsonii* (Hartig). J. Chem. Ecol. **9**:119–127.

Waldbauer, G.P. 1968. The consumption and utilization of food by insects. Adv. Insect. Physiol. **5**:229–288.

Waldbauer, G.P. & G. Fraenkel. 1961. Feeding on normally rejected plants by maxillectomized larvae of the tobacco hornworm, *Protoparce sexta* (Lepidoptera: Sphingidae). Ann. Entomol. Soc. Am. **54**:477–485.

Walker, G.P. & T.M. Perring. 1994. Feeding and oviposition behavior of whiteflies (Homoptera: Aleyrodidae) interpreted from AC electronic feeding monitor wave forms. Ann. Entomol. Soc. Am. **87**:363–374.

Waller, D.A. & C.G. Jones. 1989. Measuring herbivory. Ecol. Entomol. **14**:479–481.

Weston, P.A. & J.R. Miller. 1985. Influence of cage design on precision of tube-trap bioassay for attractants of the onion fly, *Delia antiqua*. J. Chem. Ecol. **11**:435–440.

Wolfson, J.L. & L.L. Murdock. 1987. Method for applying chemicals to leaf surfaces for bioassay with herbivorous insects. J. Econ. Entomol. **80**:1334–1336.

Wright, J.P., D.B. Fisher & T.E. Mittler. 1985. Measurement of aphid feeding rates on artificial diets using ^{3}H-inulen. Entomol. Exp. Appl. **37**:9–11.

6

Bioassay Methods for Amphibians and Reptiles

Robert T. Mason, Douglas P. Chivers, Alicia Mathis, and Andrew R. Blaustein

6.1. Introduction 272
6.2. Amphibians 273
6.2.1. Orientation and homing 273
6.2.1.1. Frogs and toads 273
6.2.1.2. Salamanders 277
6.2.2. Chemical cues in kin recognition 278
6.2.2.1. Larval amphibians 278
6.2.2.2. Post-metamorphosis 282
6.2.3. Territorial pheromones 282
6.2.3.1. Source of the chemical signal 282
6.2.2.2. Response variables 284
6.2.4. Marking behaviors 286
6.2.5. Reproductive pheromones and sexual advertisement 287
6.2.5.1. Source of the chemical signal 287
6.2.5.2. Response variables 287
6.2.5.3. Species discrimination 288
6.2.5.4. Male courtship pheromones 288
6.2.6. Predator/prey interactions 289
6.2.6.1. Chemical alarm cues 289
6.2.6.2. Antipredator responses 292
6.2.7. Noxiousness and toxicity 294
6.3. Reptiles 296
6.3.1. Olfaction and vomerolfaction 296
6.3.2. Attractants and repellents 297
6.3.3. Aggregation 299
6.3.4. Trailing 302
6.3.5. Sexual behavior 305
6.3.6. Femoral pores 307
6.3.7. Copulatory plugs 307
6.3.8. Combat behavior 308

6.3.9. Predator/prey recognition 308
6.4. Conclusions 310
6.5. Acknowledgments 311
6.6. References 311

6.1. Introduction

The study of chemical communication, semiochemicals, and pheromones in amphibians and reptiles has benefited from a great deal of attention in recent years. Amphibians and reptiles present a number of problems with regard to studying both the behaviors elicited by chemical cues, and the elucidation of the chemical signals themselves. For example, many taxa of amphibians have both life history and environmental constraints that expose a given animal to an aqueous environment as a larva, then in just a few weeks, metamorphosis imposes a terrestrial existence.

Studies have evolved from astute observations of natural history to elegantly designed experiments that elucidate the causal mechanisms responsible for mediating behavioral responses to chemical cues. The literature pertaining to this area of research is surprisingly large, but weighted to investigations of relatively few behaviors. Sources of semiochemicals in these diverse taxa vary widely as well. All amphibians possess glandular structures in their skin that keep the skin moist as well as produce chemical signals. Some of these chemical secretions are toxic in amphibians while toxic secretions from reptile skin are currently unknown. In contrast to amphibian skin, reptilian skin is a highly keratinized, relatively impermeable integument that helps to retard transcutaneous water loss.

Chemical communication is an integral facet of amphibian and reptilian life. For example, it is used in prey detection or foraging, predator detection, and pheromonally mediated behaviors such as species, kin, and individual recognition; mate choice; alarm signaling; noxiousness and toxicity, and territoriality. One of the interesting results of compiling the studies for this chapter is both how similar amphibians and reptiles can be in terms of behaviors elicited to chemical cues (e.g., aggregation), but also how unique some behaviors are to either taxa (e.g., noxiousness in amphibians and trailing in reptiles). In this chapter, representative studies are grouped according to the type of behavior, and not in a phylogenetic hierarchy. The order of discussion of the behaviors is wholly arbitrary.

This is not a review of the literature concerning chemical communication in amphibians and reptiles, but rather, focuses on the methodologies employed by investigators in documenting and elucidating the behavioral mechanisms that subserve chemical communication in these two vertebrate classes. Thus, those examples chosen serve to exemplify a diverse set of research paradigms and provide the reader with an overview of the methods that have been utilized to study chemical communication in amphibians and reptiles. This review is not all-inclusive of the literature and the reader will easily be able to cite publications that we did not include. For example, for the sake of brevity, we have omitted studies dealing with chemical cues involved in feeding. We have tried to cite reviews from the literature that are significant in their completeness and timeliness.

6.2. Amphibians

6.2.1. Orientation and Homing

Many species of amphibians return to the same sites each year to breed (recently reviewed by Sinsch 1990), and several studies have examined the cues used in homing and orientation. To examine the role of cues used in orientation, investigators often have used experiments where amphibians are transported from a home area (e.g., breeding pond) to an area away from where they were captured. They are then monitored to see if they can successfully "home" back to or orient toward their capture site after one or more of their senses are obliterated. In other studies, investigators have examined the role of chemical cues in amphibian orientation in laboratory test chambers. The basic methods used in these studies are reviewed below.

6.2.1.1. Frogs and Toads

In one homing and orientation study, Tracy and Dole (1969) captured western toads (*Bufo boreas*) from breeding sites and displaced them in light-tight jars 50–200 m to open areas where they were released. The role of olfaction in orientation of the toads was investigated by severing the olfactory nerves of some individuals and comparing their ability with nonimpaired and blinded individuals to "home" back toward the breeding sites.

Toads that were not olfactorily impaired and those that were blinded, showed a clear ability to relocate their capture site. Anosmic toads appeared to be disoriented. The authors concluded that olfactory cues provided the principal means of orientation in these animals. However, the authors also pointed out that changes in variables such as cloud cover, celestial cues, and wind direction could have influenced these results. Moreover, they suggested that severing the olfactory nerve could have "eliminated the 'desire' to return to the breeding site rather than their ability to relocate it."

Dole (1965, 1968) studied homing and orientation in leopard frogs (*Rana pipiens*) under a variety of conditions using methods similar to those described above. As in *B. boreas,* blinded *R. pipiens* could home as successfully as normal individuals displaced comparable distances, from both upwind and downwind of home.

Grubb (1970) examined homing ability and orientation in individually marked postreproductive Mexican toads (*Bufo valliceps*). Displacements were conducted by transporting toads in light-proof bags various distances from their capture sites. Toads were intact, blinded, or anosmic. Toads were made anosmic by either destroying portions of the olfactory tract or by destroying the nasal epithelium by injecting 0.02 ml of 40% formalin into the nares of toads. Several toads were also blinded and rendered anosmic with formalin.

All intact toads were recaptured within ten days after displacement. Twenty of 23 blinded toads moved toward home. Ten toads that received formalin injections also moved toward home. In 9 of 14 trials, toads with destroyed olfactory tracts also moved toward home. Control toads given the same operation procedures but not made anosmic also moved toward home. Only three of 24 toads that were both blind and anosmic moved toward home. Grubb concluded that both visual and olfactory cues are used in homing in Mexican toads.

Grubb (1973a, 1973b, 1975, 1976) has investigated olfactory orientation in a number of anuran species in the laboratory using an olfactometer (Fig. 1 in Grubb 1973a). Test animals were collected from breeding populations. The main portion of the olfactometer was made of gray-painted plywood. The top of the apparatus had glass windows for illumination by room lighting. Terminal pipes were made of PVC tubing. A fan at the exit port pulled unfiltered room air through the apparatus. Water samples from home ponds and from foreign ponds were used as stimuli. Odorants from these sources were placed in the terminal pipes out of view from the test animals. The floor of the apparatus terminated in black plastic bag drop traps downwind from the connection of the terminal pipes. All species that were tested, fell into these drop traps.

Animals were tested by placing them under an inverted wire mesh basket upwind from the fan and released when the basket was raised by a pulley. Groups of animals were tested. Group size varied with the species being tested. Different species were tested for different durations. For example, some species were tested for 10 min, whereas several tree frog species (e.g., *Hyla chrysoscelis* and *Pseudacris clarki*) were tested for at least 20 min. The testing times were arbitrarily set but were consistent within species.

A response was scored if, after the allotted time period, an animal was in either one of the drop traps or at a position at least half way to the end of one of the arms. To control for possible biases in the apparatus or positional effects, the position of the odorants were reversed after half of a sample had been tested.

In an initial series of tests using the olfactometer (Grubb 1973a), reproductively active male *B. woodhousei* and *P. clarki,* as well as both male and female *P. streckeri,* moved toward the chemical cues of water from home ponds rather than those from the water in foreign ponds. Neither *H. chrysoscelis* nor *Scaphiopus hurteri* demonstrated chemosensory orientation.

In a subsequent study using the same olfactometer and similar methods, Grubb (1973b) showed that *B. valliceps* from some populations discriminated between chemical cues from their home pond versus those from a foreign pond, but toads from at least one population of this species did not make this discrimination. Moreover, toads that were in breeding condition, but had not been exposed to their breeding habitat for at least 1 month responded positively to the chemical cues from the breeding pond. Similarly, Southern leopard frogs (*Rana ultricularia*) discriminated the chemical cues from their home pond from that of control water even after 2 months under laboratory confinement (Grubb 1975).

Grubb (1976) had also employed mazes to measure how trained *B. valliceps* escape from noxious stimuli. Two types of mazes were used in these studies, multiple T-mazes and simple T-mazes (Fig. 1 in Grubb 1976). Mazes contained visual and olfactory cues. Performance parameters were running time and ratio of correct (in the trained direction) to incorrect turns at choice points. The olfactory cues were a few drops of specific odorants (anise oil, benzaldehyde, cedarwood oil, citral, creosote beechwood, geraniol, limonene, vanillin) placed on a cotton plug within a perforated, covered plastic cup that was situated midway along each arm of the mazes against the wall opposite the entrance of the preceding arm.

In one experiment, toads were trained in simple T-mazes to move to the right in response to cedarwood oil and a square and to the left in response to citral and a circle. In another experiment, toads were trained to move toward a specific sequence of odors. In both experiments, toads oriented with respect to olfactory cues and the spatial configuration of the maze but not to visual cues. Thus, this study further supports the view that amphibians, as represented by *B. valliceps* in this study, possess the olfactory mechanisms necessary to orient with reference to an "odoriferous landscape" in nature (Grubb 1976).

In an early examination of the role of olfaction in toad orientation, Martof (1962) conditioned toads (*Bufo fowleri*) to move onto a block by repeatedly shocking them. Five of the toads that learned to move onto the block were chosen for conditioning in an olfactometer. A current of air passed through an experimental chamber where toads were placed and were subjected to an odorant. Ten seconds after this exposure toads were shocked and conditioned to move onto a block to avoid the shock. The main odorant used was 2,2,4-trimethylpentane. Prior to conditioning some toads were rendered anosmic by severing the olfactory tracts.

Intact toads exhibited a high degree of conditioning and generally moved onto the block after the odorant was introduced. The anosmic toads did not respond to the odorants. Thus, discrimination was due to olfactory stimulation.

In a second series of experiments, a T-maze was used (Fig. 1 in Martof 1962). Humidified, filtered air was divided into two equal flows each of which entered the end of the maze and exited through the third arm of the maze where frogs were released. Odors were produced by bubbling controlled amounts of filtered air through soil, muck and vegetation obtained from various habitats. Frogs were given a choice of moving towards different odors derived from natural habitats.

Pseudacris triseriata moved significantly more often toward odors from a *Pseudacris* breeding habitat than odors obtained from an upland forest. *Hyla crucifer* did not discriminate between these two types of odorants. Martof (1962) concluded that *P. triseriata* showed olfactory discrimination but *Hyla* did not. He also suggested that the maze was more appropriate for *Pseudacris* than *Hyla* because *Hyla* was more arboreal and leaped more often and thus may have left the entrance tunnels before they could have distinguished between the odors.

Tropical frogs are also able to home after being displaced (e.g., McVey et al. 1981) but until recently, the sensory basis for homing in these species was unknown. A recent study of the dart-poison frog, *Dendrobates pumilio*, a species found in tropical America, suggests that airborne chemical cues are used for recognition of home areas (Forester & Wisnieski 1991). In this study, an olfactometer (similar to Fig. 3 in Dawley 1984a) was also used.

Individuals were confined in the anterior portion of one half of a Plexiglas™ chamber. The posterior portion of the chamber was connected to a vacuum line, and the anterior end received a glass Y-maze. When the vacuum was turned on, air was pulled through the system. Airflow was regulated so that it would be the same in both arms of the Y-maze.

In one series of tests, animals were given a choice of air pulled from a communal cage that housed other frogs or an empty aquarium. These sources were hidden from the view of the test animals. In another series of experiments, animals were housed in aquariums for three weeks with potting soil, sphagnum moss, a water dish, a plastic hide box and two plants (*Rheo discolor*) and a bromeliad (*Neoregelia*). Animals were placed in the olfactometer and tested to see if they could discriminate between (1) home and empty aquarium odors, (2) home versus potted *Coleus* plant odors, (3) bromeliad versus *Coleus* odors, or (4) home versus bromeliad odors.

Animals in the first series of experiments chose to associate with odors from the communal aquarium over those from a blank aquarium. In the second series of experiments, tested animals showed a significant preference for their home cage over an empty aquarium or one containing *Coleus* plants. They also showed a significant preference for bomeliad odors over *Coleus* odors. But they did not discriminate between their home aquarium and an aquarium containing a bromeliad.

These results suggest that airborne chemical cues are involved in habitat recognition. The authors suggest that odors from vegetation play a significant role in the ability of this species to identify suitable habitat.

Orientation via chemical cues may be more important in some species than in others. For example, Oldham (1966, 1967) studied homing and orientation in American toads (*Bufo americanus*) and green frogs (*Rana clamitans*). Oldham (1966) partially destroyed the olfactory tracts of 99 American toads and used 99 intact toads treated as operational controls. An additional 112 toads were left untreated. The toads were then transported to a point of land between two pools. About the same frequencies of returns occurred to the pools. Oldham concluded that olfaction was not necessary for orientation and homing in American toads. However, since only a portion of the olfactory system was destroyed, it is possible that olfaction was still involved in homing and orientation.

Oldham (1967) conducted similar translocation experiments with blinded, intact and olfactorily impaired green frogs (*Rana clamitans*). Blind frogs were capable of relocating their home pool suggesting that olfaction was involved in orientation.

6.2.1.2. Salamanders

Classic homing and orientation studies in newts were conducted by Twitty and his colleagues (e.g. Twitty 1959; Grant et al. 1968). The methods employed by Twitty (1959) were similar to the displacement studies described above for frogs and toads. Twitty (1959) showed that when the newt *Taricha rivularis* is displaced up to 640 m, from 80–95% orient their movements back to the home site. This result was also achieved when salamanders were permanently blinded by surgical removal of both eyes. Twitty (1959) concluded that visual cues were not essential in homing and orientation in this species and that olfaction was probably the sense used. This is supported by experiments where the nares were plugged by vaseline. Newts with plugged nares could not efficiently orient toward the home area.

More evidence for olfaction as an important sense in newt homing behavior was obtained by Grant et al. (1968). *T. rivularis* that were displaced after their olfactory epithelium was injected with formaldehyde oriented almost as well as controls. However, newts with surgically removed olfactory nerves failed to home.

The northern dusky salamander (*Desmognathus fuscus*) is a nonmigratory species with a relatively small home range (Barthalamus & Bellis 1972). This species is capable of returning to its original 3-m capture sector after displacements 30 m upstream (Barthalamus & Bellis 1969). Barthalamus and Bellis (1969) suggested that homing was based on olfactory cues because of the heavy forest canopy that would not allow them to home via celestial cues.

In a follow-up study, Barthalamus and Bellis (1972) used anosmic, blinded, sham-treated, and normal salamanders and displaced them at 3-m intervals up to 30 m from a home site. Animals were blinded by cauterizing the eyes with a red-hot dissecting needle. Salamanders were rendered anosmic by cauterizing the olfactory epithelium with a red-hot #2 insect pin into the nasal cavity. The sham controls were burned between the eyes and nostrils with a red-hot dissecting pin. Each test group had a corresponding control animal.

The homing abilities of the blinded and sham-treated salamanders were not significantly different from the normal displaced animals. More of these animals returned home than accepted their release site. However, more of the anosmic animals accepted their release site than returned home. The authors concluded that olfactory cues are important in homing in this species.

Similar studies supported the suggestion that olfactory cues are important in homing and orientation in some salamander species (e.g., Forester 1979) but not others (e.g., Landreth & Ferguson 1967).

These studies are important in an historical context as well as for the methodologies utilized. Presently, investigators refer to chemical cues rather than the more generic terms of "odors" or "odorants." The reason is that there are two separate and distinct chemosensory systems in most vertebrates. The olfactory system primarily serves to detect and discriminate more volatile compounds. The term "odor" would be appropriate for referring to the olfactory system. The vomeronasal system is more specialized for relatively nonvolatile chemical cues. Cooper and Burghardt (1990) suggest the term "vomodor" to refer to those chemical cues that stimulate the vomeronasal organ. Thus, in the older literature, investigators refer generically to "odors" when they may not have specifically tested olfaction as opposed to vomerolfaction. To attempt to reinterpret these older data would inevitably lead to confusion and misinterpretation. Suffice it to say that current investigators must be much more assiduous in their experimental designs and interpretations when referring to chemical cues stimulating the chemosensory organs of either amphibians or reptiles.

6.2.2. Chemical Cues in Kin Recognition

6.2.2.1. Larval Amphibians

Many animal species can discriminate between kin and nonkin (often called kin recognition). In most cases where the sensory basis for kin recognition has been investigated, chemical cues are involved. This is true for insects, fish, amphibians, and mammals (Fletcher & Michener 1987). Chemical cues are the basis for kin recognition in all amphibians where it has been examined (Blaustein & Walls 1995).

Variations of two basic methods, choice tests and the use of a Y-maze, have been used to examine whether frog and toad larvae use chemical cues to discriminate between kin and nonkin. Choice tests were used to investigate the role of chemical cues in Cascades (*Rana cascadae*) and red-legged (*R. aurora*) frog tadpoles (Blaustein & O'Hara 1982; Blaustein et al. 1993). In standard tests, individuals were allowed to associate near members of one of two stimulus groups within a rectangular test tank (122 × 44 × 30 cm) (see Fig. 6.1). A pencil mark was drawn widthwise to delineate the two halves of the tank. To create end compartments for holding stimulus tadpoles, a partition of 1.5-mm plastic

---

Figure 6.1. Methods for assessing the role of chemical cues in kin recognition in tadpoles and recently metamorphosed frogs. Top depicts standard test aquarium where stimulus tadpoles are placed in end compartments separated with a screen mesh from test animals (in center). Test animals could use both visual and chemical cues emanating from stimulus groups to distinguish between kin and nonkin. To test the role of vision in kin recognition, mesh barriers were replaced with water-tight glass panes that impeded water flow. Middle shows chambers (12.5 × 5.5 × 8 cm) in end compartments in which stimulus tadpoles were placed. The chambers were composed of 3-mm-thick opaque plastic fronts (facing the test animals) and sides. The top, back, and bottom of the chambers were made of 1.5-mm plastic mesh. Several 1.5-mm holes were bored into the sides. These chambers allowed chemical

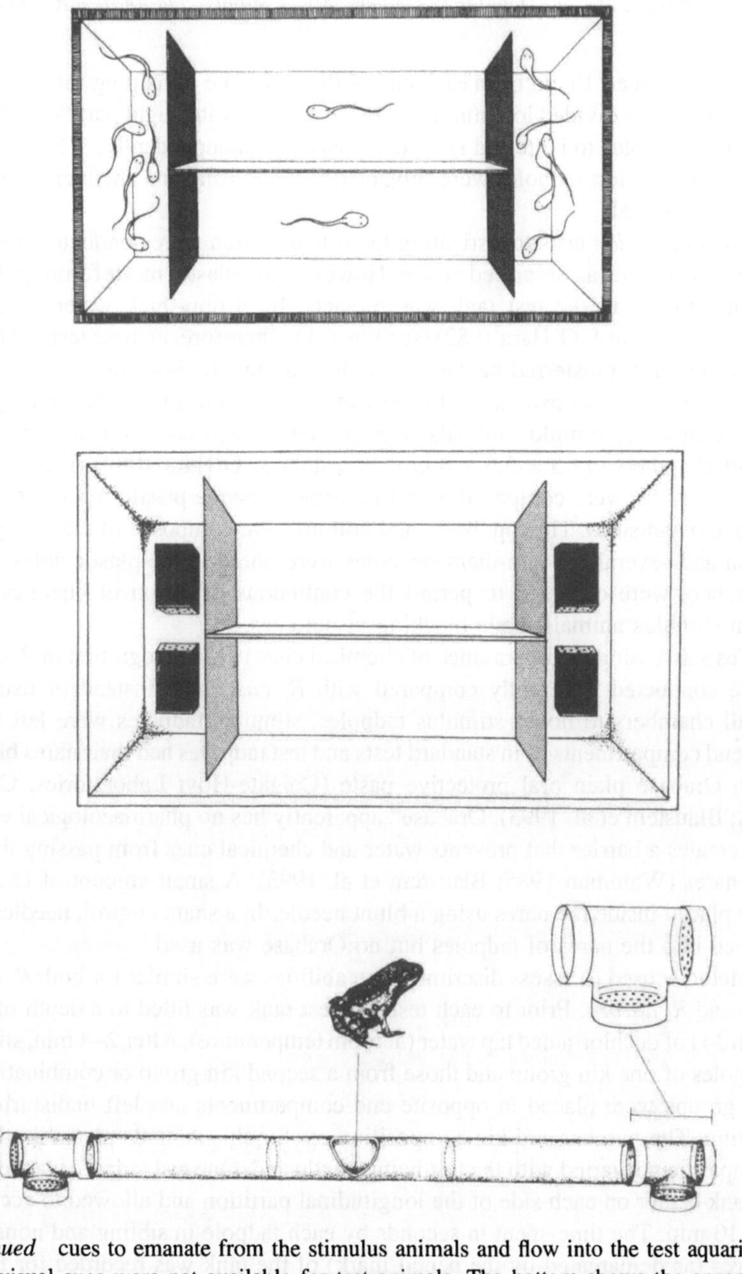

Continued cues to emanate from the stimulus animals and flow into the test aquarium, while visual cues were not available for test animals. The bottom shows the apparatus used to test kin recognition in newly metamorphosed frogs. The test individual was placed into the center of a plastic tube (84 cm long, 5.5 cm diameter). Stimulus frogs were placed into the ends of the tube and separated from the test animals by a 1.5-mm plastic mesh. There was limited visual contact and airborne chemical cues available, but there was no water contact between stimulus and test frogs. Neither tadpoles nor frogs are drawn to scale. Redrawn from Blaustein and O'Hara (1982).

mesh was placed 15 cm from each end of the tank. The remaining central portion of the tank was divided longitudinally by an opaque water-tight partition allowing two test tadpoles to be tested simultaneously (but independently). Thus, in these standard tests, test tadpoles were subjected to both visual and chemical cues from stimulus animals.

Rana cascadae tests investigating the role of vision were conducted using the same procedures as described above. However, the plastic mesh forming the end compartments in the test tank was replaced by 4-mm-thick water-tight glass panes (Blaustein & O'Hara 1982) (see Fig. 6.1). Therefore, in these tests, chemical cues were not transferred between stimulus and test individuals.

R. cascadae tests examining the role of chemical cues used the same general procedures, but stimulus animals were placed in each end compartment within small chambers (12.5 × 5.5 × 8 cm; Blaustein & O'Hara 1982) (see Fig. 6.1). The chambers were composed of a 3-mm-thick opaque plastic front (facing test animals) and sides. The top, back, and bottom were composed of 1.5-mm plastic mesh and several 1.5-mm-diameter holes were bored in the plastic sides. These chambers were designed to permit the continuous diffusion of chemical cues from stimulus animals while blocking visual cues.

Tests assessing the importance of chemical cues in kin recognition in *R. aurora* were conducted differently compared with *R. cascadae.* Instead of using the small chambers to house stimulus tadpoles, stimulus tadpoles were left within the end compartments as in standard tests and test tadpoles had their nares blocked with Orabase plain oral protective paste (Colgate-Hoyt Laboratories, Canton, MA; Blaustein et al. 1993). Orabase® apparently has no pharmacological effects, but creates a barrier that prevents water and chemical cues from passing through the nares (Waldman 1985; Blaustein et al. 1993). A small amount of Orabase® was placed inside the nares using a blunt needle. In a sham control, needles were placed into the nares of tadpoles but no Orabase was used.

Methods used to assess discrimination abilities were similar for both *R. cascadae* and *R. aurora.* Prior to each test, the test tank was filled to a depth of 6 cm with 34 l of dechlorinated tap water (at room temperature). After 2–4 min, stimulus tadpoles of one kin group and those from a second kin group or combinations of kin groups were placed in opposite end compartments and left undisturbed for 15 min. The number and kin composition of stimulus animals placed in the end compartments varied with the test being conducted. One test tadpole was released at tank center on each side of the longitudinal partition and allowed to acclimate for 10 min. The time spent in seconds by each tadpole in sibling and nonsibling halves (as demarcated by the pencil mark) of the tank was recorded for four 5-min trials (total = 1200 s) at 10-min intervals. Observations were taken from behind a plastic blind. Following each test, the tank was drained and thoroughly rinsed. The two tadpoles tested simultaneously were always members of the same sibling group. Stimulus tadpoles were alternated from one end to the other between tests. Other controls for the apparatus and testing procedures have been reported elsewhere (e.g., Blaustein & O'Hara 1981; O'Hara & Blaustein 1981). All stimulus tadpoles and test individuals were matched for body size.

The results for both *R. cascadae* and *R. aurora* showed that tadpoles discriminated between kin and nonkin based on chemical cues. Test individuals spent significantly more time near stimulus groups composed of kin than those composed of nonkin when their chemical senses were not impaired. The sham controls in the *R. aurora* tests also discriminated between kin and nonkin. Tadpoles did not discriminate between kin and nonkin when provided with visual cues alone. Sound production played no role in kin discrimination (Blaustein & O'Hara 1982).

Waldman (1985) investigated the sensory basis of kin recognition in American toad (*Bufo americanus*) tadpoles using a Y-maze. When simultaneously presented with water flowing from two containers, each containing a different kin group, test animals spent more time oriented toward their siblings than toward nonsiblings. Tadpoles whose external nares were blocked with Orabase® did not discriminate between siblings and nonsiblings. When individuals were retested after the Orabase® was removed, they did discriminate between siblings and nonsiblings. Test animals presented simultaneously with water flowing from siblings and dechlorinated water without tadpoles, did not discriminate between these two stimulus sources. When individuals were exposed to water from nonsiblings and dechlorinated water, they oriented significantly toward the side with the dechlorinated water. This suggests that *B. americanus* tadpoles avoid chemical cues from nonsiblings.

Dawson (1982) investigated chemical cues in *B. americanus* tadpoles using choice tests in an apparatus similar to, but much smaller than, the one used by Blaustein and O'Hara (1982). Dawson placed cheesecloth over the end compartments so that test tadpoles could not see stimulus tadpoles. When the stimuli were chemical cues alone, tadpoles could discriminate between siblings and nonsiblings. However, when both visual and chemical cues were available, the test tadpoles did not discriminate between siblings and nonsiblings. This visual effect was not observed in tests of other species.

Larval Arizona tiger salamanders (*Ambystoma tigrinum nebulosum*) can discriminate between kin and nonkin (Pfennig et al. 1994). In experimental tests, cannibalistic morphs may prey preferentially on unrelated individuals over related ones. The sensory basis for this discrimination was examined by covering the nares of test animals with Vetbond® (3M Animal Care Products), an *n*-butyl cyanoacrylate tissue adhesive. This material blocks the passage of water over the nasal epithelium. In tests, some animals had their nares blocked, while sham controls had the adhesive placed between their nares. Larvae whose nares were blocked consumed kin and nonkin randomly. Animals without their nares blocked consumed more nonkin than kin. These results suggest that discrimination is based on chemical cues.

Although chemical cues are obviously used in kin discrimination in larval amphibians, the nature of the kin recognition chemical signal(s) (pheromone) has not been identified in any species. The cues used in kin recognition are influenced by interacting genetic, social and environmental components (reviewed in Blaustein & Walls 1995).

6.2.2.2. Post-metamorphosis

At least one study suggests that chemical cues are important in kin recognition after metamorphosis (Blaustein et al. 1984). Newly metamorphosed Cascades froglets (*R. cascadae*) were tested to see if they could discriminate between siblings and nonsiblings in choice tests.

Single test individuals were placed in the center of a plastic tube (84 cm long; 5.5 cm diameter; Fig. 1 in Blaustein et al. 1984). Ten froglets from the test individual's sibling group and 10 froglets from a nonkin group were placed in opposite ends of partitioned end chambers. The tube was moist but there was no waterborne contact between the stimulus groups or between the test animals and stimulus groups. Test animals were separated by a plastic barrier with several small perforations. Thus, there was limited visual contact and airborne chemical contact between the stimulus and test animals.

In these tests, animals could discriminate between kin and nonkin until at least 47 days after metamorphosis. Although limited visual contact could have allowed these metamorphs to discriminate between siblings and nonkin, airborne chemical cues are implicated. Unfortunately, olfactory impairment tests were not conducted so the use of chemical cues is not clear-cut. However, a recent study by Graves et al. (1993) implies that aggregation tendencies in recently metamorphosed Great Plains toads (*Bufo cognatus*) are based at least in part on chemical cues.

6.2.3. Territorial Pheromones

To date we are aware of no studies indicating whether either anuran amphibians or caecilians advertise territories using chemical markers. Therefore, this section concentrates solely on studies of salamanders. We recommend that readers also consult reviews in Madison (1977), Jaeger (1986), and Mathis et al. (1995).

6.2.3.1. Source of the Chemical Signal

For experimental studies of terrestrial salamanders, territorial markers have been obtained in a number of ways. A common and simple collection technique is to introduce individual salamanders into small chambers lined with a damp substrate (typically either soil, paper towels, or filter paper) and allow the salamanders to mark the substrate with secretions and fecal pellets (Tristram 1977; McGavin 1978; Jaeger & Gergits 1979; Dawley 1984a; Keen et al. 1987; Anthony 1993; Graves 1994). The length of time allowed for marking is variable, but a period of several days to 1 week are common.

Simon and Madison (1984) attempted to distinguish between fecal pellets and cloacal odors as possible sources for substrate pheromones. In that study, cloacal washes were obtained by injecting 0.05-ml aliquots of distilled water into the salamanders' cloacae and immediately retracting the syringe from the cloaca. However, both cloacal washes and fecal pellets were equally effective at eliciting responses from conspecifics. A number of more recent studies have used fecal pellets alone as the stimulus (Jaeger et al. 1986; Horne & Jaeger 1988; Ovaska & Davis 1992). The advantages to using fecal pellets are that they can be frozen for long-term storage without loss of effectiveness (Jaeger et al. 1986; Horne & Jaeger 1988) and that they occur in discrete units that (1) are easy to collect and (2) can be placed easily in specified locations such as in front of burrows (Jaeger et al. 1986; Walls et al. 1989; Ovaska & Davis 1992).

When fecal pellets are used as experimental stimuli, it often is also desirable to include a "blank" control. The material that should be used as the control (i.e., neutral pseudopellet) is somewhat problematic. A number of studies have used paper toweling that is rolled into the size and shape of a fecal pellet (Jaeger et al. 1986; Horne & Jaeger 1988). As a neutral chemical stimulus, this type of surrogate pellet is ideal because it is identical to the paper toweling that provides the substrate for the test. However, this surrogate is not a perfect visual control because its cryptic coloration blends in with the substrate; fecal pellets are typically much darker than the paper towel substrate (A. Mathis, personal observation). To avoid this problem, other studies have used pseudopellets made of either clay (Ovaska & Davis 1992) or soil (Anthony 1993); these surrogates are probably better controls in a visual sense, but, unfortunately also introduce "new" chemical cues to the chamber. A more unambiguous control used by Anthony and Wicknick (1993) consisted of a 1-ml volume of sterile potting soil spread over the substrate; the experimental stimulus was an identical volume of fecal material plus sterile potting soil. The disadvantage to this technique is that it requires the loss of the fecal "pellet" as a discrete unit.

Airborne chemicals can also be effective at eliciting responses by terrestrial salamanders. Madison (1975) used an olfactometer to generate a simple choice apparatus for discrimination of airborne chemicals from neighbors versus strangers. The olfactometer is constructed so that test individuals are offered a choice between sides of a container containing separate streams of compressed air that have been passed over each of the source salamanders. This apparatus presents an effective means of discriminating between responses to substrate versus airborne chemicals, but there are some negative considerations. In olfactometer experiments, test animals are forced to move toward an airflow to make a choice. Desiccation is a significant problem for terrestrial salamanders and individuals may sometimes be hesitant to move into the stream of air. Dawley (1986a) reported that when "reluctant" individuals were stimulated to move by a dim light, they would often move rapidly through the olfactometer without stopping at the junction of the side arms of the Y-maze. Therefore, a large sample size may be required to deal with the variability of responses in these types of olfactometers.

Simons and Felgenhauer (1992) attempted to identify specific sources of pheromone production for *Plethodon cinereus* by collecting chemical stimuli from various areas of the body for use in a behavioral assay. To obtain the chemical stimuli, they rubbed cotton swabs over the following sources: the midventral surface of the base of the tail (which is known to contain serous glands), the shoulder (also containing serous glands), fecal pellets, fecal material from the colon, secretory products from the collecting (urinary) ducts, the chin, and the middorsal tail base. Stimuli from all areas except the chin and the dorsal surface of the tail elicited some response by conspecifics. Of particular interest was a behavior in response to stimuli from fecal pellets: the postcloacal press. Simons and Felgenhauer interpreted this behavior as a possible attempt at remarking by the salamander; this behavior occurred when the salamander paused with its tail base over the swab. A histological description of the glands located in the ventral postcloacal region and other sites involved in the study are included in their paper. In a subsequent study, Simons et al. (1994) used a different technique to isolate the postcloacal gland as a possible source for territorial pheromones. In this experiment, they blocked the postcloacal gland with a synthetic bilayer of rosin and an acyanocrylate glue (see Simons et al. 1993 for details of the methodology). Based on the results of this study, the author concluded that postcloacal gland markings are important for identification of the territorial resident.

6.2.3.2. Response Variables

A number of studies have assessed discrimination by quantifying the amount of time that individuals spent near different markings (Madison 1975; Parzefall et al. 1980; Jaeger et al. 1986; Ducey & Ritsema 1988; Ovaska & Davis 1992). These studies have clearly indicated that salamanders in a number of species are able to discriminate between their own chemical cues and those of conspecifics. However, further interpretation of the results of these studies is surprisingly difficult. A number of studies have reported that salamanders find their own markings more attractive than either unmarked substrates/burrows (Jaeger et al. 1986; Keen et al. 1987; Graves 1994), or the markings of conspecifics (Jaeger & Gergits 1979; Simon & Madison 1984; Jaeger et al. 1986; Horne & Jaeger 1988). In the latter case, these results could be interpreted as indicating avoidance of conspecific markings (e.g., Jaeger & Gergits 1979; Jaeger et al. 1986). However, in the critical test of responses to unmarked substrates/borrows versus those marked by conspecifics, some species appear to avoid (Keen et al. 1987; Ovaska & Davis 1992) while other species appear to be attracted to (Parzefall et al. 1980; Roudebush & Taylor 1987; Ducey & Ritsema 1988; Simons & Felgenhauer 1992; Anthony 1993) the markings of conspecifics. To add more complexity, individuals of *Plethodon cinereus* preferred unmarked burrows to burrows marked by conspecific fecal pellets (Jaeger et al. 1986) but preferred substrates marked by conspecifics to unmarked substrates (Simons et al. 1994); therefore responses may be context dependent. We recommend that future "choice" studies include at least the following three comparisons: (1) own versus blank, (2) own versus conspecific, and (3) blank versus conspecific.

Most of the salamanders that have been studied for the presence of possible territorial pheromones belong to the family Plethodontidae (Jaeger 1986; Mathis et al. 1995). A unique feature of plethodontids is the presence of nasolabial grooves running from the nostril to the upper lip (Brown 1968). Plethodontids often perform a stereotypical behavior called nose-tapping in which they press the base of the nasolabial grooves to objects (e.g., substrates: Ovaska 1989; fecal pellets: Anthony 1993; the body of other salamanders: Jaeger & Gabor 1993) (see Fig. 6.2). Dawley and Bass (1989) used radioactively labeled proline to demonstrate that chemicals are taken up through the nasolabial grooves and preferentially concentrated in the vomeronasal organ rather than the main olfactory chamber. Graves (1994) further demonstrated that the vomeronasal system is necessary for recognition of substrate marks for *Plethodon cinereus* by cauterizing the nasolabial grooves of test subjects. Cauterized salamanders preferred own-marked substrates to that of conspecifics less often than did control subjects.

Nose-taps (NTs) appear to be a good bioassay for detection of chemical cues by plethodontids. However, the functional interpretation of differences in NTs is ambiguous because it is unclear whether relatively higher rates of NTs indicate more "interest" in detection of chemical information. For example, McGavin (1978) showed that *Plethodon cinereus* performs more NTs to chemical stimuli from familiar conspecifics than to stimuli from unfamiliar conspecifics. This result could be interpreted as either (1) more interest in markings of familiar conspecifics or (2) that reduced nose tapping to unfamiliar stimuli occur because the receiver increases frequency of performance of another behavior that is inconsistent with simultaneous performance of NTs (e.g., vigilance behavior). Evidence in support of the latter interpretation is provided by Mathis and Simons (1994), who reported that males of *P. cinereus* performed more NTs in the control condition than when in the presence of swabs marked by conspecific pheromones. Jaeger and Wise (1991) reported that female *P. cinereus* nose-tapped more than males to markings from conspecific males. They interpreted these results as females attempting to gain more information from markings (i.e., for mate choice)

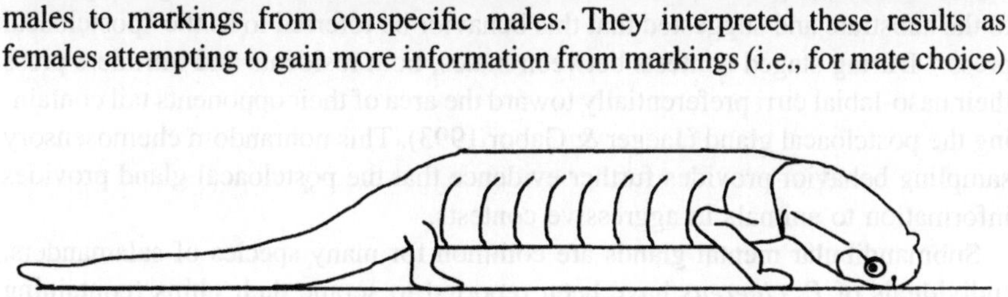

Figure 6.2. Nose-tapping by a plethodontid salamander. Salamanders from this family have nasolabial grooves that run from the nostril to the upper lip and deliver chemicals to the vomeronasal organ. Nose-taps may be directed toward the substrate, as pictured, or towards objects such as fecal pellets or the bodies of other salamanders.

than required by competing males. An alternative explanation is that males lower their rate of nose-tapping as an artifact of becoming more vigilant for the competitor. Therefore, although behaviors such as NTs may be useful as a bioassay, interpretation of the functional significance of differences between treatments should be made with caution.

The behavioral repertoire of some species of salamanders is known to include several agonistic postures. Changes in the frequency of occurrence of these postures also has been used as a bioindicator of the presence of pheromonal markings (Jaeger et al. 1986; Horne & Jaeger 1988; Mathis 1990; Mathis & Simons 1994). Additional responses have included time spent looking toward burrows marked with fecal pellets (Ovaska & Davis 1992), time "exploring" markings (Simons & Felgenhauer 1992; Mathis & Simons 1994), and the rate of production of fecal pellets (Mathis 1990). Because the interpretation of standard attraction/avoidance data can be problematic, we recommend the use of additional response variables, such as those indicated above whenever possible.

6.2.4. Marking Behaviors

Clearly, for some terrestrial species, deposition of fecal pellets can be considered as marking behavior. Mathis (1990) found that females of *Plethodon cinereus* deposited fecal pellets at a faster rate when exposed to their own chemical cues than when exposed to chemical cues of males or other females, suggesting the hypothesis that the primary function of fecal markers for females is advertisement of their own areas. Therefore, if an experiment is to last for several days, the rate of fecal pellet production can be easily quantified.

Jaeger et al. (1981) reported that individuals of *P. cinereus* that were introduced to clean chambers showed a high frequency of a behavior in which the salamanders pressed the area around the cloaca to the substrate. This behavior has been hypothesized to function in pheromonal marking for territorial advertisement (Jaeger et al. 1981; Jaeger 1984). Simons and Felgenhauer (1992) suggested that the ventral serous glands posterior to the cloaca, not the vent opening, are pressed to the substrate and suggested that this behavior be referred to as the "postcloacal press." During staged contests between males, both residents and intruders press their naso-labial cirri preferentially toward the area of their opponents tail containing the postcloacal gland (Jaeger & Gabor 1993). This nonrandom chemosensory sampling behavior provides further evidence that the postcloacal gland provides information to animals in aggressive contests.

Submandibular mental glands are common for many species of salamanders. Individuals of *P. cinereus* have been reported to scrape their chins (containing mental glands) against the substrate in what appears to be marking behavior (Jaeger et al. 1981; Jaeger 1984). However, Simons and Felgenhauer (1992) reported that swabs rubbed over the chin did not elicit a response from conspecific males of this species. It therefore seems unlikely that males are using these glands to advertise to other males. It is possible that this marking behavior may function in advertisement to females. Additional marking behaviors are very difficult to identify. An individual could theoretically release secretions from either the skin or the cloaca at any time.

6.2.5. Reproductive Pheromones and Sexual Advertisement

6.2.5.1. Source of the Chemical Signal

In general, the same techniques are used to collect stimuli for tests of pheromones used in reproduction as for tests of territorial pheromones. It is possible that in many cases the same pheromones may serve both functions: advertisement to potential competitors and to potential mates. Although the collection techniques are similar for both types of pheromones, many studies of sexual advertisement pheromones have focused on aquatic animals (primarily newts) and the techniques described above for terrestrial salamanders have been modified for use in aquatic environments. For example, test animals might be exposed to water from a chamber that had contained another salamander (Dawley 1984b) or from sponges soaked with secretions from other salamanders (Twitty 1955).

Several studies have used olfactometers for presentation of chemical stimuli to potential mates, including both aquatic (Dawley 1984b; Verrell 1985; Malacarne & Vellano 1987; Cogalniceanu 1994) and terrestrial (Dawley 1984a) systems. Studies of terrestrial salamanders have used both fecal pellets (Walls et al. 1989; Jaeger & Wise 1991) and substrate markings (Mathis 1990).

6.2.5.2. Response Variables

Not surprisingly, many studies have considered attraction/avoidance as indicators of detection of sexual advertisement pheromones (Dawley 1984a, 1984b; Walls et al. 1989; Malacarne & Vellano 1987; Cogalniceanu 1994). These responses are the easiest to quantify and interpretation of the results may be more straightforward for sexual advertisement pheromones than for territorial pheromones (see discussion above). For example, males are predicted to be attracted to females and individuals of either sex should be attracted to potential mates that they find more attractive.

A few other response variables have also been used to determine if chemical stimuli might function in sexual advertisement. Males of the territorial salamander *Plethodon cinereus* produce fecal pellets most rapidly when exposed to substrates marked by females in comparison to substrates marked by other males or by self (Mathis 1990), indicating the importance of fecal pellets as sexual advertisement for males. In this species, an interesting behavior occurs in which the salamander pushes its snout into a fecal pellet and "mashes" it (Horne & Jaeger 1988; Jaeger & Wise 1991). This "fecal squashing" behavior occurs most often in gravid (versus nongravid) females, and has never been observed for males (Jaeger &

Wise 1991), suggesting that females are gaining information (diet-related cues reflecting territory quality?; see below) that is not of immediate interest to males.

Most studies control for possible effects of diet on chemical stimuli by feeding all individuals the same prey. However, the results of one study indicate that male diet may influence sexual advertisement pheromones. Walls et al. (1989) reported that gravid female *P. cinereus* spent more time looking toward fecal pellets of males that had recently eaten high-quality diets (termites) than toward pellets of males that had recently eaten low-quality diets (ants). Females also spent more time in burrows that were marked by the "high-quality" pellets.

Studies of chemical communication under field conditions can be difficult. However, Gergits and Jaeger (1990) observed several instances of courtship under natural conditions on the forest floor for *P. cinereus*. They reported that males appeared to locate gravid females by following pheromonal trails. In all 10 courtship interactions that were observed, males initially nose-tapped the substrate with a side-to-side head motion, aligned themselves with the trail and then followed it precisely. In another study, Twitty (1955) reported that sponges soaked with water that had previously contained a female newt (genus *Taricha*), attracted large numbers of males from downstream. There is a critical need for development and application of techniques that allow for experimental study of salamander pheromones under natural conditions.

6.2.5.3. Species Discrimination

Chemical stimuli may often be used by salamanders for species discrimination during potential courtship encounters. These studies may be of particular importance because of the potential for the use of chemical stimuli as premating isolating mechanisms. Studies of species discrimination have typically used attraction/avoidance as response criteria and have used either olfactometers (Dawley 1984b, 1986a, 1986b; Roudebush & Taylor 1987; Verrell 1989; Uzendoski & Verrell 1993) or other choice experiments (Twitty 1995; Jaeger & Gergits 1979; Parzefall et al. 1980).

6.2.5.4. Male Courtship Pheromones

By definition, courtship pheromones are transferred between sexual partners during courtship. Morphological studies indicate that many species of salamanders have sexually dimorphic glands that hypertrophy during the mating season. The presence of these glands is typically associated with behaviors that occur during courtship in which the male appears to deliver the pheromones to the female. These behaviors include rubbing the gland over the females' nares or waving the tail so that water currents are delivered to the females' nares. The combination of morphological and behavioral data provide compelling evidence for the occurrence of courtship pheromones in salamanders. For more detailed information, we recommend the excellent reviews of Arnold and Houck (1982), Houck (1986), and Verrell (1988).

Experimental studies of courtship pheromones have been rare. Malacarne and Curtassa (1983) reported that female newts (*Triturus cristatus*) whose nares had been occluded with petroleum jelly were less likely to respond to courtship than were controls, providing some evidence that chemoreception may be important during courtship. A study of the salamander *Desmognathus ochrophaeus* (Houck & Reagan 1990) provides the best experimental data for the effectiveness of courtship pheromones in increasing female receptivity. In this species, as in many plethodontids, males are thought to "inject" courtship pheromones from their mental glands into the female's bloodstream by pulling his lower jaw across the female's back and scraping the path of the pheromone with his premaxillary teeth. Prior to courtship, Houck and Reagan (1990) ablated the mental glands from male subjects and created an "elixir" of supernatant from excised glands plus saline that had been homogenized and centrifuged. Prior to courtship encounters with the glandless males, females were treated with either a disk of filter paper soaked with the elixir or disks soaked in saline. Females treated with the courtship pheromone mated an average of 28% faster than females in the control condition.

6.2.6. *Predator/Prey Interactions*

The use of chemical cues in predator/prey interaction appears widespread among amphibians (Kats & Dill 1998). Chemical alarm pheromones have been examined most extensively in larval frogs and toads (Chivers & Smith 1998). However, recent experiments also suggest that caudate amphibians use chemical alarm pheromones. Numerous studies have investigated the use of chemical substances that make amphibians noxious or toxic to predators. Chemical cues from predators often induce antipredator responses in amphibian prey. In some cases, chemical cues may be the primary sensory stimuli used for predator recognition. Chemosensory-induced antipredator responses may include behavioral defenses as well as alterations in life-history characteristics.

6.2.6.1. *Chemical Alarm Cues*

Following the classical work of von Frisch (1938, 1941) on alarm reactions of European minnows (*Phoxinus phoxinus*) to stimuli from injured conspecifics, Eible-Eibesfeldt (1949), Hrbacek (1950), and Kulzer (1954) were among the first to examine alarm reactions of common toad tadpoles (*Bufo bufo*) to stimuli from injured conspecifics. In their typical experiments, a tadpole or group of tadpoles were crushed in a small volume of water and the resulting extract was slowly added to water containing other tadpoles. Experiments conducted under both field and laboratory conditions showed that the tadpoles suddenly changed directions and swam away from the extract. Alternatively, the tadpoles sank to the bottom, laid motionless for some time and then rapidly swam away. Pfeiffer (1966) employed similar techniques to examine the presence of alarm reactions in nine species of larval anurans from five different families. He found behavioral responses in two species of bufonid tadpoles (*Bufo bufo* and *B. calamita*) and speculated that alarm pheromones may be restricted to the family Bufonidae.

In a series of field experiments, Petranka (1989) tested whether tadpoles of the American toad (*Bufo americanus*) avoid stimuli from injured conspecifics. He established a series of 10- × 10-cm plots in the shallow water of a pond and photographed each plot before and after adding various stimuli to the center of the plots. His initial experiment demonstrated that stimuli from injured conspecifics decreased the number of tadpoles occupying the experimental plots. In contrast, addition of stimuli from injured hylids (*Pseudacris triseriata*) increased the number of tadpoles occupying experimental plots, while addition of a pond water control did not substantially alter the number of tadpoles occupying the plots. Subsequent tests using the same protocol showed that the greater the number of tadpoles used to produce the injured tadpole extract, the greater the magnitude of the avoidance response. As well, tadpoles simultaneously exposed to injured conspecifics and food exhibited a response intermediate between tadpoles exposed to only injured conspecifics or to only food.

Hews and Blaustein (1985) tested the western toad (*Bufo boreas*) for a response to injured conspecifics. Tests were conducted on solitary tadpoles in opaque plastic tubs (28 × 18 × 12 cm) that were filled to a depth of 4 cm with dechlorinated tap water. Stimuli from injured tadpoles was introduced into one end of the tank and then observations of the tadpoles spatial distribution and activity were taken. The tadpoles avoided the side of the tank where injured tadpole extract was introduced. Furthermore, the tadpoles increased their activity level as indicated by an increase in the frequency with which they crossed a center line drawn on the bottom of the tank. The tadpoles did not respond in the same manner to stimuli from injured Pacific treefrogs (*Hyla regilla*) or to a control of dechlorinated tap water. In a follow up experiment, Hews (1988) placed groups of toad tadpoles in rectangular 7-L containers. Each container housed a predatory dragonfly naiad in a central enclosure. Toad skin extract or a control of dechlorinated tap water was added to the containers 2 min prior to the predators being released. Her results showed that the predator's attack success rate was significantly lower when toads were "warned" with conspecific stimuli than when they were "warned" with dechlorinated tap water.

Hews and Blaustein (1985) also tested the response of Cascade frog tadpoles (*Rana cascadae*) using an experimental design similar to that described above. *R. cascadae* tadpoles did not avoid the side of the chamber where stimuli from injured conspecifics was added. However, the tadpoles did show an increase in the time they were active. Hokit and Blaustein (1995) quantified the number of times groups of nine *R. cascadae* tadpoles crossed the center line of a 100- ×

60-cm tank, in response to stimuli from injured conspecifics, stimuli from a predatory newt (*Taricha granulosa*) and a blank control. The tadpoles were less active in response to the predator cue than to either the crushed conspecifics or the control. These results do not support the earlier finding of an alarm response in *R. cascadae* tadpoles. Wilson and Lefcort (1993) demonstrated an alarm response of another ranid frog (the red-legged frog, *Rana aurora*) to stimuli from injured conspecifics. They utilized a gravitational flow through system developed by Petranka et al. (1987) (see Fig. 6.3).

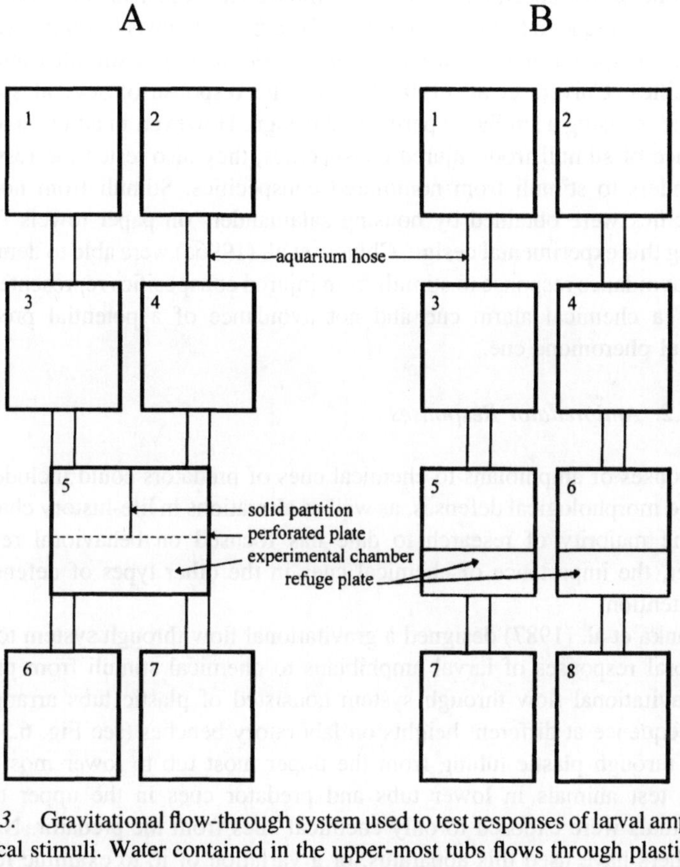

Figure 6.3. Gravitational flow-through system used to test responses of larval amphibians to chemical stimuli. Water contained in the upper-most tubs flows through plastic tubing to the lowermost tubs. In system *A,* water from tubs 3 and 4 flows along either side of the solid partition of tub 5. By placing a predator in either tub 3 or 4, a chemical gradient is established in the experimental chamber of tub 5. In this system, the behavioral responses typically quantified include avoidance of the area where the chemical stimuli enters the test chamber and overall activity level of the test animals. In system *B,* a chemical gradient is not established in the test tub. Instead, the behavioral responses typically quantified include shelter use and overall activity levels. Redrawn from Petranka et al. (1987).

Lutterschmidt et al. (1994) were the first to examine chemical alarm cues in caudate amphibians. They tested whether the adult plethodontid *Desmognathus ochrophaeus* avoided skin extracts of injured conspecifics. Injured conspecific stimuli were added to paper towels that lined one side of a petri dish (15 cm diameter) while distilled water was added to paper towels on the other side of the dish. Test salamanders were observed every 5 min for 2 h to determine whether they "avoided" stimuli from injured conspecifics. Their results show a significant avoidance of stimuli from injured conspecifics. This may represent avoidance of a chemical alarm cue or alternatively avoidance of a territorial pheromone or intraspecific predator cue. Marvin and Hutchison (1995) employed a similar technique to test the avoidance of fire-bellied newts (*Cynops pyrrhogaster*) and red-spotted newts (*Notophthalmus viridescens*) to stimuli from injured conspecifics. Chivers et al. (1996a) tested the response of several species of salamanders using a similar experimental design. However, in addition to testing avoidance of stimuli from injured conspecifics, they also tested the response of salamanders to stimuli from noninjured conspecifics. Stimuli from noninjured conspecifics were obtained by housing salamanders on paper towels for 48 h. By using this experimental design, Chivers et al. (1996a) were able to demonstrate that an avoidance response to stimuli from injured conspecifics represented avoidance of a chemical alarm cue and not avoidance of a potential predator or territorial pheromone cue.

6.2.6.2. Antipredator Responses

Responses of amphibians to chemical cues of predators could include behavioral and morphological defenses, as well as alterations in life-history characteristics. The majority of research to date has focused on behavioral responses, however, the importance of chemical cues in the other types of defenses warrants attention.

Petranka et al. (1987) designed a gravitational flow through system to test the behavioral responses of larval amphibians to chemical stimuli from predators. The gravitational flow through system consisted of plastic tubs arranged in a linear sequence at different heights on laboratory benches (see Fig. 6.3). Water flowed through plastic tubing from the upper most tub to lower most tub. By placing test animals in lower tubs and predator cues in the upper tubs, the test animals were exposed to only chemical cues from the predator. Numerous researchers have used this apparatus, or a variation of it, to examine responses of larval amphibians to chemical cues (Kats et al. 1988, 1994; Elliott et al. 1993; Lefcort & Eiger 1993; Wilson and Lefcort 1993; Kiesecker et al. 1996). The behavioral responses that the researchers quantified differed in the various studies, but typically included measurements of activity level, refuge use and/or avoidance of the area where the chemical stimuli enters the test chambers.

In one study, Kats (1988) plugged the external nares of small-mouthed salamander larvae (*Ambystoma texanum*) with Orabase® gel before testing them in a gravitational flow through system. In this experiment, the larval salamanders did not respond to cues of green sunfish (*Lepomis cyanellus*) while sham-operated individuals responded by increasing refuge use. This study indicates that chemoreception was responsible for predator recognition. In another of these studies, Kiesecker et al. (1996) demonstrated that western toad tadpoles (*Bufo boreas*) respond to chemical stimuli from predatory insects, but did not respond to the same predators using only visual cues. This study indicates that chemical cues may be the primary sensory system used in predator recognition (see also Stauffer & Semlitch 1993). Using the gravitational flow-through system, Wilson and Lefcort (1993) showed that *Rana aurora* tadpoles responded to chemical cues of rough-skinned newts (*Taricha granulosa*) that were previously fed *R. aurora* tadpoles, but not to newts fed insect prey. This study emphasizes the importance of the predator's diet in chemosensory-based predator recognition studies (for further discussion see Mathis & Smith 1993; Chivers et al. 1996b, 1997b).

Several studies have examined the use of chemical cues by larval amphibians under field conditions. For example, Feminella and Hawkins (1994) tested the response of tailed frog tadpoles (*Ascaphis truei*) to chemical cues from various predators. They quantified activity of tadpoles in the presence and absence of fish and salamander predators that were visually isolated from the test animals in upstream enclosures. In response to giant salamanders (*Dicamptodon* spp.), cutthroat trout (*Salmo clarki*), and brook trout (*Salvelinus fontinalis*), tadpoles decreased activity two-, three- and sixfold respectively. In another study, Sih et al. (1992) manipulated the presence of green sunfish in stream pools and monitored several behavioral responses of larval small-mouthed salamanders (*Ambystoma barbouri*). They demonstrated that larvae were less active when they first entered fish pools when compared to fishless pools. Furthermore, the larvae shifted activity periods from daytime to nighttime when in fish pools compared to fishless pools. In these experiments, the fish were placed in 10-L buckets that were pierced with numerous small holes. The top of the buckets were covered with dark fiberglass screen. This design would likely not allow detection of visual or mechanical cues of the predator and hence, likely represents a response to chemical cues alone. In a follow-up study, Sih and Kats (1994) dropped streamside salamander larvae into laboratory aquariums that had previously contained green sunfish or control aquariums that lacked fish cues. They then recorded the time taken for the salamanders to reach the bottom. This technique simulated salamanders drifting into stream pools. Their results showed that chemical cues were responsible for the adaptive "sink-to-the-bottom" response. Furthermore, they showed that the response was not dependent on previous experience and that the response was not age-dependent.

Few studies have examined the responses of terrestrial amphibians to chemical cues from predators. In one study, two plethodontids (*Plethodon dorsalis* and *P. richmondi*) were tested for an avoidance response to chemical cues from ringneck snakes (*Diadophis punctatus*) (Cupp 1994). In his first experiment, Cupp (1994) lined half of a series of rectangular plastic containers (31 × 23 cm) with moistened paper towels that had been conditioned by a snake for 3 days. The other half of the containers were lined with moistened paper towel that did not come from a snake housing. In his second set of experiments, Cupp (1994) lined half of the plastic containers with paper towels from a snake housing and the other half of the same experimental containers with paper towel that previously housed a different species of salamander. In both sets of experiments the salamanders' position was observed every 3 min for 2 h, and in both sets of experiments both *P. dorsalis* and *P. richmondi* avoided the cues of the snake. Another plethodontid, *Desmognathus ochrophaeus* similarly avoided chemical cues of ringneck snakes over cues of another salamander species (Cupp 1994).

Only one study has demonstrated that amphibians can exhibit an induced morphological defense against predators. In this study, *Hyla chrysoscelis* tadpoles developed larger, deeper tailfins which aid in escape from predators (McCollum & Van Buskirk 1996). The importance of chemical cues in inducing this change remains unknown. However, given that behavioral studies indicate that chemical stimuli may be of prime importance in predator recognition (e.g. Kiesecker et al. 1996), the importance of chemical cues in morphological defenses deserves exploration.

Few studies have investigated the ways in which amphibians alter their life history characteristics in response to predation. However, one such study implicates the importance of chemical cues. Sih and Moore (1993) incubated salamander eggs (*Ambystoma barbouri*) in direct contact with flatworms (*Phagocottus gracilis*) and isopods (*Lirceus fontinalis*). Flatworms are potential predators on hatchling salamander larvae but isopods are not. Sih and Moore (1993) demonstrated that salamanders delayed hatching in response to flatworms but not in response to isopods. By delaying hatching, salamanders reach a developmental stage where they are less susceptible to predation by flatworms. In a follow-up study, they used partitions that blocked visual and mechanical but not chemical cues of the flatworms from reaching the salamander eggs. In response to chemical cues alone, the salamanders exhibited the same delayed hatching response.

6.2.7. Noxiousness and Toxicity

Numerous species of amphibians contain substances in their skin that render them noxious or toxic to potential predators. The active substances in the skin include amines (norepinephrine and histamine), piperidines, steroidal alkaloids, bufodienolides, and tetrodotoxin (reviews in Daly et al. 1978; Daly 1982).

Early studies of the noxious and toxic properties of amphibians have their basis in field observations that indicate that predators appear to avoid consuming particular prey species. Following these observations researchers typically attempted to ascertain the noxiousness of a particular species by comparing a predator's "willingness" to eat several different types of prey, including the species that is suspected of being noxious (Voris & Bacon 1966; Walters 1975; Brodie et al. 1978; Kruse & Stone 1984; Peterson & Blaustein 1991, 1992). In most experiments the researchers simultaneously introduce two species of prey (that are matched for size and developmental stage) into a container with the predator and then quantify the proportion of each prey type captured over a given period of time. In addition, several studies quantify the number of each prey type that are captured and then rejected before being consumed. Peterson and Blaustein (1991), for example, established the unpalatability of *Bufo boreas* tadpoles by comparing whether newts (*Taricha granulosa*) were less likely to eat *Bufo* tadpoles over *Hyla regila* tadpoles offered simultaneously. They also demonstrated that newts preferentially captured and ate *Rana cascadae* tadpoles over *Bufo* tadpoles offered simultaneously. During these observations, Peterson and Blaustein (1991) documented that 13 of 23 *Bufo* tadpoles captured during the experiments were rejected. In contrast, only 1 of 18 *Rana* tadpoles was rejected. In this case, the *Rana* tadpole was captured simultaneously with a *Bufo* tadpole and was released when the *Bufo* tadpole was released.

From the standpoint of methodology for future experiments, several important points must be made. For example, the hunger level of the predator must be carefully controlled in palatability tests. Kruse and Stone (1984) satiated large-mouth bass (*Micropterus salmoides*) and then starved them for 1 to 7 days. They found that bass starved for only 1 day almost totally abstained from consuming *Bufo americanus* and *B. woodhousei* tadpoles. However, as the fasting period extended, significantly more *Bufo* tadpoles were consumed. The developmental stage of tadpoles must also be considered in palatability studies. Brodie et al. (1978) tested the palatability of *B. americanus* tadpoles to predatory beetles (*Dytiscus verticalis*) and giant water bugs (*Lethocerus americanus*) and found that metamorphosing *Bufo* were not eaten while *Bufo* tadpoles in earlier developmental stages were eaten. In this experiment, the predators did not avoid eating other species of metamorphosing anurans. Formanowicz and Brodie (1982) provided another example of stage-dependent unpalatability. As a final cautionary note, it is important to test noxiousness of a particular species to a variety of different types of predators. Predators that swallow prey whole or invertebrate predators that pierce the skin and suck body fluids are less affected by chemical defences than predators that masticate, bite, or somehow taste the prey (e.g., Peterson & Blaustein 1991, 1992).

One common technique used to examine the toxicity of compounds in the skin of amphibians is to isolate the toxic compounds and inject them into laboratory mice to assay their toxicity. Daly et al. (1978), for example, used this technique to determine the toxicity of dendrobatid frogs. In another study, Brodie et al. (1974) calculated that a single newt (*Taricha granulosa*) contains enough taricha-toxin to kill approximately 25,000 white mice.

Finally, Wassersug (1971) tested the palatability of eight species of Costa

Rican tadpoles by having 11 human volunteers taste and rank the species on a palatability scale.

6.3. Reptiles

The study of chemical communication in reptiles began in earnest in the 1920s and 1930s with some of the pioneering work of Baumann and Noble. Baumann studied life history patterns in the European adder (*Vipera aspis*) (Baumann 1927, 1929). At approximately the same time, G.K. Noble was conducting complementary studies in North American colubrids (Noble 1937). The discipline has grown rapidly since and now references number in the thousands. Many substantive and informative reviews have been written in the interim (Evans 1961; Carpenter & Ferguson 1977; Madison 1977; Burghardt 1970, 1980; Simon 1983; Mason 1992; Halpern 1992; Cooper 1994). As a result, the field of chemical communication in reptiles has been well and thoroughly covered in the recent past. This part of the review also focuses on those methodologies commonly used in studies of chemical communication in reptiles.

6.3.1. Olfaction and Vomerolfaction

As Cooper (1994) stated, tongue-flicking is the quintessential squamate behavior. It serves as an environmental sampler and delivery device to the vomeronasal or Jacobson's organ found in the roof of the mouth or in the nasal passages. The vomeronasal organ is common to most vertebrate taxa but is most highly developed in squamates (Halpern 1992). The olfactory sense, although well developed in reptiles, has not been extensively studied. Gustation has not been studied systematically to date, although taste buds do exist in squamates (Schwenk 1985). Tongue-extrusion, or tongue-flicking in reptiles is associated with vomeronasal organ chemoreception. Because of the ease of observing and quantifying tongue-flicks in social contexts, these behaviors are commonly reported in investigations of chemical communication in reptiles.

The reception and perception of chemical cues from the environment by the vomeronasal system has been termed vomerolfaction (Cooper & Burghardt 1990). Thus, vomodors are those chemical cues detected by the vomeronasal system. Definitive evidence of the vomeronasal organ's role in mediating socially important chemical cues has been obtained by means of complicated physiological experiments in which the vomeronasal nerves are either anesthetized, ablated, or severed. Sectioning the olfactory nerves and vomeronasal nerves of male garter snakes leads to deficits in courtship behavior (Kubie et al. 1978; Halpern & Kubie 1980) and aggregation behavior (Heller & Halpern 1982b). Those animals with olfactory nerve lesions alone continued to court attractive females. Those males with sectioned vomeronasal nerves failed to recognize and court attractive females. Identical results were found in the European adder (*Vipera berus*)

(Andrén 1982). The vomeronasal organ is also essential for detecting prey, at least in squamates. Impairment of the vomeronasal system, but not the olfactory system, renders garter snakes (*Thamnophis*) unable to recognize earthworms as a food source (Halpern & Frumin 1979). This is also the case in lizards (Graves & Halpern 1989, 1990; Cooper & Alberts 1991).

Turtles have received far less study in this regard, but the vomeronasal organ of turtles does respond to general odorants and it has been postulated that the vomeronasal system may sense chemical stimuli of a social nature (Hatanaka et al. 1988; Shoji & Kurihara 1991). Crocodilians do not possess a functional vomeronasal organ.

A cautionary note must be made regarding the functional interpretation of the tongue-flick in studies with reptiles. Often the differences in tongue-flicks or tongue-flick rate are taken as evidence of "interest" in a given chemical cue. However, reptiles will respond with increased tongue-flick rates to virtually any novel stimulus in their environment. To avoid ambiguity, experiments should be designed that include additional behaviors whenever possible such that the functional significance of tongue-flicking is interpretable in a broader behavioral context. Such experiments are difficult to design, however, the results will be valuable in clarifying the role of chemical cues in behaviors such as aggregation and predator/prey responses.

6.3.2. Attractants and Repellents

All snakes possess at least a pair of cloacal scent glands that are under voluntary control. On being disturbed or handled, snakes often express foul-smelling fluids from these glands into the cloaca and into the environment. Handling often causes snakes to writhe and smear this malodorous secretion onto the handler. It has been suggested that these secretions serve either a defensive or alarm function or perhaps both. Cloacal gland secretions removed from Texas blind snakes (*Leptotyphlops dulcis*) repelled ants of the species *Labidus coecus*, *Neivamyrmex nigrescens*, and *Solenopsis geminata* (Gehlbach et al. 1968). In a two-choice test where animals were scored on how many times a snake entered either half of an aquarium and how much time was spent on each side, conspecific blind snakes were attracted to the cloacal gland secretions. Individuals of other snake species that compete for ant and termite larvae with Texas blind snakes were significantly repelled by these secretions.

Cloacal gland secretions in snakes may also serve an alarm function in addition to their defensive role. Prairie rattlesnakes (*Crotalus viridis*) experience a significant rise in their heart rate after exposure to a threatening stimulus in the presence of cloacal gland materials from conspecifics (Graves & Duvall 1988). Those animals not exposed to conspecific cloacal gland material and subsequently threatened responded with significantly lower heart rates. Presumably, animals not being threatened would have lower heart rates.

Lizards also have glands in their lower gastrointestinal tract and cloaca. Lipid components from the urodaeal gland of male broad-headed skinks (*Eumeces laticeps*) and black-lined plated lizards (*Gerrhosaurus nigrolineatus*) elicit significantly more tongue-flicks than do water controls (Cooper et al. 1986; Cooper & Trauth 1992). When the gland was removed and macerated and presented to courting males, tongue-flick rates to the homogenate were significantly higher than those directed toward cloacal lavages, suggesting that the courtship pheromone in these skinks is contained primarily in this gland. Solvent extracts of these glands indicate that the sex pheromone is probably a lipid (Cooper & Garstka 1987). Male broad-headed skinks (*E. laticeps*) tongue-flick female-derived cloacal cues significantly more than those of males or controls (Cooper & Vitt 1984). Males of this species can also recognize the cloacal cues of females of their own species when compared to female cloacal cues from two closely related species (Cooper & Vitt 1986a). Exogenous estrogen injections to female broad-headed skinks render their cloacal chemical cues more attractive to breeding male skinks than do cloacal cues from sham-injected females (Cooper 1995). Furthermore, males with brighter orange head coloration, which may indicate higher circulating androgen concentrations, respond with more tongue-flicks to the cloacal cues of estrogen-treated females than to sham-injected females.

Turtles have several specialized glands that secrete aromatic odoriferous semiochemicals into the environment. Inguinal, axillary, and Rathke's glands all lie along the angle formed by the plastron and carapace and possess pores that empty into the environment. Almost all turtle families possess these glands. In the stinkpot turtle (*Sternotherus odoratus*), the contents of their musk glands are ejected when the turtles are handled or molested. Voluntary control of these malodorous glands implies a defensive role for the semiochemicals (Eisner et al. 1977). The secretions deter feeding by small fish, but tests with larger predators are needed (Eisner et al. 1977). The long-necked turtle (*Chelodina longicollis*) also possesses inguinal musk glands, but the secretions although malodorous, do not deter feeding by potential predators including mammals, birds, and reptiles (Kool 1981).

Crocodilians possess two sets of actively secreting semiochemical glands. Paired mandibular glands in the throat region and cloacal scent glands become active and secretory during the breeding season, especially in males. Males can be observed rubbing their mandibular glands across the snouts of females during courtship, perhaps introducing a courtship pheromone (McIlhenny 1935; Evans 1961; Burrage 1965; Garrick 1978).

The cloacal glands in alligators are thought to play a role in territorial behavior among breeding bulls. Secretions from these glands are quite odoriferous. During the breeding season, bull alligators are thought to use these glands to mark territories and ward off competing males (Evans 1961). Experiments conducted under controlled conditions are needed before conclusions regarding the role of these glands in chemical communication can be reached.

6.3.3. Aggregation

Aggregation behavior is a widespread and relatively well studied phenomenon in reptiles (Gillingham 1987; Gregory et al. 1987). Temperate zone snakes are commonly found aggregating in overwintering or hibernation sites (Parker & Brown 1973; Aleksiuk 1976; Gregory 1982; Duvall et al. 1985), or during gestation (Gregory 1975; Graves et al. 1986; Graves & Duvall 1995). Although aggregation is a commonly observed behavior in many reptilian taxa, its functional significance is still a subject of considerable debate.

Experimental evidence documenting aggregation behavior has focused on the deposition of chemical cues on the substrate. However, the conclusions and interpretations drawn from these investigations are extremely variable and no consensus has been reached on the generality of this behavior.

Aggregation behavior was examined in snakes by placing two groups of eight male eastern garter snakes (*Thamnophis s. sirtalis*) in a large arena with a hide box in each corner (Heller & Halpern 1982a) (see Fig. 6.4). In these experiments, snakes aggregated under the hide boxes and distributed themselves nonrandomly. Once aggregated under a shelter, snakes returned to these shelters responding only to the presence of chemical cues left on the paper under the shelter. In a similar study, prairie ring-neck snakes (*Diadophis punctatus arnyi*) also aggregated under shelters in a nonrandom pattern study (Dundee & Miller 1968). When the substrate under one of the hide boxes was exchanged with another, a significant number of snakes relocated to the "new" hide box. The authors termed the deposition of chemical cues "habitat conditioning," although they could not conclude whether the source of the chemical cues was integumentary, fecal, cloacal gland material, or something else. Similar results were obtained with neonatal individuals of brown snakes (*Storeria dekayi*) and eastern garter snakes (*Thamnophis sirtalis*) (Burghardt 1983). Both species formed aggregations with conspecifics. Changing the hideboxes did not affect aggregation tendencies, but rotating the aquarium did, implying that visual spatial cues may be an important corequisite to chemical cues.

Aggregations of different species of snakes are commonly encountered in the field. Diamondback water snakes (*Nerodia r. rhombifera*) and plains garter snakes (*Thamnophis radix*) were examined for their aggregation preferences (Porter & Czaplicki 1974). In these experiments, animals chose to aggregate on one side of an aquarium or the other. Chemical cues from bedding of conspecifics repelled the water snakes, while in similar studies, the garter snakes were attracted to the side with the chemical cues of conspecifics. Both species ignored the chemical cues left by the other species. In prairie rattlesnakes (*Crotalus v. viridis*), animals initially chose to aggregate on the side of the aquarium marked by conspecifics, but after 1 day, they avoid the conspecific marked side of the aquarium (King et al. 1983). When chemical cues from bull snakes (*Pituophis melanoleucus*

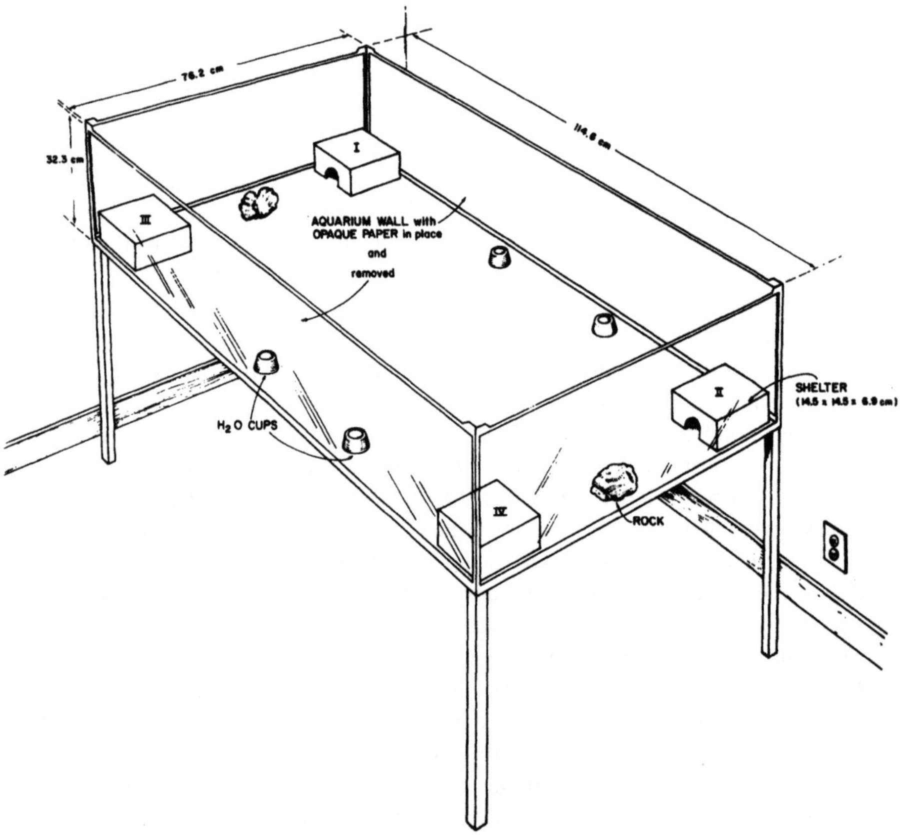

Figure 6.4. Aquarium used for aggregation experiments. Chemical cues left on papers under the shelters can be moved to different shelters to test for aggregation tendencies to pheromones. Redrawn from Heller and Halpern (1982a).

deserticola) are used, rattlesnakes show no preference, apparently ignoring chemical cues from heterospecifics.

Aggregation with, or avoidance of, conspecifics based on integumental chemicals has also been suggested. However, Ford and Holland (1990) showed no significant tendency for adult male checkered garter snakes (*Thamnophis marcianus*), corn snakes (*Elaphe guttata*), or copperheads (*Agkistrodon contortrix*) to avoid aggregating on substrates marked by conspecific males.

To control for species, age, sex, and methodological differences, Allen et al. (1984) compared the aggregation tendencies of three sympatric species of water snake (*Nerodia*). One half of the test aquarium had a clean gravel substrate, and the other gravel was soiled by conspecifics. The results were somewhat equivocal.

The diamondback water snake (*N. rhombifera*) avoided the half of the aquarium containing conspecific chemical cues, the Mississippi green water snake (*N. cyclopion*) preferred the side soiled or marked by conspecifics, while banded water snakes (*N. fasciata*) also preferred the conspecific marked substrate. Males of the species avoid conspecific chemical cues altogether.

The source of the chemical cues involved in aggregation behavior in snakes may include integumental cues, cloacal gland material, and fecal material. When methylene chloride extracts of skin lipids from eastern garter snakes (*Thamnophis s. sirtalis*) and plains garter snakes (*T. radix*) were compared to extracts from corn snakes (*Elaphe guttata*) there was no aggregation preference for heterospecifics (Graves & Halpern 1988). Skin lipid extracts were placed on a filter paper under a hide box on one side of an aquarium. The other side had the solvent control. Neonate plains garter snakes (*T. radix*) preferred to aggregate on the side marked by conspecifics. Thus, skin lipids alone are sufficient to serve as a chemical attractant eliciting aggregation behavior in at least one species of snake.

In further studies, the ability of juvenile eastern garter snakes (*Thamnophis s. sirtalis*) to discern bedding marked by chemical cues from themselves, conspecifics, or blanks were compared (Halpin 1990). As in the previous studies, the garter snakes in this study responded to the odor of conspecifics significantly more than to the blank condition. Responses measured were tongue-flicks and general activity. Interestingly, animals in these tests responded to chemical cues of themselves at an intermediate level. The author concluded that individual recognition may be operating by means of chemical cues in these snakes. However, further investigation is needed to document this claim.

Not all investigators have documented juvenile snakes aggregating with conspecifics. Waye and Gregory (1993) conducted aggregation experiments with neonatal wandering garter snakes, *Thamnophis elegans vagrans*. Neonates were offered a choice of aggregating under a shelter with a paper towel on the floor containing fecal cues from conspecifics, heterspecifics, and heterogeneric snakes. Interestingly, unlike most other aggregation studies, the authors found that neonatal wandering garter snakes preferred to aggregate with the chemical cues of a congeneric in preference to a conspecific. In addition, the neonates showed no discrimination between the chemical cues of a heterogeneric and those of conspecifics or congenerics or even blanks. Finally, the neonates did show a strong preference for the chemical cues of neonate conspecifics over those of conspecific adults, perhaps reflecting their familiarity with the chemical cues of juveniles.

Aggregation behavior has also been reported for a number of lizard species (for review see Mason 1992). During the breeding season male chameleons (*Chameleo chameleo*) aggregate but still appear to maintain territories (Ogilvie 1966). These territories may be maintained by chemical cues rubbed onto the substrate from a temporal pouch. These glands secrete an odoriferous material, especially during the breeding season. Hatchling green iguanas (*Iguana iguana*) use chemical cues to aggregate (Burghardt et al. 1977), and perhaps even for kin recognition based on fecal deposits and body odors (Werner et al. 1987).

Tortoises are known to aggregate in the field and chemical cues have been implicated in this behavior (Patterson 1971). Urine from male and female desert tortoises (*Gopherus agassizi*) does not affect the aggregation tendencies of conspecifics. However, when urine from male Texas tortoises (*G. berlandieri*) was spread in the aggregation area, a significant number of both males and females abandoned the aggregation site. Further, the spreading of fecal pellets of unfamiliar tortoises and especially the fecal pellets of dominant males, caused subordinate animals to abandon aggregation sites. Perhaps there are pheromone-producing glands in the cloacas of tortoises that warrant investigation.

As noted earlier, interpretation of attraction/avoidance results as applied to aggregation behavior can be equivocal. Additional behavioral response variables that support tongue-flicking data should be included whenever possible.

6.3.4. Trailing

The trailing ability of reptiles, especially the squamate snakes and lizards is well known. Snakes and many, but not all, lizards rely on chemical cues to trail prey items and this vast literature has been extensively reviewed (Burghardt 1970, 1980; Ford 1986; Mason 1992; Cooper 1994). Reptiles will also trail conspecifics to aggregation sites, overwintering hibernacula, and for access to mates. The methodology of choice in examining trailing behavior now seems to have settled on the use of mazes, especially the Y-maze. In these experiments, chemical cues are laid down on the floor of the maze on removable paper including one arm of the maze. The other arm is usually left blank or contains solvent controls.

Trailing to overwintering hibernacula was examined in neonatal timber rattlesnakes (*Crotalus horridus*) (Brown & MacLean 1983). Neonates successfully trailed adult conspecifics in the maze. Integumental cues and cloacal chemical cues almost certainly play a role. Shed skins and skin lipids may serve as chemical "sign posts" guiding juvenile prairie rattlesnakes back to their dens (Graves et al. 1986). Burger (1989) followed-up on these leads and also examined the ability of juvenile snakes to trail adults. In this case, juvenile pine snakes (*Pituophis m. melanoleucus*) were able to distinguish and trail chemical cues from adult conspecifics. This ability would confer a selective advantage to those individuals using this behavior to successfully locate overwintering hibernacula.

Adult garter snakes (*Thamnophis sirtalis*) were able to trail chemical cues deposited by conspecifics, but not very accurately (Costanzo 1989). In this study, a modified maze was used that enabled the investigator to produce a stimulus trail by leading a snake down the maze and through 1 of 17 gates. The results, however, indicated that snakes strongly preferred to pass through gates nearest the walls of the maze. Snakes are known to be thigmotactic and these results may reflect this desire of snakes to move along surfaces. Unfortunately, this is a confounding variable that was not controlled for in the experimental design.

Chemical cues undoubtedly play a role in den location by snakes. However, other cues, such as celestial, solar, and visual cues, as well as topographic landmarks, have been implicated (reviewed by Gregory 1982; Gregory et al. 1987).

More often trailing studies focus on the role of trailing in mate location and recognition. A number of studies have investigated this phenomenon, again relying primarily on the Y-maze methodology. There remains a great deal of ambiguity when examining the trailing abilities of the sexes in a given species. The motivation for males following females, especially during the breeding season, is clear. The other combinations of trail producer and trail follower have yielded somewhat equivocal results and a clear pattern cannot be ascertained. For example, in one of the earliest studies, Noble and Clausen (1936) examined trailing abilities of brown snakes (*Storeria dekayi*) in the fall to chemical cues from conspecifics. In this study, the trailing animals did not distinguish sex. Both sexes trailed chemical cue from males and females. If the trailing animal were migrating back to a den, this behavior would be advantageous. When chemical cues from the integument and cloacae of these animals were collected and tested in the spring, males significantly preferred to follow the trails of females. Females did not prefer the trails of either sex. During the breeding season, chemical cues from the integument of female eastern garter snakes (*Thamnophis sirtalis*) were compared with chemical cuese from the female cloaca in their ability to elicit trailing from males. Males significantly preferred trails from integumental cues over those from the cloaca. Similar results have been obtained with the European viper (*Vipera berus*) (Andrén 1982, 1986).

Trailing of chemical cues has also been tested both within and between species. The eastern garter snake (*Thamnophis s. sirtalis*), the red-sided garter snake (*T. s. parietalis*), the western aquatic garter snake (*T. couchi*), and the western terrestrial garter snake (*T. elegans*) were tested for their ability to trail either conspecific or congeneric individuals in a Y-maze experiment (Heller & Halpern 1981). Interestingly, snakes in these experiments tended to follow the trail of the animal immediately preceding it in the maze. Both conspecific and heterospecific individuals were followed with similar frequencies. The results suggest a mechanism by which animals of different species can locate and utilize group hibernacula.

The trailing abilities of garter snakes of the genus *Thamnophis* have been extensively studied by Ford and his colleagues. In initial experiments, a five-armed maze was used to demonstrate that male garter snakes are quite successful at following trails of conspecific females, and not those trails of heterospecifics (Ford 1978, 1981). In studies with a Y-maze, male garter snakes still tended to follow the trails of their own females, but in addition, in this study they also followed trails of heterospecifics rather than a blank arm (Ford 1982; Ford &

Schofield 1984; Ford & O'Bleness 1986). The results of these experiments are somewhat conflicting and point to the need for additional experiments to tease apart the causal mechanisms underlying this important behavior in snakes.

Directionality of a trail is a critical piece of information that the follower needs to determine. Experiments in a test arena with removable pegs elucidated a probable mechanism by which snakes determine the direction of a trail (Ford & Low 1983) (see Fig. 6.5). When a stimulus snake deposited a trail in the test box, snakes were able to trail in the correct direction. Then, when the direction of the pegs was removed and reversed, the trailing snake would also reverse direction. Under natural conditions, snakes deposit chemical cues in a similar fashion on objects in the environment that they have crawled over and past. Lizard trailing behavior has also been observed in Y-mazes. Male broad-headed skinks (*Eumeces laticeps*) can distinguish and follow the trails of females by detecting chemical cues through tongue-flicking and tongue-touching the substrate (Cooper & Vitt 1986b, 1986c).

Future studies should investigate the robustness of this behavior under natural or field conditions. Laboratory studies can elucidate behaviors, however, these results need to be verified under the more realistic conditions actually experienced by animals in their native environment.

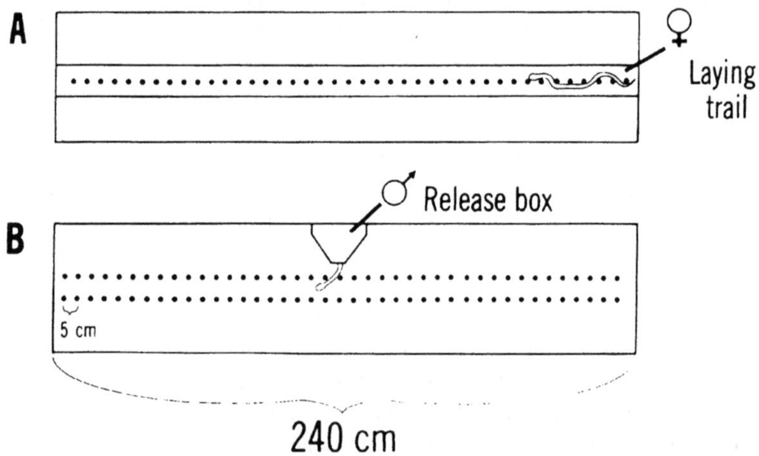

Figure 6.5. Arena used to test direction determining ability of garter snakes to female pheromone trails. (**A**) Arena setup with a single row of pegs showing the female laying a trail between two removable boards. (**B**) Arena showing the male release. Males are able to trail females by detecting pheromone cues left on the pegs. Skin lipids left in the environment by females are detected by male tongue-flicking. Males follow females during the breeding season using trailing behavior to find suitable mates. Redrawn from Ford and Low (1983).

6.3.5. Sexual Behavior

In most species of snakes, sexual behavior is characterized by the male investigating the female with rapid tongue-flicking to her dorsal surface. The male then presses his chin onto the female's dorsal surface, rubbing forward toward her head while continuing to rapidly tongue-flick. The occurrence of these two behaviors simultaneously is characteristic of courtship and sex behavior. Noble (1937) first hypothesized that males detected an odor or pheromone in the skin of females that elicited male courtship. Estrogen seems to play a role in the production and/or expression of the pheromone in garter snakes (*Thamnophis sirtalis*) (Crews 1976; Kubie et al. 1978).

Field and laboratory studies have identified the sex pheromone of the red-sided garter snake (*Thamnophis sirtalis parietalis*) (Mason et al. 1989, 1990; Mason 1993). Hexane extracts of skin lipids from sexually attractive females were isolated and tested in a bioassay. Liquid aliquots of fractions from these skin lipids were applied to paper towels or filter paper. Positive responses in the bioassay were scored when courting male garter snakes exhibit significantly increased tongue-flicks to the sample in conjunction with the chin-rubbing behavior that is exclusive to courtship behavior. In additional tests, the latency to court, number of tongue-flicks elicited, and tongue-flick rate were examined (Mason et al. 1990). The long-chain saturated and monounsaturated methyl ketones identified are the sex pheromone of the red-sided garter snake. Similar studies using the same experimental paradigm have identified saturated and mono- and diunsaturated methyl ketones in another colubrid, the brown tree snake of Guam (*Boiga irregularis*) (Murata et al. 1991).

Licking behaviors are common to many species of lizards and may be analogous to tongue-flicking in snakes when observed in a reproductive context. Male western banded geckos (*Coleonyx variegatus*) licked females repeatedly before taking a neck grip and initiating copulation attempts (Greenberg 1943). Male leopard geckos (*Eublepharis macularius*) are aggressive and territorial. Males routinely licked all individuals that they contacted. If the animal contacted was a female in breeding condition, she was courted. If the animal contacted was a male, a fight would quickly ensue. Sex recognition cues seem to be related to integumental lipids as shedding caused territorial males to misinterpret the sex of the stimulus animals. Shedding caused territorial males to bite females as if they were males (Mason & Gutzke 1990).

When geckos (*E. macularius* and *Paroedura pictus*) were placed in an empty cage that was previously occupied by a conspecific, both sexes responded with higher tongue-flick rates than to a cage that was formerly empty or occupied by the test animal itself (Brillet 1990). Further, male *Eublepharis* responded to the chemical cues of female *Eublepharis* with significantly more tongue-flicks as well as tail vibrations which are only observed in a reproductive context. Males did not respond to females that were out of sight when chemical cues were only presented airborne. Finally, substrate borne female chemical cues were more effective at eliciting courtship behavior from males than visual cues alone.

Licking behaviors in lizards are also frequently directed toward the cloacal region, especially during the breeding season. These behaviors have been noted in side-blotched lizards, *Uta stansburiana* (Ferguson 1966; Tinkle 1967); Mallee dragons, *Amphibolurus fordi* (Cogger 1978); horned lizards, *Phrynosoma platyrhinos* and *P. coronatum* (Tollestrup 1981); desert iguanas, *Dipsosaurus dorsalis* (Glinski & Krekorian 1985); and *Tropidurus delanonis* (Werner 1978). Indeed, behavioral responses to integumental cues, cloacal cues, substrate rubbing and licking are widespread in lizards (see Mason 1992 for review).

Olfactory cues have been studied in both five-lined skinks (*Eumeces fasciatus*) and ground skinks (*Scincella lateralis*) by means of an olfactometer (Duvall et al. 1980). Airborne chemical cues from conspecific males and females showed that these animals respond to the odors with tongue-flicks and snout-dips (Duvall et al. 1980). Chemical cues from heterospecific animals were ignored.

An olfactometer was also used to investigate responses to chemical cues from alligators (*Alligator mississippiensis*) (Johnsen & Wellington 1982). Alligators were scored for the amount of gular pumping observed in response to several synthetic chemical stimuli as well as secretions from the cloacal glands, mandibular glands, and blanks. Gular pumping seems to be correlated with the reception of chemical cues from the environment, presumably because it serves to pass odorants over the chemosensory organs (Johnsen & Wellington 1982). Alligators in this study responded most strongly to the cloacal gland secretions of male alligator donors.

In turtles, the mental glands of desert tortoises (*Gopherus* spp.) have been extensively studied with regard to pheromone production. Both males and females possess a large mental gland on the head that hypertrophies during the breeding season and secretes odorous fluids (Auffenberg 1966). Both male and female gopher tortoises (*G. polyphemus*) rub their forearms against their chin glands and wave their forearms only at males during courtship (Auffenberg 1969, Weaver 1970). Presumably, males are challenging other males while females may be soliciting matings from conspecific males.

Gopher tortoises challenge one another by confronting an approaching individual with a head challenge (Auffenberg 1964). If the challenge is not returned (female response), the male proceeds to the posterior of the animal and sniffs its cloacal area. If the second individual is a sexually attractive female, he mounts. Thus, terrestrial tortoises probably have at least two significant sources of pheromones: the mental glands, which may be most important in male-male combat, and the cloaca, which provides sex pheromones that are important in mating behavior.

In aquatic turtles, courtship behavior seems to rely on pheromones produced and expressed from the cloaca. There are no experimental studies specifically investigating the existence of these chemical cues, however, anecdotal reports do suggest their presence. In Florida redbelly turtles (*Pseudemys nelsoni*) (Kramer & Fritz 1989), painted turtles (*Chrysemys picta*) (Ernst 1971), Suwanee river cooters (*Chrysemys concinna suwanniensis*) (Jackson & Davis 1972), map turtles (*Graptemys* spp.) (Ernst 1974), Florida east coast terrapin (*Malaclemys terrapin tequesta*) (Seigel 1980), mud turtles (*Kinosternon* spp.) and stinkpot turtles (*Sternotherus* spp.) (Mahmoud 1967), courting males chase females and sniff their cloacal areas, implying that chemical cues important in coordinating mating behavior are expressed by females.

Clearly, experimental evidence is needed to document and explore the role of chemical cues in mediating agonistic behaviors in turtles.

6.3.6. Femoral Pores

Many lacertilian reptiles possess epidermal structures on the ventral surface of the thigh or the precloacal abdominal area. These glands, or femoral pores, exude a waxy secretion, are usually larger in males, and seem to respond to rising androgen concentrations in the blood, especially during the breeding season (Chiu & Maderson, 1975). Behavioral responses to the femoral pore secretions were investigated in the desert iguana (*Dipsosaurus dorsalis*) (Alberts 1989). Lizards tongue-touched tiles coated with femoral pore secretions significantly more than paraffin-coated tiles or controls. Because there were no consistent differences between male and female secretions, the femoral pore secretions may be more important in individual recognition than sex recognition (Alberts 1989).

Although technically not femoral pores, the precloacal pores of the amphisbaenian *Blanus cinereus* were investigated for their possible role as a pheromone source (Cooper et al. 1994). Cotton swabs were immersed in water as a control, or rubbed over the precloacal pores of male and female *B. cinereus,* then presented to males. Latency to tongue-flick as well as total numbers of tongue-flicks were recorded. Male *B. cinereus* tongue-flick significantly more to female precloacal pore chemical cues than to those of males or water blanks. Tests with integumental chemical cues revealed that males respond equally to male and female skin cues. Thus, they do not appear to be able to distinguish sex differences based on integumental cues alone. Rather, precloacal pore chemical cues are necessary for correct sex identification to occur. Cooper et al. (1994) raise the possibility that olfaction, as well as vomerolfaction, may be involved in sex recognition in this species.

6.3.7. Copulatory Plugs

In garter snakes (*Thamnophis*), males leave a gelatinous plug in the cloaca of the female immediately following mating. Sexually active males refuse to court mated females immediately after the female has mated and up to 48 h later (Ross

& Crews 1977). In addition, a pheromone associated with the copulatory plug in the plains garter snake (*Thamnophis radix*) renders captive males sexually refractory for periods of 24–72 h after exposure (Ross & Crews 1978). Thus, there appear to be pheromones in the copulatory plug that not only render mated females temporarily unattractive, but also cause sexually active males to become sexually quiescent, ceasing courtship behavior after exposure to this pheromone. Similar studies in the Swedish viper (*Vipera berus*) did not yield any behaviors similar to those exhibited by garter snakes (Nilson & Andrén 1982).

6.3.8. Combat Behavior

Texas tortoises (*Gopherus berlandieri*) respond to the mental gland secretions of conspecific males with combat behavior (Rose 1970). Plaster models of tortoises painted with mental gland secretions from conspecific male tortoises elicited head bobbing and ramming of the model by males and females. Females responded primarily with head bobbing, a courtship behavior.

6.3.9. Predator/Prey Recognition

A common defensive behavior exhibited by over 30 taxa of pit vipers and some colubrids to snake-eating (ophiophagous) is a posture known as body bridging (Weldon 1982). In this posture, a body loop can be used to strike at the attacking snake, thus disabling or driving it off. In tests examining the source of the chemical compounds eliciting this behavior, cloacal cues and lipids removed from the ventrum of ophiophagous king snakes failed to elicit body bridging from rattlesnakes (Bogert 1941). However, skin lipids removed from the dorsal surface of the king snakes prompted body bridging from rattlesnakes even if they were blindfolded (Bogert 1941). In further studies, methanol-soaked cotton balls were used to extract chemical cues from king snakes and elicited body bridging from several species of pit vipers and colubrids (Weldon & Burghardt 1979). Juvenile garter snakes can distinguish between the skin lipids of the ophiophagous common king snake (*Lampropeltis getulus*) and two nonophiophagous snakes, the western hognose snake (*Heterodon nasicus*) and plains garter snake (*Thamnophis radix*) by exhibiting increased tongue-flicks to the ophiophage skin extracts (Weldon 1982). However, this is not always the case as the prairie rattlesnake (*Crotalus viridis*) and cottonmouth (*Agkistrodon piscivorus*) give the opposite response and did not show increased tongue-flick rate when exposed to chemical cues from a king snake (*Lampropeltis* spp.) and a nonophiophagous hognose snake (*Heterodon* spp.) (Chiszar et al. 1978). These conflicting results demonstrate the need for studies in which additional behaviors are needed to supplement the tongue-flick data. These additional behaviors need to more directly address the motivational state of the animal. Tongue-flick data are often only a sign of increased activity of the animal. In any case, they are strengthened by including additional response variables.

Y-mazes have also been utilized to study the response of snakes to chemical cues from potential predators. Juvenile pine snakes (*Pituophis m. melanoleucus*) significantly avoid the arm of a maze treated with skin lipids from an ophiophagous predator, the common king snake (*Lampropeltis getulus*) (Burger 1989). In further studies, soiled bedding from the cages of nonophiophagous rat snakes (*Elaphe obsoleta*), corn snakes (*Elaphe guttata*), and ophiophagous milk snakes (*Lampropeltis triangulum*) and common king snakes (*L. getulus*) were deposited on one arm of a Y-maze (Burger 1990; Burger et al. 1991). Juvenile pine snakes (*P. melanoleucus*) were able to discern and avoid the odors of the ophiophagous snakes and either ignored the nonophiophagous snakes, or were attracted to them, while preferring to follow the trails of conspecifics. These responses would be adaptive to young pine snakes who need to avoid syntopic snake-eating snakes. However, it benefits them to be able to distinguish those chemical cues of predators from those of conspecifics and nonophiophagous snakes like corn snakes that occasionally hibernate with pine snakes. Shortly after hatching, pine snakes need to migrate to overwintering hibernacula. The ability to follow chemical cues from conspecifics and corn snakes would thus be highly desirable. Corn snakes (*E. guttata*) apparently can also distinguish between ophiophagous versus nonophiophagous snakes by responding with more tongue-flicks to the ophiophagous snake's skin lipids (Weldon et al. 1990).

Cloacal scent gland secretions are commonly thought to act as predator deterrents. Several mammalian predators avoid food treated with the cloacal scent gland secretions of the desert king snake (*Lampropeltis getulus splendida*) (Price & LaPointe 1981). Cloacal scent gland secretions from western diamondback rattlesnakes (*Crotalus atrox*) cause coyotes and dogs to roll and rub the substrate around these secretions, although the reasons for this behavioral response are not discussed by the authors. The secretions may distract the predators long enough for the snake to escape predation (Weldon & Fagre 1989).

Lizards will also respond to chemical cues from snake predators with increased tongue-flick rates (Thoen et al. 1986). When the viviparous lizard (*Lacerta vivipara*) was housed in the cage of lizard-eating snakes (adders, *Vipera berus*, and smooth snakes, *Coronella austriaca*), the lizard responded with higher tongue-flick rates than those in response to the chemical cues from a non-lizard-eater, the grass snake (*Natrix* spp.). The lizards also altered their movement patterns by moving more slowly, vibrating their tails and shaking their feet.

To determine whether viviparous lizards (*L. vivipara*) show innate responses to chemical cues from predatory snakes from birth, or learn them through experience, the preceding experiments were duplicated (Van Damme et al. 1995). Seventeen juvenile common lizards (*L. vivipara*) were exposed to a clean cage or to a cage that had previously housed the lizard-eating adder (*V. berus*) or smooth snake (*C. austriaca*). Juveniles that had never been exposed to predator chemical cues before behaved like the animals used in the previous study by Thoen et al. (1986). One difference in the behavior of juvenile and adult lizards is that juveniles exposed to predator chemical cues almost completely refrain from basking behavior, while adults do not alter their time spent in this behavior after exposure to the snake predator chemical cues (Van Damme et al. 1995). Two subspecies of Iberian wall lizard (*Podarcis hispanica*) are able to identify the snake predator (Lataste's viper, *Vipera latastei*) solely through detection of chemical cues left in a previously occupied cage (Van Damme & Castilla 1996). In this particular experiment, one of the subspecies has been isolated from predation by the viper for over 100 years. The second subspecies of lizard still occurs sympatrically with the snake. The results indicated that even after 100 years of isolation, the lizards still respond to the chemical cues of the snake in the same fashion as did the sympatric subspecies. The behaviors measured were a series of motor behaviors associated with arousal in these lizards.

White-throated monitor lizards (*Varanus albigularis*) eat nonvenomous snakes, while venomous snakes prey on the monitor lizards. When presented with the nonvenomous sand snake (*Psammophis leightoni*), monitor lizards attack and consume the snakes (Phillips & Alberts 1992). However, when the same lizards were presented with two venomous snakes, the spitting cobra (*Naja nigricollis*) and the horned adder (*Bitis caudalis*) (all are known to eat monitor lizards), the lizards never attacked these snakes. Finally, when insect prey were covered with the skins of the venomous snakes, the lizards did not attack them after investigating with tongue-flicking. However, if the insect prey were covered with the sand snake skin, or they were not investigated with tongue-flicking, then the lizards attacked and devoured the insect. Thus, chemical cues from predatory snakes are recognized by these lizards and discriminated from those of nonpredatory snakes, even if the lizards were naive to snake predation.

6.4. Conclusions

Clearly, the study of chemical signals in amphibians and reptiles holds great promise for elucidating complex behaviors in these animals. In this review, we have concentrated on those methodologies that seem to be generally applicable to the investigation of specific behaviors elicited by chemical cues. We have not attempted to cover all the literature, but rather, have tried to cover fewer topics in greater detail. In so doing, we have discovered that there is little overlap in the similarity of behaviors exhibited in response to chemical cues among amphibians and reptiles. Thus, in many respects, the grouping of the rather disparate groups of amphibians and reptiles is more an historic construct than one based on similarity of behaviors. Further, some very promising areas of research dealing with chemical signalling and pheromones have scarcely been investigated.

We can point out to the reader areas of research that seem to be particularly worthy of further study. Thus, the mechanisms responsible for kin recognition have been well studied in amphibians, especially anuran amphibians; however, few studies have investigated these behaviors in reptiles. Further, territorial behavior based on chemical cues has been rigorously studied in salamanders and some orders of reptiles, but not all. This seems to be a particularly tractable behavior for studies in the future. In contrast to the salamanders and most reptiles, very little is known regarding the role of chemical cues in territoriality and mating behavior in frogs and toads. Amphibians also pose the interesting problem of life-history constraints. Thus, social behaviors in larvae of amphibians may be entirely different from those of adults. For example, aggregation studies with anuran tadpoles are far more tractable than such studies with adults. In reptiles, there are no larval stages, however there are behaviors where responses of juveniles differ significantly from adults (e.g., trailing, aggregation).

Amphibians and reptiles continue to provide robust models for the study of chemical communication. The methodologies developed for quantifying behavioral responses to chemical cues are critically important for continuing studies of the chemical ecology of these two groups. To elucidate the chemical cues responsible for eliciting social behaviors, investigators must have rigorous, unequivocal bioassays. Amphibians and reptiles exhibit behaviors in response to chemical cues that are relatively simple, discriminating, and tractable. They can also be extremely sensitive to chemical cues, perhaps more so than any other vertebrates. By bridging both the aquatic and terrestrial environments, amphibians and reptiles occupy a number of widespread habitats. Comparisons of both the similarities and differences between chemically mediated behaviors in these groups may lead to general principles that are applicable to studies of similar behaviors in other vertebrates, such as mammals, where behaviors often result from a multicomponent sensory input.

6.5. Acknowledgments

The authors would like to thank the editors, Jocelyn Millar and Ken Haynes, for their valuable insights and suggestions for improving the chapter. Brent Graves provided an in-depth critique of the chapter and we are indebted to him. The work of the authors was supported by the National Science Foundation INT-9114567, IBN-9357245; The Whitehall Foundation; and the U.S. Fish and Wildlife Service to R.T.M.; NSERC of Canada to D.P.C.; Southwest Missouri State University Faculty Research Development grant to A.M.; and the National Science Foundation DEB-9423333 to A.R.B.

6.6. References

Alberts, A.C. 1989. Ultraviolet visual sensitivity in desert iguanas: implications for pheromone detection. Anim. Behav. **38**:129–137.

Aleksiuk, M. 1976. Reptilian hibernation: evidence of adaptive strategies in *Thamnophis sirtalis parietalis.* Copeia **1976**:170–178.

Allen, B.A., G.M. Burghardt & D.S. York. 1984. Species and sex differences in substrate preference and tongue-flick rate in three sympatric species of water snakes, *Nerodia.* J. Comp. Psychol. **98**:358–367.

Andrén, C. 1982. The role of the vomeronasal organs in the reproductive behavior of the adder, *Vipera berus.* Copeia **1982**:148–157.

Andrén, C. 1986. Courtship, mating and agonistic behaviour in a free-living population of adders, *Vipera berus* (L.). Amphibia-Reptilia **7**:353–383.

Anthony, C.D. 1993. Recognition of conspecific odors by *Plethodon caddoensis* and *P. ouachitae.* Copeia **1993**:1028–1033.

Anthony, C.D. & J.A. Wicknick. 1993. Aggressive interaction and chemical communication between adult and juvenile salamanders. J. Herpetol. **27**:261–264.

Arnold, S.J. & L.D. Houck. 1982. Courtship pheromones: evolution by natural and sexual selection. *In:* Biochemical Aspects of Evolutionary Biology, ed. M. Nitecki, pp. 173–211, University of Chicago Press, Chicago.

Auffenberg, W. 1964. Notes on the courtship of the land tortoise *Geochelone travancorica* (Boulenger). J. Bombay Nat. Hist. Soc. **61**:247–253.

Auffenberg, W. 1966. On the courtship of *Gopherus polyphemus.* Herpetologica **22**:113–117.

Auffenberg, W. 1969. Social behavior of *Geochelone denticulata.* Q. J. Fla. Acad. Sci. **32**:50–58.

Barthalamus, G.T. & E.D. Bellis. 1969. Homing in the northern dusky salamander, *Desmognathus fuscus fuscus* (Rafinesque). Copeia **1969**:148–153.

Barthalamus, G.T. & E.D. Bellis. 1972. Home range, homing and the homing mechanism of the salamander, *Desmognathus fuscus.* Copeia **1972**:632–642.

Baumann, F. 1927. Experimente über den Geruchsinn der Viper. Rev. Suisse Zool. **34**:173–184.

Baumann, F. 1929. Experimente über den Geruchsinn und den Beuteerwerb der Viper (*Vipera aspis*). Z. vergl. Physiol. **10**:36–119.

Blaustein, A.R. & R.K. O'Hara. 1981. Genetic control for sibling recognition? Nature **290**:246–248.

Blaustein, A.R. & R.K. O'Hara. 1982. Kin recognition cues in *Rana cascadae* tadpoles. Behav. Neural. Biol. **36**:77–87.

Blaustein, A.R. & S.C. Walls. 1995. Aggregation and kin recognition. *In:* Amphibian Biology, eds. H. Heatwole & B.K. Sullivan, pp. 568–602, Surrey Beatty & Sons, Chipping Norton, NSW, Australia.

Blaustein, A.R., R.K. O'Hara, & D.H. Olson. 1984. Kin preference behaviour is present after metamorphosis in *Rana cascadae* frogs. Anim. Behav. **32**:445–450.

Blaustein, A.R., T. Yoshikawa, K. Asoh & S.C. Walls. 1993. Ontogenetic shifts in tadpole kin recognition: loss of signal and perception. Anim. Behav. **46**:525–538.

Bogert, C.M. 1941. Sensory cues used by rattlesnakes in their recognition of ophidian enemies. Ann. N.Y. Acad. Sci. **41**:329–344.

Brillet, C. 1990. Role des informations olfactives et visuelles dans la discrimination du sexe chez deux especes de geckos nocturnes: *Eublepharis macularius* et *Paroedura pictus.* Biol. Behav. **15**:1–22.

Brodie, E.D. Jr., J.L. Hensel Jr. & J.A. Johnson. 1974. Toxicity of the urodele amphibians *Taricha, Notophthalmus, Cynops* and *Paramesotriton* (Salamandridae). Copeia **1974**:506–511.

Brodie, E.D. Jr., D.R. Formanowicz, & E.D. Brodie III. 1978. The development of noxiousness of *Bufo americanus* tadpoles to aquatic insect predators. Herpetologica **34**:302–306.

Brown, C.W. 1968. Additional observations on the function of the nasolabial grooves of plethodontid salamanders. Copeia **1968**:728–731.

Brown, W.S. & F.M. MacLean. 1983. Conspecific scent-trailing by newborn timber rattlesnakes, *Crotalus horridus.* Herpetologica **39**:430–436.

Burger, J. 1989. Following of conspecific and avoidance of predator chemical cues by pine snakes (*Pituophis melanoleucus*). J. Chem. Ecol. **15**:799–806.

Burger, J. 1990. Response of hatchling pine snakes (*Pituophis melanoleucus*) to chemical cues of sympatric snakes. Copeia **1990**:1160–1163.

Burger, J., W. Boarman, L. Kurzava & M. Gochfeld. 1991. Effect of experience with pine (*Pituophis melanoleucus*) and king (*Lampropeltis getulus*) snake odors on Y-maze behavior of pine snake hatchlings. J. Chem. Ecol. **17**:79–87.

Burghardt, G.M. 1970. Chemical perception in reptiles. *In:* Communication by Chemical Signals, eds. J.W. Johnston, D.G. Moulton & M. Turk, pp. 241–308, Appleton-Century-Crofts, New York.

Burghardt, G.M. 1980. Behavioral and stimulus correlates of vomeronasal functioning in reptiles: Feeding, grouping, sex, and tongue use. *In:* Chemical Signals. Vertebrates and Aquatic Invertebrates, eds. D. Müller-Schwarze & R.M. Silverstein, pp. 275–301, Plenum Press, New York.

Burghardt, G.M. 1983. Aggregation and species discrimination in newborn snakes. Z. Tierpsychol. **61**:89–101.

Burghardt, G.M., H.W. Green & A.S. Rand. 1977. Social behavior in hatchling green iguanas: life at a reptile rookery. Science **195**:689–691.

Burrage, B.R. 1965. Copulation in a pair of *Alligator mississippiensis.* Br. J. Herp. **3**:207–208.

Carpenter, C.C. & G.W. Ferguson. 1977. Variation and evolution of stereotyped behavior in reptiles. *In:* Biology of the Reptilia, eds. C. Gans & D.W. Tinkle, pp. 335–554, Academic Press, New York.

Chiszar, D., K.M. Scudder, L. Knight & H.M. Smith. 1978. Exploratory behavior in prairie rattlesnakes (*Crotalus viridis*) and water moccasins (*Agkistrodon piscivorus*). Psychol. Rec. **28**:363–368.

Chiu, K.W. & P.F.A. Maderson. 1975. The microscopic anatomy of epidermal glands in two species of gekkonine lizards, with some observations on testicular activity. J. Morphol. **147**:23–40.

Chivers, D.P. & R.J.F. Smith. 1998. Chemical alarm signalling in aquatic predator-prey systems: a review and prospectus. Ecoscience. In press.

Chivers, D.P., J.M. Kiesecker, M.T. Anderson, E.L. Wildy & A.R. Blaustein. 1996a. Avoidance response of a terrestrial salamander (*Ambystoma macrodactylum*) to chemical alarm cues. J. Chem. Ecol. **22**:1709–1716.

Chivers, D.P., B.D. Wisenden & R.J.F. Smith. 1996b. Damselfly larvae learn to recognize predators from chemical cues in the predator's diet. Anim. Behav. **52**:315–320.

Chivers, D.P., J.M. Kiesecker, E.L. Wildy, M.T. Anderson, & A.R. Blaustein. 1997a. Chemical alarm signalling in terrestrial salamanders: intra- and interspecific responses. Ethology **103**:599–613.

Chivers, D.P., E.L. Wildy & A.R. Blaustein. 1997b. Eastern long-toed salamander (*Ambystoma macrodactylum columbianum*) larvae recognize cannibalistic conspecifics. Ethology **103**:187–197.

Cogalniceanu, D. 1994. The relative importance of vision and olfaction in mate recognition in male newts (genus *Triturus*). Herpetologica **50**:344–349.

Cogger, H.G. 1978. Reproductive cycles, fat body cycles and socio-sexual behaviour in the Mallee Dragon, *Amphibolurus fordi* (Lacertilia: Agamidae). Aust. J. Zool. **26**:653–672.

Cooper, W.E. 1994. Chemical discrimination by tongue-flicking in lizards: a review with hypotheses on its origin and its ecological and phylogenetic relationships. J. Chem. Ecol. **20**:439–487.

Cooper, W.E. 1995. Effects of estrogen and male head coloration on chemosensory investigation of female cloacal pheromones by male broad-headed skinks (*Eumeces laticeps*). Physiol. Behav. **58**:1221–1225.

Cooper, W.E. & A.C. Alberts. 1991. Tongue-flicking and biting in response to chemical food stimuli by an iguanid lizard (*Dipsosaurus dorsalis*) having sealed vomeronasal ducts: vomerolfaction may mediate these behavioral responses. J. Chem. Ecol. **17**:135–146.

Cooper, W.E. & G.M. Burghardt. 1990. Vomerolfaction and vomodor. J. Chem. Ecol. **16**:103–105.

Cooper, W.E. & W.R. Garstka. 1987. Lingual responses to chemical fractions of urodaeal glandular pheromone of the skink, *Eumeces laticeps*. J. Exp. Zool. **242**:249–253.

Cooper, W.E. & S.E. Trauth. 1992. Discrimination of conspecific male and female cloacal chemical stimuli by males and possession of a probable pheromone gland by females in a cordylid lizard, *Gerrhosaurus nigrolineatus*. **48**:229–236.

Cooper, W.E. & L.J. Vitt. 1984. Conspecific odor detection by the male broad-headed skink, *Eumeces laticeps:* effects of sex and site of odor source and of male reproductive condition. J. Exp. Zool. **230**:199–209.

Cooper, W.E. & L.J. Vitt. 1986a. Interspecific odour discrimination by a lizard *Eumeces laticeps*. Anim. Behav. **34**:367–376.

Cooper, W.E. & L.J. Vitt. 1986b. Interspecific odour discriminations among syntopic congeners in scincid lizards (genus *Eumeces*). Behaviour **97**:1–9.

Cooper, W.E. & L.J. Vitt. 1986c. Tracking of female conspecific odor trails by male broad-heded skinks (*Eumeces laticeps*). Ethology **71**:242–248.

Cooper, W.E., W.R. Garstka & L.J. Vitt. 1986. Female sex pheromone in the lizard *Eumeces laticeps*. Herpetologica **42**:361–366.

Cooper, W.E., P. Lopez & A. Salvador. 1994. Pheromone detection by an amphisbaenian. Anim. Behav. **47**:1401–1411.

Costanzo, J. 1989. Conspecific scent trailing by garter snakes (*Thamnophis sirtalis*) during autumn: further evidence for use of pheromones in den location. J. Chem. Ecol. **15**:2531–2538.

Crews, D. 1976. Hormonal control of male courtship behavior and female attractivity in the garter snake (*Thamnophis sirtalis parietalis*). Horm. Behav. **7**:451–460.

Cupp, P.V. Jr. 1994. Salamanders avoid chemical cues from predators. Anim. Behav. **48**:232–235.

Daly, J.W. 1982. Alkaloids of neotropical poison frogs (Dendrobatidae). *In:* Progress in the Chemistry of Organic Natural Products, Vol. 41, eds. W. Herz, H. Grisebach & G.W. Kirby, pp. 205–340, Springer-Verlag, Vienna.

Daly, J.W., G.B. Brown, M. Mensah-Dwumah & C.W. Myers. 1978. Classification of skin alkaloids from neotropical dart-poison frogs (Dendrobatidae). Toxicon **16**:163–188.

Dawley, E.M. 1984a. Recognition of individual, sex and species odour by salamanders of the *Plethodon glutinosus-P. jordani* complex. Anim. Behav. **32**:353–361.

Dawley, E.M. 1984b. Identification of sex through odors by male red-spotted newts, *Notophthalmus viridescens*. Herpetologica **40**:101–105.

Dawley, E.M. 1986a. Behavioral isolating mechanisms in sympatric terrestrial salamanders. Herpetologica **42**:156–164.

Dawley, E.M. 1986b. Species discrimination by chemical cues in a unisexual-bisexual complex of salamanders. J. Herpetol. **20**:114–116.

Dawley, E.M. & A.H. Bass. 1989. Chemical access to the vomeronasal organs of a plethodontid salamander. J. Morphol. **200**:163–174.

Dawson, J.T. 1982. Kin recognition and schooling in the American toad (*Bufo americanus*). Ph.D. thesis, State University of New York at Albany.

Dole, J.W. 1965. Summer movements of adult leopard frogs, *Rana pipiens* Schreber, in northern Michigan. Ecology **46**:237–255.

Dole, J.W. 1968. Homing in leopard frogs, *Rana pipiens*. Ecology **49**:386–399.

Ducey, P.K. & P. Ritsema. 1988. Intraspecific aggression and responses to marked substrates in *Ambystoma maculatum* (Caudata:Ambystomatidae). Copeia **1988**:1008–1013.

Dundee, H.A. & M.C. Miller III. 1968. Aggregative behavior and habitat conditioning by the prairie ringneck snake, *Diadophis punctatus arnyi*. Tulane Stud. Zool. Bot. **15**:41–58.

Duvall, D., R. Herskowitz & J. Trupiano-Duvall. 1980. Responses of five-lined skinks

(*Eumeces fasciatus*) and ground skinks (*Scincella lateralis*) to conspecific and interspecific chemical cues. J. Herp. **14**:121–127.

Duvall, D., M.B. King & K.J. Gutzwiller. 1985. Behavioral ecology and ethology of the prairie rattlesnake. Nat. Geog. Res. **1985**:80–111.

Eible-Eibesfeldt, I. 1949. Über das Vorkommen von Schreckstoffen bei Erdkrötenquappen. Experientia **5**:236.

Eisner, T., W.E. Connor, K. Hicks, K.R. Dodge, H. Rosenberg, T.H. Jones, M. Cohen & J. Meinwald. 1977. Stink of stinkpot turtle identified: w-phenylalkanoic acids. Science **196**:1374–1349.

Elliott, S.A., L.B. Kats & J.A. Breeding. 1993. The use of conspecific chemical cues for cannibal avoidance in California newts (*Taricha torosa*). Ethology **95**:186–192.

Ernst, C.H. 1971. Observations of the painted turtle, *Chrysemys picta.* J. Herpetol. **5**:216–220.

Ernst, C.H. 1974. Observations on the courtship of male *Graptemys pseudogeographica.* J. Herpetol. **8**:377–378.

Evans, L.T. 1961. Structure as related to behavior in the organization of population in reptiles. *In:* Vertebrate Speciation, ed. W.F. Blair, pp. 148–178, University of Texas Press, Austin.

Feminella, J.W. & C.P. Hawkins. 1994. Tailed frog tadpoles differentially alter their feeding behavior in response to non-visual cues from four predators. J. North Am. Benthol. Soc. **13**:310–320.

Ferguson, G.W. 1966. Releasers of courtship and territorial behaviour in the side-blotched lizard, *Uta stansburiana.* Anim. Behav. **14**:89–92.

Fletcher, D.J.C. & C.D. Michener. eds. 1987. Kin Recognition in Animals. John Wiley, Chichester.

Ford, N.B. 1978. Evidence for species specificity of pheromone trails in two sympatric garter snakes, *Thamnophis.* Herpetol. Rev. **9**:10–11.

Ford, N.B. 1981. Seasonality of pheromone trailing behavior in two species of garter snake, *Thamnophis* (Colubridae). Southwestern Nat. **26**:385–388.

Ford, N.B. 1982. Species specificity of sex pheromone trails of sympatric and allopatric garter snakes. Copeia **1982**:10–13.

Ford, N.B. 1986. The role of pheromone trails in the sociobiology of snakes. *In:* Chemical Signals in Vertebrates IV. Ecology, Evolution and Comparative Biology, eds. D. Duvall, D. Müller Schwarze, & R.M. Silverstein, pp. 261–278, Plenum Press, New York.

Ford, N.B. & D. Holland. 1990. The role of pheromones in the spacing behaviour of snakes. *In:* Chemical Signals in Vertebrates, 5, eds. D.W. Macdonald, D. Müller-Schwarze & S.E. Natynczuk, pp. 465–472, Oxford University Press, Oxford.

Ford, N.B. & J.R. Low, Jr. 1983. Sex pheromone source location by snakes: a mechanism for detection of direction in nonvolatile trails. J. Chem. Ecol. **10**:1193–1199.

Ford, N.B. & M.L. O'Bleness. 1986. Species and sexual specificity of pheromone trails of the garter snake, *Thamnophis marcianus.* J. Herpetol. **20**:259–262.

Ford, N.B. & C.W. Schofield. 1984. Species specificity of sex pheromone trails in the plains garter snake, *Thamnophis radix.* Herpetologica **40**:51–55.

Forester, D.C. 1979. Homing to the nest by female mountain dusky salamanders (*Desmognathus ochrophaeus*) with comments on the sensory modalities essential to clutch recognition. Herpetologica **35**:330–335.

Forester, D.C. & A. Wisnieski. 1991. The significance of airborne olfactory cues to the recognition of home area by the dart-poison frog (*Dendrobates pumilio*). J. Herpetol. **25**:502–504.

Formanowicz, D.R. & E.D. Brodie, Jr. 1982. Relative palatabilities of members of a larval amphibian community. Copeia **1982**:91–97.

Garrick, L.D. 1978. Love among the alligators. Animal Kingdom **78**:2–8.

Gehlbach, F.R., J.F. Watkins & H.W. Reno. 1968. Blind snake defensive behavior elicited by ant attacks. BioScience **18**:784–785.

Gergits, W.F. & R.G. Jaeger. 1990. Field observations of the behavior of the red-backed salamander (*Plethodon cinereus*): courtship and agonistic interactions. J. Herpetol. **24**:93–95.

Gillingham, J.C. 1987. Social behavior. *In:* Snakes: Ecology and Evolutionary Biology, eds. R.A. Seigel, J.T. Collins & S.S. Novak, pp. 184–209, Macmillan, New York.

Glinski, T.H. & C.O. Krekorian. 1985. Individual recognition in free-living adult male desert iguanas, *Dipsosaurus dorsalis.* J. Herpetol. **19**:541–544.

Grant, D., O. Anderson & V. Twitty. 1968. Homing orientation by olfaction in newts (*Taricha rivularis*). Science **160**:1354–1356.

Graves, B.M. 1994. The role of nasolabial grooves and the vomeronasal system in recognition of home area by red-backed salamanders. Anim. Behav. **47**:1216–1219.

Graves, B.M. & D. Duvall. 1988. Evidence of an alarm pheromone from the cloacal sacs of prairie rattlesnakes. Southwestern Nat. **33**:339–345.

Graves, B.M. & D. Duvall. 1995. Aggregation of squamate reptiles associated with gestation, oviposition, and parturition. Herpetol. Mono. **9**:102–119.

Graves, B.M. & M. Halpern. 1988. Neonate plains garter snake (*Thamnophis radix*) are attracted to conspecific skin extracts. J. Comp. Psych. **102**:251–253.

Graves, B.M. & M. Halpern. 1989. Chemical access to the vomeronasal organs of the lizard *Chalcides ocellatus.* J. Exp. Zool. **249**:150–157.

Graves, B.M. & M. Halpern. 1990. Roles of vomeronasal organ chemoreception in tongue flicking, exploratory and feeding behaviour of the lizard, *Chalcides ocellatus.* Anim. Behav. **39**:692–698.

Graves, B.M. & M. Halpern. 1991. Discrimination of self from conspecific chemical cues in *Tiliqua scincoides* (Sauria:Scincidae). J. Herpetol. **25**:125–126.

Graves, B.M., D. Duvall, M.B. King, S.L. Lindstedt & W.A. Gern. 1986. Initial den location by neonatal prairie rattlesnakes: functions, causes, and natural history in chemical ecology. *In:* Chemical Signals in Vertebrates IV. Ecology, Evolution, and Comparative Biology, eds. D. Duvall, D. Müller-Schwarze & R.M. Silverstein, pp. 285–304, Plenum Press, New York.

Graves, B.M., C.H. Summers & K.L. Olmstead. 1993. Sensory mediation of aggregation among postmetamorphic *Bufo cognatus*. J. Herpetol. **27**:315–319.

Greenberg, B. 1943. Social behavior of the Western banded gecko, *Coleonyx variegatus baird*. Physiol. Zool. **16**:110–122.

Gregory, P.T. 1975. Aggregation of gravid snakes in Manitoba, Canada. Copeia **1975**:185–186.

Gregory, P.T. 1982. Reptilian hibernation. *In:* Biology of the Reptilia, Vol. 13, ed. C. Gans & F.H. Pough, pp. 53–154. Academic Press, New York.

Gregory, P.T., J.M. Macartney & K.W. Larsen. 1987. Spatial patterns and movements. *In:* Snakes: Ecology and Evolutionary Biology, eds. R.A. Seigel, J.T. Collins & S.S. Novak, pp. 366–395, Macmillan, New York.

Grubb, J.C. 1970. Orientation in post-reproductive Mexican toads, *Bufo valliceps*. Copeia **1970**:674–680.

Grubb, J.C. 1973a. Olfactory orientation in *Bufo woodhousei fowleri, Pseudacris clarki* and *Pseudacris streckeri*. Anim. Behav. **21**:726–732.

Grubb, J.C. 1973b. Olfactory orientation in breeding Mexican toads, *Bufo valliceps*. Copeia **1973**:490–497.

Grubb, J.C. 1975. Olfactory orientation in southern leopard frogs, *Rana ultricularia*. Herpetologica **31**:219–221.

Grubb, J.C. 1976. Maze orientation by Mexican toads, *Bufo valliceps* (Amphibia, Anura, Bufonidae), using olfactory and configurational cues. J. Herpetol. **10**:97–104.

Halpern, M. 1992. Nasal chemical senses in reptiles: structure and function. *In:* Biology of the Reptilia: Hormones, Brain, and Behavior, eds. C. Gans & D. Crews, pp. 423–523, University of Chicago Press, Chicago.

Halpern, M. & N. Frumin. 1979. Roles of the vomeronasal and olfactory systems in prey attack and feeding in adult garter snakes. Phyiol. Behav. **22**:1183–1189.

Halpern, M. & J.L. Kubie. 1980. Chemical access to the vomeronasal organ of the garter snakes. Physiol. Behav. **24**:367–371.

Halpin, Z.T. 1990. Responses of juvenile eastern garter snakes (*Thamnophis sirtalis sirtalis*) to own, conspecific and clean odors. Copeia **1990**:1157–1160.

Hatanaka, T., T. Shibuya & J. Inouchi. 1988. Induced wave responses of the accessory olfactory bulb to odorants in two species of turtle, *Pseudemys scripta* and *Geoclemys reevesii*. Comp. Biochem. Physiol. **91A:**377–385.

Heller, S. & M. Halpern. 1981. Laboratory observations on conspecific and congeneric scent trailing in garter snakes (*Thamnophis*). Behav. Neural Biol. **33**:372–377.

Heller, S.B. & M. Halpern. 1982a. Laboratory observations of aggressive behavior of garter snakes, *Thamnophis sirtalis*. J. Comp. Physiol. Psych. **96**:967–983.

Heller, S.B. & M. Halpern. 1982b. Laboratory observations of aggressive behavior of garter snakes, *Thamnophis sirtalis:* roles of the visual, olfactory, and vomeronasal senses. J. Comp. Physiol. Psych. **96**:984–989.

Hews, D.K. 1988. Alarm response in larval western toads, *Bufo boreas:* release of larval chemicals by a natural predator and its effect on predator capture efficiency. Anim. Behav. **36**:125–133.

Hews, D.K., & A.R. Blaustein. 1985. An investigation of the alarm response in *Bufo boreas* and *Rana cascadae* tadpoles. Behav. Neural Biol. **43**:47–57.

Hokit, D.G. & A.R. Blaustein. 1995. Predator avoidance and alarm-response behaviour in kin discriminating tadpoles (*Rana cascadae*). Ethology **101**:280–290.

Horne, E.A. & R.G. Jaeger. 1988. Territorial pheromones of female red-backed salamanders. Ethology **78**:143–152.

Houck, L.D. 1986. The evolution of salamander courtship pheromones. *In:* Chemical Signals in Vertebrates IV. Ecology, Evolution, and Comparative Biology, eds. D. Duvall, D. Müller-Schwarze & R.M. Silverstein, pp. 173–190, Plenum Press, New York.

Houck, L.D. & N.L. Reagan. 1990. Male courtship pheromones increase female receptivity in a plethodontid salamander. Anim. Behav. **39**:729–734.

Hrbacek, J. 1950. On the flight reaction of tadpoles of the common toad caused by chemical substances. Experientia **6**:100–101.

Jackson, C.G. & J.D. Davis. 1972. Courtship display behavior of *Chrysemys concinna suwanniensis.* Copeia **1972**:385–387.

Jaeger, R.G. 1984. Agonistic behavior of the red-backed salamander. Copeia **1984**:309–314.

Jaeger, R.G. 1986. Pheromonal markers as territorial advertisement by terrestrial salamanders. *In:* Chemical Signals in Vertebrates IV. Ecology, Evolution, and Comparative Biology, eds. D. Duvall, D. Müller-Schwarze & R.M. Silverstein, pp. 191–203, Plenum Press, New York.

Jaeger, R.G. & C.R. Gabor. 1993. Intraspecific chemical communication by a territorial salamander via the postcloacal gland. Copeia **1993**:1171–1174.

Jaeger, R.G. & W.F. Gergits. 1979. Intra- and interspecific communication in salamanders through chemical signals on the substrate. Anim. Behav. **27**:150–156.

Jaeger, R.G. & S.E. Wise. 1991. A reexamination of the male salamander "sexy faeces hypothesis." J. Herpetol. **25**:370–373.

Jaeger, R.G., R.G. Joseph, & D.E. Barnard. 1981. Foraging tactics of a terrestrial salamander: sustained yield in territories. Anim. Behav. **29**:1100–1105.

Jaeger, R.G., J.M. Goy, M. Tarver, & C.E. Marquez. 1986. Salamander territoriality: pheromonal markers as advertisement by males. Anim. Behav. **34**:860–864.

Johnsen, P.B., J.L. Wellington. 1982. Detection of glandular secretions by yearling alligators. Copeia **1982**:705–708.

Kats, L.B. 1988. The detection of certain predators via olfaction by small-mouthed salamander larvae (*Ambystoma texanum*). Behav. Neural Biol. **50**:126–131.

Kats, L.B. & L.M. Dill. 1998. The scent of death: chemosensory assessment of predation risk by prey animals. Ecoscience. In press.

Kats, L.B., J.W. Petranka & A. Sih. 1988. Antipredator defenses and the persistence of amphibian larvae with fishes. Ecology **69**:1865–1870.

Kats, L.B., J.A. Breeding, K.M. Hanson & P. Smith. 1994. Ontogenetic changes in California newts (*Taricha torosa*) in response to chemical cues from conspecific predators. J. North. Am. Benthol. Soc. **13**:321–325.

Keen, W.H., M.G. McManus & M. Wohltman. 1987. Cover site recognition and sex differences in cover site use by the salamander, *Desmognathus fuscus.* J. Herpetol. **21**:363–365.

Kiesecker, J.M., D.P. Chivers & A.R. Blaustein. 1996. The use of chemical cues in predator recognition by western toad tadpoles. Anim. Behav. **52**:1237–1245.

King, M., D. McCarron, D. Duvall, G. Baxter & W. Gern. 1983. Group avoidance of conspecific but not interspecific chemical cues by prairie rattlesnakes (*Crotalus viridis*). J. Herpetol. **17**:196–198.

Kool, K. 1981. Is the musk of the long-necked turtle, *Chelodina longicollis,* a deterrent to predators? Aust. J. Herpetol. **1**:45–53.

Kramer, M. & U. Fritz. 1989. Courtship of the turtle, *Pseudemys nelsoni.* J. Herpetol. **23**:84–86.

Kruse, K.C. & B.M. Stone. 1984. Largemouth bass (*Micropterus salmoides*) learn to avoid feeding on toad (*Bufo*) tadpoles. Anim. Behav. **32**:1035–1039.

Kubie, J.L., A. Vagvolgyi & M. Halpern. 1978. The roles of the vomeronasal and olfactory systems in the courtship behavior of male snakes. J. Comp. Physiol. Psychol. **92**:627–641.

Kulzer, E. 1954. Untersuchungen über die Schreckreaktion bei Erdkrötenkaulquappen (*Bufo bufo* L.). Z. Tierpsychol. **36**:443–463.

Landreth, H.F. & D.E. Ferguson. 1967. Newt orientation by sun-compass. Nature **215**:516–518.

Lefcort, H. & S.M. Eiger. 1993. Antipredatory behaviour of feverish tadpoles: implications for pathogen transmission. Behaviour **126**:13–27.

Lutterschmidt, W.I., G.A. Marvin & V.H. Hutchinson. 1994. Alarm responses by a plethodontid salamander (*Desmognathus ochrophaeus*): conspecific and heterospecifc "Schreckstoff." J. Chem. Ecol. **20**:2751–2759.

Madison, D.M. 1975. Intraspecific odor preferences between salamanders of the same sex: dependence on season and proximity of residence. Can. J. Zool. **53**:1356–1361.

Madison, D.M. 1977. Chemical communication in amphibians and reptiles. *In:* Chemical Signals in Vertebrates, eds. D. Müller-Schwarze & M.M. Mozell, pp. 135–168, Plenum Press, New York.

Mahmoud, I.Y. 1967. Courtship behavior and sexual maturity in four species of kinosternid turtles. Copeia **1967**:314–319.

Malacarne, G. & R. Curtassa. 1983. Sexual selection in the crested newt. Anim. Behav. **31**:1256–1257.

Malacarne, G. & C. Vellano. 1987. Behavioral evidence of a courtship pheromone in the crested newt, *Triturus cristatus carnifex* Laurenti. Copeia **1987:** 245–287.

Martof, B.S. 1962. Some observations on the role of olfaction among Salientian Amphibia. Physiol. Zool. **35**:270–272.

Marvin, G.A. and V.H. Hutchison. 1995. Avoidance response by adult newts (*Cynops pyrrhogaster* and *Notophthalmus viridescens*) to chemical alarm cues. Behaviour **132**:95–105.

Mason, R.T. 1992. Reptilian pheromones. *In:* Biology of the Reptilia: Hormones, Brain, and Behavior, eds. C. Gans and D. Crews, pp. 114–228, University of Chicago Press, Chicago.

Mason, R.T. 1993. Chemical ecology of the red-sided garter snake. *Thamnophis sirtalis parietalis.* Brain, Behav. Evol. **41**:261–268.

Mason, R.T. & W.H.N. Gutzke. 1990. Sex recognition in the Leopard gecko, *Eublepharis macularius* (Sauria: Gekkonidae): possible mediation by skin-derived semiochemicals. J. Chem. Ecol. **16**:27–36.

Mason, R.T., H.M. Fales, T.H. Jones, L.K. Pannell, J.W. Chinn & D. Crews. 1989. Sex pheromones in snakes. Science **245**:290–293.

Mason, R.T., T.H. Jones, H.M. Fales, L.K. Pannell & D. Crews. 1990. Characterization, synthesis, and behavioral response to the sex attractiveness pheromones of the red-sided garter snake, *Thamnophis sirtalis parietalis.* J. Chem. Ecol. **16**:27–36.

Mathis, A. 1990. Territorial salamanders assess sexual and competitive information via chemical signals. Anim. Behav. **44**:953–962.

Mathis, A. & R.J.F. Smith. 1993. Fathead minnows, *Pimephales promelas,* learn to recognize northern pike, *Esox lucius,* as predators on the basis of chemical stimuli from minnows in the pike's diet. Anim. Behav. **46**:645–656.

Mathis, A. & R.R. Simons. 1993. Size-dependent responses of resident male red-backed salamanders to chemical stimuli from conspecifics. Herpetologica **50**:335–344.

Mathis, A., R.G. Jaeger, W.H. Keen, P.K. Ducey, S.C. Walls, & B.W. Buchanan. 1995. Aggression and territoriality by salamanders and a comparison with the territorial behavior of frogs. *In:* Amphibian Biology, Vol. 2, Social Behaviour, eds. H. Heatwole & B.K. Sullivan, pp. 633–676, Surrey Beatty and Sons, Chipping Norton, NSW, Australia.

McCollum, S.A. & J. Van Buskirk. 1996. Costs and benefits of a predator-induced polyphenism in the gray treefrog *Hyla chrysoscelis.* Evolution **50**:583–593.

McGavin, M. 1978. Recognition of conspecific odors by the salamander *Plethodon cinereus.* Copeia **1978**:356–358.

McIlhenny, E.A. 1935. The Alligator's Life History, Christopher Publishing House, Boston.

McVey, M.E., R.C. Zahary, D. Perry & J. MacDougal. 1981. Territoriality and homing behavior in the poison dart frog (*Dendrobates pumilio*). Copeia **1981**:1–8.

Murata, Y., T.H. Jones, L.K. Pannell, H. Yeh, H.M. Fales & R.T. Mason. 1991. New ketodienes from the integumental lipids of the Guam brown tree snake, *Boiga irregularis.* J. Nat. Products **54**:233–240.

Nilson, G. & C. Andrén. 1982. Function of renal sex secretion and male hierarchy in the adder, *Vipera berus,* during reproduction. Horm. Behav. **16**:404–413.

Noble, G.K. 1937. The sense organs involved in the courtship of *Storeria, Thamnophis,* and other snakes. Bull. Am. Mus. Nat. Hist. **73**:673–725.

Noble, G.K. & H.J. Clausen. 1936. The aggregation behavior of *Storeria dekayi* and other snakes, with especial reference to the sense organs involved. Ecol. Monogr. **6**:271–316.

O'Hara, R.K. & A.R. Blaustein. 1981. An investigation of sibling recognition in *Rana cascadae* tadpoles. Anim. Behav. **29**:1121–1126.

Ogilvie, P.W. 1966. An anatomical and Behavioral Investigation of a Previously Undescribed Pouch Found in Certain Species of the Genus *Chameleo*. Ph.D. dissertation, University of Oklahoma, Norman.

Oldham, R.S. 1966. Spring movements in the American toad, *Bufo americanus*. Can. J. Zool. **44**:63–100.

Oldham, R.S. 1967. Orienting mechanisms of the green frog, *Rana clamitans*. Ecology **48**:477–491.

Ovaska, K. 1989. Pheromonal divergence between populations of the salamander *Plethodon vehiculum* in British Columbia. Copeia **1989**:770–775.

Ovaska, K. & T.M. Davis. 1992. Faecal pellets as burrow markers: intra- and interspecific odour recognition by western plethodontid salamanders. Anim. Behav. **43**:931–939.

Parker, W.S. & W.S. Brown. 1973. Species composition and population changes in two complexes of snake hibernacula in northern Utah. Herpetologica **29**:319–326.

Parzefall, J., J.P. Durand & B. Richard. 1980. Chemical communication in *Necturus maculosus* and his cave-living relative *Proteus anguinus* (Proteidae, Urodela). Z. Tierpsychol. **53**:133–138.

Patterson, R. 1971. Aggregation and dispersal behavior in captive *Gopherus agassizi*. J. Herpetol. **5**:214–216.

Peterson, J.A. & A.R. Blaustein. 1991. Unpalatability in anuran larvae as a defense against natural salamander predators. Ethol. Ecol. Evol. **3**:63–72.

Peterson, J.A. & A.R. Blaustein. 1992. Relative palatabilities of anuran larvae to natural aquatic insect predators. Copeia **1992**:577–584.

Petranka, J.W. 1989. Response of toad tadpoles to conflicting chemical stimuli: predator avoidance versus "optimal" foraging. Herpetologica **45**:283–292.

Petranka, J.W., L.B. Kats & A. Sih. 1987. Predator-prey interactions among fish and larval amphibians: use of chemical cues to detect predatory fish. Anim. Behav. **35**:420–425.

Pfeiffer, W. 1966. Die Verbreitung der Schreckreaktion bei Kaulquappen und dei Herkunft des Schreckstoffes. Z. vergl. Physiol. **52**:79–98.

Pfennig, D.W., Sherman, P.W. & J.P. Collins. 1994. Kin recognition and cannibalism in polyphenic salamanders. Behav. Ecol. **5**:225–232.

Phillips, J.A. & A.C. Alberts. 1992. Naive ophiophagus lizards recognize and avoid venomous snakes using chemical cues. J. Chem. Ecol. **18**:1775–1783.

Porter, R.H. & J.A. Czaplicki. 1974. Responses of water snakes (*Natrix r. rhombifera*) and garter snakes (*Thamnophis sirtalis*) to chemical cues. Anim. Learn. Behav. **2**:129–132.

Price, A.H. & J.L. LaPointe. 1981. Structure-functional aspects of the scent gland in *Lampropeltis getulus splendida*. Copeia **1981**:138–146.

Rose, F.L. 1970. Tortoise chin gland fatty acid composition: behavioral significance. Comp. Biochem. Physiol. **32**:577–580.

Ross, P. & D. Crews. 1977. Influence of the seminal plug on mating behavior in the garter snake. Nature **267**:344–345.

Ross, P. & D. Crews. 1978. Stimuli influencing mating behavior in the garter snake, *Thamnophis radix.* Behav. Ecol. Sociobiol. **4**:132–142.

Roudebush, R.E. & D.H. Taylor. 1987. Chemical communication between two species of desmognathine salamanders. Copeia **1987**:744–748.

Schwenk, K. 1985. Occurrence, distribution, and functional significance of taste buds in lizards. Copeia **1985**:91–101.

Seigel, R.A. 1980. Courtship and mating behavior of the diamondback terrapin *Malaclemys terrapin tequesta.* J. Herpetol. **14**:420–421.

Shoji, T. & K. Kurihara. 1991. Sensitivity and transduction mechanisms of responses to general odorants in turtle vomeronasal system. J. Gen. Physiol. **98**:909–919.

Sih, A. & L.B. Kats. 1994. Age, experience, and the response of streamside salamander hatchlings to chemical cues from predatory sunfish. Ethology **96**:253–259.

Sih, A. & R.D. Moore. 1993. Delayed hatching of salamander eggs in response to enhanced larval predation risk. Am. Nat. **142**:947–960.

Sih, A., L.B. Kats & R.D. Moore. 1992. Effects of predatory sunfish on the density, drift, and refuge use of stream salamander larvae. Ecology **73**:1418–1430.

Simon, C.A. 1983. A review of lizard chemoreception. *In:* Lizard Ecology: Studies of a Model Organism, eds. R.B. Huey, E.R. Pianka & T.W. Schoener, pp. 119–133, Harvard University Press, Cambridge, MA.

Simon, G.S. & D.M. Madison. 1984. Individual recognition in salamanders: cloacal odours. Anim. Behav. **32**:1017–1020.

Simons, R.R. & B.E. Felgenhauer. 1992. Identifying areas of chemical signal production in the red-backed salamander, *Plethodon cinereus.* Copeia **1992**:776–781.

Simons, R.R., B.E. Felgenhauer & R.G. Jaeger. 1993. A technique to explore the effects of integumental glands on the behavior of salamanders. J. Herpetol. **27**:99–100.

Simons, R.R., B.E. Felgenhauer & R.G. Jaeger. 1994. Salamander scent marks: site of production and their role in territorial defence. Anim. Behav. **48**:97–103.

Sinsch, U. 1990. Migration and orientation in anuran amphibians. Ethol. Ecol. Evol. **2**:65–79.

Stauffer, H.-P. & R.D. Semlitsch. 1993. Effects of visual, chemical and tactile cues of fish on the behavioural responses of tadpoles. Anim. Behav. **46**:355–364.

Thoen, C., D. Bauwens & R.F. Verheyen. 1986. Chemoreceptive and behavioral responses of the common lizard *Lacerta vivipara* to snake chemical deposits. Anim. Behav. **34**:1805–1813.

Tinkle, D.W. 1967. The life and demography of the side-blotched lizard, *Uta stansburiana.* Misc. Publ. Mus. Zool. Univ. Mich. **132**:1–182.

Tollestrup, K. 1981. The social behavior and displays of two species of horned lizards, *Phrynosoma platyrhinos* and *Phrynosoma coronatum.* Herpetologica **37**:130–141.

Tracy, C.R. & J.W. Dole. 1969. Orientation of displaced California toads, *Bufo boreas,* to their breeding sites. Copeia **1969**:693–700.

Tristram, D.A. 1977. Intraspecific olfactory communication in the terrestrial salamander *Plethodon cinereus.* Copeia **1977**:597–600.

Twitty, V.C. 1959. Migration and speciation in newts. Science **130**:1735–1743.

Twitty, V.C. 1955. Field experiments on the biology and genetic relationships of the California species of *Triturus.* J. Exp. Zool. **129**:129–148.

Uzendoski, K. & P. Verrell. 1993. Sexual incompatibility and mate-recognition systems: a study of two species of sympatric salamanders (Plethodontidae). Anim. Behav. **46**:267–278.

Van Damme, R. & A.M. Castilla. 1996. Chemosensory predator recognition in the lizard *Podarcis hispanica:* effects of predation pressure relaxation. J. Chem. Ecol. **22**:13–22.

Van Damme, R., D. Bauwens, C. Thoen, D. Vanderstighelen & R.F. Verheyen. 1995. Responses of naive lizards to predator chemical cues. J. Herpetol. **29**:38–43.

Verrell, P.A. 1985. Male mate choice for large, fecund females in the red-spotted newt, *Notophthalmus viridescens:* how is size assessed? Herpetologica **41**:382–386.

Verrell, P.A. 1988. The chemistry of sexual persuasion. New Sci. **118**:40–43.

Verrell, P.A. 1989. An experimental study of the behavioral basis of sexual isolation between two sympatric plethodontid salamanders, *Desmognathus imitator* and *D. ochrophaeus.* Ethology **80**:274–282.

von Frisch, K. 1938. Zur Psychologie des Fisch-Schwarmes. Naturwissenschaften **26**:601–606.

von Frisch, K. 1941. Über einen Schreckstoff der Fischhaut und seine biologische Bedeutung. Z. vergl. Physiol. **29**:46–145.

Voris, H.K. & J.P. Bacon. 1966. Differential predation on tadpoles. Copeia **1966**:594–598.

Waldman, B. 1985. Olfactory basis of kin recognition in toad tadpoles. J. Comp. Physiol. A **156**:565–577.

Walls, S.C., A. Mathis, R.G. Jaeger & W.F. Gergits. 1989. Male salamanders with high quality diets have faeces attractive to females. Anim. Behav. **38**:546–548.

Walters, B. 1975. Studies of interspecific predation within an amphibian community. J. Herpetol. **9**:267–279.

Wassersug, R.J. 1971. On the comparative palatability of some dry-season tadpoles from Costa Rica. Am. Midl. Nat. **86**:101–109.

Waye, H.L. & P.T. Gregory. 1993. Choices of neonate *Thamnophis elegans vagrans* between conspecific, congeneric, and heterogeneric odors. J. Herpetol. **27**:435–441.

Weaver, W.G. 1970. Courtship and combat behavior in *Gopherus berlandieri.* Bull. Fla. State Mus. Biol. Sci. **15**:1–43.

Weldon, P.J. 1982. Responses to ophiophagous snakes by snakes of the genus *Thamnophis.* Copeia **1982**:788–794.

Weldon, P.J. & G.M. Burghardt. 1979. The ophiophage defensive response in Crotaline snakes: extension to new taxa. J. Chem. Ecol. **5**:141–151.

Weldon, P.J. & D.B. Fagre 1989. Responses by canids to scent gland secretions of the western diamondback rattlesnake (*Crotalus atrox*). J. Chem. Ecol. **15**:1589–1604.

Weldon, P.J., B.S. Dunn, Jr., C.A. McDaniel & D.I. Werner. 1990. Lipids in the femoral gland secretions of the green iguana (*Iguana iguana*). Comp. Biochem. Physiol. **95B**:541–543.

Werner, D.I. 1978. On the biology of *Tropidurus delanonis,* Baur (Iguanidae). Z. Tierpsych. **47**:337–395.

Werner, D.I., E.M. Baker, E. del C. Gonzalez & I.R. Sosa. 1987. Kinship recognition and grouping in hatchling green iguanas. Behav. Ecol. Sociobiol. **21**:83–89.

Wilson, D.J. & H. Lefcort. 1993. The effects of predator diet on the alarm response of red-legged frog, *Rana aurora,* tadpoles. Anim. Behav. **46**:1017–1019.

7

Bioassays for Mammals and Birds

Dale L. Nolte and J. Russell Mason

7.1.	Introduction	327
7.2.	Chemical senses	327
	7.2.1. Olfaction	327
	7.2.2. Vomeronasal chemoreception	328
	7.2.3. Trigeminal chemoreception	328
	7.2.4. Terminal nerve	329
	7.2.5. Septal organ	329
	7.2.6. Taste	329
7.3.	Test paradigms	330
	7.3.1. Naturalistic observations	331
	7.3.2. Orientating responses	331
	7.3.3. Choice tests	331
	7.3.4. Flavor-avoidance learning	333
	7.3.5. Operant conditioning	335
	7.3.6. Single-subject designs	336
7.4.	Experimental apparatus	337
	7.4.1. Feed trays	337
	7.4.2. Drinking tubes	337
	7.4.3. Arena	338
	7.4.4. Mazes	339
7.5.	Intraspecific behaviors	339
	7.5.1. Detection and recognition	342
	7.5.1.1. Mammal bioassays	342
	7.5.1.2. Bird bioassays	346
	7.5.2. Reproduction	349
	7.5.3. Dominance	351
	7.5.4. Territorial behaviors	352
7.6.	Interspecific behaviors	355
	7.6.1. Resource exploitation	356
	7.6.1.1. Mammal bioassays	356

7.6.1.2. Fetal mammal bioassays 360
7.6.1.3. Lactating mammal bioassays 362
7.6.1.4. Bird bioassays 363
7.6.2. Predator avoidance (defense) 365
7.6.3. Community or mutualistic interactions 368
7.7. Case studies 369
7.7.1. Chemical correlates and sensory perception 369
7.7.2. Responses of birds to aversive stimuli 370
7.7.3. Relating repellency and chemical structure 371
7.7.4. Conspecifics response to scent marks 373
7.7.5. Aversive components of predator odors 374
7.7.6. Chemically mediated foraging by black bears 376
7.8. Summary 378
7.9. Acknowledgments 378
7.10. References 378

7.1. Introduction

The best possible outcome of a bioassay is the elicitation of a specific behavior pattern by a defined chemical stimulus (Müller-Schwarze 1977). This chapter presents methods to assess the role of chemical stimuli in the behavior of birds and mammals. An overview of the chemosensory capacities of birds and mammals to perceive stimuli is provided first, followed by a description of test paradigms that can be incorporated into bioassays. The preponderance of the chapter describes examples of these paradigms incorporated into bioassays to assess the role of chemical signals in inter- and intra-specific behaviors. The examples, like past studies in mammalian and avian chemical ecology, are disproportionately represented by rodents. Most of the described procedures, however, could be modified for other species, provided the apparatuses and the response variables can be altered to accommodate the different subject and stimuli.

7.2. Chemical Senses

Olfaction is typically used as a label for all of the chemical senses with the exception of taste. This is an excessively general use of the term, and it leads to the misinterpretation of results. For example, mislabeling trigeminally mediated responding as olfactory is an error similar in magnitude and implication to mislabeling visually mediated behavior as auditory. As we discuss below, a variety of other sensory systems in the nasal and oral cavities respond to chemical stimuli and mediate different categories of behavior.

7.2.1. Olfaction

Olfaction is commonly called the sense of smell, and the system is designed to detect relatively low molecular weight (i.e., volatile) substances. This sense can perform quality coding, and receptors located in the nose make monosynaptic connections with the olfactory bulbs of the forebrain (Shepherd 1985). Unlike other sensory receptors, olfactory neurons of birds and mammals are replaced every 30–60 days (Graziadei & Monti-Graziadei 1980). Despite this rapid replacement, perception apparently remains unaffected and olfactory receptors respond to innumerable odors, sometimes with exposure-induced changes in sensitivity (Wysocki & Beauchamp 1991). Because of these features, some have speculated olfaction and the immune system may share certain similarities (Yamazaki et al. 1991). Perhaps because the olfactory bulbs project directly to midbrain structures, odor memories are more emotional than memories mediated by other sensory systems (Herz & Engen 1996).

The principal biological function of olfaction for most mammals appears to be the discovery of foods and the avoidance of odors associated with negative consequences. However, the system probably has other roles. For example, odors such as androsterone may have sexual significance for a variety of mammals including humans (Wysocki & Beauchamp 1991).

7.2.2. Vomeronasal Chemoreception

Vomeronasal chemoreception controls a variety of biologically important behaviors. Like olfaction, the vomeronasal system performs quality coding. Typically, stimuli are of a social nature, and vomeronasal activity induces hormonal changes, affects the success of pregnancy, alters the course of puberty, modulates female cyclicity and ovulation, elicits courtship and attraction, and modulates reproductive behavior and aggression (Wysocki & Meredith 1987). Depending on the species, the vomeronasal organ opens into the nasal cavity, the roof of the mouth, or the nasopalatine canal (Wysocki & Meredith 1987). The organ detects and discriminates among relatively high molecular weight (i.e., chiefly nonvolatile) substances, and molecules as large as 60,000 Da are effective stimulants (Wysocki et al. 1985). Chemicals may be transported to the organ via specialized structures (e.g., the nasopalatine canals), and in some species, a vascular pump can be used to suck materials into the organ and then expel them (Meredith 1982). An important physiological difference between mammals and birds is that none of the avian species examined to date have had a vomeronasal organ (Kare & Mason 1986).

7.2.3. Trigeminal Chemoreception

Trigeminal chemoreception mediates the perception of chemical pain, as well as thermal, tactile, and proprioceptive cues (Green et al. 1990). This review is limited to the discussion of chemical pain, the kind of stimulation that humans describe as burning, itching, or tingling (Rozin et al. 1982). Trigeminal nociceptors are free nerve endings that can be found in the mouth and nose, as well as exposed mucus membranes throughout the body (Silver 1987). Although the system most commonly responds to high concentrations of chemicals (Green et al. 1990), it can respond to substances presented at concentrations well below those that stimulate olfaction or taste (Tucker 1971). Trigeminal stimuli may be volatile (e.g., ammonia) or nonvolatile (e.g., capsaicin); lipophyllicity is important. Virtually any substance will stimulate trigeminal receptors if it is presented at a sufficiently high concentration (e.g., Prah & Benignus 1984). This and other evidence suggests that unlike olfaction, vomeronasal chemoreception, and taste, the trigeminal pain receptors respond to stimulus quantity rather than stimulus quality (Silver et al. 1989). Trigeminal stimulation can elicit a variety of protective physiological reflexes that minimize exposure to noxious chemicals. Such reflexes are among the strongest in the body (James & Daly 1969). Obviously, because trigeminal reflexes are strong and invariant, they can influence the responding of animals and confound the interpretation of experimental results.

Recent studies suggest that strong trigeminal stimuli are substances that mimic endogenous chemicals (e.g., bradychinins) released during tissue damage (Clark 1995). Curiously, there are profound taxonomic differences between avian and mammalian trigeminal chemoreception. Mammalian irritants are rarely aversive to birds, and vice versa (Mason et al. 1991b).

7.2.4. *Terminal Nerve*

Terminal nerve (nervus terminalis) receptors are chemically receptive, and present in both birds and mammals. The biological function of this system remains obscure, although it probably mediates some reproductive behaviors (Wysocki & Meredith 1987). In some fishes, electrical stimulation of the terminal nerve causes sperm release (Demski & Northcutt 1983). Because the terminal nerve is severed in ablation studies of vomeronasal function, it is conceivable that some behaviors attributed to the vomeronasal system are actually mediated by the terminal nerve (Wysocki & Meredith 1987).

7.2.5. *Septal Organ*

Septal organ function, like that of the terminal nerve, is obscure. The organ is apparently present in both mammals and birds. Little information is available on the biological importance of this system, although it may serve to trigger sniffing behavior in canines (D.A. Marshall, personal communication, 1985).

7.2.6. *Taste*

Taste detects and discriminates among four to six qualities (i.e., sweet, starch, sour, salty, bitter, umami or protein-taste) depending on the species under examination (Kuwamura & Kare 1987; Sclafani 1991). Obligate mammalian carnivores often lack the ability to detect sweet mono- and disaccharide sugars. Apparently the sweet receptor in these animals is modified for the detection of free amino acids. Birds show poor taste sensitivity when compared with mammals (Kare & Mason 1986). Although birds have a higher salivary flow rate than mammals, they have relatively fewer taste buds.

7.3. Test Paradigms

Regardless of the field of study, a principal difference between humans and other vertebrates is that humans can use language to describe the sensory and hedonic properties of stimuli. To understand the sensory universe of other animals, inferences are drawn from learned and unlearned behaviors exhibited in relation to stimulus presentations. Clearly, such inferences are conditional because behavior is context dependent. The method or methods selected to address a sensory question invariably will determine the quality and generalizability of the answer obtained. As an illustration of this point, birds are notoriously insensitive to mammalian irritants including capsaicin (Norman et al. 1992). In the absence of training, birds will ingest feed containing capsaicin as though the irritant was undetected. If consumption, however, is followed by lithium chloride-induced malaise, capsaicin adulterated food is subsequently avoided, showing that the substance is detected but not perceived as an irritant (Mason & Clark 1995, 1996).

The complexity of a testing paradigm influences the generalizability of the results. In simple laboratory tests, a variety of chemicals appear at least moderately aversive (Mason et al. 1989). Under field conditions, however, even weak avoidance of the same materials may not be observed (M.L. Avery, personal communication, 1988). Most likely, the laboratory setting lacks critical features present under more natural conditions. Variables that influence the behavior of captive subjects include differences among species (Schafer & Brunton 1971; Schafer et al. 1983; Kononen et al. 1986, 1987; Espaillat & Mason 1990; Mason & Bonwell 1993; Mason et al. 1993), sex (Espaillat & Mason 1990), age (Hakim et al. 1996), group size (Kononen et al. 1986), social interactions (Mason & Reidinger 1982), previous experience (Starr et al. 1964; Greig-Smith 1987), type of treated food and prior food deprivation (Thompson et al. 1981), color of the treated food (Greig-Smith & Rowney 1987; Mason & Reidinger 1983), the type of alternative untreated food available (Avery et al. 1995; Rogers 1974), the number of choices available (Bennett 1989a, 1989b), and ambient temperature (Avery et al. 1994). The presence of grit may also be important to birds that require it to process foods (Gionfriddo & Best 1995).

More broadly, the evaluation of chemical cues is inherently more difficult than the evaluation of other sensory stimuli because the stimuli are difficult to control. Relative to visual or auditory cues, chemicals can not be applied with high precision (Lawless 1987).

A variety of methods are available for the study of chemosensory abilities.

These range from naturalistic observations in field studies and large enclosures (e.g., Avery & Decker 1991; Cummings et al. 1991) to laboratory feeding tests (e.g., Mason et al. 1989) and studies involving operant and classical conditioning (Passe & Walker 1985). Conn (1993) provides an overview of paradigms for the study of behavior.

7.3.1. Naturalistic Observations

Naturalistic observations can be useful when animals respond in a stereotypic manner to the presentation of stimuli (Shallenberger 1975). In open field tests, animals are placed in a rectangular or circular arena. Circular arenas are preferable, especially when using species that are prone to linger in corners of the field. Often, the space within the arena is divided into a grid, and the animal's movement around the field can be monitored by video camera, photoelectric beams, or motion detectors. If marking behaviors (e.g., the deposition of urine to signal territory boundaries) are being studied, then inspecting the apparatus with a black light is useful because urine will fluoresce under ultraviolet illumination. Open field tests are typically used to study territorial and aggressive behaviors with rodents, and they can be applied to the study of place avoidance or preference, for example, in response to predator odors (Epple et al. 1995). While open field tests are easy to perform, and relatively quick, control of stimulus presentations is difficult. Open field testing is not recommended if information about the precise sensory capacities of an animal is desired.

7.3.2. Orientating Responses

Other methodologies use alerting or orienting responses as measures of stimulus detection and discrimination (Wenzel 1972). Changes in respiratory volume or heart rate can be recorded before and during stimulus presentations, and such changes during exposure to stimuli is taken as an indication of detection. The principal disadvantage to the use of orienting responses is that subjects habituate after multiple stimulus presentations (i.e., neophobia and consequent orientation to the formerly novel stimuli tend to diminish). Accordingly, the gradually increasing failure to respond as a result of stimulus habituation can be confused with failure to respond as a result of nondetection. Habituation can sometimes be reduced by increasing the time intervals (e.g., from hours to days or weeks) between test sessions.

7.3.3. Choice Tests

Choice tests are a simple extension of naturalistic observations. Feeding, drinking, or other behaviors can be used as dependent measures. For the evaluation of taste, drinking tests are more sensitive than feeding tests for at least two reasons. First, tastants in fluid are more accessible to receptors than tastants in food.

Second, the presence of extraneous taste cues is easier to control when stimuli are presented in a fluid vehicle. For the evaluation of odors, feeding tests are often superior to drinking tests. This is because volatile stimuli are more easily expressed in the typical feeding context (e.g., a covered dish) than in a drinking context (e.g., stimulus presentation in a drinking tube with a small opening).

The characteristics of stimuli must be considered when deciding between drinking and feeding tests as a method for sensory evaluation. Hydrophobic substances may be easier to use in feeding tests because they are easier to mix thoroughly with a diet. Such substances are difficult to keep in solution or suspension for drinking tests without the use of emulsifying agents (e.g., dimethyl sulfoxide, Tween®-80). Emulsifiers typically have aversive (e.g., bitter) taste characteristics that interfere with the evaluation of stimuli (Silver et al. 1985).

Regardless of whether drinking or feeding tests are used, a variety of protocols exist. The simplest is the one-choice test. Ingestion (feeding or drinking) is the typical dependent measure. Animals are presented with a sample of food or fluid for a period of time, and consumption is indirectly measured as the difference between the starting and ending weight or volume of stimulus presented during the trial. Spillage is sometimes measured in addition to consumption, although spillage typically tracks consumption, per se (Mason et al. 1991c). Animals are generally deprived prior to test to ensure consumption, with the length of deprivation dependent on the species. Comparison between pretreatment and treatment consumption is taken as a measure of the response to the stimuli. Short duration tests (15 min to 1 h) are taken to reflect sensory events, while longer-duration tests (2–6 h) involve both sensory and postingestive effects. In all likelihood, even short-duration tests involve some postingestive feedback (Feldman & Richardson 1986; Provenza 1995).

Two-choice tests involve the presentation of two stimuli, either a treated and an untreated sample, or less commonly, two treated samples. Again, after a period of time, consumption (and/or spillage) is measured. When a series of treated samples are being compared in two-choice tests, completely counterbalanced or Latin-square designs are used. Counterbalancing is important because experience with successive stimuli influences the results. Preference relationships among stimuli in pairwise comparisons may be intransitive, i.e., stimulus "a" may be preferred to stimulus "b," and stimulus "b" may be preferred to stimulus "c," but it is not necessarily true that stimulus "a" is preferred to stimulus "c." The results of two-choice tests often are expressed as "preference ratios" or "difference scores." With the former, consumption of one food is divided by total consumption of both foods in a test session. A preference ratio of 0.5 represents indifference for the food in the numerator of the equation, while quotients of 1.0 and 0.0 represent absolute preference and complete rejection, respectively (Mason et al. 1989). With difference scores, mean consumption of one food is subtracted from mean consumption of the other. A score of 0 indicates indifference between foods, while scores larger than 0 reflect preference for the food in the numerator of the ratio (Mason & Otis 1990). When avoidance responding is measured, the term "suppression ratio" is substituted for the term "preference ratio" (Mason et al. 1989).

Whether to employ two-choice or one-choice tests depends on the question. Two-choice tests are more sensitive for detecting differences. If the question is whether a stimulus exerts strong effects (and not merely if it is detected), then one-choice designs may be better. If possible, both one- and two-choice tests should be performed to elucidate the characteristics of the response of the animal to the substances under examination (e.g., Mason et al. 1989).

There may be instances in which it is desirable to simultaneously investigate more than two items. When three stimulus substances are compared, triangle tests may be the best option. In this design, items are placed in containers and the containers are presented in a spatially triangular configuration. Over the course of testing, each of the stimulus substances is presented in each of the containers. This ensures that all items are sampled and guards against position biases that might otherwise confound results.

It is unwise to present more than three stimuli simultaneously because animals may fail to sample all of the items. A better design option is to use a series of counterbalanced two-choice tests, as described above.

A serious drawback of naturalistic observations and simple one- and two-choice tests is that these methods depend on untrained responding. Simply because an animal fails to respond is not evidence that it has failed to detect a stimulus. For this reason, training procedures have been developed. Typically, these methods employ classical or operant conditioning. With the former, the presentation of a stimulus is paired with the delivery of a reinforcer. For example, presentation of an odor is paired with the delivery of electric shock. The shock leads to an unconditioned increase in heart rate or a measurable change in some other physiological response. After several pairings, presentations of odor alone lead to conditional or conditioned increases in heart rate. The concentration of the odor stimulus can then be manipulated, and questions can be asked about minimum detectable concentrations etc. Not infrequently, greater stimulus sensitivities are measured when trained animals are used than when data are collected from untrained animals in simple choice tests.

7.3.4. *Flavor-Avoidance Learning*

A related methodology is food- or flavor-avoidance learning. Animals are presented with a food or flavor (conditional or conditioned stimulus), and ingestion is paired with sickness (unconditional or unconditioned stimulus). Following this aversive experience, animals avoid the conditional stimulus, and this learned avoidance tends to generalize to other foods or flavors, depending on the degree to which these other substances resemble the material ingested before the onset of malaise (Stewart et al. 1983). The strength of the conditional response can be influenced by a number of factors, including the novelty of the food or flavor (the greater the novelty, the greater the avoidance response), the magnitude of the induced malaise (the greater the sickness, the greater the avoidance response), and the amount of time intervening between ingestion and the onset of sickness (the shorter the period, the greater the avoidance response). Expression is also affected by the subsequent use of one- versus two-choice tests (Dragoin et al. 1971). Aversions are expressed more strongly and for longer periods of time when the latter test paradigm is employed.

The use of food-avoidance learning can be extended from chemosensory evaluations to the evaluation of visual perception with birds (Reidinger & Mason 1983). Unlike mammals, birds tend to treat the visual characteristics of food like tastes. For mammals, avoidance of any characteristic of the conditional stimulus other than taste (e.g., odor, visual cues, texture) is relatively weak, and tends to be a function of higher-order conditioning (Garcia & Hankins 1977). That is, avoidance of these other cues depends on their association with the conditioned taste. While gastrointestinal sickness may be the most effective means of inducing food avoidance learning, any substance that induces malaise or has a psychogenic consequence (e.g., amphetamines, alcohol) can elicit food or flavor avoidance. The reader is referred to extensive published bibliographies on food avoidance learning for further information (e.g., Riley & Clarke 1977; Riley & Tuck 1985, Garcia & Riley, in press).

Examination of flavor-avoidance learning can be coupled with physiological measurements to provide additional useful information. Mason et al. (1985) paired grooming of saccharin-flavored carboxymethylcellulose gel with lithium chloride injections to induce sickness, to demonstrate that rodents ingest substances from their fur while grooming that they would otherwise not eat or drink. Subsequent to conditioning, voles refused to drink saccharine solutions or groom saccharin-flavored gel from cage mates postconditioning, but they continued to ingest flavored gel while grooming their own fur. Blood samples were then collected from all animals. The failure of avoidance in autogrooming was associated with elevated blood corticosteroids. Rodents tend to increase grooming in stressful contexts (Cohen & Price 1979). Apparently, the voles became trapped in a positive feedback loop during autogrooming (i.e., saccharin detection during grooming induced stress, and stress in turn induced further grooming).

Food preferences also can be learned, and, like learned aversions, conditioned preferences can be used to evaluate chemosensory abilities. Typically, preferences are induced by pairing the ingestion of a novel flavor with calories (Bolles et al. 1981; Messier & White 1984; Booth 1985; Mehiel & Bolles 1984, 1988; Simbayi et al. 1985), recovery from nutritional deficiency (Garcia et al. 1967; Zahorik et al. 1974), and recovery from malaise (Green & Garcia 1971; Zahorik 1977; Sherman et al. 1983). The novel flavor is subsequently preferred, as are other flavors that subjects perceive as similar to the conditioned stimulus. However, because conditioned food preferences are more ephemeral than conditioned food aversions, they are employed less frequently in studies of sensory function.

7.3.5. Operant Conditioning

In operant conditioning, animals are trained to show a voluntary response or operant (e.g., a bar or treadle press) to obtain positive reinforcement (e.g., food or water delivery) or avoid negative reinforcement (e.g., electric shock). Once the basic operant is established, so-called contingencies are implemented such that reinforcement is delivered only when the operant is exhibited in the presence of another stimulus (e.g., odorant, light, tone). Responding in the absence of this stimulus typically leads to a delay in the delivery of the reinforcer (i.e., the animal is "encouraged" to respond only when it detects the presence of the reinforced stimulus). For a detailed explanation of free-operant learning, the reader is referred to the discussions in Mackintosh (1974).

A critical feature in successful operant conditioning is the selection of a response that the species under study readily performs (Breland & Breland 1961, 1966). In general, the selected response should be exhibited at a reasonable rate in the absence of training. Such behaviors are much easier to shape and use as operants. Perch-hopping, for example, is a usable response for various passerine birds. Key-pecking, on the other hand, is a response that is difficult to establish. Key-pecking is useful with poultry and pigeons (perhaps because these birds peck while feeding), while lever-pressing is rapidly acquired by rats and dogs. Choosing an operant that an animal is "unprepared" to make or that is exhibited at a low frequency leads to unnecessarily difficult training and response maintenance. The classic studies of "prepared" versus "unprepared" responding are those of Breland & Breland (1961, 1966). In attempting to train animals to perform various tricks, they discovered that although a response could be established, the behavior of animals became progressively less efficient despite the maintenance of the instrumental contingency. When pigs were trained to pick up a token and drop it into a container (i.e., insert a coin into a piggy bank), the animals would drop the token, root it along the way, pick it up and flip it into the air, drop it, root it some more, etc. Raccoons trained to perform the same task would spend time rubbing the tokens together instead of inserting them into the container. Needless to say, these diversionary behaviors persisted in spite of substantial delays in reinforcement. In other cases, aspects of a desired response could not be shaped, apparently because of species-selective constraints. Breland and Breland (1966) reported that although a cow could be trained to chase a man and engage in a bullfight, no amount of food deprivation would induce cows to run (rather than walk). Limits on response acquisition have been termed "instinctive drift." Learned behavior is said to "drift" toward instinctual or unlearned behavior. In the context of learning theory, Mackintosh (1974) argues that the reinforcer used in any instrumental context is also a potent unconditioned stimulus. A set of reflexive responses is elicited by this unconditioned stimulus and these reliably accompany its presentation (i.e., the token used in the experiments with pigs and raccoons was both a conditioned and an unconditioned stimulus).

The valence of the reinforcement (positive or alimentary, negative or defensive; Pavlov 1927) also affects the instrumental response. Rate of response and measurable sensory acuity will vary, depending on whether positive (e.g., food, water, access to a sexual partner, pleasurable electrical stimulation of the brain) or negative (electric shock, airpuffs to the cornea, aversive electrical brain stimulation, bright light) reinforcement is used. For some animals, physiology prohibits the use of certain categories of reinforcer (Mackintosh 1974). Electric shock cannot be used with amphibians and reptiles, because responding degrades rapidly until animals become unresponsive to presentations of the unconditioned stimulus. Likewise, food reinforcers cannot be used with animals (e.g., snakes) that feed infrequently (Passe & Walker 1985).

Conditioned suppression is an operant paradigm that is frequently used in studies of sensory function. In this paradigm, animals are trained to exhibit a response, and then presentations of odor or some other stimulus are paired with the delivery of electric shock. In fairly short order, animals cease responding in the presence of the stimulus that signals shock delivery. The intensity of the signaling stimulus is then reduced until the animal fails to suppress responding. Response rates for an equal period of time preceding odor presentation are compared to response rates during odor presentation in a suppression ratio (SR). The ratio is calculated by subtracting pretrial responses from trial responses, and then dividing the result by the pretrial response (Passe & Walker 1985). Threshold for stimulus detection is defined as that concentration or value corresponding to an SR of 0.5. A special advantage of conditioned suppression is that it involves what Mackintosh (1974) describes as preparatory conditioned responding. Such preparatory responses appear to condition more rapidly than so-called consummatory conditioned responses (e.g., salivation, licking) and with longer delays between the onset of the conditioned stimulus and onset of the unconditioned stimulus.

7.3.6. Single-Subject Designs

Although not test paradigms in themselves, single-subject designs are an experimental option when limited resources (e.g., stimuli) or inaccessible subjects (i.e., species generally found only as single individuals in zoos) prohibit a replicated study. Kazdin (1982) provides a detailed overview for the use and limitations of single case research designs. Saudergrass and Drummer (1996) describes the application of single-subject research designs with zoo animals. Briefly, a baseline assessment of the existing level of performance is extrapolated as the predicted performance of the animal in the future, provided that no intervention is applied

(Kazdin 1982). Inferences are then made by comparing changes in response when the subject is exposed to different conditions over time. The validity of inference is enhanced when objective measures (e.g., time spent grooming) are applied; performance is assessed repeatedly at baseline and under experimental conditions; nonexperimental activity levels are stable; and interventions produce obvious, marked changes in behavior (Kazdin 1982). Single subject experiments are not replicated, thus it is inappropriate to extrapolate the responses of individuals to broader patterns characteristic of the species.

7.4. Experimental Apparatus

The choice of experimental apparatus is dictated by the question and experimental design, and the test species. An appropriate apparatus will enable the stimuli to be maintained in its original state, allow normal responses by the subject, and permits unobstructed monitoring of responses. Simple is usually better. The literature describes an array of apparatuses for specific tests with specific subjects. This section describes a few inexpensive test apparatuses for simple chemosensory bioassays.

7.4.1. Feed Trays

Test trays should provide ready access to foods and be sufficiently large to hold enough food for a complete test. Trays also need to be easy to clean between trials or disposable. A variety of trays are commercially available. We have used weigh boats thumbtacked to small blocks of wood as inexpensive but durable trays for test (Fig. 7.1). Weigh boats are available in a variety of sizes and styles, and can be changed easily to accommodate the subject. The tray is attached to the wood to provide stability and to reduce tipping. Costs are minimal, thus new trays can be constructed when contamination is a concern. A small boat tacked inside a larger boat helps to control spillage. Food pulled out of the first falls into the second, and when properly spaced, the lip of the first tray makes food spilled into the second inaccessible. For olfactory trials, subjects are prevented from making direct contact with stimuli by placing the test substance on an absorbent paper folded inside a small perforated plastic container (e.g., histo-prep capsule) tacked in the center of the tray.

7.4.2. Drinking Tubes

Calibrated drinking tubes can be constructed from a syringe and a sipper tube attached with Tygon™ tubing (Fig. 7.2). Size varies with subject, but a 10-ml syringe is generally adequate for mice, while 30- or 60-ml syringes are better for rats and guinea pigs. First, drill out the center of the syringe. This is most easily done on a press drill with a ⅜-in. bit. Next, connect a sipper tube to the

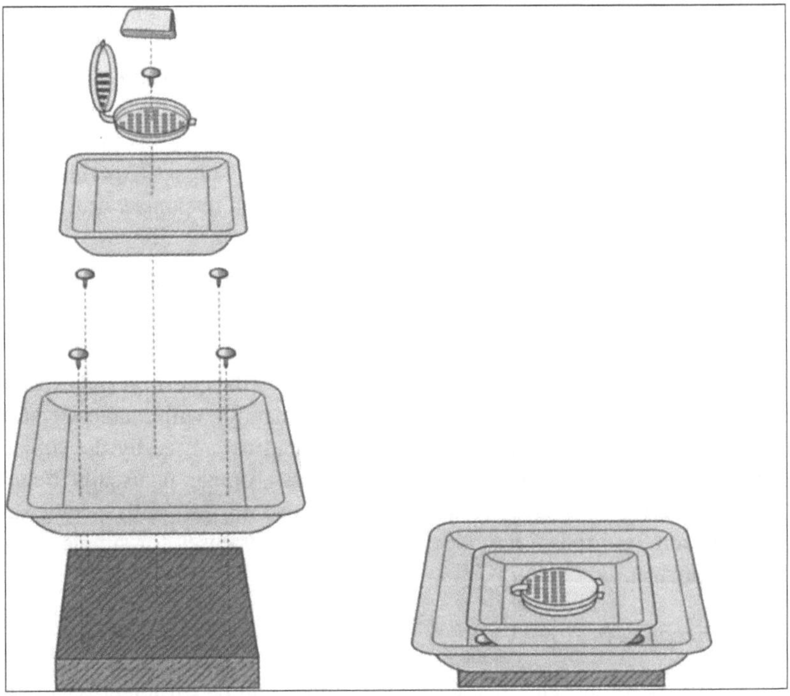

Figure 7.1. Food tray, including an inner tray to reduce spillage and placement of a histo-prep capsule to deliver odor stimuli.

syringe with a piece of Tygon™ tubing. To fill the syringe, insert the sipper tube into the liquid and retract the plunger. The syringe will not fill with liquid the first time because of the added volume of the sipper tube. Turn the syringe upright, depress the plunger until liquid emerges from the sipper tube and then repeat the process.

7.4.3. Arena

Arenas can be made to fit any subject and constructed from a variety of barriers. The need to clean them thoroughly between tests, however, generally restricts their construction to easily washable materials. Paper linings or high gloss oil-based paint helps to ease the burden of cleaning. Rectangular glass tanks found

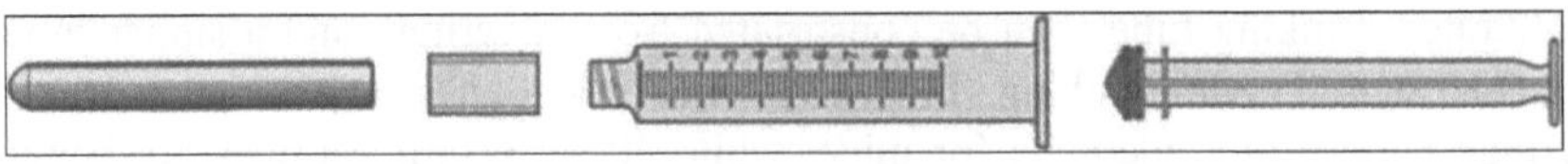

Figure 7.2. Materials to construct a calibrated drinking tube created from a sipper tube attached to a syringe with Tygon™ tubing.

in pet stores are useful for smaller animals. Ten-gallon aquariums are commonly used with small rodents. Tape can be applied to the underside of the tank to delineate halves or to plot a grid. Replace the bottom glass with screen mesh to permit odors to be introduced from below. Observing activities without disturbing the subject can be made easier if the tank is raised onto blocks and a mirror is set at a 45° angle beneath the glass floor. Observations then can be taken from the side rather than looking down into the arena. Alternatively, a video camera can be used for more remote observations. A beaker placed over the subject in the center of the arena can serve as a simple start box. The subject is observed until it orientates itself in the proper direction and then the beaker is lifted to release the animal.

7.4.4. Mazes

Clear PVC pipe and appropriate connectors can be used to construct mazes in which subjects are easily watched. These mazes are easily disassembled for cleaning, or for reconfiguring to suit other experimental designs. Pipe sizes vary, but small rodents (e.g., pocket gophers, voles, and mice) readily move through a 5-cm-diameter pipe; larger subjects, such as mountain beaver require at least a 10-cm and preferably a 15-cm diameter. Plastic kitchen storage containers with sealable lids serve well as starting points and goal boxes. Pipes can be attached to these boxes by fitting them through a hole cut in the side of the container. Split Tygon™ tubing fitted around the edge of the hole helps to seal the connection. Odors can be placed within perforated containers inside the tubes or goal boxes, or allowed to permeate through holes drilled in the wall of the pipe. A slight vacuum will draw odors through the system in the desired direction (e.g., from a goal, exiting through the rear of the start box). The design of the maze obviously depends on the question being addressed. We use a 1-m main run with 1-m arms to create a T-maze to test pocket gopher responses to food flavors (Fig. 7.3A). To investigate responsiveness to aversive odors, we connect the maze arms directly to the start box (Fig. 7.3B). This allows the subject to avoid the stimuli before entering the maze; otherwise, the subject must first pass through possibly aversive odors until it reaches the choice point. When desirable, the complexity of the system can be increased with plastic tee and elbow joints. For example, we developed a system with 120 m of pipe and multiple nest and food caches to assess pocket gopher invasion of territories contaminated with predator odors (Fig. 7.4).

7.5. Intraspecific Behaviors

Chemical signals are crucially important in the social behavior of mammals, and they mediate parental and conspecific recognition, as well as sexual, aggressive, and territorial behaviors (Johnston 1990). Bioassays to assess the role of the

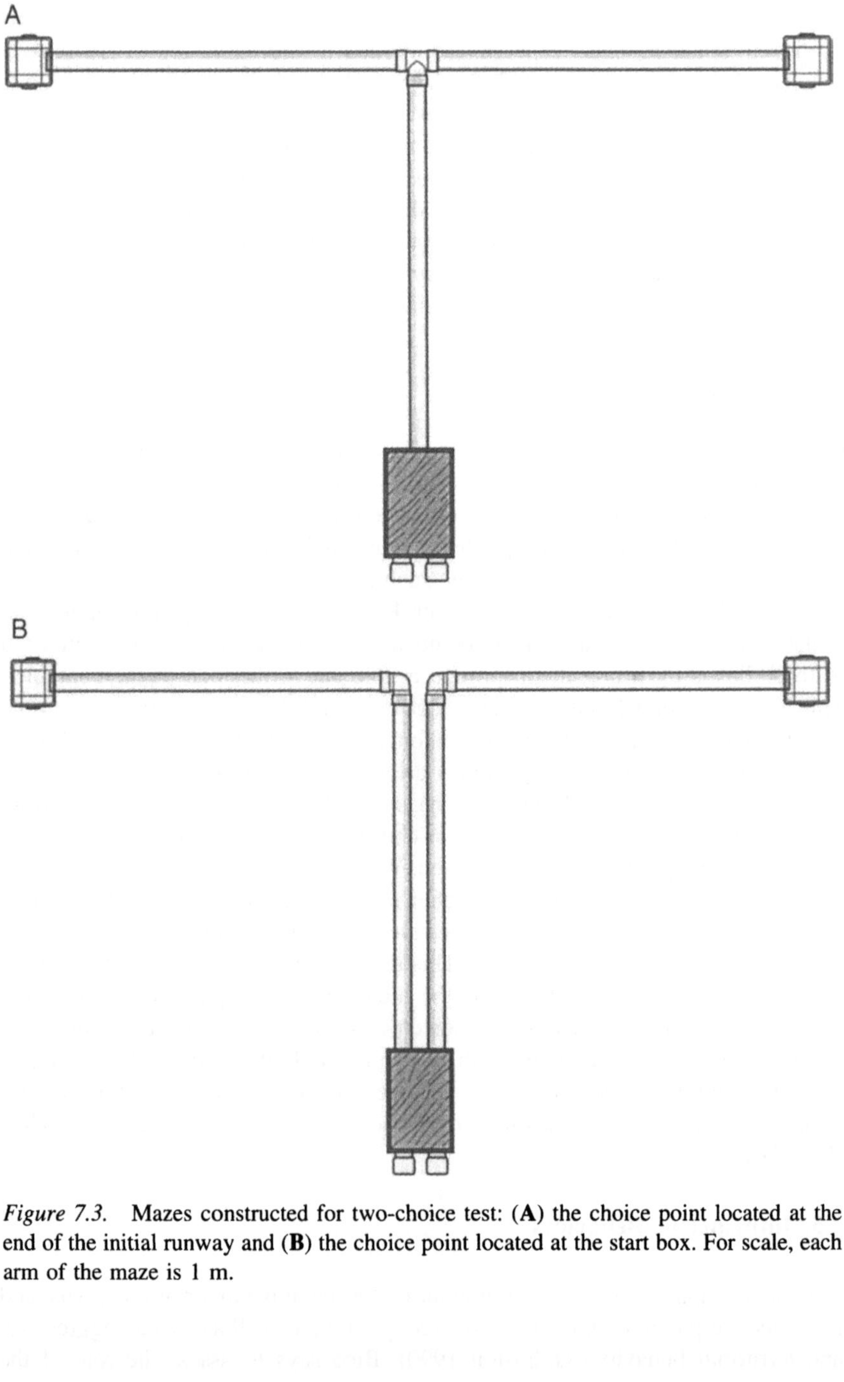

Figure 7.3. Mazes constructed for two-choice test: (**A**) the choice point located at the end of the initial runway and (**B**) the choice point located at the start box. For scale, each arm of the maze is 1 m.

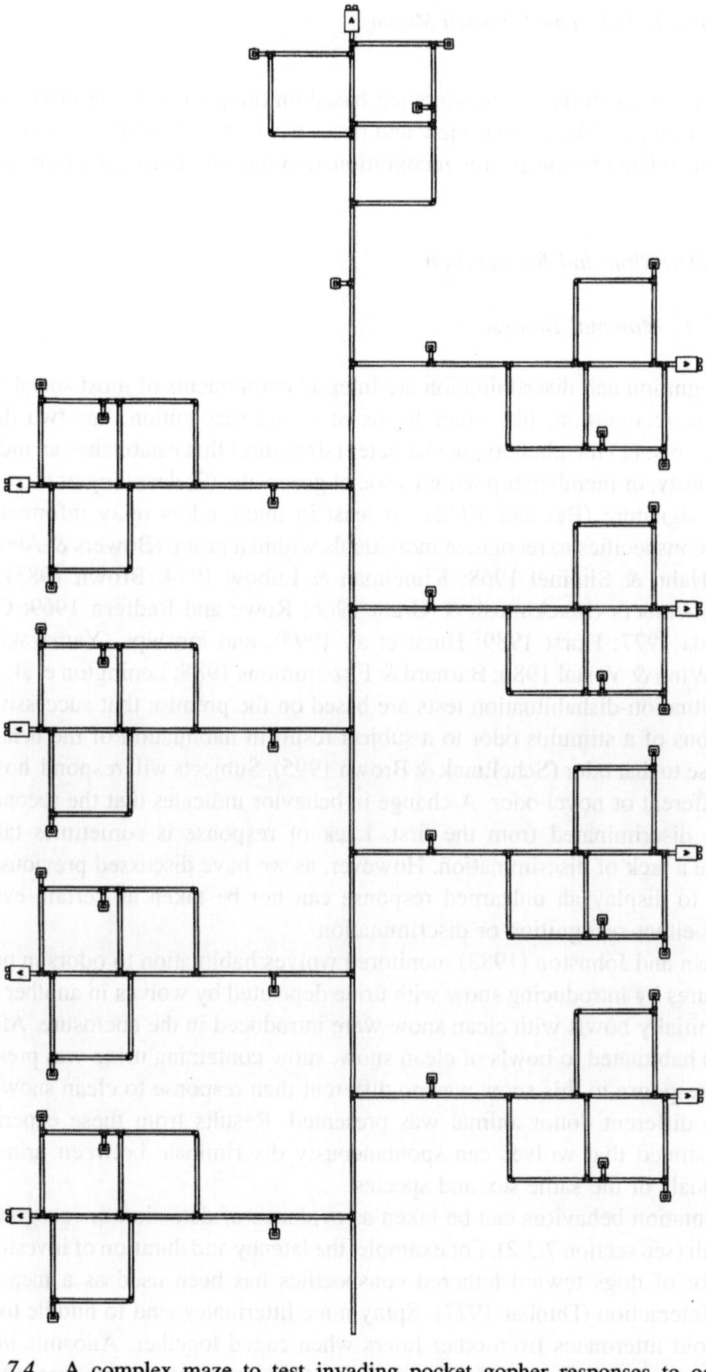

Figure 7.4. **A complex maze to test invading pocket gopher responses to odors of conspecific and predator odors. Odors can be placed either in the burrows or in the attached boxes. Number and location of odorized containers depends on the test paradigm. For scale, the distance between each connection is 1 m.**

chemical senses in these behaviors are based on the paradigms described above. This section provides an overview and describes test procedures used to address questions related to conspecific recognition, dominance, sexual, and territorial behavior.

7.5.1. Detection and Recognition

7.5.1.1. Mammal Bioassays

Recognition and discrimination are integral components of most social behaviors. Kin recognition, like other forms of social recognition, has two discrete components: (1) the phenotypic character (signature) that establishes an individual's identity, or membership within a social group and (2) decoding or recognition of that signature (Beecher 1982). At least in mice, odors relay information to enable conspecifics to recognize individuals within a group (Bowers & Alexander 1967; Hahn & Simmel 1968; Kimelman & Lubow 1974; Brown 1985), other group members (Mackintosh & Grant 1966; Rowe and Redfern 1969; Connor & Lynds 1977; Hurst 1989; Hurst et al. 1993), and kinships (Yamazaki et al. 1979; Winn & Vestal 1986; Barnard & Fitzsimmons 1988; Lenington et al. 1988).

Habituation-dishabituation tests are based on the premise that successive presentations of a stimulus odor to a subject result in habituation of the behavioral response to that odor (Schellinck & Brown 1995). Subjects will respond, however, to a different or novel odor. A change in behavior indicates that the second odor can be discriminated from the first. Lack of response is sometimes taken to indicate a lack of discrimination. However, as we have discussed previously, the failure to display an unlearned response can not be taken as certain evidence against either recognition or discrimination.

Brown and Johnston (1983) monitored wolves habituation to odors in outdoor enclosures by introducing snow with urine deposited by wolves in another enclosure. Initially bowls with clean snow were introduced in the enclosure. After the wolves habituated to bowls of clean snow, snow containing urine was presented. When response to this snow was no different than response to clean snow, urine from a different donor animal was presented. Results from these experiments demonstrated that wolves can spontaneously discriminate between urine from individuals of the same sex and species.

Orientation behaviors can be taken as evidence of detection or recognition of a stimuli (see section 7.3.2). For example, the latency and duration of investigatory behavior of dogs toward tethered conspecifics has been used as a measure of social interaction (Dunbar 1977). Spiny mice littermates tend to huddle together and avoid littermates from other litters when caged together. Anosmia induced by intranasal infusion of $ZnSO_4$ abolish affiliation, suggesting that olfactory signatures are the basis for the selective interactions among littermates as well as the avoidance of unfamiliar conspecifics (Porter 1986). Loughry & McDonough

(1994) tested the ability of infant nine-banded armadillos to discriminate among conspecific odors by placing the subject in circular tanks and recording the time spent in respective quadrants treated with odors. These results again showed that some animals use odors for discrimination of kin and that odors used in discrimination may be individually distinct.

Hurst (1989) developed a test apparatus to examine olfactory communication in populations of wild house mice. A grid on interchangeable tiles (0.15 × 0.15 m) on the floor of an enclosure (1.2 × 1.2 m) allowed experimental manipulation of the substrate once it had been marked by the resident mice. A variety of test odors could also be incorporated temporarily into the marked substrate by replacing a home area tile with marked test tiles. Individual responses to each test odor were measured in random order and without interference from other family members by isolating a test area (0.3 × 1.2 m) within each enclosure immediately before each test. To test responses to the marks of specific individuals, test tiles were marked artificially with urine collected from individual family members, using 15 µl per test tile to provide a strong mark signal that imitated the largest streaks deposited naturally by resident mice. These results showed that mice differentiate between odors deposited by kin and marks made by a member of an unfamiliar family.

Progression through a maze can reflect responses to a sensory signal. Random movements suggest an indifference, while consistent responses indicate attraction or avoidance. Naked mole rats response to conspecific urines were tested in a Perspex™ tunnel system that comprised a T-maze type choice point within a maze that permitted the subject to return to the start box rather than have the experimenter remove the subject from the test system (Heth & Todrank 1995). The initial runway ended in a T-junction where the animal had to turn to the left or to the right to continue through the tunnels. The two tunnels converged into a common tunnel that led back to the original start box. Sliding doors in all parts of the system kept the animal running in the same direction to ensure that they always approached the stimulus from the start box runway tunnel. A sliding floor in the tunnels to the right and left of the choice point allowed stimuli to be presented either as an odor through small perforations or as a liquid that could be both smelled and touched with the nose in small grooves in the tunnel floor.

Animals can also be trained to select for specific odors. Consistent correct responses means the subject is discriminating between cues. Subjects trained to approach conspecific odors using water reinforcement repeatedly select the arm of a Y-maze that contains that odor (Yamazaki et al. 1979; Singer et al. 1993). A slight vacuum draws odors toward the choice point. Responses are recorded as correct or incorrect. Complexity of the choice can be increased with additional arms added to the maze. Voznessenskaya et al. (1992a) trained rats to move through a five-arm maze. Response measurements included subject movements within the box and their response latencies to reach the correct maze arm and time to enter each of the incorrect entrances.

Mammalian neonates typically display preferences for select odors immediately after birth (Schaal et al. 1995). As neonates emerge to a new environment, chemical signals provide the most salient cues, particularly for altricial neonates born with sealed eyes and ears, who must therefore rely on haptic and chemical signals. Chemical signals on which parental-offspring recognition rely are generally cues that the neonates experienced *in utero*. Nursing behavior of some neonates is directed by these stimuli. Orientation appears to be innate or at least unlearned.

Basic paradigms for testing neonate responsiveness to significant chemical cues are similar to tests described above. However, lack of subject mobility may require adaptations of the experimental apparatus. Gerling and Yahr (1982) used a Plexiglas™ rectangle (40 × 28 × 22 cm) with a wire mesh screen as the floor. This frame rested on triangular blocks so that the mesh floor was suspended 2 cm above the counter. The mesh floor was further supported from underneath by a black Plexiglas™ bar, that when viewed from above, divided the floor into equal sections. Stimulus was placed beneath the screen on each side of the support bar so that it could be smelled but not contacted. Neonates were initially placed lengthwise over the center support bar, and tests began when the neonate moved off the center line. Preference for an odor was determined as the amount of time spent on a side during the course of a 5-min test. Sides on which pups spent more than 150 s were operationally defined as preferred. Attraction to maternal nest odors was tested in an open arena, covering one side of the arena with bedding from the pup's home cage and the other side with bedding from virgin females (Porter & Ruttle 1975). Tests with sensory impaired neonates indicated that the choices were directed by chemosensory cues. Similar tests have been conducted with parental bedding or cotton swabs with odors spaced equal distances from the center line starting point. Neonates are placed perpendicular to the choices, and observed until they move one direction or the other, or until 2 min have elapsed without a demonstrated choice (Porter & Doane 1976; Porter et al. 1978).

Precocial neonates are more mobile than altricial young and thus test procedures can take advantage of more elaborate response repertoires. Schaal et al. (1995) tested olfactory preferences in newborn lambs. Immediately after birth, lambs were placed in an isolation box (Plexiglas™ 50 × 35 × 40 cm), that prevented them from reaching the maternal udder but permitted typical maternal behavior by the ewe (sensory access and licking the offspring). As soon as the lambs were able to stand, they were given a 5-min two-choice test. The testing device was a Plexiglas™ box (50 × 45 × 30 cm) with a top at the approximate height of an ewe's udder. Cotton pads were attached to the top and along the upper front wall on opposite sides of the chamber. The top of the box was heated by two rubber hot water bottles and by an infrared bulb hung (20 cm) above the device. The temperatures of both cotton pads were equal during tests though among tests they varied from 26 to 32°C. Recordings of low-pitched bleats emitted by an unfamiliar, recently parturient ewe and the background sound of a lambing pen were played through a speaker located in a central position behind the odor stimuli. The maternal vocalizations were highly arousing for most lambs. The whole box was covered with paper to attenuate and equalize the brightness of the light on both pads. Prior to trials, the lamb was introduced into the center of the box, aided to stand, and then the experimenter oriented the lamb's muzzle toward each odor pad, and then toward the medial neutral zone (Vince & Stanier 1991). Subsequent orientation toward one of the pads was defined as any direct contact with a pad or any directed movement that resulted in the lamb's muzzle being within 2 cm of a pad.

Altricial neonates use chemical signals to locate maternal milk. For example, chemical stimuli distributed across the belly in an increasing strength guides offspring toward the nipples of a lactating rabbit (Keil et al. 1990). Tests of this behavior were conducted by placing pups in a stand made from a Styrofoam™ block with holes cut diagonally into it, the diameter and depth were adjustable to hold pups of different ages (Keil et al. 1990). Stimuli were presented on a small glass rod, the tip roughened so to absorb approximately 30 μl of liquid when dipped into a sample. Stimuli were offered to pups by holding the tip 5 mm in front of the muzzle and a positive response recorded if the presentation elicited clear, searchlike movements and/or grasping of the rod within 10 s. Hudson and Distel (1983) addressed a similar question by restraining does on their backs in a U-shaped trough, the ends of which were closed by dividers to form an arena enclosing the six rear nipples. Shaving the belly of the doe had no impact on pup response, but when the shaved belly was covered with adhesive tape, leaving only the nipples bare, pups did not search and only attached to nipples when they encountered them by chance.

Effect of prenatal experiences with chemical signals on the first suckling episode of rats were assayed in similar tests on pups exposed to flavors *in utero* (Pederson & Blass 1982). Fetuses were exposed to flavors injected into the fetal environment on day 20 of gestation. After anesthetizing the female, the uterine horns were exposed in a midline laparotomy. Access to individual amniotic sacs were gained by passing a 30-gauge needle through the distended clear uterus and seated in the space between the fetus's jaw and chest. Then 0.2 ml of sterile citral warmed to body temperature was injected. Litters were delivered via caesarean section on day 22. Uterine horns were removed and placed on a heated tray covered with paper towels saturated with warm isotonic saline. Amniotic sacs were removed, and each pup was dried with gauze pads and placed in a styrofoam cup in a warm incubator for the next hour. All rats were stroked for the next hour with a soft artist's brush to mimic postnatal maternal stroking. Responsiveness to odors painted on the nipples of recently parturient females was then assessed in an approach similar as that described for rabbits above.

7.5.1.2. Bird Bioassays

Though generally neglected by workers in the chemical senses, birds offer several distinct anatomical advantages over mammals. Birds, unlike mammals, lack a vomeronasal system, and thus, the question of which system mediates a response is simplified. Also, the anatomical arrangement of the olfactory system in birds allows olfactory fibers to be selectively cut (e.g., Mason & Silver 1983), and this simplifies the problem of isolating olfactory and trigeminal components of odor perception. Knowledge of olfactory physiology, combined with well-designed behavioral tests, can be used to evaluate the consequences of surgical intervention. Measures of changes in consumption (feeding and drinking); heart and respiratory rates; and operant discrimination procedures, including conditioned suppression, have all been applied in studies of chemosensory systems in birds.

An unusually good example of the use of an orienting response to describe sensory function was performed by Wenzel & Sieck (1972). In their studies, pigeons, shearwaters, ducks, turkey vultures, chickens, and other birds were lightly wrapped in a stretch bandage and supported in a quiet unlighted enclosure. The birds' heads were placed into a tunnel into which odorants or clear filtered air could be presented. Heart and respiratory rates were recorded during the presentation of stimuli. In general, odorant produced greater changes in heart rate than respiratory rate.

Several important control procedures were used in these experiments. First, the experimenters severed the olfactory nerves in some of their subjects to assess olfactory contributions to responding. Second, responses to presentations of filtered air were recorded in addition to responses to presentations of odors. This permitted investigators to distinguish responses mediated by changes in airflow (reflecting the operation of the apparatus) from responses elicited by odor stimuli. Third, at the end of each session, the darkened test chamber was suddenly illuminated to measure the responsiveness of birds to a stimulus with known arousal value.

Shallenberger (1975) used procedures and an apparatus similar to those described by Wenzel and Sieck (1972), with the exception that natural odors were used rather than artificial scents. In his experiments, bottled air passed through activated charcoal for drying and deodorization. The clean dry air stream then split and passed through one of two cylinders or a bypass. A valve permitted selection among the airflows, and rotameters were used to assure that responding was not a result of pressure differences among air flows, per se. Pairs of wedge-tailed shearwaters served as subjects. One member of each pair was placed in a darkened test chamber, with monitoring electrodes attached, and left for 20–30 min to adapt to the experimental context. When the birds's heart and respiratory rates had stabilized, air from the chamber containing the other member of the pair was diverted into the test chamber. Odor trials were interspersed with trials in which only filtered air passed through the chamber and changes in heart and respiratory rate were recorded. For some birds, respiratory rates increased in response to presentations of odorized air; for others, respiratory rate decreased. Any change was taken as evidence of detection, and the authors reported that changes in respiratory rate was more reliable than changes in heart rate. When orienting responses are used as a measure of detection, idiosyncracies in the display of attention are likely to occur. Rigid operational definitions of the orientation can result in the failure to collect important information. In the work with the shearwaters, had only increases in respiratory rate been recorded, it is likely that markedly different conclusions would have been drawn by Shallenberger (1975).

There are many excellent examples of operant test procedures with birds. A number of these studies have employed odors. The use of taste cues as stimuli has been less frequent, owing to inherent difficulties in presenting nonvolatile stimuli to birds in an operant context.

Michelsen (1959) used an olfactometer to measure threshold sensitivities in pigeons. Birds were trained to peck a response key to initiate a flow of air through a cylindrical breathing chamber. The bird then put its head into the airstream and pecked one of two keys. Pecking one key in the presence of an odor stimulus caused the delivery of a grain reinforcer. In the absence of the stimulus, pecking the other key delivered reinforcement. Thus, symmetrical reinforcement contingencies were used to obtain reports from birds about both the presence and the absence of odor. Incorrect responses terminated the trial and prevented access to grain. Odor and no-odor trials were presented randomly, except when the bird responded incorrectly. In that case, the same stimulus response condition was presented in the next trial to prevent response perseveration, which would have permitted the animal to obtain grain 50% of the time by simply responding on the same key. To determine whether the odorant delivery mechanism produced extraneous cues, control trials were conducted in which only nonodorized or only odorized air was delivered regardless of the key pecked. Also, in subsequent studies of the same birds, bilateral section of the olfactory nerves was performed in an attempt to separate the contribution of olfaction from trigeminal chemoreception to performance.

More recently, Walker (1983) used a similar procedure to study detection thresholds in pigeons. A restrained, food-deprived pigeon was positioned with its head in a breathing chamber that contained two response keys. The onset of each trial was signaled by the illumination of both keys. The bird was trained to peck one key when clean air was present and another when odor was present. Responding to the odorized key resulted in the delivery of food. Responding to the other key terminated the trial and led to a time-out period in which the test cubicle darkened. Likewise, a failure to respond when the keys were illuminated led to a time-out. Thus, the experimental contingencies required a response on each trial. Threshold was defined as the odorant concentration was elicited 25%

more "correct" than "incorrect" responses. Olfactometry is a discipline to itself, and considerable research has been devoted to their development and construction. An excellent discussion of olfactometry and the construction of olfactometers is provided by Dravnieks (1974, 1975).

Another approach to the operant evaluation of olfaction was taken by Stattelman and his colleagues (Stattelman et al. 1975). Here, the responsiveness of pigeons, chickens, and quail was examined. Subjects were placed in a chamber through which filtered air passed at a constant rate. Odorants were injected into the airstream with a gas syringe, and odorant concentrations in the airstream were evaluated by gas chromatography. Subjects were trained to peck at a bar in the presence of an odor. Concentrations were then reduced until pecking stopped. The lowest concentration that reliably elicited pecking was defined as threshold.

This study was notable because the evaluation of several species with the same apparatus facilitated interspecific comparisons. However, since responses were only reinforced in odor trials, one might expect a response bias toward reporting the presence of an odor when there was none.

Conditioned suppression is another operant method that has been applied to the study of avian olfaction and trigeminal chemoreception (Passe & Walker 1985). Henton et al. (1969) were the first to employ a conditioned suppression method to demonstrate odorant detection in pigeons. Subsequently, Henton (1969) also used the method to demonstrate that birds could discriminate between odorants, and Shumake et al. (1969) showed that pigeons could perform odor intensity discriminations. Initially, pigeons were trained to detect amyl acetate at 7% of vapor saturation. Subsequently, a second odor was introduced at 1% of vapor saturation. The concentration of the second odor was gradually increased until the subject could no longer discriminate between the two intensities. Transection of the olfactory nerves resulted in the loss of responding to amyl acetate suggesting the importance of that chemosensory system to the detection of this chemical.

Conditioned suppression also has been used to monitor recovery of sensory function following surgical transection of the olfactory nerves. Henton et al. (1969) trained pigeons to suppress key-pecking in the presence of amyl acetate. After response acquisition, birds were subjected to bilateral resection of the olfactory nerves. Testing within 24 h of surgery showed that the suppression response was abolished, although it could be reestablished in the presence of considerably higher (3 versus 10%) odorant concentrations. These latter findings suggest the importance of trigeminal chemoreception. Olfactory nerves will regenerate, so care should be taken to test birds before regrowth occurs which can be within as little as 75 days (Tucker et al. 1975).

A series of final examples are provided by studies of homing pigeons and their use of odor cues. The reader is referred to Benvenuti et al. (1992) for a review of these studies. Homing is a strong unlearned behavior and when studied in combination with various physiological manipulations (e.g., bilateral resection of the olfactory nerves; Baldaccini et al. 1975), considerable information about the importance of olfaction to navigation by pigeons has been obtained. These studies underscore the importance of invariant behaviors for the study of sensory processing in the absence of training.

A variety of methods have been applied to the study of the chemical senses in birds. These include simple feeding and drinking studies, as well as classical and operant conditioning. Generally, birds exhibit greater apparent sensitivity after training, and responding is strongly contextual. Because birds lack some of the chemosensory systems present in other vertebrates, and because the systems they possess are organized simply (relative to mammals), birds represent a useful model for the development of basic chemosensory information.

7.5.2. Reproduction

Many mammalian species make extensive use of chemical signals in sociosexual communication in priming of reproductive functions (Epple et al. 1989). Signals may serve as long-range attractants or as short-range promoters of sexual behavior (Nyby 1983). Cues emitted by conspecifics or other species may serve either to suppress, or conversely, to initiate reproductive status in females. Onset or delay of puberty or changes in the adult female reproductive cycle are common measures for these signals. Precopulatory signals to entice perspective mates or to discern receptive partners vary among species and can be complex. Likewise, test response variables vary and need to be well defined. Signals that promote copulatory behaviors are generally recorded as attempted or successful matings.

Multiple factors influence the kinds and quantity of reproductive behaviors that are ultimately displayed. Responses may require overlapping or consecutive auditory, visual, tactile and chemical cues. Prairie voles are more likely to exhibit estrous odor preferences when odor presentation are paired with stimuli from other modalities (Taylor & Dewsbury 1988). The test paradigm, therefore, may need to incorporate other contextual cues to avoid false negative responses. Weak or inconsistent responses may also occur if stimuli are inadequately presented. Vaginal secretion applied to a anesthetized, castrated golden hamster will induce males to exhibit sexual behaviors, whereas the same secretions on a clay model do not (Johnston 1974). Relative preference for or attraction to stimuli may change with reproductive status. In Djungarian hamsters four scents provide information about the female's reproductive state (Ferkin & Johnston 1995a, 1995b). Males are attracted to one of these scents only on the day before receptivity, two scents are primarily attractive during receptivity, and their attraction toward the fourth scent increases as receptivity approaches.

Stringent standardization of test conditions will reduce inherent variability (see Cherry 1993). Factors such as the strain of the animal, age, physiological condition, and social and sexual experiences of the subject, should be held constant. Social status of donor and test subjects affects responses. Assays should be conducted at consistent times across days. Reversed light/dark cycles are convenient when studying nocturnally active animals. Habituating subjects to the test arena prior to test reduces anxiety and increases overall performance. Photoperiod affects the reproductive cycles of some species and tests conducted during a nonreproductive period will be futile or provide misleading results.

Influence of chemical signals on the sexual maturation of subjects can be assessed by comparing animals under different exposure regimes. Werner and Clarke (1995) exposed weanling females, matched for age and body weight, to a sexually experienced male, a naive male, or no male by placing a male within a cage ($16 \times 16 \times 8$ cm; mesh 5 mm^2) inside a female's cage ($44 \times 28 \times 21$ cm). Females were sacrificed after 3 weeks and reproductive systems and pituitary glands were extracted. Ovarian follicular development was classified according to Pederson and Peters (1968). A similar experiment was conducted to isolate the effects of urine without the male presence. Procedures were similar but stimuli were painted on the dorsal surface of the female nose and in the region of the nostrils, using a water-color paint brush. The presence of chemical stimuli in urine that delayed puberty in field mice at high population densities was demonstrated by presenting urine collected in the field to albino mice in the laboratory (Coppola & Vandenbergh 1987).

Choice test in a reproductive context are generally simple. For example, subjects may be presented with odors on opposite ends of microscope slides and observers record the amount of time subjects spend investigating each end of the slide (Ferkin et al. 1994; Ferkin & Johnson 1995a, 1995b). A series of two-choice tests can elicit responses from subjects to stimuli collected from donor animals at various physiological stages. Petrulis and Johnston (1995) used a neutral arena ($60 \times 60 \times 27$ cm) with two wire-covered openings (11×9 cm) on either end of the apparatus where odors could be placed. Similar apparatuses can be used to test other species. Doty and Dunbar (1974) constructed a plywood kennel ($60 \times 75 \times 100$ cm) with two holes (7.7 cm) located 25 cm apart and 34 cm off the floor to test dogs. Stimuli were placed in Plexiglas™ boxes ($14 \times 15 \times 30$ cm) that were attached to the kennels over the holes. Photocells and a cumulative clock were used to record the duration and number of times a dog inserted it's nose through an opening.

Onset or increased incidence of some behaviors can indicate if animals detect reproductive stimuli. Copulation is the most obvious behavior to consider, but other responses may be informative. For example, voles often engage in autogrooming when exposed to odors of a female, with response strength affected by the reproductive condition of the donor (Ferkin et al. 1996). The flehman response meets the specificity and quantitative requirements as a reliable response in a bioassay to access the activity of compounds for sexual communication in Asian elephants (Rasmussen et al. 1993; Perrin & Rasmussen 1994). Specific samples can be collected from conspecifics or subjects can be allowed direct contact with donor animals. Donor animals can be anesthetized, confined to cages, or tethered to decrease displays of nonchemosensory signals. Bodyak & Surov (1995) observed the behavioral responses of golden hamster males toward anesthetized females and males, and toward anesthetized males painted with vaginal discharge or fresh Harderian gland homogenate from females. Behavioral categories evaluated included intermale aggression (flank-gland marking, digging, nonspecific movements of the rostrum, washing, grooming, placing rostrum close to stimulus animal's flank) or heterosexual courtship (flatten, pulling stimulus animal's skin, mounts, intromissions, ejaculation). Lepri (1994) measured the amount of time females spend in close proximity to inexperienced males to assess whether impaired vomeronasal function altered female attraction to males. Influence of conspecifics odors on the reproductive success of arctic foxes was tested by placing urine samples in den boxes and then monitoring sexual activity and reproductive success (Bartos & Rodl 1990).

Animals can also be trained to respond in the presence of selected stimuli. Michael and Bonsall (1977) used operant conditioning and reversible anosmia to assess the effect of olfactory signals on the sexual behavior of rhesus monkeys. Males were first trained to repeatedly press a lever for food and then to press the lever to gain access to a receptive female. A task males readily performed when in the presence of ovariectomized females that had been estrogenized but not in the presence of nonestrogenized ovariectomized females. Subsequently, males rendered anosmic (plugs of gauze impregnated with bismuth-iodoform-paraffin paste were inserted above the superior turbinate bones) did not respond to estrogenized females. When their olfactory acuity was returned by removing the plugs the males returned to pressing the lever. A similar study was conducted to assess the inhibitory effects of progesterone on female attractiveness (Michael & Bonsall 1977). Treating females with progesterone (50 mg/day) normally abolishes the sexual activity of rhesus monkeys. Temporarily anosmic males, however, continued to press for access and mate females until their olfactory acuity was returned.

Subjects trained to identify reproductive odors need not be conspecifics. Rats were trained to detect the estrus-specific odor in cow urine (Dehnhard & Claus 1988). The olfactory discrimination apparatus consisted of a teflon cylinder attached to a test-chamber. Air flow (18 L/h) was guided through gas-washing bottles containing urine samples and then into a Teflon® test chamber with the rat. Rats were rewarded for responding in the presence of estrous but not diestrous urine. Once trained these rats were used to differentiate among fractions that contained chemical constituents specific to estrus urine.

7.5.3. Dominance

Chemical signals are a means to identify a dominant individual within a group and to prime responses by conspecifics when encountered (Gosling & McKay 1990; Hurst 1993). Aggressive encounters between conspecifics increase when their ability to detect signals is impaired. Reproductive success is also disproportionately higher among dominant individuals because they gain differential access to mates (Dewsbury 1982).

Dominance can be interpreted from behavioral displays (e.g., submissive postures, investigation, aggression). Because aggression may cause stress and possible injury to experimental animals, efforts to minimize the risk of protracted or injurious encounters should be taken. Hurst et al. (1994) group housed males at weaning before aggressive behavior developed. Attempts to introduce unfamiliar males can result in considerable and sustained fighting (Rowe & Redfern 1969; Poole & Morgan 1975). Regardless, means to separate subjects or disrupt aggressive behaviors should be considered well before an experiment is implemented.

Patterns of scent marking with urine by bank voles are distinctly different between dominant and subordinate males (Desjardins et al. 1973; Rozenfeld et al. 1987). Dominant males leave fine-tract urine marks especially in locations where conspecifics are likely to encounter them. Subordinate animals, on the other hand urinate in large puddles, usually in locations away from dominate individuals (Rozenfeld & Rasmont 1991). Horne and Ylonen (1996) used this criteria to assess dominance between male pairs in an arena (60 × 40 × 34 cm) longitudinally divided by a wire mesh partition. Brown craft paper, placed on the arena floor, changed color on contact with urine, which made urination marks and puddles easy to detect.

Conspecific response to odors from dominantes can be tested as previously described in discussion of other discrimination testing. Open fields can be used to investigate whether subjects avoid substrates marked with odors from dominante animals (Jones & Nowell 1974; Koyoma 1995). Also direct interactions between or among individuals can be studied (Hurst et al. 1993, 1994).

Hafer (1990) tested the effect of dominance in mice on the Bruce effect. The Bruce effect is the termination of pregnancy in recently mated females exposed to the odor of an unfamiliar male (Bruce 1960). Urinary cues of male house mice greatly affects the reproductive status of females and signals emitted by dominant males are more likely to elicit a response than cues from subordinates. Dominance in males can be ranked by exposing mice to one another in a small aquarium for 10 min, and scoring chase, attack and fights. These males then encountered females in estrus in one of the following test regimes: a control group where females were mated with a stud male that remained with the female throughout the experiment; an experimental group where a dominant stud male was replaced by a subdominate male; and another experimental group where the subordinate stud male was replaced by a dominant male. Pregnancy was blocked (no births) and estrus resumed rather quickly when subordinate males were replaced by dominant males, but not when dominant males were replaced by subordinate males.

7.5.4. Territorial Behaviors

Many animals defend a specified area of their habitat against other members of their species and often mark it chemically in a specific manner (Gosling 1982).

The intruder's response to marked territories depends on the resource benefit paired with the risk of detection along with the competitive ability of the intruder (Gosling 1990). Gosling et al. (1996) demonstrated that large house mice are less likely to avoid marked territories than are smaller conspecifics. Subjects were tested in a tunnel shaped as a modified Y-maze. The floor of the maze was covered with white "benchcote" paper. One randomly chosen arm of the Y and half the crossbar were marked with spots of donor male urine. The other arm and half of the crossbar were marked with spots of distilled water. Marks were 5 μl delivered by an Eppendorf® pipette and placed at 3-cm intervals in two alternating rows along the floor. Fans were used to generate gentle airflow through the system and away from the nest box. This reduced the amount of odor the subject encountered in the stem of the Y before it made its right/left choice. Mice were allowed 5 min in the nest box at the base of the Y to adjust to its surroundings before each test. The amount of time taken to cross a start line after leaving the nest box was recorded, along with which arm of the Y was entered first and the time spent in each section of the maze for each subsequent minute of the trial.

Response to scent marks may also vary with experience in establishing and defending a territory. Heise and Hurst (1995) used an open arena to assess the effect of territorial experience on the response of house mice to marked territories. Nonterritorial mice were kept isolated from outside groups, while territorial mice were permitted olfactory and visual contact through a wire screen. Subjects were then introduced to an arena with each half treated with a different scent (i.e., small spots of urine applied to filter paper on the floor). Behavioral patterns recorded in each minute for 15 min included the initial side of the field visited; the frequency and duration of subsequent visits; movements around the edge of the field; escape attempts; and time spent investigating the stimuli, floor, and walls along with time spent questing, grooming, and resting.

A common practice among mammals is to deposit scents in a place previously marked by a conspecific (Ralls 1971; Johnson 1973; Thiessen & Rice 1976). Johnson et al. (1994) relied on habituation to investigate golden hamster recognition of overmarked scents. Subjects were presented with glass plates with simulated overmarks (i.e., the secretion of one donor painted or drizzled over the secretion of another donor), in a series of 5-min trials conducted at 15-min intervals. Time spent investigating scents declined over time. Half the subjects were then presented with plates marked with the bottom stimulus paired with a novel scent, while the remaining subjects were presented the overmark (top) stimulus paired with a novel scent. Subjects spent more time investigating the novel scent when offered the top scent, but spent equal time investigating the novel and bottom scents. Thus, overmarking by golden hamsters masks the scent of their conspecifics.

Overmarking can be a reliable assay to assess the recognizable components of a scent mark in some species. Adult tree shrews reliably overmark the urine of a strange fertile male conspecific by a movement called chinning (Holst 1985). Stralendorff (1987) used this natural response to identify the active fraction of urine. Isolated fractions were applied to the front of an animal's home cage and the number of times the sample was overmarked by chinning within 2 min after application was recorded. Control substances (e.g., female urine or urine from an immature male) were applied either 30 min before or 30 min after the test stimulus to ensure that chinning occurred only in the presence of scent from strange fertile males.

Response to conspecific scent marks can be assessed in the field by monitoring trap success. For example, if urine-contaminated traps catch fewer animals than clean traps, avoidance can be inferred, and vice versa. One caution is that all traps should be equally accessible. Temporal comparisons need to be conducted in such a manner that changes in trap success reflect responses to cues and are not the result of a change in the environment or the physiological status of the target animals. An overview of potential trapping regimes and special considerations for selected species is provided by Davis (1991). Drickamer (1995) monitored the status of individuals captured first relative to the status of individuals captured next to study the social biology of house mice. Clean traps were substituted for contaminated traps to assess the response of voles to conspecific odors in traps (Boonstra & Krebs, 1976). Clean traps were placed exactly as the contaminated trap had been set. Trap success for clean traps was compared with trap success for contaminated traps. Heske (1987) set two longworth live traps with their entrances ~ 5-cm apart in a runway or at a burrow opening such that they would be encountered simultaneously by an approaching vole. Clean pine shavings were placed in the nest box and in front of the entrance of one trap, while an equal amount of soiled pine shavings was used to bait the other.

Field observations can sometimes provide a perspective of subject responses under natural conditions. Species that are not easily disturbed by intruders can be followed and their behaviors recorded. Richardson (1990) followed aardwolfs in a jeep at a distance of 30 m to describe their dispersion pattern of scent markings and the response of conspecifics. Meaney (1990) followed pikas on foot, locating cheek marks and recording pika locations to the nearest stake in a 5 × 5-m grid established across the study site. Less direct observations are necessary for most species.

Animal responses to artificial scent marks created by the investigator also can be monitored. Svendsen and Huntsman (1988) treated mounds of mud and debris along a stream with fractions of beaver castoreum. Only beavers that swam within 10 m of these mounds were included in the study. Stimuli were considered active if a beaver made a detectable change in direction or altered its behavior in relation to it. A similar approach has been used to assess woodchuck responses to oral-gland scents (Meier 1991). Scents were applied to cotton balls affixed to stakes near the entrance of an occupied burrow.

Gorman and Stone (1990) used a combination of bioassays to assess the avoidance behavior of European moles. Consistent responses across both bioassays provided confidence that the behaviors were not influenced by the methodology. One strategy compared the catching efficiency of clean Friesian traps with that of traps that had contained a captive male mole for 3–4 h just prior to use. Overall an equal trapping effort was made with each trap type. Only one trap was set in any mole territory and was left in place for up to 72 h, with daily resetting if blocked off with soil. The second strategy was to place cotton balls, clean or smeared with preputial secretion, into tunnels of radio-tagged moles 10–15 min before a mole was likely to visit the spot. Particular sites which moles visited on a regular basis, and which contained shallow surface tunnels, were identified prior to the assay. The reaction of a mole to a stimulus was scored as positive if the mole retreated down the tunnel and negative if the mole passed the stimulus and continued along the tunnel.

Mykytowycz et al. (1976) demonstrated that scent marks deposited by rabbits serve to increase their own territorial confidence. The tests (10 min) were carried out in an indoor 5 m^2 arena. Before each test the floor of the pen was covered with a sheet of clean paper on which fecal pellets or eight filter pads impregnated with the given odor stimulus were scattered randomly. The two rabbits were then simultaneously introduced to the pen. When a rabbit was tested in the presence of its own odor it was referred to as the home animal. Behaviors recorded to indicate the territorial confidence of an animal were the time to first physical contact between animals, frequency and duration of approaches, aggressive behavior, and attempts to escape from the pen. Tests were conducted once in the presence of the odor familiar to animal A and then again in the presence of an odor stimulus familiar to animal B. Chemical stimuli were collected from perspective subjects prior to test. Chin glands of individual rabbits were massaged to induce secretory pores to emit drops of secretion on the surface of the skin. The secretion was then collected into glass capillaries and diluted to 1:1000 immediately prior to the test. Fecal pellets become coated with anal gland secretion when passing through the end part of the rectum and hence were the natural vehicle for its dispersion. Urine was collected overnight from beneath the rabbits' cages, and precautions were taken to avoid contamination with fecal pellets. Small quantities of sebum from the inguinal pouches were collected with a cotton swab.

7.6. Interspecific Behaviors

Chemical signals also play a role in interactions among species. Cues signal palatable foods, predators, and plant toxins. Though less common, or at least less documented, interspecific signals may also influence physiology. For example, the reproductive status of some voles is mediated by chemical cues in plants (Berger et al. 1981) and predator odors can delay the reproductive status of rats (Voznessenskaya et al. 1992b).

7.6.1. Resource Exploitation

Bioassays can be used to investigate the relationship between chemical signals and nutrition. Clearly, the importance of cues in this context varies among species. Odors emitted by a mouse signifies something different to a weasel than to a cow. Likewise, succulent plants are desirable to gazelles but of limited value to lions. Yet other species (e.g., bears) readily exploit the nutritional benefits of both mice and plants. Animals' responses to dietary cues, though not absolute, generally reflect their foraging guild. Predators rely on chemical signals to detect prey presence. Search-and-capture behaviors are the usual response of predators when they encounter cues that denote prey. Once captured prey items most likely will be ingested. The choice to ingest is perhaps certain because of the limited nutritional variation among meats. However, nutritional variation among and within plants is common. Consequently herbivores rely on chemical signals to identify which plants or plant parts can be ingested; capture is reasonably ensured. Thus, the decision point for carnivores is whether to pursue prey when it is detected, while for herbivores it is whether to ingest prey. Thus, search or attack behaviors are appropriate to assess responses to transient prey (e.g., rodents), while intake is a more appropriate response variable for animals encountering stationary prey (e.g., plants).

7.6.1.1. Mammal Bioassays

Choice tests are the most commonly used method to evaluate food acceptance and preferences (see section 7.3.3). Flavors can be added to a readily accepted carrier or base food to control for consequences, or odor stimuli can be placed adjacent to a desirable resource. Flavors that reduce intake relative to a baseline are defined as aversive, while those that increase intake are taken to show a preference.

Food-deprived subjects generally provide more consistent results in bioassays. Preliminary studies may be needed to determine an optimal deprivation schedule. Where optimum is defined as a level that reliably induces consumption during test sessions without inducing a severe reduction in body weight. Nolte et al. (1993b, 1993c) evaluated flavored water intake by mice after an 18-h deprivation period. Response to flavors was tested for 3 h and then plain water was provided for an additional 3 h before the deprivation schedule was reinstated. Mice that have been adapted to this schedule drank very little during the second 3-h period if the test substance was not aversive (i.e., sufficient water was consumed during stimulus presentation to induce satiety).

Mountain beaver do not adapt well to food deprivation, but they can be trained to respond to a preferred food (e.g., apple). Therefore, mountain beaver can be accustomed to a feeding schedule that includes apples and then this highly preferred food is treated with stimuli or placed adjacent to test odors during test trials (Epple et al. 1993, 1995; Nolte et al. 1995a, 1995b).

A two-choice test in a laboratory setting basically consists of providing the subject with the desired choice whole maintaining all other variables constant. Drinking tests conducted with mice (Nolte et al. 1993b, 1993c) and food tests with mountain beaver (Nolte et al. 1995a, 1995b) are suitable examples. Mice were randomly assigned to treatment groups and adapted to a water deprivation schedule as described above. On each of 4 pretreatment days, water was presented at 0900 h, and water intake between 0900 and 1200 h was determined to the nearest 0.1 ml. Drinking tubes used in these trials are described in section 7.4.2. Between 1200 and 1500 h, animals were provided *ad libitum* access to water. A 4-day treatment period immediately followed pretreatment. Treatment procedures were identical to those used during the pretreatment period except that each group was given its respective treatment during the 3-h measurement period.

Feeding tests with mountain beaver were similar in that animals were first adapted to the experimental regime (Nolte et al. 1995a, 1995b). Mountain beaver were presented with 20 untreated apple cubes in weigh-boats (see section 7.4.1) spaced at even intervals around the perimeter of their pens for 4 consecutive days. The number of apple cubes in each weigh boat and within the immediate area (<30 cm) was counted after 24 h. All remaining cubes were then removed and another 20 fresh cubes were placed in the weigh boats. A 4-day treatment period immediately followed pretreatment. Treatment procedures were identical except that apple cubes were treated with their respective stimuli.

Distance between the offered choices can influence the ease or rapidity with which animals switch from one stimulus to another. As distance increases, so does the cost in time and energy to switch choices. This generally results in more pronounced consumption of one alternative within a single test. Another strategy to increase the cost associated with a choice is to block access to alternatives once a food is selected. Hosoi et al. (1995) trained sheep and goats to select foods in a maze. During test trials, blocks were placed behind animals as they entered one arm or the other of the maze, effectively eliminating the option of back tracking to select an alternative food.

Choice assays can be conducted in the field with slight modifications of the laboratory procedure. Bean and Mason (1995) used a simple successive choice procedure to investigate the attractiveness of liquid baits to white-tailed deer. One-liter polyethylene bottles (housed inside PVC pipe as armor), with a metal single-ball sipper tube attached, were suspended in metal holders about 1 m above ground level. Infrared motion detectors were mounted nearby at an equivalent height and 3–4 m away from the dispenser. Bait holders were positioned so that there were large trees behind them to block extraneous recordings. Detectors were tuned so that they only recorded visits by objects ≥60 cm in diameter within 0.5 m of the dispenser. (See Zielinski & Kucera 1995 for suggestions on using motion detectors to monitor animal activity.) Test liquids were randomly assigned to dispensers and the number of visits and consumption from each dispenser (loss of liquid) was measured over a 2-week period.

One measure of the attractiveness of scents to carnivores is the number of approaches to a scent station. Typically, a scent station is a scented stake surrounded by a raked patch of dirt. Bullard et al. (1983) placed scent stations near known coyote locations to assess relative efficacy of materials for use as attractions. Approaches to the scents were then calculated by counting the number of tracks and deposited scent mounds (scats) in the raked area after a specified interval.

Prior experience with foods affects an animal's dietary habits and this can impact the outcome of bioassays. Carnivore recognition of prey improves with experience, and the efficiency of search behaviors increases (Apfelbach 1992). Likewise, herbivore diet selection is modified and expanded as new foods are encountered (Provenza 1995). Even flavors encountered *in utero* (Stickrod et al. 1982; Smotherman 1982a, 1982b, Hepper 1989; Nolte & Mason 1995) and by young in mother's milk can produce significant effects (Galef & Sherry 1973; Capretta & Rawls 1974; Nolte & Provenza 1991).

The paradigm to condition animals to avoid flavors is described in section 7.3.4. Basically the hedonic value of a flavor paired with negative gastrointestinal consequences declines, while a flavor's hedonic value increases if associated with positive consequences. Numerous studies with a variety of species have been conducted and are described elsewhere (see reviews by Barker et al. 1977; Zahorik 1977, Zahorik and Houpt 1981; Braveman & Bronstein 1985; Provenza & Balph 1988, 1990; Provenza 1995).

Two examples of flavor avoidance conditioning provide a perspective on procedures. Clapperton et al. (1996) conditioned brushtail possums to avoid a novel raspberry-flavored food. Possums were fed their normal basal mash modified by adding sucrose and a raspberry flavor. Food consumption was monitored to ensure all animals showed measurable consumption. Possums were assigned to groups and given either an interperitoneal injection of sodium chloride (414 mg/kg, 0.35 M), injections of 4 mg/kg NaCN (0.204 M), or injections of 200 mg/kg LiCl. The amount of LiCl administered was based on previously reported doses used to induce aversion in other mammals (Du Toit et al. 1991). The procedure was repeated once a week for 3 consecutive weeks. Intake of the flavored mash was then measured at weekly intervals for another 8 weeks.

A more elaborate approach was employed to condition rats to prefer flavors paired with intragastric infusions of protein (Perez et al. 1996). Intragastric infusions were according to procedures described by Elizalde & Sclafani (1990). Catheters attached to surgically implanted gastric cannulas were connected to a dual-channel infusion swivel, which in turn was connected to two peristaltic infusion pumps. The pumps were activated automatically by drinkometer circuits whenever the rat drank from sipper tubes. A flavor paired with an infusion of 10% calcium caseinate or a flavor paired with water infusions was presented to rats on alternating days for 6 days. On days 7 and 8, rats were given a choice between the flavors.

Nutritional information is transferred among conspecifics by chemical signals. Rats can detect what foods their conspecifics have been eating by odors contained in breath (Galef & Stein 1985). Food volatiles, when paired with carbon disulfide elicits transfer of diet preference similar to that produced by direct exposure to a conspecific (Galef et al. 1988). Apparently, the social transfer of a food preference requires both a food cue and an endogenous (e.g., CS_2) cue produced by the signaling animal. Bean et al. (1988) compared attraction and intake of food pellets treated with increasing concentrations of carbon disulfide. Treated pellets were weighed and placed in one of two bait enclosures in an arena. The other enclosures contained either plain food pellets or pellets scented with butanol. Food deprived mice then were placed individually in the arena and observed for 20 min. Photoelectric circuits recorded entries and departures from each bait enclosure. At the end of the observation period, consumption was recorded. Mice were attracted to carbon-disulfide-scented food and the attraction increased with increasing CS_2 concentrations (Bean et al. 1988).

Chemical signals may enable rats to identify feeding sites used by conspecifics (reviewed by Galef 1986), much as ants deposit chemical trails that lead nestmates to food. Galef and Buckley (1996) observed conspecific responses to trails deposited by rats trained to run through a maze. A food cup was located at the end of each arm of the maze and a guillotine door at the arm entrances permitted control of subjects access to either arm. Initially leader rats were permitted to make 10 trips between the start box and the food cup in a selected arm of the maze. After the 10th trip the leader was removed and the block removed so both arms of the maze were accessible. A naive subject was introduced into the start box and the maze arm it selected was recorded. Any subject that failed to enter an arm within the 5-min trial period was eliminated from the experiment. Subsequently, Galef and Buckley (1996) used several variations of this test paradigm to show that trails are deposited as rats moved away from food and not toward food; and that the attractiveness of a trail is affected by the number of times it has been used, but not by the amount of food a rat ingested before using it.

Although visual and auditory stimuli are important stimuli for hunting behavior by ferrets, food searching behavior is reliably exhibited only in the presence of prey odors with which the ferret is familiar. Apfelbach (1978, 1986, 1992) demonstrated that ferrets develop olfactory search images for specific prey odors experienced early in life. Response to odors was tested in a circular chamber subdivided into four equal sections for rapid identification of ferret location during trials (Apfelbach 1978). For each trial, clean air was channelled to the chamber for 7 min, followed by the scented air for an additional 3 min. Reaction to an odor was recorded as the amount of time a ferret spent in the quadrant closest to the source of a smell. Preferences for food items were tested by placing different prey items within a ferret's cage and the item taken first was considered to be preferred (Apfelbach 1986). Also, changes in heart rate in response to odors were recorded by fitting subjects with a body harness containing electrodes (Apfelbach 1986). Baseline rates were recorded during a 1-min interval prior to stimulus onset. This baseline was compared with the heart rate during a 2-min exposure period and the rate during a 3-min postexposure period. Apfelbach (1992) used a Y-maze to measure ferret attraction to odors and subsequent approach behaviors, such as latency time in responding, decision time, and locomotory speed.

Operant conditioning procedures are useful in assessing minimum detection concentrations of stimuli, as well as in quantifying the amount of effort or time delay a subject will endure to obtain a preferred food. Procedures and considerations for operant conditioning are presented in section 7.3.5. Considerations and reinforcement schedules for testing canines food preferences with operant conditioning are presented by Beidler et al. (1984). Similar approaches are feasible for tests with other species. Rashotte and Smith (1984) note that the advantages of operant conditioning are that it separates consumption from preferences because performance is measured by work and not ingestion and also, that it minimizes ingestion, thereby minimizing postingestional effects on performances. Finally, because work is measured and not feeding per se, operant methods permit comparisons of foods that differ in size, moisture content, textures, etc.

Moulton (1977) used operant conditioning to determine the minimum detectable thresholds of dogs for several odors. Dogs were first trained to insert their noses into an airstream to obtain water as positive reinforcement. Next, odorant was added to the airstream as a response contingency, that is reinforcement was presented only in the presence of the odor. Incorrect responses were punished with time-outs (i.e., the test was interrupted for a period of time). Stimuli concentrations were successively lowered until the subject's performance dropped to a level of chance.

7.6.1.2. Fetal Mammal Bioassays

Olfactory and gustatory stimuli cross the placental barrier (Smotherman & Robinson 1987; Keeler 1988) and are present in the fetal blood and amniotic fluid (Hauser et al. 1985; Nolte et al. 1992). These substances can stimulate olfactory and taste receptors (Bradley & Beidler 1970; Maruniak et al. 1983), stimulate ingestive behaviors (Nolte & Mason 1995), and possibly influence receptor development (Bradley & Mistretta 1980).

Fluids from the fetal environment can be collected for chemical assays from sacrificed or anesthetized subjects. Nolte et al. (1992) collected samples of amniotic fluid, allantoic fluid, fetal blood and maternal blood to ascertain whether garlic crossed the placental barrier from mother to fetus. Briefly, a uterine horn and fetus were exteriorized from a ewe at approximately 110 days of gestation, under halothane general anesthesia. Samples were aspirated with a syringe before and at 10-min intervals for 2.5 h after the ewe was gavaged with 6 ml of Egyptian garlic oil. Amniotic and allantoic fluids were aspirated from the exteriorized uterus. Fetal blood was drawn from a fetal artery near a cotyledon and maternal blood was obtained from the jugular vein. Blood samples were centrifuged and serum placed in separate vials. All samples were frozen until assayed. A panel of judges detected garlic odor in samples of allantoic fluid, fetal blood, and maternal blood collected 50, 100, and 150 min after the ewe was given garlic and in samples of amniotic fluid collected 100 min after treatment.

Stimuli can be directly or indirectly introduced to the fetal environment by feeding them to the mother. Female rats fed garlic during gestation gave birth to pups with olfactory preferences for garlic (Hepper 1988). Ortho-aminoacetophenone (OAP), a trigeminal irritant, is more readily consumed by pups if their mother's drank OAP during gestation (Nolte & Mason 1995). Stimuli introduced directly (e.g., injection) into the fetal environment provide better control of the onset and concentration of stimuli exposed to the fetus. Rat pups develop preferences for flavors injected into the amniotic sac (Smotherman 1982a), and aversions to flavors paired with aversive conditioning (Stickrod et al. 1982; Smotherman 1982b). Surgical procedures to inject flavors are described in detail elsewhere (Stickrod et al. 1982; Smotherman 1982a, 1982b). Briefly, both horns of the uterus are exposed and elevated through a midline laparotomy. Solutions (0.04 ml) are then injected into the amniotic fluid adjacent to the nose and mouth area of fetuses using a 30-gauge needle inserted through the uterine wall. Pups are conditioned to avoid flavors by injecting them with lithium chloride (0.02 ml of a 0.15-M solution). Fetuses are manipulated in the uterus so intraperitoneal injections can be made from a dorsal lateral approach. To prevent exposures to extraneous chemical cues, pups are taken by cesarean section on gestational day 22 and fostered to adoptive mothers (Grota 1968).

Bioassays to assess the effects of flavors in utero depend on the response capacity of the subject. Robinson et al. (1993) used an artificial nipple to elicit oral grasping by fetal rats. Nipple attachment may be a feasible response variable for neonates. Smotherman (1982a) trained 11-day-old rat pups to transverse a runway to reach their mother and attach to a nipple. If pups failed to attach to a nipple within 100 s, the trials were interrupted. Pup responding to odors was tested by introducing odors at the goal end of the apparatus. The relative time to move through the maze and the number of interrupted trials were used as indicators of neonate response to odor. Movements within an arena relative to an odor source also have been used to indicate preferences. Finally, it may be possible to paint nipples with flavors and then record attachment by young as a method of evaluating preference (Stickrod et al. 1982).

Drinking tests are feasible when testing older subjects. Nolte and Mason (1995) used a single-choice drinking test to assess pup response to trigeminal irritants they were exposed to in utero. First, pups were adapted to an 18-h water deprivation schedule. Pups were presented with tap water in a 10 ml graduated sipper tube at 0900 h. At 1500 h, the tubes were removed until the following morning. Adaptation was followed by 4 days of pretreatment. Water intake between 0900 and 1200 h was measured during pretreatment. Treatment was similar to pretreatment, except the sipper tubes contained the respective test stimuli.

Grace and Russek (1969) rated the reactions of canine pups, on the day of birth, to drops of taste stimuli placed on their tongue. Distilled water and test solutions were introduced in random order with drop pipettes, while two observers independently recorded the response. Their response categories were (1) clear rejection (spitting, vomiting, or movement away from the drop pipette); (2) weak rejection (less intense aversive reaction with occasional movement toward pipette); (3) indifference (no evident reaction); (4) weak acceptance (swallowing and movement toward pipette with an occasional movement away; and (5) clear acceptance (rapid swallowing, obvious movement toward the pipette) (Jacobs 1964).

7.6.1.3. *Lactating Mammal Bioassays*

Nursing offspring can experience flavors of foods ingested by the mother through flavors transmitted to the milk (Dougherty et al. 1962; Bassettte et al. 1986). Offspring develop preferences for flavors in milk that reflect gustatory cues in the diet of their mother (Galef & Clark 1972; Galef & Henderson 1972; Galef & Sherry 1973; Capretta & Rawls 1974; Mainardi et al. 1989; Nolte & Provenza 1991; Altbacker et al. 1995). These pups later seek and consume food with similar flavors even if the food is normally unpalatable (Galef & Henderson, 1972). Chemical constituents within milk can be determined through samples collected from lactating females. Offspring experience with flavors in mother's milk can be manipulated by feeding or inserting (e.g., gavage) flavors into the mother. Flavors in milk also can be presented to a nursing offspring through bottle feeding. Most mammals readily adapt to bottle-feeding after a short training period. Stimuli concentration and extent of exposure can be easily controlled. Bottle-feeding also isolates the feeding experience to chemical stimulation.

Milk delivery can be used as a reinforcer for classical conditioning of an odor preference (Johanson & Teicher 1980; Brake 1981). This provides a means to test otherwise less responsive young subjects. Johanson and Teicher (1980) used this strategy to demonstrate that 3-day-old rats are capable of appetitive associative learning. Milk was delivered through a fine intraoral cannula installed in the back of the pups' mouths. Odor exposure started 15 s before milk delivery and continued through a 5-s infusion of milk. Conditioning was successful only when training occurred while pups were nutritionally deprived and when the milk and odor pairings occurred in a warm ambience. Similar trials with pups exposed to odors while suckling their mothers indicated that suckling in itself was rewarding, but stronger preferences were formed when odors were paired with milk delivery than when odors were paired with suckling without milk delivery (Brake 1981).

Figure 7.5. Apparatus to deliver milk during one-choice tests with suckling neonates. Intake of milk is measured first during a pretreatment period and then when flavors are added during a subsequent treatment period. Two-choice tests can be conducted by presenting two nipples connecting to containers with different flavors.

Young rat pups (10 days) also form taste aversions to flavors in milk. Preweanling rats, however, will not form aversions if flavors are presented while they are attached to a nipple (Martin & Alberts 1979).

Test paradigms to assess the effect of suckling experiences, need to consider the developmental capacities of the subject. Young altricial animals can orient themselves toward an odor within a maze or arena. Even 1-day-old rat pups can learn to probe upward into a paddle if their efforts are rewarded with milk (Johanson & Hall 1979). These pups also can learn to differentiate between paddles paired with different odors if only one delivers milk. For larger animals, a nipple bar attached to the side of the pen can permit easy access for young animals (Fig. 7.5). Milk temperature can be controlled by placing containers inside an ice chest with a tube connecting the nipple. Effort needed to receive milk can be manipulated by changing the relative height of the nipples to the container.

7.6.1.4. Bird Bioassays

Classic studies of free-ranging birds were published by Brower (1960). In her work, pastries were adulterated with quinine hydrochloride (a bittering agent) and/or color (green or red). Pastries containing quinine and color were defined as "models," while pastries containing color alone were defined as "mimics." Both pastry types were then distributed at different densities on lawns where the feeding behavior of birds could be observed. By manipulating the number of pastries containing quinine, she was able to determine the minimum number of quinine-adulterated baits necessary to keep birds from foraging on pellets treated with color alone. These data were used to develop a model of mimicry in which the number of models (quinine and color) needed to confer measurable protection on mimics (color alone) could be specified.

On a smaller scale, Avery and his colleagues (e.g., Decker & Avery 1990) showed the value of naturalistic observations in large outdoor aviaries. In their studies, groups of red-winged blackbirds were released into a 0.25-ha outdoor enclosure. The enclosure contained two paddies, one seeded with clay-coated rice and the other seeded with plain rice. Foraging times and the number of birds feeding within each paddy were recorded. The results showed that the clay increased foraging times and decreased the number of feeding birds.

A third example of the value of careful observation of free-ranging birds is provided by Wenzel (1972). She demonstrated the olfactory ability of the brown kiwi in the following way. Baited and unbaited feeding tubes were buried under 3 cm of dirt and covered with nylon screening. The birds, who have poor eyesight and feed at night, usually broke into the baited stations and not into the unbaited ones. This was taken as evidence that the birds used odor cues to find food.

Food avoidance learning has been used to examine birds' responsiveness to the color (Mason & Reidinger 1983), taste (Westbrook et al. 1980), and odor (Mason & Silver 1983). In each of these cases, similar procedures were used; Mason and Reidinger (1983) is used here as an example.

Red-winged blackbirds were individually caged under an 8:16 light-dark cycle selected to maximize consumption during the light period. Birds were visually isolated from one another to control for the influence of social facilitation on consumption (Mason & Reidinger 1982) and to prevent birds from seeing visual stimuli presented to conspecifics. After a 5-day pretreatment period during which baseline food consumption was evaluated, birds were assigned to two different groups, and then each group was presented with a sample of food in a colored cup. Different groups were presented with red or green cups and after 60 min, each bird received an oral gavage of methiocarb. Methiocarb is a carbamate insecticide and bird repellent that produces gastrointestinal sickness like a gavage or an injection of lithium chloride. During the posttreatment period, both groups were presented with both red and green cups in two-choice tests, and consumption was evaluated. The results showed that birds avoided food in the cup color experienced just prior to gavage.

Standard protocols are available for avian toxicity testing and the avoidance of pesticides. Because the details of these protocols have been published elsewhere (e.g., Schafer et al. 1983; Bennett & Schafer 1988; Grau et al. 1992; Kononen et al. 1986, Kononen 1988, Minneau et al. 1996), they are not presented in detail here.

7.6.2. Predator Avoidance (Defense)

The olfactory detection of predators by their prey is an important aspect of mammalian chemical signaling (Sullivan & Crump 1986). Mammalian predators typically produce pungent odors (Macdonald 1985) that can be recognized by prey (Ylonen et al. 1992). Odor detection can lead to the display of a variety of defensive behaviors including fight, immobilization, and increased wariness (Blanchard et al. 1990). The anticipated response depends on the species of prey, the species of predator, and the context in which stimuli are presented (Lima & Dill 1990). A natural response, for example flight, may not always be possible under laboratory conditions. Observed behaviors in the field also may vary depending on the immediate amount of cover in which to hide and the relative distance to a known protected hideaway (Merkins et al. 1991). Further, an appropriate prey response may depend on the perceived predator (Jedrzejewski et al. 1993).

Kavaliers et al. (1992) demonstrated that predator odors can elicit analgesia in mice. Temperature sensitivity was determined 15 min prior to, immediately after, and 15 min after exposure to predator odors for either 30 s or 15 min. Sensitivity was measured as the latency of a foot-licking response to a thermal stimulus (50° hot plate). Animals were significantly less sensitive (i.e., greater delay in response) when tested after odor exposures than when tested in the absence of predator odors.

Dell'Omo et al. (1994) recorded the amount of time mice spent in proximity to fox or rabbit odors and the frequency of several behaviors, including grooming, rearing, sniffing, freezing, and defensive burying. Mice were individually introduced to an arena, similar to their holding box, with a perforated container (35-mm film canister) containing either fox or rabbit faeces for 3 min. Activity was video recorded from above and subsequently scored. Two perpendicular lines divided the video screen into four equal sections, and time spent in the quadrant with the odor was a direct measure of avoidance.

Jederzejewski et al. (1993) developed an elaborate three-chambered arena with twigs for climbing and tubes for hiding to provide bank voles various escape options when presented with predator odors. The test apparatus consisted of a glass terrarium (1.5 × 1.7 m) divided into three equal sections with access between sections via a small hole cut in the dividers. Each section contained four plastic tubes (4 × 12 cm), four rubber tubes (3 × 20 m), four vertical twigs, and *ad libitum* access to oats and water. The tubes simulated rodent tunnels and the twigs provided arboreal escape routes during trials. Bank voles were introduced the night before a trial and allowed to acclimatize to the apparatus. Voles were removed from the test chamber the next morning while a predator was introduced to one of the peripheral sections to dispense its odor. Voles then were reintroduced to the chamber and their activity recorded for 10-min periods at intermittent intervals. Voles avoided the contaminated chamber and escape responding varied dependent on the type of predator. They were more likely to climb in the presence of mustelid odors than odors from canids, and avoided tubes simulating tunnels in the presence of mustelids.

The role of chemoreception in antipredator defenses can be demonstrated by comparing the response of intact and chemosensory impaired subjects to predator scents. Laboratory studies indicate that subjects with chemosensory impairments are less responsive to predators than are intact subjects. Naive intact rats presented with cat scent exhibited freezing and micturition more frequently than did rats whose olfactory bulbs had been removed (Sieck et al. 1974). Mollenauer et al. (1974) placed intact and bulbectomized rats in a cage adjacent to a cat separated by a wire mesh. Olfactory-impaired rats exhibited fewer signs of fear, defecating less and exploring more than did intact rats. A similar approach could be implemented in the field, and survivorship of intact subjects could be compared to that with individuals whose chemical senses have been impaired (Weldon 1990).

Response to food can be used as an indirect measure of prey response to predator scents. Stimuli can be placed either directly on the food or in close proximity to the food. For example, choice-tests were used to assess the avoidance of lion fecal odors by rabbits (Boag & Mlotkiewicz 1994) and deer (Abbot et al. 1990). In both cases, subjects were offered a choice between treated and untreated pelleted food and relative intake was taken to reflect avoidance. Likewise, arena tests have been used to demonstrate that domestic livestock will investigate but reduce their ingestion of feed in the presence of predator odors (Pfister et al. 1990). Odors also can be applied to natural forage to assess whether target species avoid the treated plants (Sullivan et al. 1988; Calder & Gorman 1991). Epple et al. (1995) monitored caching behavior to assess the response of mountain beaver to food resources associated with predator odors.

Merkins et al. (1991) compared the impact of cover on the intake of food by voles in the presence of predator odors. Feeding stations were established at a 2×2 m spacing in a diagonal pattern across both halves of three enclosures (6×11 m). Four nest boxes were located along the midline of the enclosures with straw strewn between them to provide cover. Prior to each test, vegetation within each enclosure was mowed to a height of 1 cm above ground level and all burrows and runways previously constructed by voles were destroyed. Experimental conditions consisted of cover without predator odor, cover with predator odor, and predator odor without cover. Treatments were applied to one half of an enclosure and they were always paired with an untreated half, no cover and no repellent. Predator odors were presented in a PVC tube impregnated with 1:1 mixture of 2-propylthietane and 3-propyl-1,2-dithiolane. A device was attached to each feeding station within the treated portions of the enclosures.

Cover was provided by dispersing straw to a depth of 5–15 cm. One female vole was then introduced to each nest box and given 4 days to acclimatize. Starting on the fifth day, food taken from each feeding station was measured by weighing the remaining food and subtracting it from the known starting weight. Voles avoided predator odors in the absence of cover, however, when cover was present, they preferred to feed on the side with cover, regardless of whether or not predator odors were present.

Relative capture success of clean and predator contaminated traps can be an indicator of avoidance. Stoddart (1976) assessed the effects of weasel odors on trap success in a mixed community of rodents. Two traps were placed at 10-m intervals on a 6 × 6 grid. Paired traps were placed as close as possible to each other and a pretreatment period indicated either trap was as likely as the other to capture a rodent. One trap in each pair then was randomly selected to be treated with weasel and gland secretion during the treatment period. Robinson (1990) found that trapping patterns made little difference in small mammal response to mink odors. Tests results were similar regardless of whether the traps were placed in pairs immediately adjacent to each other or in rows at 10-m intervals. Tobin et al. (1995) compared the capture rate of rats in traps known to have previously captured a mongoose to that of clean traps.

Relative capture rate has been used as an indicator of pocket gopher avoidance of predator odor fractions relative to areas without predator odors (Sullivan & Crump 1986). Animals were introduced to a neutral zone (35 × 6.5 cm) from which they had equal access to two arenas (35 × 50 cm). The floor of the neutral area and the arenas was covered in wax paper, to provide easy removal of cues that might linger from one trial to the next. Longworth live-traps located directly opposite the arena entrances simulated burrow system entrances. Trials were conducted under dim red light to stimulate darkness and enhance subject activity. On capture, a subject was removed, the test apparatus cleaned, and the wax paper replaced. Subsequently, the number of gophers captured in each arena was tallied to assess the impact of each compound on avoidance behavior. Similar field assays were conducted by comparing the relative success of traps containing a capillary tube filled with one of several possible predator odor fractions. The open end of the tube faced the door of the trap such that a pocket gopher could not enter and be caught in a trap without first being exposed to the odor of the respective compound.

Predator avoidance by potential prey can be inferred in the field by recording changes in prey populations or rates of invasion on treated versus untreated sites. For example, a capture-and-release trap program was used to assess pocket gopher avoidance of stoat and ferret anal-gland scents (Sullivan et al. 1988). Gophers were live-trapped at 3-week intervals on checkerboard grids established across treatment sites. Captured gophers were ear-tagged and the point of capture recorded. Conspecific invasion on treated sites was similarly monitored on sites where all pocket gophers had been removed. Pocket gopher mound counts can also serve as indicators of relative activity (Sullivan et al. 1990). Davis (1991) provides an overview of census methods for terrestrial vertebrates.

7.6.3. Community or Mutualistic Interactions

Chemical cues emitted by one species have been shown to alter the reproduction status of another species. Similar paradigms discussed above for intraspecific interactions can be used to assess whether behavioral patterns are altered because of interspecific odors.

Ronkaninen and Ylonen (1994) assessed changes in the behavior patterns of bank voles in the presence of mustelid odors. Mating pairs in reproductive condition were housed together for 3 days and then exposed to mustelid odors in an arena. The odors were introduced to the arena by placing a stoat in it for 3 h immediately prior to introducing the mated pairs. On subsequent days, stoat odor was sprayed on sawdust in the nest. General activity was measured by counting the number of passes into the nest chamber. Interactive behavior among pairs was observed at 10-min intervals and recorded as approaching, amicable interactions, copulations, threats, attacks and flights.

Predation risk was simulated in a similar experiment by placing a weasel in a cage adjacent to the paired voles for 2–3 h a day (Koskela et al. 1996). Females were considered to be estrus if they showed lordosis in response to mounting attempts by a male during a 10-min observation period.

The vaginal smear technique is another commonly used means to determine the estrus cycle in female rodents (Whitten & Champlin 1978). Collection of vaginal smears, however, can induce ovulation (MacFarlane & Taylor 1982) or cause abnormally long estrus cycles (Whitten & Champlin 1978) in some species. The lordosis response to males is a reliable indicator of estrus in at least voles and does not disturb the natural cycle (Carter et al. 1990).

Estrus in rats can be delayed by exposing females in metestrus to predator odors (Voznessenskaya et al. 1992b). Mink anal sac secretion was applied to the bedding of test animals and the onset of estrus compared to that of control females. Subsequently, they demonstrated this delay to be mediated by signals detected by the vomeronasal organ. Tests were repeated using rats that had either their vomeronasal organ removed or whose main olfactory system was blocked by 5% $ZnSO_4$. Delays in estrus continued in rats with blocked olfactory system but not in animals without a vomeronasal organ.

The reproductive status of some species can be enhanced by or requires intraspecific chemical signals. Berger et al. (1981) demonstrated the reproductive effort of montaine voles was triggered by chemical signals found in their food through a field trial. Rolled oats adulterated with 6-methoxybenzoxazolinone (6-MBOA), normally found in food resources, was provided to sexually inactive voles. Reproductive activity of voles captured from this site were then compared to voles captured from another population that had been given supplemental food without the 6-MBOA.

7.7. Case Studies

7.7.1. Chemical Correlates and Sensory Perception

Good examples of one- and two-choice feeding tests are provided in the experiments of Mason et al. (1989). These investigations were designed to evaluate the responsiveness of European starlings to a series of anthranilate derivatives. Starlings were selected as test subjects because they readily adapt to laboratory conditions, are easily obtained, and show good olfactory and taste sensitivity (Espaillat & Mason 1990; Clark & Mason 1987). Both methyl and dimethyl anthranilate are repellent to birds, and Mason and his colleagues wanted to explore the sensory systems mediating the avoidance response, as well as physical and chemical characteristics of anthranilates that might explain repellency.

Starlings were individually caged under a 6:18 light-dark cycle chosen to maximize the feeding rate during daylight without reducing the total amount consumed (Mason et al. 1989). Water was freely available, and before experiments began, birds were permitted *ad libitum* access to food (Purina Flight Bird Conditioner) and grit.

One week before the experiment began, birds were adapted to a food deprivation regime. Deprivation involved removing food and water from the cages just before dark. Just after light on the following day, each bird was presented with two cups (preceding two-choice tests) or one cup (preceding one-choice tests), each containing 20 g of food. For two-choice tests, the cups were bound together with a rubber band. This assured the left-right position of cups in cages, and minimized the number of cups upset by feeding birds. Consumption was measured after 2 h. Birds were then left undisturbed with free access to food and grit for the remaining hours of light. The 2-h feeding trial procedure followed by 4-h free access to food and then overnight food deprivation remained in effect throughout the experiments. In subsequent two-choice tests, birds were presented with two cups of food, one of which was adulterated with an anthranilate derivative. Tests were repeated on 5 successive days, and on each day, birds were presented with a different anthranilate derivative. All of the derivatives were presented at the same concentration. The data were evaluated in a two factor (anthranilates, food cups) repeated measures analysis of variance (ANOVA). In addition, suppression scores were calculated by dividing consumption of anthranilate treated feed by total consumption. These scores were correlated with the molecular weights and boiling points of the chemicals.

For one-choice tests, different groups of birds were tested with each of the various anthranilates. Over-days, each substance was presented at five concentrations. The data were evaluated in a two-factor (anthranilates, concentrations)

ANOVA. As in two-choice tests, rank-order correlation coefficients were used to associate behavioral suppression scores with the physicochemical properties of the derivatives. For this purpose six properties (molecular weight, boiling point, number of carbon atoms, percentage oxygen, percentage nitrogen, percentage nonanthranilate carbons) were initially considered in a factor analysis to reduce the number of variables. The factor scores (Norusis 1986) were used to compute the rank correlation.

Subsequently, the investigators performed surgical bisections of the olfactory nerves, and then surgical bisections of both the olfactory and nasal trigeminal nerves. Two- and one-choice testing was then repeated, and statistically evaluated as previously described.

Using simple one- and two-choice feeding tests, Mason and his coworkers were able to answer several fundamental questions. First, two-choice tests were used to investigate whether each of the anthranilates was detected, and suppression scores calculated from these data were used in an attempt to differentiate among the repellencies of the various substances. One-choice tests were used to decrease the likelihood of differential responding, and thereby increase the probability of finding differences among the derivatives. Repeating the simple test procedures after surgical interventions allowed the investigators to examine the relative contribution of two sensory systems to the observed behavioral performance.

7.7.2. Responses of Birds to Aversive Stimuli

There may be situations in which it is either difficult or inappropriate to present stimulus samples in cups. One solution to this problem is to present stimuli on the substrate in such fashion that stimuli can be retrieved and separated at the end of the test session. Mason et al. (1993) determined whether methyl anthranilate could be used to eliminate or reduce the consumption of pelleted pesticides. Such pesticides present so-called nontarget hazards to birds that ingest them for grit (Gionfriddo & Best 1995). Methyl anthranilate was incorporated into pelleted baits and the pellets were presented to cowbirds in two laboratory experiments and an outdoor aviary evaluation. Cowbirds were chosen as test subjects because they are a species potentially at risk from pelleted pesticide ingestion. In laboratory tests, birds were permitted a 2-week period to adapt to captivity. During this period, the birds were individually caged and given free access to feed presented on top of 500 ml of sand in plastic tubs inserted into the cages. Sand particles were uniformly smaller than U.S. Standard sieve size 18. Birds were food-deprived overnight to encourage feed and pellet consumption during testing. During the pretreatment period that followed adaptation, birds were given 30 g of feed and 1 g of control pellets in plastic tubs containing 500 ml of sand, as in adaptation. While the feed was presented in a metal cup positioned in the center of the tub, the pellets were evenly spread over the sand surface. At the conclusion of each 3-h session the amount of feed remaining was weighed and differences from the beginning of the trial were defined as consumption. To evaluate pellet ingestion pellet size was standardized prior to each trial as particles that passed through a #16 sieve but were caught on a #18 sieve. This pellet size range encompassed most pellets but was also the same size as feed granules; hence the separate presentation.

At the end of each session, tubs of sand were poured through the #18 sieve, and pellets trapped on the sieve were weighed. Visual inspection of sand indicated that this procedure trapped all of the pellets (easily seen against the white background). To control for moisture absorption or loss during test sessions, 1-g samples of pellets were left in the test room during test periods. Mean percentage changes in the weight of these samples were used to adjust the weight of pellet samples recovered from the tubs. Because moisture associated changes in feed weights were negligible in pilot tests, no similar controls for moisture absorption or loss were implemented.

At the end of the pretreatment period, birds were weighed and ranked on the basis of consumption of feed. These rankings were then used to assign birds to groups that were balanced with respect to feed intake. Pellet consumption was not used to assign birds to groups because measurable ingestion was low. The pretreatment period was then followed by treatment trials in which birds were presented with pellets that either did or did not contain methyl anthranilate, and consumption was estimated as described above.

In outdoor aviary tests, birds were released into 16 large ($6 \times 3 \times 3$ m) rectangular enclosures ($n = 12$ birds/enclosure). The ground underneath each enclosure had been tilled. During a 1-week pretreatment period, and throughout the treatment period, all birds were given free access to sunflower seeds, cracked corn, and wheat seeds that were liberally spread on the ground within the enclosures. Water was freely available from two bowls. On the day following the last day of pretreatment, the enclosures were randomly assigned to various treatments, and different pellet types were evenly spread on the floors of enclosures. After these treatment applications, the number of pellets inside 929-cm^2 sampling plots ($n = 6$ plots/enclosure) were counted. At the end of the treatment period, the number of pellets within sampling plots was again counted, and the difference from the number initially present was taken as a measure of consumption (differences were adjusted by counting the number of pellets within a control enclosure without birds). Two-factor (treatment condition, date) analysis of variance were used to evaluate the data.

7.7.3. *Relating Repellency and Chemical Structure*

Relative to feeding tests and their superiority for simple olfactory testing, drinking tests may be more useful for the measurement of taste sensitivity. There are several reasons for this. Among the most important is that stimuli are presented in solution, and this enhances receptor accessibility. An illustration of this advantage is the relatively greater sensitivity of birds to acetophenone derivatives in drinking (Clark 1997; Clark & Shah 1991) versus feeding (Mason et al. 1991a) tests. Because the drinking tests devised by Clark and Shah (1991) were both elegant and effective, the procedures are described in detail below.

Clark and Shah (1991) tested isomers and moieties of acetophenones for their effectiveness as bird repellents to better understand the nature of repellency in birds. Starlings served as test subjects. Birds were individually caged under a 12:12 light:dark cycle. Food and grit were available ad libitum, and before experiments began, birds were permitted free access to tapwater. Because acetophenones are generally insoluble in water, stimuli were prepared by mixing each substance with water under low heat to yield a saturated emulsion of 0.5% (mass/vol or vol/vol). Lower concentrations (0.25, 0.125, 0.0625, 0.0313) were prepared by serial dilutions.

In one-choice tests, starlings were given 3 days of pretreatment during which water consumption was measured after 6 h. At the end of this period, individuals whose variance around the 3-day mean was greater than one standard deviation of the population variance were excluded from further observation. Those birds with stable daily water consumption were ranked according to mean water consumption and assigned to treatment groups. The bird with the highest consumption was assigned to the 0.5% treatment group, the bird with the second highest consumption was assigned to the 0.25% treatment group, and so forth, until all of the birds were assigned to a group (n = 4 birds/group).

After group assignment, birds were given a 1-day drinking trial. Birds had free access to feed and tapwater overnight. At 0930 h, water was replaced with the preassigned concentration of chemical, and drinking was recorded at 2-h intervals for 6 h. At the end of the test period, birds were again given free access to tap water. Consumption of tapwater was monitored overnight and for a 6-h period the next day. Overnight consumption was measured to determine whether birds compensated for any water deficits resulting from the experimental treatments. The within-group 6-h posttreatment-day water consumption was compared with the mean within group 6-h pretreatment water consumption to determine whether consumption had returned to pretreatment levels. If not, testing of the next compound was postponed until diurnal water consumption returned to pretreatment levels. Once this condition was met, the birds were tested with the next substance, with groups of birds receiving different concentration according to a counterbalanced predetermined schedule; i.e., test groups initially were randomly assigned to one of the six concentrations to be tested. On subsequent tests, groups were assigned the next highest concentration. The order in which acetophenones were tested was arbitrary.

Naive birds were not used for each test because it was not practical to capture the 216 starlings that would have been required to complete the experiment. If an individual's 6-h interexperiment water consumption was within one standard error of its pretreatment value, the bird was used in the next experiment (i.e., testing of the next compound), otherwise a new bird was substituted. Throughout the experiment, only two birds were replaced, and this was because the birds developed the habit of jumping up and down on the drinking tubes, causing spillage.

To test for carryover effects due to treatment, Clark and Shah (1991) examined the intertreatment tapwater consumption of all birds with a repeated measures one-way analysis of variance. They compared each posttreatment day's water consumption by partition using an *a priori* contrast with pretreatment water consumption (Norusis 1986). Each substance was considered separately.

Two *a priori* hypotheses were tested. First, did consumption of treated water differ from a theoretical value of zero consumption? The analysis required a slight modification in calculation of the treatment sums of squares, where the grand mean was replaced by zero and the degrees of freedom reflected the number of treatments (six) considered in the experiment. Estimates of the error term remained the same as in a standard analysis of variance. *Post hoc* comparisons were made using a modification of Dunnett's *t*-test, again using a theoretical value of zero rather than the mean, and comparing the resulting t to critical values in Dunnett's calculated distribution with P set at <0.05. The second question was whether mean water consumption differed among treatments (concentration). A one-factor analysis of variance was used to compare group means, and Scheffe's *post hoc* tests were used to identify significant ($P < 0.05$) differences among means.

A least-squares fit for a nonlinear function of the form $y(x) = a(\exp - sx)$ with the Marquardt-Levenberg algorithm, where a is the consumption at 0% concentration, s is the rate of change, and x is the concentration of the compound, to fit curves for the dose-response relationship. The two-parameter model was used because it provided the best fit for most of the compounds tested over the broadest range of concentrations.

Because care was taken to ensure that carryover effects were not impacting results, Clark and Shah (1991) were able to achieve a high degree of precision. They concluded that the basicity of a substituted phenyl ring, as defined by the electron-donating substituent, was an important feature influencing acetophenone repellency; more basic substituents resulted in greater repellency. Isomeric position of the electron-donating substituent, which leads to resonance of lone pairs of electrons, also was important. An ancillary contributory factor was the presence of intramolecular hydrogen bonds. Combining these features resulted in the greatest aversiveness.

7.7.4. Conspecifics Response to Scent Marks

The role of chemical communication in the social behavior of mountain beavers is mostly unknown (Brown 1985). The use of scent marks, however, was suspected because although solitary, mountain beavers often live close to conspecifics

(Martin 1971). Further, nest sites are a valuable resource, rarely vacated voluntarily, and a vacated nest is quickly occupied by other mountain beavers (Campbell et al. 1988).

Field observations of mountain beaver is difficult if not impossible because they are fossorial and predominately nocturnal. Therefore, Nolte et al. (1993a) created an artificial system to monitor mountain beaver activity. Each system consisted of an open cage (204 × 92 × 64 cm) with an assortment of tunnels and chambers below. Burrows were constructed of clear PVC pipe (15 cm diameter) and chambers consisted of polyethylene buckets (30 cm diameter × either 27 or 35 cm) with Plexiglas™ windows in the lids. Free access to food and water was provided in the open cage and aspen shavings covered the cage floor.

Mountain beaver were introduced to the artificial burrow system by moving the nest bucket attached to their individual cage to the burrow system. Thus, each introduced mountain beaver had the option to retreat to its own nest. Introduced animals emerged into a small cage. Access between this and the larger enclosure was controlled by a sliding panel.

Interactions between mountain beavers were observed in the morning at the beginning of the light period. Mountain beavers were introduced to the burrow system by attaching their nest buckets from individual cages as described above. Trials were concluded when animals returned to their respective nests and failed to emerge again for 15 min. Encounters were always between animals unfamiliar with each other. An intruder was never reintroduced to a system if it had previously encountered the resident animal.

These observations provided insight into the role of chemical communication among mountain beaver. Mountain beaver appeared to mark burrows outside their nest with odors. They alternated between grooming and rubbing against chamber entrances and burrows, and while grooming they frequently put their feet to their mouths. Subsequently, marked areas were repeatedly sniffed by both the intruder and established animal. Although mountain beaver marked burrows with scents, these marks were not sufficient to prevent invasion by conspecifics. In the absence of residents the intruders repeatedly explored marked systems, rapidly exploiting any opportunity to establish themselves. Encounters between conspecifics, however, were invariably agonistic. Established animals invariably retained possession of the burrow system and familiarity does not fully explain this dominance. The current resident animal ejected an intruding conspecific after living in a burrow system for only 72 h, even if the intruder had previously occupied the same system for 30 days and only 72 h earlier had successfully ejected the current resident. We concluded that scent marking may serve to enhance the territorial confidence of mountain beaver.

7.7.5. Aversive Components of Predator Odors

Predator odors are generally aversive to potential prey species. Avoidance appears to be mediated, at least in part, by urinary constituents, which are not species-specific. Such compounds may constitute a generalized meat eater cue (Abbott et al. 1990; Epple et al. 1993), or a predator "leitmotiv" (Stoddart 1980). We conducted a series of experiments to assess whether diet manipulations affect repellency of urine, and to investigate the contribution of sulfurous compounds to its repellency (Nolte et al. 1994).

The first experiment assessed whether diet manipulations affect the repellency of a predator urine to several prey species. Four rodents served as subjects: mountain beaver, house mice, deer mice and guinea pigs. Urine was collected from coyotes maintained on cantaloupe (CU) for 2 weeks and then from the same coyotes fed minced raw meat (MU) for two weeks. Test procedures were similar for all species, though food and deprivation schedules varied. On each of 2 pretreatment days, animals were given their respective foods in cups containing perforated containers with a piece of absorbent paper treated with 1 ml of tap water. On the 2 treatment days that followed, the animals were given the same foods, but the absorbent paper was treated with 1 ml of either CU or MU. Urine samples (1 ml) were pipetted onto pieces of absorbent paper placed inside small (38 mm diameter × 8 mm) perforated plastic containers. For all individuals of each species, the left-right position of CU urine samples was randomly determined on day 1, and then reversed on day 2. The data for each species was evaluated separately in a two-factor repeated measures ANOVA. In each case, urine type was the main effect, with the animals nested within urine type and the repeated measure was days.

Animals avoided MU more than they did CU in the first experiment. These differences, however, may have reflected dilution effects. Therefore, we conducted a second experiment with deer mice using CU and MU samples that had been lyophilized and then rehydrated to a common concentration. On each of 4 pretreatment days, food deprived mice (18 h) were presented with two weigh-boats, each containing 40 sunflower seeds of known weight, and a perforated plastic container with an enclosed absorbent paper treated with 1 ml of tap water. After 2 h, the weigh-boats were removed and the number of seeds remaining were assessed. On the basis of ingestion, animals were then assigned to two counterbalanced groups, and tested for 4 days. One group was offered a choice between sunflower seeds associated with CU or tap water for 2 h. The other group was offered a choice between sunflower seeds associated with MU or tap water for 2 h. Differences scores were calculated by subtracting CU or MU ingestion from control (tap water scented) ingestion by individual deer mice. These data were evaluated in a two-factor repeated measures ANOVA.

Diet manipulation experiments showed that the presence of meat in a donor coyote's diet increased the repellency of the urine. This result led us to hypothesize that byproducts of meat digestion in urine (e.g., sulfur compounds) might contribute to aversiveness. To test this hypothesis, we removed sulfur-containing substances from MU by precipitation with mercuric chloride. The responses of mountain beaver to urine stimuli with and without sulfur compounds were then investigated. Counterbalanced two-choice tests were used to assess ingestion of apple cubes associated with either tap water or MU, and apple cubes associated with either tap water or sulfur removed (SR). For each test, apple cubes were available for 12 h on 2 consecutive days. Difference scores were calculated by subtracting the number of apple cubes associated with MU or SR ingested by mountain beaver from the number of apple cubes associated with tap water ingested during the same trial. Subsequently, these data were evaluated in a two-factor repeated measures ANOVA.

Overall, the results were consistent with the hypothesis that (as yet unknown) sulfurous odors associated with meat digestion are important for the repellency of predator urines. A plausible speculation is that prey use such odors as cues to the diet composition of donors. It may also be that predators attend to sulfurous urinary odors and interpret them as indicators of diet composition; this provides an explanation of the attractiveness of predator urines to these species (Bullard et al. 1983).

7.7.6. Chemically Mediated Foraging by Black Bears

Following winter dormancy, black bears commonly forage on Douglas fir trees. This foraging appears to be selective as not all timber stands are utilized, and within stands, one tree may be stripped while its neighbor is ignored or merely sampled. An effort was implemented to determine whether bear selection is influenced by the chemical constituents present in the trees.

Kimball et al. (1998) developed a method to investigate bear forage selection in the field because of the difficulty in obtaining stimuli, as well as subjects that adequately reflect conditions present under normal conditions. First, samples of vascular tissue were collected from preferred trees and rejected trees. Trees where bears had removed more than 2500 cm^2 of the bark were considered to be preferred, while trees with less than 250 cm^2 of the bark removed were considered to have been sampled and then rejected. Trees with no evident foraging marks were not sampled since it could not be ascertained whether they had been encountered by a foraging bear. Chemical assays then were conducted to quantify the carbohydrates and terpenoids within these samples.

The first study revealed that trees selected by bears were higher in carbohydrates and lower in terpenoids than adjacent rejected trees. Although these results suggested a correlation between chemical constituents and bear foraging it was not known whether bears actually selected forage based on this criteria. Therefore, a choice test was designed to determine if bear would select for carbohydrates and avoid terpenoids in foods.

At each test site, four feeders were placed in close proximity (10 m) to one another and filled with an untreated pellet (Fig. 7.6). Feeders were arranged in

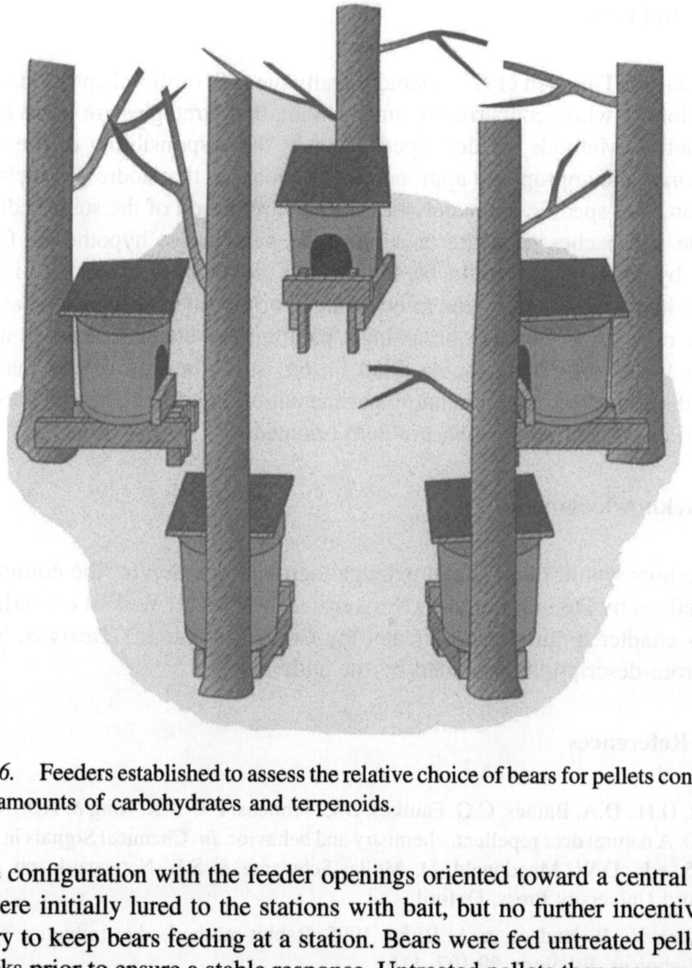

Figure 7.6. Feeders established to assess the relative choice of bears for pellets containing variable amounts of carbohydrates and terpenoids.

a square configuration with the feeder openings oriented toward a central focus. Bears were initially lured to the stations with bait, but no further incentive was necessary to keep bears feeding at a station. Bears were fed untreated pellets for two weeks prior to ensure a stable response. Untreated pellets then were replaced by test diets. The diets reflected the four combinations of high and low carbohydrates and high and low terpenoids identified in trees. Test diets were formulated to closely mimic the carbohydrate and terpenoid constituents and their respective concentrations. Intake of feed was indirectly monitored by measuring the amount of feed taken every 3 days for 3 weeks.

Bears ate more of the foods with high- as opposed to low-carbohydrate concentrations. Pellets with a high-terpenoid concentration were ingested less than pellets low in terpenoids. Thus, intake of the pelleted diets reflected the corelational data from the first study investigating chemical constituents in preferred and rejected trees. These combined results give credence to the hypothesis that bear foraging on trees is at least in part mediated by chemical constituents.

10.8 Summary

Paraphrasing Thiessen (1977), paradoxically the task of this chapter is to describe methodology while concurrently emphasizing that strategies are more important than tactics. Methods are described, but it is the responsibility of the reader to incorporate the appropriate approach into a bioassay that addresses their specific question. The specific characteristics of the stimuli and of the subject dictate the feasible approaches to refute or support the perspective hypotheses. Creativity fueled by knowledge should be foremost in developing bioassays. Bioassays, along with the apparatus, need to be changed or modified as questions are refined or new ones arise. Fixation on a single paradigm or attachment to an apparatus at best limits opportunities. As Platt (1964) said "beware of the man of one method or one instrument, either experimental or theoretical, he tends to become method orientated rather than problem oriented."

7.9. Acknowledgments

The authors would like to acknowledge their appreciation for the comments and suggestions by Drs. Frederick D. Provenza and Michael W. Fall on earlier drafts of this chapter. Figures were created by Carl Yoshihara (Y-Designs, San Jose, CA) from descriptions provided by the authors.

7.10. References

Abbott, D.H., D.A. Baines, C.G. Faulkes, D.C. Jennens, P.C.Y.K. Ning & A.J. Tomlinson. 1990. A natural deer repellent: chemistry and behavior. *In*: Chemical Signals in Vertebrates 5, eds. D.W. Macdonald, D. Müller-Schwarze & S.E. Natynczuk, pp. 599–609, Oxford University Press, Oxford.

Altbacker, V., R. Hudson & A. Bilko. 1995. Rabbit-mother's diet influence pups' later food choices. Ethology **99**:107–116.

Apfelbach, R. 1978. A sensitive phase for the development of olfactory preferences in ferrets (*Mustela putorius*). Z. Saugetierkd. **43**:289–295.

Apfelbach, R. 1986. Imprinting on prey odours in ferrets (*Mustela putorius F. Furo* L.) and its neural correlates. Behav. Proc. **12**:363–381.

Apfelbach, R. 1992. Ontogenetic olfactory experience and adult searching behavior in the carnivorous ferret. *In*: Chemical Signals in Vertebrates 6, eds. R.L. Doty & D. Müller-Schwarze, pp. 155–159, Plenum Press, New York.

Avery, M.L. & D.G. Decker. 1991. Repellency of fungicidal rice seed treatments to red-winged blackbirds. J. Wildl. Manage. **55**:327–334.

Avery, M.L., D.G. Decker & D.L. Fischer. 1994. Cage and flight pen evaluation of avian repellency and hazard associated with imidachloprid-treated rice seed. Crop Protect. **13**:535–540.

Avery, M.L., D.G. Decker, J.S. Humphrey, E. Aronov, S.D. Linscombe & M.O. Way. 1995. Methyl anthranilate as a rice seed treatment to deter birds. J. Wildl. Manage. **59**:50–56.

Baldaccini, N.E., S. Benvenuti, V. Fiaschi & F. Papi. 1975. New data on the influence of olfactory deprivation on the homing behavior of pigeons. *In*: Olfaction and Taste V, eds. D.A. Denton & J.P. Coghlan, pp. 351–354. Academic Press, New York.

Barker, L.M., M.R. Best & M. Domjam. 1977. Learning Mechanisms in Food Selection. Baylor University Press, Waco, TX.

Barnard, C.J. & J. Fitzsimmons. 1988. Kin recognition and mate choice in mice: the effects of kinship, familiarity and social interference on intersexual interaction. Anim. Behav. **36**:1078–1090.

Bartos, L. & P. Rodl. 1990. Effects of conspecific's urine on reproduction in arctic foxes. *In*: Chemical Signals in Vertebrates 5, eds. D.W. Macdonald, D. Müller-Schwarze & S.E. Natynczuk, pp. 353–359, Oxford University Press, Oxford.

Bassette, R., D.T.C. Fung & V.R. Manthar. 1986. Off-flavors in milk. CRC Crit. Rev. Food Sci. Nutr. **24**:1–52.

Bean, N.J. & J.R. Mason. 1995. Attractiveness of liquid baits containing natural and artificial sweeteners to white-tailed deer. J. Wildl. Manage. **59**:610–613.

Bean, N.J., B.G. Galef & J.R. Mason. 1988. The effect of carbon disulfide on food consumption by house mice. J. Wildl. Manage. **52**:502–507.

Beecher, M.D. 1982. Signature systems and kin recognition. Amer. Zool. **22**:477–490.

Beidler, L.M., M.E. Rashotte, J.C. Smith & R.W. Griffin. 1984. Neuroscience and Biobehavioral Reviews: Studies in Canine Olfaction, Taste and Feeding. Ankho International, Inc. Fayetteville, New York.

Bennett, R.S. 1989a. Role of dietary choices in the ability of bobwhite to discriminate between insecticide-treated and untreated food. Environ. Toxicol. Chem. **8**:731–738.

Bennett, R.S. 1989b. Factors influencing discrimination between insecticide-treated and untreated food by northern bobwhite. Arch. Environ. Contam. Toxicol. **18**:697–705.

Bennett, R.S. & D.W. Schafer. 1988. Procedure for evaluating the potential ability of birds to avoid chemically contaminated food. Environ. Toxicol. Chem. **7**:359–362.

Benvenuti, S., P. Ioale & F. Papi. 1992. The olfactory map of homing pigeons. *In*: Chemical Signals in Vertebrates VI, eds. R.L. Doty & D. Müller-Schwarze, pp. 429–434, Plenum Press, New York.

Berger, P.J., N.C. Negus, E.H. Sanders & P.D. Gardner. 1981. Chemical triggering of reproduction in *Microtus montanus*. Science. **214**:69–70.

Blanchard, R.J., C.D. Blanchard, J. Rodgers & S.M.Weiss. 1990. The characterization and modelling of antipredator defensive behavior. Neurosci. Biobehav. Rev. **14**:463–472.

Boag, B. & J.A. Mlotkiewicz. 1994. Effect of odor derived from lion faeces on behavior of wild rabbits. J. Chem. Ecol. **20**:631–637.

Bodyak, N.D. & A.V. Surov. 1995. The females' Harderian gland and pheromonal communication in golden hamsters. *In*: Chemical Signals in Vertebrates VII, eds. R. Apfelbach, D. Müller-Schwarze, K. Reutter & E. Weiler, pp. 325–331, Pergamon, Oxford.

Bolles, R.C., L. Haywood, & C. Crandall. 1981. Conditioned taste preferences based on caloric density. J. Exp. Psychol. **7**:59–69.

Boonstra, R. & C.J. Krebs. 1976. The effect of odour on trap response in *Microtus townsendii*. J. Zool. Lond. **180**:467–476.

Booth, D.A. 1985. Food-conditioned eating preferences and aversions with introceptive elements: conditioned appetites and satieties. *In*: Experimental Assessments and Clinical Applications of Conditioned Food Aversions, eds. N.S. Braveman & P. Bronstein, pp. 22–41, New York Academy of Science, New York.

Bowers, J.M. & B.K. Alexander. 1967. Mice: individual recognition by olfactory cues. Science **158**:1208–1210.

Bradley, R.M. & L.M. Beidler. 1970. Investigations of intravenous taste using the perfused rat tongue. Fed. Proc. **29**:A522.

Bradley, R.M. & C.M. Mistretta. 1980. Developmental changes in neurophysiological taste responses from the medulla in sheep. Brain Res. **191**:21–34.

Brake, S.C. 1981. Suckling infant rats learn a preference for a novel olfactory stimulus paired with milk delivery. Science **211**:506–508.

Braveman, N.S. and P. Bronstein. 1985. Experimental Assessments and Clinical Applications of Conditioned Food Aversions. N. Y. Academy of Science, New York.

Breland, K. & M. Breland. 1961. The misbehavior of organisms. Am. Psych. **16**:661–664.

Breland, K. & M. Breland. 1966. Animal Behavior, MacMillian Publishing, New York.

Brower, J.V.Z. 1960. Experimental studies of mimicry. Part IV. The reactions of starlings to different proportions of models and mimics. Am. Natural. **44**:271–281.

Brown, R.E. 1985. The rodents II: suborder Myomopha. *In*: Social Odours in Mammals. Vol. 1, eds. R.E. Brown & D.W. Macdonald, pp. 345–457, Clarendon Press, Oxford.

Brown, D.S. & R.E. Johnston. 1983. Individual discrimination on the basis of urine in dogs and wolves. *In*: Chemical Signals in Vertebrates 3, eds. D. Muller-Schwarze & R.M. Silverstein pp. 343–346, Plenum, New York.

Bruce, H.M. 1960. A block to pregnancy in the mouse by proximity of strange males. J. Reprod. Fertil. **1**:96–103.

Bullard, R.W., F.J. Turkowski & S.R. Kilburn. 1983. Responses of free-ranging coyotes to lures and their modifications. J. Chem. Ecol. **9**:877–888.

Calder, C.J. & M.L. Gorman. 1991. The effects of red fox *Vulpes vulpes* faecal odours on the feeding behavior of orkney voles *Microtus arvalis*. J. Zool. **224**:599–606.

Campbell, D.L., J.D. Ocheltree & M.G. Carey. 1988. Adaptation of mountain beaver (*Aplodontia rufa*) to removal of underground nests. Northw. Sci. **62**:75.

Capretta, P.J. & L.H. Rawls. 1974. Establishment of a flavor preference in rats: importance of nursing and weanling experience. J. Comp. Physiol. Psychol. **88**:670–673.

Carter, C.S., D.M. Witt, S.R. Manock, K. A. Adams, J.A. Bahr & K. Carlstead. 1990. Hormonal correlates of sexual behavior and ovulation in male-induced and postpartum estrus of female prairie vole. Physiol. Behav. **46**:941–948.

Cherry, J.A. 1993. Measurements of sexual behavior: control of variables. *In*: Paradigms for the Study of Behavior, ed. P.M. Conn, pp. 3–15, Academic Press, New York.

Clapperton, B.K., L.R. Matthews, M.S. Fawkes & A.J. Pearson. 1996. Lithium and cyanide-induced conditioned food aversions in brushtail possums. J. Wildl. Manage. **60**:195–201.

Clark, L. 1995. Modulation of avian responsiveness to chemical irritants: effects of prostaglandin E1 and analgesics. J. Exp. Zool. **271**:432–440.

Clark, L. 1997. A review of the bird repellent effects of 117 carbocyclic compounds. *In*: Repellents in Wildlife Management, ed. J.R. Mason, in press, Colorado State University Press, Fort Collins.

Clark, L. & J.R. Mason. 1987. Olfactory discrimination of plant volatiles by the European starling. Anim. Behav. **35**:227–235.

Clark, L. & P.S. Shah. 1991. Nonlethal bird repellents: in search of a general model relating repellency and chemical structure. J. Wildl. Manage. **55**:538–545.

Cohen, J.A. & E.O. Price. 1979. Grooming in the Norway rat: Displacement activity or "boundary-shift"? Behav. Neural Biol. **26**:177–188.

Conn, P.M. 1993. Methods in Neuroscience: Paradigms for the Study of Behavior. Academic Press, New York.

Connor, J.L. & P.G. Lynds. 1977. Mouse aggression and the intruder-familiarity effect: evidence for multiple-factor determination. J. Comp. Physio. Psychol. **91**:270–280.

Coppola, D.M. & J.G. Vandenbergh. 1987. Induction of a puberty-regulating chemosignal in wild mouse populations. J. Mammal. **68**:86–91.

Cummings, J.L., J.R. Mason, D.L. Otis & J.M. Heisterberg. 1991. Evaluation of dimethyl and methyl anthranilate as Canada goose repellents on grass. Wildl. Soc. Bull. **19**:184–190.

Davis, D.E. 1991. CRC Handbook of Census Methods for Terrestrial Vertebrates. CRC Press, Boca Raton, FL.

Decker, D.G. & M.L. Avery. 1990. Reducing blackbird damage to newly planted rice with a nontoxic clay-based seed coating. Proc. Vert. Pest Conf. **14**:327–331.

Dehnhard, M. & R. Claus. 1988. Reliability criteria of a bioassay using rats trained to detect estrus-specific odor in cow urine. Theriogenology **30**:1127–1138.

Dell'Omo, G., M. Fiore & E. Alleva. 1994. Strain differences in mouse response to odours of predators. Behav. Proc. **32**:105–116.

Demski, L.S. & R.G. Northcutt. 1983. The terminal nerve: a new chemosensory system in vertebrates. Science **220**:435–437.

Desjardins, C., J.A. Maruniak & F.H. Bronson. 1973. Social rank in house mice: differentiation revealed by ultraviolet visualization of urinary marking patterns. Science **182**:939–941.

Dewsbury, D.A. 1982. Dominance rank, copulatory behavior, and differential reproduction. Q. Rev. Biol. **57**:135–159.

Doty, R.L. & I. Dunbar. 1974. Attraction of beagles to conspecific urine, vaginal and anal sac secretion odors. Physiol. Behav. **12**:825–833.

Dougherty, R.W., W.F. Shipe, G.V. Gudnason, R.A. Ledford, R.D. Peterson & R. Scarpellino. 1962. Physiological mechanisms involved in transmitting flavors and odors to milk. 1. contribution of eructated gases to milk flavor. J. Dairy Sci. **45**:472–476.

Dragoin, W.B., G.E. McCleary & P.A. McCleary. 1971. A comparison of two methods of measuring conditioned taste aversions. Behav. Res. Methods and Instr. **3**:309–310.

Dravnieks, A. 1974. Instruction Manual: Dynamic Dilution Binary Scale Olfactometer. American Society for Testing and Materials, Philadelphia.

Dravnieks, A. 1975. Instrumental aspects of olfactometry. *In*: Methods in Olfactory Research, ed. D.G. Moulton, A. Turk & J.W. Johnston, pp. 1–62, Academic Press, N.Y.

Drickamer, L.C. 1995. Odors in traps: does most recent occupant influence capture rates in house mice? J. Chem. Ecol. **21**:541–556.

Dunbar, I.F. 1977. Olfactory preferences in dogs: the response of male and female beagles to conspecific odors. Behav. Biol. **20**:471–481.

Du Toit, J.T., F.D. Provenza & A. Nastis. 1991. Conditioned taste aversions: how sick must a ruminant get before it learns about toxicity in foods? Appl. Anim. Behav. Sci. **30**:35–46.

Elizalde, G. & A. Sclafani. 1990. Flavor preferences conditioned by intragastric polycose: a detailed analysis using an electronic esophagus preparation. Physiol. Behav. **47**:63–77.

Epple, G., A. Belcher, K.L. Greenfield, I. Kuderling, K. Nordstrom & A.B. Smith. 1989. Scent mixtures used as social signals in two primate species: *Saguinus fuscicollis* and *Saguinus oedipus*. *In*: Perception of Complex Smells and Tastes, eds. D.E. Laing, W.S. Cain, R.L. McBride & B.W. Aclre, pp. 1–25, Academic Press, Australia.

Epple, G., J.R. Mason, D.L. Nolte & D.L. Campbell. 1993. Effects of predator odors on feeding in the mountain beaver (*Aplodontia rufa*). J. Mammal. **74**:715–722.

Epple, G., J.R. Mason, E. Aronov, D.L. Nolte, R.A. Hartz, R. Kaloostian, D.L. Campbell & A.B. Smith. 1995. Foraging responses to natural and synthetic predator cues in *Aplodontia rufa*, the North American mountain beaver. Ecol. Appl. **5**:1163–1170.

Espaillat, J.E. & J.R. Mason. 1990. Differences in taste preference between red-winged blackbirds (*Agelaius phoeniceus*) and European starlings (*Sturnus vulgaris*). Wilson Bull. **102**:292–299.

Feldman, M. & C.T. Richardson. 1986. The cephalic phase of gastric secretion. *In*: Interaction of the Chemical Senses and Nutrition, eds. M.R. Kare & J.G. Brand, pp. 181–192, Academic Press, New York.

Ferkin, M.H. & R.E. Johnston. 1995a. Effects of pregnancy, lactation and postpartum oestrus on odour signals and the attraction to odours in female meadow voles. Anim. Behav. **49**:1211–1218.

Ferkin, M.H. & R.E. Johnston. 1995b. Meadow voles, *Microtus pennsylvanicus*, use multiple sources of scent for sex recognition. Anim. Behav. **49**:37–44.

Ferkin, M.H., F.H. Ferkin & M. Richmond. 1994. Sources of scent used by prairie voles, *Microtus ochrogaster*, to convey sexual identity to conspecifics. Can. J. Zool. **72**:2205–2209.

Ferkin, M.H., E.S. Sorokin & R.R. Johnston. 1996. Self-grooming as a sexually dimorphic communicative behaviour in meadow voles, *Microtus pennsylvanicus*. Anim. Behav. **51**:801–810.

Galef, B.G. 1986. Olfactory communication among rats of information concerning distant diets. *In*: Chemical Signals in Vertebrates IV: Ecology Evolution and Comparative Biology, eds. D. Duvall, D. Müller-Schwarze & R.M. Silverstein, pp. 487–505, Plenum Press, New York.

Galef, B.G. & L.L. Buckley. 1996. Use of foraging trails by Norway rats. Anim. Behav. **51**:765–771.

Galef, B.G. & K.M. Clark. 1972. Mother's milk and adult presence: two factors determining initial dietary selection by weaning rats. J. Comp. Physiol. Psychol. **78**:220–223.

Galef, B.G. & P.W. Henderson. 1972. Mother's milk: a determinant of feeding preferences of weaning rats pups. J. Comp. Physiol. Psychol. **78**:213–219.

Galef, B.G. & D.F. Sherry. 1973. Mother's milk: a medium for transmission of cues reflecting the flavor of mother's diet. J. Comp. Physiol. Psychol. **83**:374–378.

Galef, B.G. & M. Stein. 1985. Demonstrator influence on observer diet preference: analyses of critical social interactions and olfactory signals. Anim. Learn. Behav. **13**:31–38.

Galef, B.G., J.R. Mason, G. Preti & N.J. Bean. 1988. Carbon disulfide: a semiochemical mediating socially-induced diet choice in rats. Physiol. Behav. **42**:119–124.

Garcia, J. & W.G. Hankins. 1977. On the origin of food aversion paradigms. *In*: Learning Mechanisms in Food Selection, eds. L.M. Barker, M.R. Best & M. Domjan, pp. 3–22, Baylor University Press, Waco, TX.

Garcia, J. & A.L. Riley. In press. Conditioned taste aversions. *In*: Encyclopedia of Comparative Psychology, eds. G. Greenberg & M. Haraway, pp. in press, Academic Press, New York.

Garcia, J., F.R. Ervin, C.H. Yorke & R.A. Koelling. 1967. conditioning with delayed vitamin injections. Science **155**:716–718.

Gerling, S. & P. Yahr. 1982. Maternal and paternal pheromones in gerbils. Physiol. Behav. **28**:667–673.

Gionfriddo, J.P. & L.B. Best. 1995. Grit use by house sparrows: effects of diet and grit size. Condor **97**:57–67.

Gorman, M.L. & R.D. Stone. 1990. Mutual avoidance by European moles *Talpa europaea*. *In*: Chemical Signals in Vertebrates 5, eds. D.W. Macdonald, D. Müller-Schwarze & S.E. Natynczuk, pp. 367–377, Oxford University Press, Oxford.

Gosling, L.M. 1982. A reassessment of a function of scent marking in territories. Z. Tierpsychol. **60**:89–118.

Gosling, L.M. 1990. Scent-marking by resource holders: alternative mechanisms for advertising the costs of competition. In: Chemical Signals in Vertebrates 5, eds. D.W. Macdonald, S. Natynczuk, & D. Müller-Schwarze, pp. 315–328, Oxford University Press, Oxford.

Gosling, L.M. & H.V. McKay. 1990. Competitor assessment by scent matching: an experimental test. Behav. Ecol. Sociobiol. **26**:415–420.

Gosling, L.M., N.W. Atkinson, S.A. Collins, R.J. Roberts & R.L. Roberts. 1996. Avoidance of scent-marked areas depends on the intruder's body size. Behav. **113**:491–502.

Grace, J. & M. Russek. 1969. The influence of previous experience on the taste behavior of dogs toward sucrose and saccharin. Physiol. Behav. **4**:553–558.

Grau, R., W. Pfluger & R. Schmuck. 1992. Acceptance studies to assess the hazard of pesticides formulated as dressings, baits, and granules, to birds. Proc. Brighton Crop Protec. Conf. Pests Diseases **1**:787–792.

Graziadei, P.P.C. & G.A. Monti-Graziadei. 1980. Plasticity of connections in the olfactory sensory pathway: transplantation studies. *In*: Olfaction and Taste VII, ed. H. van der Starre, pp. 155–158, IRS Press, Ltd. London.

Green, B.G., J.R. Mason & M.R. Kare. 1990. Chemical Senses: Vol. 2, Irritation. Marcel Dekker, New York.

Green, K.F. & J. Garcia. 1971. Recuperation from illness: flavor enhancement in rats. Science **173**:749–751.

Greig-Smith, P.W. 1987. Aversions of starlings and sparrows to unfamiliar, unexpected, or unusual flavours and colours in food. Ethology **74**:155–163.

Greig-Smith, P.W. & C.M. Rowney. 1987. Effects of color on the aversions of starlings and house sparrows to five chemical repellents. Crop Prot. **6**:402–409.

Grota, L.J. 1968. Factors influencing the acceptance of Cesarean delivered offspring by foster mothers. Physiol. Behav. **3**:265–269.

Hafer, A.A. 1990. Dominance status and its influence on the Bruce Effect in house mice. *In*: Chemical Signals in Vertebrates 5, eds. D.W. Macdonald, D. Müller-Schwarze & S.E. Natynczuk, pp. 270–275, Oxford University Press, Oxford.

Hahn, M.E. & E.C. Simmel. 1968. Individual recognition by natural concentrations of olfactory cues in mice. Psychon. Sci. **12**:183–184.

Hakim, S., W.J. McShea, & J.R. Mason. 1996. Attractiveness of liquid baits to fawns and adult white-tailed deer. J. Wildlife Diseases. **32**:395–398.

Hauser, G.J., D. Chitayet, L. Berns, D. Brauer & B. Muhlbauer. 1985. Peculiar odours in newborns and maternal prenatal ingestion of spicy foods. Eur. J. Pediatr. **144**:403.

Heise, S & J.L. Hurst. 1995. Territorial experience causes a shift in the responsiveness of female house mice to odours from dominant males. *In*: Chemical Signals in Vertebrates VII, eds. R. Apfelbach, D. Müller-Schwarze, K. Reuter & E. Weiler, pp. 291–296, Pergamon, Oxford.

Henton, W.W. 1969. Conditioned suppression to odorous stimuli in pigeons. J. Exp. Anal. Behav. **12**:175–185.

Henton, W.W., J.C. Smith & D. Tucker. 1969. Odor discrimination in pigeons following section of the olfactory nerves. J. Comp. Physiol. Psych. **69**:317–323.

Hepper, P.G. 1988. Adaptive fetal learning: parental exposure to garlic affects postnatal preferences. Anim. Behav. **36**:935–936.

Hepper, P.G. 1989. Foetal learning: implications for psychiatry? Br. J. Psych. **155**:289–293.

Herz, R.S. & T. Engen. 1996. Odor memory: review and analysis. Psychon. Bull. Rev. **3**:300–313.

Heske, E.J. 1987. Responses of a population of California voles, *Microtus californicus*, to odor-baited traps. J. Mamm. **68**:64–72.

Heth, G. & J. Todrank. 1995. Assessing chemosensory perception in subterranean mole rats: different responses to smelling versus touching odorous stimuli. Anim. Behav. **49**:1009–1016.

Holst, V.D. 1985. The primitive eutherians II: a case study of the tree shrew, *Tupia belangeri*. *In*: Social Odours in Mammals, eds. R.E. Brown & D. Macdonald, pp. 155–216, Clarendon Press, Oxford.

Horne, T.J. & H. Ylonen. 1996. Female bank voles prefer dominant males; but what if there is no choice? Behav. Ecol. Sociobiol. **38**:401–405.

Hosoi, E., D.M. Swift, L.R. Rittenhouse & R.W. Richards. 1995. Comparative foraging strategies of sheep and goats in a t-maze apparatus. Appl. Anim. Behav. Sci. **44**:37–46.

Hudson, R. & H. Distel. 1983. Nipple location by newborn rabbits: behavioral evidence for pheromonal guidance. Behaviour **85**:260–275.

Hurst, J.L. 1989. The complex network of olfactory communication in populations of wild house mice *Mus domesticus* Rutty: urine marking and investigation within family group. Anim. Behav. **37**:705–725.

Hurst, J.L. 1993. The priming effects of urine substrate marks on interactions between male house mice, *Mus musculus domesticus*. Anim. Behav. **45**:55–81.

Hurst, J.L., J. Fang & C.J. Barnard. 1993. The role of substrate odours in maintaining social tolerance between male house mice, *Mus musculus domesticus*. Anim. Behav. **45**:997–1006.

Hurst, J.L., J.M. Fang & C.J. Barnard. 1994. The role of substrate odours in maintaining social tolerance between male house mice, *Mus musculus domesticus*: relatedness, incidental kinship effects and the establishment of social status. Anim. Behav. **48**:157–168.

Jacobs, H.L. 1964. Observations on the ontogeny of saccharin preference in the neonate rat. Psychon. Sci. **1**:105–106.

James, J.E.A. and M.B. Daly. 1969. Nasal reflexes. Proc. Royal Soc. Med. **62**:1287–1293.

Jedrzejewski, W., L. Rychlik & B. Jedrzejewski. 1993. Response of bank voles to odours of seven species of predators: experimental data and their relevance to natural predator-vole relationships. Oikos **68**:251–257.

Johanson, I.B. & W.G. Hall. 1979. Appetitive learning in 1-day-old rat pups. Science **205**:419–421.

Johanson, I.B. & M.H. Teicher. 1980. Classical conditioning of an odor preference in 3-day-old rats. Behav. Neural Biol. **29**:132–136.

Johnson, R.P. 1973. Scent marking in mammals. Anim. Behav. **21**:521–535.

Johnston, R.E. 1974. Sexual attraction function of golden hamster vaginal secretion. Behav. Biol. **12**:111–117.

Johnston, R.E. 1990. Chemical communication in golden hamsters: from behavior to molecules and neural mechanisms. *In*: Contemporary Issues in Comparative Psychology, ed. D.A. Dewsbury, pp. 381–409, Sinauer Press, New York.

Johnston, R.E., G. Chiang & C. Tung. 1994. The information in scent over-marks of golden hamsters. Anim. Behav. **48**:323–330.

Jones, R.B. & N.E. Nowell. 1974. A comparison of the aversive and female attractant properties of urine from dominant and subordinate male mice. Anim. Learn. Behav. **2**:141–144.

Kare, M.R. & J.R. Mason. 1986. The special senses. *In*: Avian Physiology, ed. P. Sturkie, pp. 59–73, Cornell University Press, Ithaca, NY.

Kavaliers, M., D. Innes & K.P. Ossenkopp. 1992. Predator-odor analgesia in deer mice: neuromodulatory mechanisms and sex differences. *In*: Chemical Signals in Vertebrates VI, eds. R.L Doty & D. Müller-Schwarze, pp. 529–535, Plenum Press, New York.

Kazdin, A.E. 1982. Single Case Research Designs. Oxford University Press, New York.

Keeler, R.F. 1988. Livestock models of human birth defects, reviewed in relation to poisonous plants. J. Anim. Sci. **66**:2414–2427.

Keil, W., F.V. Stralcndorff & R. Hudson. 1990. A behavioral bioassay for analysis of rabbit nipple-search pheromone. Physiol. Behav. **47**:525–529.

Kimball, B.A., D.L. Nolte, R.M. Engeman, J.J. Johnston & F.R. Stermitz. 1998. Chemically mediated foraging preference of free ranging black bears (*Ursus americanus*). J. Mammal. **79**: in press.

Kimelman, B.R. & R.E. Lubow. 1974. The inhibitory effect of preexposed olfactory cues on intermale aggression in mice. Physio. Behav. **12**:919–922.

Kononen, D.W. 1988. Two methods for evaluating avian food avoidance data. Bull. Environ. Contam. Toxicol. **40**:619–625.

Kononen, D.W., J.R. Hochstein & R.K. Ringer. 1986. A quantitative method for evaluating avian food avoidance behavior. Environ. Toxic. Chem. **5**:823–830.

Kononen, D.W., J.R. Hochstein & R.K. Ringer. 1987. Avoidance behavior of mallards and northern bobwhites exposed to carbofuran-contaminated food and water. Environ. Toxic. Chem. **40**:619–625.

Koskela, E., T.J. Horne, T. Mappes & H. Ylonen. 1996. Does risk of small mustelid predation affect the oestrus cycle in the bank vole? Anim. Behav. **51**:1159–1164.

Koyoma, S. 1995. Differential response to the odor of familiar intruder mice in male mice. J. Ethol. **13**:47–56.

Kuwamura, Y. & M.R. Kare. 1987. Umami: A Basic Taste. Marcel Dekker, New York.

Lawless, H.T. 1987. Gustatory psychophysics. *In*: Neurobiology of Taste and Smell, eds. T.E. Finger & W.L. Silver, pp. 401–422, Wiley Interscience, New York.

Lenington, S., K. Egid & J. Williams. 1988. Analysis of a genetic recognition system in wild house mice. Behav. Genet. **18**:549–564.

Lepri, J.J. 1994. The vomeronasal organ coordinates reproduction in some, but not all prairie voles. *In*: Chemical Signals in Vertebrates VII, eds. R. Apfelbach, D. Müller-Schwarze, K. Reutter & E. Weiler, pp. 273–280, Pergamon, Oxford.

Lima, S.L. & L.M. Dill. 1990. Behavioral decisions made under the risk of predation: a review and prospectus. Can. J. Zoo. **68**:619–640.

Loughry, W.J. & C.M. McDonough. 1994. Scent discrimination by infant nine-banded armadillos. J. Mammal. **75**:1033–1039.

Macdonald, D.W. 1985. The carnivores: order Carnivora. *In*: Social Odours in Mammals, eds. E.R. Brown & D.W. Macdonald, pp. 619–622, Oxford Science Publications, Oxford.

MacFarlane, J.D. & J.M. Taylor. 1982. Nature of estrus and ovulation in *Microtus townsendii* (Bachman). J. Mammal. **63**:104–109.

Mackintosh, N.S. 1974. The Psychology of Animal Learning. Academic Press, New York.

Mackintosh, J.H. & E.C. Grant. 1966. The effect of olfactory stimuli on the agnostic behavior of laboratory mice. Z. Tierpsychol. **23**:584–587.

Mainardi, M., M. Poli & P. Valsecchi. 1989. Ontogeny of dietary selection in weaning mice: effects of early experience and mother's milk. Biol. Behav. **14**:185–194.

Martin, P. 1971. Movements and activities of the mountain beaver *Aplodontia rufa*. J. Mammal. **52**:717–723.

Martin, L.T. & J.R. Alberts. 1979. Taste aversions to mother's milk: the age-related role of nursing in acquisition and expression of a learned association. J. Comp. Physiol. Psych. **93**:430–445.

Maruniak, J.A., W.L. Silver & D.G. Moulton. 1983. Olfactory receptors respond to blood-borne odorants. Brain Res. **265**:312–316.

Mason, J.R. & W.R. Bonwell. 1993. Evaluation of turpentine as a bird repellent seed treatment. Crop Prot. **12**:453–457.

Mason, J.R. & L. Clark. 1995. Mammalian irritants as chemical stimuli for birds: the importance of training. Auk **112**:511–514.

Mason, J.R. & L. Clark. 1996. Capsaicin detection in trained European starlings: the importance of olfaction and trigeminal chemoreception. Wilson Bull. **107**:165–169.

Mason, J.R. & D.L. Otis. 1990. Effectiveness of six potential irritants on consumption by red-winged blackbirds (*Agelaius phoeniceus*) and starlings (*Sturnus vulgaris*). *In*: Chemical Senses: Vol. 2, Irritation, eds. B.G. Green, J.R. Mason & M.R. Kare, pp. 309–324, Marcel Dekker, New York.

Mason, J.R. & R.F. Reidinger. 1982. Observational learning of food aversions in red-winged blackbirds (*Agelaius phoeniceus*). Auk **99**:548–554.

Mason, J.R. & R.F. Reidinger. 1983. Importance of color for methiocarb-induced food aversions in red-winged blackbirds. J. Wildl. Manage. **47**:383–393.

Mason, J.R. & W.L. Silver. 1983. Trigeminally mediated odor aversions in starlings. Brain Res. **269**:196–199.

Mason, J.R., Y. Katz & R.F. Reidinger. 1985. Contextual and hormonal factors influence expression of conditioned taste aversions by pine voles. Physiol. Behav. **35**:979–983.

Mason, J.R., M.A. Adams & L. Clark. 1989. Anthranilate repellency to starlings: chemical correlates and sensory perception. J. Wildl. Manage. **53**:55–64.

Mason, J.R., M.L. Avery, J.F. Glahn, D.L. Otis, R.E. Matteson & C.O. Nelms. 1991a. Evaluation of methyl anthranilate relative to starch-plated dimethyl anthranilate as a bird repellent livestock feed additive. J. Wildl. Manage. **55**:182–187.

Mason, J.R., N.J. Bean, P.S. Shah & L. Clark. 1991b. Taxon-specific differences in responsiveness to capsaicin and several analogues: correlates between chemical structure and behavioral aversiveness. J. Chem. Ecol. **17**:2539–2550.

Mason, J.R., L. Clark & P.S. Shah. 1991c. Ortho-aminoacetophenone repellency to birds: perceptual and chemical similarities to methyl anthranilate. J. Wildl. Manage. **55**:334–340.

Mason, J.R., L. Clark & T.P. Miller. 1993. Evaluation of a pelleted bait containing methyl anthranilate as a bird repellent. Pest. Sci. **39**:299–304.

Meany, C.A. 1990. Spatial utilization and scent mark location in a territorial lagomorph. *In*: Chemical Signals in Vertebrates 5, eds. D.W. Macdonald, D. Müller-Schwarze & S.E. Natynczuk, pp. 388–393, Oxford University Press, Oxford.

Mehiel, R. & R.C. Bolles. 1984. Learned flavor preferences based on caloric outcome. Anim. Learn. Behav. **12**:421–427.

Mehiel, R. & R.C. Bolles. 1988. Learned flavor preferences based on calories are independent of initial hedonic value. Anim. Learn. Behav. **16**:383–387.

Meier, P.T. 1991. Response of adult woodchucks (*Marmota monax*) to oral-gland scents. J. Mammal. **72**:622–624.

Meredith, M. 1982. Stimulus access and other processes involved in nasal chemosensory function: Potential substrates for neuronal and hormonal influence. *In*: Olfaction and Endocrine Regulation, ed. W. Breipohl, pp. 223–236, Information Retrieval Ltd., London.

Merkins, M., A.S. Harestad & T.P. Sullivan. 1991. Cover and efficacy of predator based repellents for Townsend's vole *Microtus townsendii*. J. Chem. Ecol. **17**:401–412.

Messier, C. & N.M. White. 1984. Contigent and non-contigent actions of sucrose and saccharine reinforcers: effects on taste preference and memory. Physiol. Behav. **32**:195–203.

Michael, R.P. & R.W. Bonsall. 1977. Chemical signals and primate behavior. *In*: Chemical Signals in Vertebrates, eds. D. Müller-Schwarze & M.M. Mozell, pp. 251–271, Plenum Press, New York.

Michelsen, W.J. 1959. Procedure for studying olfactory discrimination in pigeons. Science **130**:630–631.

Minneau, P., B.T. Collins & A. Baril. 1996. On the use of scaling factors to improve interspecies extrapolation of acute toxicity in birds. Reg. Toxicol. Pharm. **24**:24–29.

Mollenauer, S., R. Plotnik & E. Snyder. 1974. Effects of olfactory bulb removal on fear responses and passive avoidance in the rat. Physiol. Behav. **12**:141–144.

Moulton, D.G. 1977. Minimum odorant concentrations detectable by the dog and their implications for olfactory receptor sensitivity. *In*: Chemical Signals in Vertebrates, eds. D. Müller-Schwarze & M.M. Mozell, pp. 455–464, Plenum Press, New York.

Müller-Schwarze, D. 1977. Complex mammalian behavior and pheromone bioassay in the field. *In*: Chemical Signals in Vertebrates, eds. D. Müller-Schwarze & M.M. Mozell, pp. 413–433, Plenum Press, New York.

Mykytowycz, R., E.R. Hesterman, S. Gambale & M.L. Dudzinski. 1976. A comparison of the effectiveness of the odors of rabbits, *Oryctolagus cuniculus*, in enhancing territorial confidence. J. Chem. Ecol. **2**:13–24.

Nolte, D.L. & J.R. Mason. 1995. Maternal ingestion of ortho-aminoacetophenone during gestation affects intake by offspring. Physiol. Behav. **58**:925–928.

Nolte, D.L. & F.D. Provenza. 1991. Food preferences in lambs after exposure to flavors in milk. Appl. Anim. Behav. Sci. **32**:381–389.

Nolte, D.L., F.D. Provenza, R. Callan & K.E. Panter. 1992. Garlic in the ovine fetal environment. Physiol. Behav. **52**:1091–1093.

Nolte, D.L., G. Epple, D.L. Campbell & J.R. Mason. 1993a. Response of mountain beaver (*Aplodontia rufa*) to conspecifics in their burrow system. Northw. Sci. **67**:251–255.

Nolte, D.L., J.R. Mason & L. Clark. 1993b. Avoidance of bird repellents by mice (*Mus musculus*). J. Chem. Ecol. **19**:427–432.

Nolte, D.L., J.R. Mason & L. Clark. 1993c. Nonlethal rodent repellents: differences in chemical structure and efficacy from nonlethal bird repellent. J. Chem. Ecol. **19**:2019–2027.

Nolte, D.L., J.R. Mason, G. Epple, E. Aronov & D.L. Campbell. 1994. Why are predator urines aversive to prey? J. Chem. Ecol. **20**:1505–1516.

Nolte, D.L., B.A. Kimball, K.L. Kelly, Z. Zhang & D.L. Campbell. 1995a. Herbivore avoidance of a simple digitalis extract. J. Agric. Food Chem. **43**:830–832.

Nolte, D.L., K.L. Kelly, B.A. Kimball & J.J. Johnston. 1995b. Herbivore avoidance of digitalis extracts is not mediated by cardiac glycosides. J. Chem. Ecol. **21**:1447–1455.

Norman, D.M., J.R. Mason & L. Clark. 1992. Capsaicin effects on consumption of feed by cedar waxwings, and house finches. Wilson Bull. **104**:549–551.

Norusis, M.J. 1986. Advanced Statistics SPSS/PC+. SPSS, Inc., Chicago, IL.

Nyby, J. 1983. Volatile and nonvolatile chemosignals of female rodents: differences in hormonal regulation. *In*: Chemical Signals in Vertebrates 3, eds. D. Müller-Schwarze & R.M. Silverstein, pp. 179–193, Plenum Press, New York.

Passe, D.H. & J.C. Walker. 1985. Odor psychophysics in vertebrates. Neurosci. Biobehav. Rev. **9**:431–467.

Pavlov, I.P. 1927. Conditioned Reflexes. Oxford University Press, Oxford.

Pederson, P.E. & E.M. Blass. 1982. Prenatal and postnatal determinants of the 1st suckling episode in albino rats. Psychol. Biol. **15**:349–356.

Pederson, T. & H. Peters. 1968. Proposal for a classification of oocytes and follicles in the mouse ovary. J. Reprod. Fert. **17**:555–557.

Perez, C., K. Ackroff & A. Sclafani. 1996. Carbohydrate- and protein-conditioned flavor preferences: effects of nutrient preloads. Physiol. Behav. **59**:467–474.

Perrin, T.E. & L.E.L. Rasmussen. 1994. Chemosensory response of female Asian elephants (*Elephas maximus*) to cyclohexane. J. Chem. Ecol. **20**:2857–2866.

Petrulis, A. & R.E. Johnston. 1995. A reevaluation of dimethyl disulfide as a sex attractant in golden hamsters. Physiol. Behav. **57**:779–784.

Pfister, J.A., D. Müller-Schwarze & D.F. Balphs. 1990. Effects of predator fecal odors on feed selection be sheep and cattle. J. Chem. Ecol. **16**:573–584.

Platt, J.R. 1964. Strong inference: certain systematic methods of scientific thinking may produce much more rapid progress than others. Science **146**:347–353.

Poole, T.B. & H.D.R. Morgan. 1975. Aggressive behavior of male mice (*Mus musculus*) towards familiar and unfamiliar opponents. Anim. Behav. **23**:470–479.

Porter, R.H. 1986. Chemical signals and kin recognition in spiny mice *Acomys cahirinus*). *In*: Chemical Signals in Vertebrates 4, Ecology, Evolution and Comparative Biology. eds. D. Duvall, D. Müller-Schwarze & R.M. Silverstein, pp. 397–411, Plenum Publishing, New York.

Porter, R.H. & H.M. Doane. 1976. Maternal pheromone in the spiny mouse (*Acomys cahirinus*). Physiol. Behav. **16**:75–78.

Porter, R.H. & K. Ruttle. 1975. The response of one-day old *Acomys cahirinus* pups to naturally occurring chemical stimuli. Z. Tierpsychol. **38**:154–162.

Porter, R.H., H.M. Doane & S.A. Cavallaro. 1978. Temporal parameters of responsiveness to maternal pheromone in *Acomys cahirinus*. Physiol. Behav. **21**:563–566.

Prah, J.D. & V.A. Benignus. 1984. Trigeminal sensitivity to contact chemical stimulation: a new method and some results. Percep. Psychophys. **35**:65–68.

Provenza, F.D. 1995. Postingestive feedback as an elementary determinant of food preference and intake in ruminants. J. Range. Manage. **48**:2–17.

Provenza, F.D. & D.F. Balph. 1988. The development of dietary choice in livestock on rangelands and its implications for management. J. Anim. Sci. **66**:2356–2368.

Provenza, F.D. & D.F. Balph. 1990. Applicability of five diet-selection models to various foraging challenges ruminants encounter. *In*: Behavioral Mechanisms of Food Selection, ed. R.N. Hughes, pp. 423–459, Springer-Verlag, Berlin.

Ralls, K. 1971. Mammalian scent marking. Science **171**:443–449.

Rashotte, M.E. & J.C. Smith. 1984. Operant conditioning methodology in the assessment of food preferences: introductory comment. *In*: Neuroscience and Biobehavioral Reviews: Studies in Canine Olfaction, Taste and Feeding, eds. L.M. Beidler, M.E. Rashotte, J.C. Smith & R.W. Griffin, pp. 211–215, Ankho International, Inc., Fayetteville, NY.

Rasmussen, L.E.L., T.D. Lee, G.D. Daves & M.J. Schmidt. 1993. Female-to-male sex pheromones of low volatility in the Asian elephant, *Elephas maximus*. J. Chem. Ecol. **19**:2115–2128.

Reidinger, R.F. & J.R. Mason. 1983. Evaluation and exploitation of weaknesses in behavioral defenses against dietary poisoning. *In*: Test Methods for Vertebrate Pest Control and Management Materials, ed. D.E. Kaukienen, pp. 20–39, American Society for Testing and Materials, Philadelphia.

Richardson, P.R.K. 1990. Scent marking and territoriality in the aardwolf. *In*: Chemical Signals in Vertebrates 5, eds. D.W. Macdonald, D. Müller-Schwarze & S. Natynczuk, pp. 378–387, Oxford University Press, Oxford.

Riley, A.L. & C.M. Clarke. 1977. Conditioned taste aversions: a bibliography. *In*: Learning Mechanisms in Food Selection, eds. L.M. Barker, M.R. Best & M. Domjan, pp. 593–615, Baylor University Press, Waco, TX.

Riley, A.L. & D.L. Tuck. 1985. Conditioned food aversions: a bibliography. *In*: Experimental Assessments and Clinical Applications of Conditioned Food Aversions, Vol. 443, eds. N.S. Braveman & P. Bronstein, pp. 381–437, Annals of the New York Academy of Sciences, New York.

Robinson, I. 1990. The effect of mink odour on rabbits and small mammals. *In*: Chemical Signals in Vertebrates 5, eds. D.W. Macdonald, D. Müller-Schwarze & S. Natynczuk, pp. 566–572, Oxford University Press, Oxford.

Robinson, S.R., J.M. Arnold, N.E. Spear & W.P. Smotherman. 1993. Artificial nipple promotes conditioned opiod activity in the rat fetus. Develop. Psychobiol. **26**:375–387.

Rogers, J.G. 1974. Responses of caged red-winged blackbirds to two types of repellents. J. Wildl. Manage. **38**:418–423.

Ronkaninen, H. & R. Ylonen. 1994. Behavior of cyclic bank voles under the risk of mustelid predation: do females avoid copulations? Oecologia **97**:377–381.

Rowe, F.P. & R. Redfern. 1969. Aggressive behaviour in related and unrelated wild house mice (*Mus musculus* L.). Ann. Appl. Biol. **64**:425–431.

Rozenfeld, F.M. & R. Rasmont. 1991. Odor cue recognition by dominant male bank voles, *Clethrionomys flareolus*. Anim. Behav. **41**:839–850.

Rozenfeld, F.M., E. Le Boulang & R. Rasmont. 1987. Urine marking by male bank voles *Clethrionomys flareolus* Schreber, 1788: *Microtidae, Rodentia*) in relation to their social rank. Can. J. Zool. **65**:2594–2601.

Rozin, P., L. Ebert & J. Schull. 1982. Some like it hot: a temporal analysis of hedonic responses to chili pepper. Appetite **3**:13–22.

Saudergrass, R.A. & L.C. Drummer. 1996. Single subject research designs and zoo research. Zoo Biol. **15**:173–182.

Schaal, B., P. Orgeur & C. Arnould. 1995. Olfactory preferences in newborn lambs: possible influence of prenatal experience. Behaviour **132**:351–366.

Schafer, E.W. & D.B. Brunton. 1971. Chemicals as bird repellents: two promising agents. J. Wildl. Manage. **35**:569–572.

Schafer, E.W., W.A. Bolles & J. Hurlbut. 1983. The acute oral toxicity, repellency, and hazard potential of 988 chemicals to one or more species of wild and domestic birds. Arch. Environ. Contam. Toxicol. **12**:355–382.

Schafer, H.M. & R.E. Brown. 1995. Methodological questions in the study of the rat's ability to discriminate between the odors of individual conspecifics. *In*: Chemical Signals in Vertebrates VII, eds. R. Apfelbach, D. Müller-Schwarze, K. Reuter & E. Weiler, pp. 427–436, Pergamon, Oxford.

Sclafani, A. 1991. The hedonics of sugar and starch. *In*: The Hedonics of Taste, ed. R.C. Bolles, pp. 59–88, Lawrence Erlbaum and Associates, Hillsdale, NJ.

Shallenberger, R.J. 1975. Olfactory use in the wedge-tailed shearwater (*Puffinus pacificus*) on Manana Is. Hawaii. *In*: Olfaction and Taste V, eds. D.A. Denton & J.P. Coghlan, pp. 355–359, Academic Press, New York.

Shepherd, G.M. 1985. Are there labelled lines in the olfactory pathway? *In*: Taste, Olfaction and the Central Nervous System, ed. D.W. Pfaff, pp. 307–321, Rockefeller University Press, New York.

Sherman, J.E., C.F. Hickis, A.G. Rice, K.W. Rusiniak & J. Garcia. 1983. Preferences and aversions for stimuli paired with ethanol in hungry rats. Anim. Learn. Behav. **11**:101–106.

Shumake, S.A., J.C. Smith & D. Tucker. 1969. Olfactory intensity-difference thresholds in the pigeons. J. Comp. Physiol. Psych. **67**:64–69.

Sieck, M.H., H.C. Baumbach, B.L. Gordon & J.F. Turner. 1974. Changes in spontaneous, odor modulated and shock induced behavior patterns following discrete olfactory system lesions. Physiol. Behav. **13**:427–439.

Silver, W.L. 1987. The common chemical sense. *In*: Neurobiology of Taste and Smell, eds. T.E. Finger & W.L. Silver, pp. 65–88, Wiley Interscience, New York.

Silver, W.L., J.R. Mason, J.A. Maruniak & D.A. Marshall. 1985. Capsaicin desensitization abolishes aversion to a strong nasal irritant: behavioral and electrophysiological evidence that smell and taste remain intact. Brain Res. **333**:45–54.

Silver, W.L., A.H. Arzt & J.R. Mason. 1989. Evidence against odorant discrimination by trigeminal chemoreceptors. J. Comp. Physiol. **164**:55–66.

Simbayi, L.C., R.A. Boakes & M.J. Burton. 1985. Acquired preferences for flavors mixed with nutritive and non-nutritive sweet solutions. Neurosci. Lett. **22**:S158.

Singer, A.G., H. Tsuchiya, J.L. Wellington, G.K. Beauchamp & K. Yamazaki. 1993. Chemistry of odor types in mice: fractionation and bioassay. J. Chem. Ecol. **19**:569–579.

Smotherman, W.P. 1982a. In utero chemosensory experience alters taste preferences and corticosterone responsiveness. Behav. Neural Biol. **36**:61–68.

Smotherman, W.P. 1982b. Odor aversion learning by the rat fetus. Physiol. Behav. **29**:769–771.

Smotherman, W.P. & S.R. Robinson. 1987. Prenatal influences on development: behavior is not a trivial aspect of fetal life. Dev. Behav. Pediatr. **8**:171–175.

Starr, R.I., J.F. Besser & R.B. Brunton. 1964. A laboratory method for evaluating chemicals as bird repellents. Agric. Food Chem. **12**:342–344.

Stattelman, A.J., R.B. Talbot & D.B. Coulter. 1975. Olfactory thresholds of pigeons (*Columbia livia*), quail (*Colinus virginianus*), and chickens (*Gallus gallus*). Comp. Biochem. Physiol. **50A**:807–809.

Stewart, C.N., R.F. Reidinger & J.R. Mason. 1983. Method for determining the taste qualities of rodenticides to rodents. *In*: Test Methods for Vertebrate Pest Control and Management Materials, ed. D.E. Kaukiened, pp. 155–164, American Society for Testing and Materials, Philadelphia.

Stickrod, G., D.P. Kimble & W.P. Smotherman. 1982. In utero taste/odor aversion conditioning in the rat. Physiol. Behav. **28**:5–7.

Stoddart, D.M. 1976. Effect of the odour of weasels (*Mustela nivalis* L.) on trapped samples of their prey. Oecologia **22**:439–441.

Stoddart, D.M. 1980. Some responses of a free living community of rodents to the odors of predators. *In*: Chemical Signals: Vertebrates and Aquatic Invertebrates, eds. D. Müller-Schwarze & R.M. Silverstein, pp. 1–10, Plenum Press, New York.

Stralendorff, F.V. 1987. Partial chemical characterization of urinary signaling pheromone in tree shrews (*Tupia belangeri*). J. Chem. Ecol. **13**:655–679.

Sullivan, T.P. & D.R. Crump. 1986. Avoidance of pocket gophers (*Thomomys talapoides*) to mustelid anal gland compounds. *In*: Chemical Signals in Vertebrates 4, Ecology, Evolution, and Comparative Biology. eds. D. Duvall, D. Müller-Schwarze & R.M. Silverstein, pp. 514–531, Plenum, New York.

Sullivan, T.P., D.R. Crump & D.S. Sullivan. 1988. Use of predator odor as repellents to reduce feeding damage by herbivores IV. northern pocket gophers (*Thomomys talapoides*). J. Chem. Ecol. **14**:379–389.

Sullivan, T.P., D.R. Crump, H. Wieser & E.A. Dixon. 1990. Response of pocket gopher (*Thomomys talapoides*) to an operational application of synthetic semiochemicals of stoat (*Mustela erminea*). J. Chem. Ecol. **16**:941–949.

Svendsen, G.E. & W.D. Huntsman. 1988. A field bioassay of beaver castoreum and some of its components. Am. Midl. Natural. **120**:144–149.

Taylor, S.A. & D.A. Dewsbury. 1988. Effects of experience and available cues on estrous versus diestrus preferences in male prairie voles (*Microtus ochrogaster*). Physiol. Behav. **42**:379–388.

Thiessen, D.D. 1977. Methodology and strategies in the laboratory. *In*: Chemical Signals in Vertebrates, eds. D. Müller-Schwarze & M.M. Moxell, pp. 391–412, Plenum Press, New York.

Thiessen, D. & M. Rice. 1976. Mammalian scent gland marking and social behavior. Psychol. Bull. **45**:1141–1151.

Tobin, M.E., R.M. Engeman & R.T. Sugihara. 1995. Effects of mongoose odors on rat capture success. J. Chem. Ecol. **21**:635–639.

Thompson, R.D., C.V. Grant & D.J. Elias. 1981. Factors affecting red-winged blackbird response to methiocarb, an avian repellent. Pest. Biochem. Physiol. **15**:166–171.

Tucker, D. 1971. Non-olfactory responses from the nasal cavity: Jacobsen's organ and the trigeminal system. *In*: Handbook of Sensory Physiology, Vol. IV, Chemical Senses, Part 1, Olfaction, ed. L.M. Beidler, pp. 151–181, Springer-Verlag, Berlin.

Tucker, D., P.P.C. Graziadei & J.C. Smith. 1975. Recovery of olfactory function in pigeons after bilateral transection of the olfactory nerves. *In*: Olfaction and Taste V, eds. D.A. Denton & J.P. Coghlan, pp. 369–373, Academic Press, New York.

Vince, M.A. & M.W. Stanier. 1991. The effect of food intake on young Soay and Clun Forst lamb's response to touch on the face. Behaviour **89**:117–127.

Voznessenskaya, V.V., V.M. Parfyonova & E.P. Zinkevich. 1992a. Individual odortypes. *In*: Chemical Signals in Vertebrates VI, eds. R.L. Doty & D. Müller-Schwarze, pp. 503–508, Plenum Press, New York.

Voznessenskaya, V.V., C.J. Wysocki & E.P. Zinkevich. 1992b. Regulation of the rat estrus cycle by predator odors: the role of the vomeronasal organ. *In*: Chemical Signals in Vertebrates VI, eds. R.L. Doty & D. Müller-Schwarze, pp. 281–284, Plenum Press, New York.

Walker, J.C. 1983. An operant procedure for testing olfactory capacities in restrained pigeons. Physiol Behav. **30**:165–168.

Weldon, P.J. 1990. Responses of vertebrates to chemicals from predators. *In*: Chemical Signals in Vertebrates 5, eds. D.W. MacDonald, D. Müller-Schwarze & S.E. Natynczuk, pp. 500–521, Oxford University Press, Oxford.

Wenzel, B.M. 1972. Olfactory sensation in the kiwi and other birds. Ann. N.Y. Acad. Sci. **188**:183–193.

Wenzel, B.M. & M.H. Sieck. 1972. Olfactory perception and bulbar electrical activity in several avian species. Physiol. Behav. **9**:287–924.

Werner, I. & J.R. Clarke. 1995. Male sexual experience affects the sexual maturation in female field voles. *In*: Chemical Signals in Vertebrates VII, eds. R. Apfelbach, D.TMüller-Schwarze, R. Reutter & E. Weiler, pp. 281–289, Pergamon, Oxford.

Westbrook, R.F., J.C. Clarke & S. Provost. 1980. Long-delay learning in the pigeon: flavor, color, and flavor-mediated color aversions. Behav. Neural Biol. **28**:398–407.

Whitten, W.K. & A.K. Champlin. 1978. Pheromones, estrus, ovulation, and mating. *In*: Methods in Mammalian Reproduction, ed. J.C. Daniel, pp. 403–417, Academic Press, London.

Winn, B.E. & B.M. Vestal. 1986. Kin recognition and choice of males by wild female house mice (*Mus musculus*). J. Comp. Psychol. **100**:72–75.

Wysocki, C.J. & G.K. Beauchamp. 1991. Individual differences in human olfaction. *In*: Chemical Senses, Vol. 3, Genetics of Perception and Communication, eds. C.J. Wysocki & M.R. Kare, pp. 353–374, Marcel Dekker, New York.

Wysocki, C.J. & M. Meredith. 1987. The vomeronasal system. *In*: Neurobiology of Taste and Smell, eds. T.E. Finger & W.L. Silver, pp. 125–150, John Wiley and Sons, New York.

Wysocki, C.J., G.K. Beauchamp, R.F. Reidinger, & J.L. Wellington. 1985. Access of large and non-volatile molecules to the vomeronasal organ of mammals during social and feeding behaviors. J. Chem. Ecol. **11**:1147–1159.

Yamazaki, K., M. Yamaguchi, L. Baranoski, J. Bard, E.A. Boyse & L. Thomas. 1979. Recognition among mice. Evidence from the use of a Y-maze. Differently scented by congenic mice of different major histocompatibility types. J. Exp. Med. **150**:755–760.

Yamazaki, K., G.K. Beauchamp, J. Bard, E.A. Boyse & L. Thomas. 1991. Chemosensory identity and immune function in mice. *In*: Chemical Senses: Genetics of Perception and Communication, eds. C.J. Wysocki & M.R. Kare, pp. 211–225, Marcel Dekker, Hillsdale.

Ylonen, H., B. Jedrzejewski, W. Jedrzejewski & J. Heikkila. 1992. Antipredatory behavior in *Clethrionomys* voles—'David and Goliath' arms race. Ann. Zool. Fenn. **29**:207–216.

Zahorik, D.M. 1977. Associative and non-associative factors in learned food preferences. *In*: Learning Mechanisms in Food Selection, eds. L.M. Barker, M.R. Best & M. Domjan, pp. 181–200, Baylor University Press, Waco, TX.

Zahorik, D.M. & K.A. Houpt. 1981. Species differences in feeding strategies, food hazards, and the ability to learn food aversions. *In*: Foraging Behavior, eds. A.C. Kamil and T.D. Sargent, pp. 289–310, Garland, New York.

Zahorik, D.M., S.F. Maier & R.W. Pies. 1974. Preferences for tastes paired with recovery from thiamine deficiency in rats: appetitive conditioning or learned safety? J. Comp. Physiol. Psychol. **87**:1083–1091.

Zielinski, W.J. & T.E. Kucera. 1995. American Marten, Fisher, Lynx, and Wolverines: Survey Methods for Their Detection. Gen. Tech. Rep. PSW-GR-157. USDA Publication, Albany, CA.

Index

Abalone, 87
Abracis flafolineata, 243
Acanthrophora, 57
Acetophenones, 372
Acheta domesticus, 164
Achromobacter aquamarinus, 12
Acinetobacter calcoaceticus, 194
Acinetobacter sp., 25
Acropora formosa, 117
Adders, 309
Additive effects, 55
Adenostoma fasciculatum, 180, 188
Agar plate diffusion assays, 156
Aggregation behavior, 301
Aggregation pheromone, 225
  reptiles, 299
Agkistrodon contortrix, 300
Agkistrodon piscivorus, 308
Aiptasia pallida, 67, 81, 88
Alarm cues, 289
Alarm pheromone, 226
  amphibian, 289
  Hymenoptera, 227
  termites, 227
Alarm signal, 93
  amphibian larvae, 94
  cloacal gland secretions, 297
  damaged individuals, 93
  opisthobranch mollusk, 97
  skin extract of the fish, 95
  tadpoles, 289
Alcyonium paessleri, 25, 30
Alexandrium tamarense, 23
Allelopathy, 117, 163, 179
  alcyonacean corals, 117
  alga, 118
  allelochemical transport, 189
  alternate explanations, 180, 184
  bioassays of soils, 196
  bracken fern, 184
  brown alga, 119
  California chaparral, 188
  combinations of compounds, 202
  coral, 118
  cover crop, 191
  crowberry, 198
  density-dependent effects, 182
  density-dependent phytotoxicity, 200
  environmental stress factors, 197
  filamentous algae, 120
  Florida scrub, 195
  germination, 198
  greenhouse experiments, 192
  inhibited radicle growth, 187
  inhibition zone, 190
  Koch's postulates, 183, 194
  leachates, 187
  leaf extracts, 198
  marine invertebrates, 117
  mechanism(s) of inhibition, 182
  monoterpenes, 189
  nutrient levels, 180
  petri dish bioassays, 192
  pitfalls, 181
  radicle elongation, 192, 198
  reef anemone, 120
  role of environmental factors, 182
  rye, 191
  soil transformation, 194
  sponge bioassays, 187

terpenes, 188
transformation, 197
transformation of plant allelochemicals, 182
transport mechanism, 182
vegetation pattern, 184
weed control, 191
Alligator mississippiensis, 306
Allomyces macrogynus, 144
Allylglucosinolate, 257
Alternaria sp., 14
Alteromonas luteo violacea, 27
Alteromonas sp., 15
Ambystoma barbouri, 293, 294
Ambystoma texanum, 293
Ambystoma tigrinum nebulosum, 281
2-Amino-3H-phenoxazin-3-one, 194
Amphibolurus fordi, 306
Amphora coffeaeformis, 21
Andropogon gyrans, 196
Anosmia, 342
Antialgal effects, 12
Antibiotic, 5, 160
Antibiotic disk susceptibility test, 7
Antifouling, 11, 24, 28, 29, 117
Antifouling assays, 112
microscope slide assay, 25
soft coral, 119
Antifungal agents, 15, 158
minimal inhibitory concentration, 159
Antimicrobial assay, 5
agar disk-diffusion assay, 7
agar plate assay, 13
bioautography, 10
cell count assays, 11
competitive interaction assays, 7
diatom, 11
direct challenge assays, 7
filamentous marine fungi, 14
pharmaceutical industry, 5
spectrophotometric methods, 11
viral plaque formation assay, 13
Aonidiella aurantii, 233
Aphytis melinus, 233
Aphytis yanonensis, 236
Aplysia, 77
Aplysilla glacialis, 28
Apo-fucoxanthinoids, 68
Arctostaphylos glandulosa, 188
Arenariomyces trifurcatus, 15
Artemisia californica, 188
Artificial seaweed, 61
Ascaphis truei, 293
Ascidians, 76, 81
Ascoglossans, 82
Ascomycetes, 148
Aspergillus nidulans, 148
Assimilation efficiency, 80
Associational chemical defense, 15
Atrazine, 200
Attachment assay
conidia, 161
zoospores, 161

Basidiomycetes, 148
Batatasin III, 199
Beauveria bassiana, 157
Benzofurans, 259
Bioassay-guided fractionation, 43, 100, 193
allelopathic interactions, 197
kairomone, 236
Bioassays for volatile chemicals, 216
Bird repellents, 372
Birds, 327
olfaction, 327
taste, 329
Bitis caudalis, 310
Black bears, 376
Blanus cinereus, 307
Boiga irregularis, 305
Brassica napus, 200
Bromus rigidus, 186, 187, 190
Bruce effect, 352
Bryopsis sp., 19
Bufo americanus, 276, 281, 290, 295
Bufo boreas, 273, 290, 293, 295
Bufo bufo, 289
Bufo fowleri, 275
Bufo valliceps, 273
Bugula neritina, 115

Caffeic acid, 236
Calamintha ashei, 195, 197
Callsobruchus maculatus, 233
Camphor, 189
Candida, 159
Cembranolide eunicin, 23
Ceratiola ericoides, 195, 196, 197
Chameleo chameleo, 301
Chelodina longicollis, 298
Chemotaxis, 16, 19
Adler method, 17
capillary tube, 17

capillary tube assay, 17
chamber method, 20
dual-well slide method, 21
marine gametes, 21
Miller method, 22
motion analyzer, 22
Müller method, 21
negative chemotaxis, 20
Choice test, 331
amphipods, 104
European starlings, 369
five-arm maze, 343
kin recognition, 278
limitation, 219
mazes, 339
multiarm olfactometers, 218
naked mole rats, 343
olfactometers, 217
snakes, 309
statistical analysis, 215
T-maze, 275, 343
triangle tests, 333
trigeminal irritants, 361
two-arm olfactometer, 218
two-choice tests, 332
Y-maze, 49, 95, 302, 309, 353
Y-tube design, 104
Chromenes, 259
Chrysemys picta, 307
Chrysoma pauciflosculosa, 195, 197
Ciguatoxins, 77, 79
1,8-Cineole, 189
Ciona intestinalis, 23
Ciona savignyi, 23
Circatidal rhythms, 105
Cladina, 195
Cladonia, 195
Cladophora, 58
Claviceps, 163
Clavularia frankliniana, 25
Cnidarians, 91
Coccidioides immitis, 157
Codium sp., 19
Coelomomyces, 145
Coleonyx variegatus, 305
Colletotrichum, 152
Combat behavior, 308
Texas tortoises, 308
Condylactis gigantea, 120
Conradina canescens, 195, 196, 197
Copepods, 83
Copulatory plugs, 307
Corals, 67, 76
Coronella austriaca, 309
Coscinodiscus sp., 30
Courtship pheromones, 288
mental glands, 289
salamanders, 288
Crocodilians, 298
courtship pheromone, 298
Crotalus atrox, 309
Crotalus horridus, 302
Crotalus v. viridis, 299
Culmorin, 15
Cylindrotheca fusiformis, 10
Cymopolia, 52
Cynops pyrrhogaster, 292

Dascyllus aruanus, 109
Daucus carota, 229
Decapod crabs, 105
Defensive metabolites, 74
Deleya marina, 27
Dendrobates pumilio, 276
Desacetylcalaminthone, 197
Desmognathus fuscus, 277
Desmognathus ochrophaeus, 289, 292, 294
Deterrents, 237, 51
Diadema antillarum, 99
Diadophis punctatus, 294
Diadophis punctatus arnyi, 299
Diarrhetic shellfish toxins, 77
DIBOA, 193
Dicamptodon spp., 293
Dictyota, 115
Dictyota menstrualis, 114, 90
Didemnins, 76, 81, 88
Digestibility reducers, 53
Digestive efficiency, 53, 80
Dihydrochalcone ceratiolin, 197
Dinoflagellate toxins, 77
Dipentene, 189
Diplodus holbrooki, 81
Dipsosaurus dorsalis, 306, 307
Disk-diffusion assay
bacteria, 9
microalgae, 9
yeasts, 9
Dominance, 351
bank voles, 352
Bruce effect, 352
urine marks, 352

Dose-response, 247
  operant conditioning, 360
Dose-response curve, 217
Dunaliella tertiolecta, 19
Dytiscus verticalis, 295

Ecological relevance, 3, 121
  assay organisms, 54
  choice of assay organism, 9
  concentration, 4, 41, 49, 56, 118, 181, 217
  high levels of odors, 100
  larval density, 116
  natural exudates, 95
  naturally occurring metabolite, 76
  organismal homogenates, 111
  organismal washes, 111
  still-water assays, 112, 114
  surface swabbings, 115
  timing of assays, 116
  turbulence, 50
  water flow, 50
Ectomycorrhizal fungi, 162
Elaphe guttata, 300, 301, 309
Elaphe obsoleta, 309
Electronic feeding monitors, 246
Empetrum hermaphroditum, 198
Encelia, 259
Endophytic fungi, 163
Enteromorpha, 58, 86
Epifluorescence microscopy, 25
Epilachna varivestris, 254
Epi-thunbergol, 21
Escherichia coli, 17
Eublepharis, 305
Eublepharis macularius, 305
Eumeces fasciatus, 306
Eumeces laticeps, 298, 304
(+)-Evodone, 197
Experimental design, 192
  Latin-square designs, 332
  randomized complete block, 192, 193
  single-subject designs, 336
  sources of variation, 214
  standardization of test conditions, 349

Favella ehrenbergii, 23
Feeding assay
  amphipod, 89
  artificial diet, 240
  cannulation, 258
  carnivore, 358
  choice tests, 356
  concentration of the marker, 254
  consumption, 252
  copepods, 68
  crustaceans, 66
  deterrence, 258
  digitizer, 245
  electronic feeding monitor, 246
  estimating area consumed, 244
  feed trays, 337
  fitness, 89
  food deprivation, 244, 356
  free-ranging birds, 363
  gravimetry, 254
  isopod, 90
  length of the trial, 244
  mussels, 69
  one-choice test, 332
  phlorotannins, 87
  plant tissue substrates, 241
  prior experience, 244
  radioisotopes, 246
  range of concentrations, 243
  sea anemones, 67
  sea stars, 64
  toxicity, 258
  uric acid method, 254
  video, 245
Feeding deterrents, 240, 73
Feeding history, 51
Feeding inhibitors, 237
Feeding stimulants, 52, 70
  black bears, 376
Festuca megalura, 186
Fishes, 45
  field studies, 46
  foraging, 45
Flavor-avoidance learning, 333
Flume experiments, 50
Folin-Denis procedure, 52
Food avoidance learning, 364
Food consumption, 248
Foraging cues, 42
  macerated tissues, 44
  natural exudates of prey, 44
  prey extracts, 44
Fractionation, 55
Fucus vesiculosus, 12
Fungal-insect associations, 162
Fungi, 160
  toxins, 163
Fusarium, 163

Gallic acid, 197
Gammarus pseudolimnaeus, 89
Gastropods, 70
Gerrhosaurus nigrolineatus, 298
Gersemia antarctica, 25, 30
Glandular trichomes, 197
Gonyaulax excavata, 78
Gopherus agassizi, 302
Gopherus berlandieri, 308
Gopherus spp., 306
Gracilaria, 52, 57
Graptemys spp., 307
Gravitational flow through system, 292
Growth assay
  ANCOVA, 251
  secondary metabolites, 86
  weight gain, 253

Habitat recognition, 276
Habituation-dishabituation tests, 342
Halimeda goreaui, 74
Halocyamine A, 13
Halocynthia roretzi, 12, 13
Helicoverpa zea, 259
Helminthosporium, 152
Hemacytometer, 11, 159
Herbicides, 181
Heterocapsa triquetra, 23
Heterodon nasicus, 308
High-speed video camera, 84
Homing behavior, 106
  amphibians, 273
  anosmic, 277
  anosmic toads, 273
  decapod crustaceans, 108
  limpets, 108
  mazes, 275
  odoriferous landscape, 275
  olfactometer, 274
  salamanders, 277
  salmonid fish, 106
  sea urchins, 108
  tropical frogs, 276
  ultrasonic transmitters, 107
  Y-shaped stream, 106
Host recognition, 233
  parasitoid, 233
Host-marking pheromones, 237
HPLC, 41
Hunger level, 51
Hydrocinnamic acid, 197
Hydroquinone, 197
β-Hydroxybutyric acid, 193
17-Hydroxygrindelic acid, 197
Hyla chrysoscelis, 274, 294
Hyla crucifer, 275
Hyla regila, 290, 295
Hypochoeris glabra, 190

Identification of plant pathogens, 152
Iguana iguana, 302
Imprinting, 106
  artificial chemical cue, 107
Indole-3-acetic acid, 163
2,3-Indolinedione, 15
Induced responses, 99
  bryozoan, 100, 101
  cladocerans, 101
  protective spines, 100
  seaweeds, 101
Interference competition, 14
  marine fungus, 14
Isatin, 15

Jacobson's organ, 296

Kairomone, 236
Kin recognition, 278, 342
  airborne chemical cues, 282
  American toad, 281
  Arizona tiger salamanders, 281
  choice tests, 278
  larval amphibians, 278
  mammal, 342
  tadpoles, 280
Kinosternon spp., 307
Kira's law, 200
Koch's postulates, 183
  pathogens, 152

Labidus coecus, 297
Laboulbeniales, 162
Lacerta vivipara, 309
Lacertilian reptiles, 307
  femoral pores, 307
  individual recognition, 307
Lactating mammal bioassays, 362
Lactuca sativa, 197
Lagenidium callinectes, 15
Lagenidium giganteum, 146, 157
Lagodon rhomboides, 46, 81
Lampropeltis getulus, 308, 309
Lampropeltis triangulum, 309
LD_{50}, 156, 247, 259, 78

Leaf disks, 243
Learned aversion, 76, 77
  flavor avoidance conditioning, 358
Learning, 57, 233, 244
  avian olfaction, 348
  birds, 347
  fetal environment, 361
  flavor-avoidance learning, 333
  food preferences, 334
  food-avoidance, 334
  operant conditioning, 335
  secondary metabolites, 88
  toad orientation, 275
Lepomis cyanellus, 293
Leptinotarsa decemlineata, 254
Leptochloa dubia, 197
Leptoglossus zonatus, 226
Leptogorgia setacea, 12
Leptosphaeria oraemaris, 15
Leptotyphlops dulcis, 297
Lethocerus americanus, 295
Ligands for fungal recognition, 159
Lindra thalassiae, 14
Lirceus fontinalis, 294
Lizard trailing behavior, 304
Lobophora variegata, 119
Lobophytum crassum, 21
Lobophytum pauciflorum, 119
Lordosis, 368
Lulworthia sp., 14
Lycopersicon esculentum, 200
Lycopersicon hirsutum, 259
Lymantria dispar, 228

Malaclemys terrapin tequesta, 307
Mammals, 327
  dominance, 352
  olfaction, 327
  taste, 329
Manduca sexta, 244
Manitol, 187
Marine fungi, 13
Marking behavior, 286
Mating types, 147
Mazes, 339
Media sativa, 190
Melampsora, 155
Melanin biosynthesis inhibition, 156
Metabolite instability, 55
6-Methoxy-2,3-benzoxazolinone, 194
Methyl anthranilate, 370
Mice, 352
Microalgae, 69, 83
Microalgal attachment assay, 27
Microbial biomass, 150
  hydrolysis of fluorescein diacetate, 150
  spore-fall method, 151
Microdeutopus gryllotalpa, 104
Micropterus salmoides, 295
Microvideo, 84
Monodictys pelagica, 14
Monoterpenes, 189, 197
Montipora digitata, 21
Motility assay, 23
Mountain beavers, 373
Multiple-arm olfactometers, 219
Muricea fruticosa, 11
Muricins, 11
Mutagenicity assays, 156

Naja nigricollis, 310
Natrix spp., 309
Naturalistic observations, 331
Navanax inermis, 97
Navenones, 97
Navicula salinicola, 12
Navicula sp., 11
Neivamyrmex nigrescens, 297
Neomeris annulata, 73
Neoregelia, 276
Nerodia, 300
Nerodia r. rhombifera, 299
Neurospora crassa, 147, 148
Notophthalmus viridescens, 292
Nudibranchs, 82
Nutritional indices, 249
  approximate digestibility, 250
  calorimetry, 255
  data acquisition, 252
  efficiency of conversion, 250
  herbivore biomass, 250
  measurements of excreta, 253
  plant defensive chemicals, 256
  relative growth rate, 250
  respirometry, 255
Nutritional quality, 73

Ochtodes, 65
Odor plume, 46
Olfactometer, 217
  activated charcoal, 218

Olfactory orientation, 274
Olfactory system, 278
  immune system, 328
  navigation by pigeons, 349
Oncopeltus fasciatus, 164
Oogoniol, 146
Operant conditioning, 335, 351, 360
Oviposition bioassay, 228, 233
  swallowtail butterflies, 229
Oviposition-deterring pheromones, 231
Ovipositions, 231
Oviposition stimulants, 228
2,2′-Oxo-1,1′-azobenzene, 193
17-Oxogrindelic acid, 197
Oysters, 71

Pachydictyol A, 52, 81
Palaemon macrodactylus, 15
Palatability, 55, 295
  calcification, 73
Papilio demoleus, 229
Papilio polyxenes, 229, 257
Papilio protenor, 229
Paralytic shellfish poisoning, 77
  gull chicks, 77
  otters, 79
Parasites, 151
Parasitic fungi, 161
Parental-offspring recognition, 344
  mammalian neonates, 344
  maternal milk, 345
Paroedura pictus, 305
Paspalum notatum, 196
Pathogens, 151
Peridroma saucia, 259
Phaeodactylum tricornutum, 11, 68
Phagocottus gracilis, 294
Phagostimulants, 237, 70
  Avicel® plate bioassay, 71
Phenolic acid, 236
Phenolics, 53
Phenols, 160, 197
Pheromone
  aggressive encounters, 103
  Coelomomyces, 145
  copulatory plug, 308
  fungus, 144
  larval release, 105
  oomycetes, 145
  recognition of shoalmates, 102
  sexual maturation, 350
  sporulating areas, 147
  steroid, 145
  yeasts, 147
β-Phenyl-lactic acid, 193
Phlorotannin, 40, 52, 80
Photosensitization experiments, 157
Phoxinus phoxinus, 289
Phrynosoma platyrhinos, 306
Phytoalexin, 155
Phytoalexin production, 160, 164
Phytophthora, 152
Phytotoxic effect, 184
Phytotoxins, 188
Pieris rapae, 257
α-Pinene, 189
Pinus palustris, 195
Pinus sylvestris, 198
Piper, 243
Pituophis m. melanoleucus, 302, 309
Pituophis melanoleucus deserticola, 299
Plethodon cinereus, 284, 285, 286, 287
Plethodon dorsalis, 293
Plocamium hamatum, 118
Pocillopora damicornis, 109
Polygonella myriophylla, 195
Polyphenolics, 80, 86
Populus tremula, 198
Postcloacal gland, 286
Postingestive bioassays, 247
Predator avoidance, 96, 365
  American toad, 290
  amphibian larvae, 96
  analgesia in mice, 365
  bank voles, 365
  capture-and-release trap program, 367
  choice-tests, 366
  deer mice, 375
  gastropods, 98
  gravitational flow through system, 292
  guinea pigs, 375
  house mice, 375
  lizards, 309
  monitor lizards, 310
  mountain beaver, 375
  relative capture success, 367
  salamanders, 292
  salmon, 95
  sea urchin, 99
  snakes, 308
  tadpoles, 290
  urinary constituents, 374

western toad, 290
Y-mazes, 309
Predator/prey interactions, 289
amphibians, 289
Predator/prey recognition, 308
Prey capture, 90
cone snails, 91
envenomation, 91
ferrets, 359
hydrozoans, 91
jellyfish, 91
octopi, 91
reptiles, 302
toxins, 90, 92
Prostaglandins, 76
Protoplast microassays, 155
Psammophis leightoni, 310
Pseudacris clarki, 274
Pseudacris triseriata, 275, 290
Pseudemys nelsoni, 307
Pseudomonas fluorescens, 18
Pseudomonas perfectomarinus, 12
Pseudomonas spp., 18
Pseudopterogorgia americana, 12
Psila rosae, 224
Pteridium aquilinum, 184
Puccinia, 155
Pycnogonids, 82
Pythium, 152

Quercus laevis, 195
Quinones, 157

Rana aurora, 291, 293
Rana cascadae, 278, 280, 290, 295
Rana clamitans, 276, 277
Rana pipiens, 273
Rana ultricularia, 275
Recognition pheromones, 227
Red algae, 77
Reptiles, 296
tongue-flicking, 296
vomeronasal organ, 296
Rhagoletis cerasi, 237
Rhagoletis pomonella, 223
Rheo discolor, 276
Rhithropanopeus harissii, 105
Rhizophydium littoreum, 19
Rhizopus suinis, 163
Rudbeckia hirta, 197

Saccharomyces cerevisiae, 147
Salamander, 285
agonistic postures, 286
nose-tapping, 285
postcloacal gland, 286
Salmo clarki, 293
Salvelinus fontinalis, 293
Salvia leucophylla, 188
Saprophytes, 151
Scaphiopus hurteri, 274
Schizachyrium scoparium, 197, 198
Schizophyllum commune, 148
Schizosaccharomyces pombe, 147
Schreckstoff, 93
Scincella lateralis, 306
Scintillation counting, 18
Sea stars, 64
Sea urchin, 73
Secale cereale, 191
Septal organ, 329
Serratia sp., 18
Sesquiterpenes, 73
Settlement assay, 24
agar matrix assay, 26
barnacle, 113
bryozoan, 115
coral-reef fish larvae, 109
field studies, 28
flume experiments, 112
microalgal assays, 29
microscope slide assay, 29
planktonic larvae, 110
spectrophotometric methods, 26
tile assay, 28
Sex pheromone, 104, 225
aquatic turtles, 306
crustaceans, 104
elephants, 350
flehman response, 350
hamster, 349
lizards, 305
monkeys, 351
salamander, 287
snakes, 305
species discrimination, 288
tortoises, 306
vaginal secretion, 349
Sinularia cruciata, 118
Siphonodictyon, 117
Sirenin, 144
Snakes, 297
cloacal scent glands, 297

Solenopsis geminata, 297
Sphaeroma quadridentata, 90
Spirulina powder, 85
Spodoptera eridania, 257
Spodoptera exigua, 158
Spodoptera frugiperda, 164
Spodoptera litura, 240
Spore germination, 152
  carbon dioxide, 161
  inoculation, 153
  insect cuticle, 158
  microhumidity chamber, 153
Statistical analyses, 42
  ANCOVA, 251
  ANOVA, 216, 247, 369, 376
  chi-square, 224
  Dunnett's *t*-test, 373
  Fisher's exact test, 216, 247
  G-test, 216, 237, 247
  inadequate controls, 83
  Kruskal-Wallis test, 216, 247
  least-squares fit, 373
  nonparametric, 216
  probit analyses, 247
  pseudoreplication, 42, 75, 86, 99
  rank-order correlation coefficients, 370
  t-test, 216, 247
  Wilcoxon two-sample test, 216, 247
Sternotherus odoratus, 298
Sternotherus spp., 307
Still-air bioassay, 223
  limitation, 223
  pitfall traps, 224
Stimulus presentation, 24
  agar, 27
  agar media, 154
  agar-based food, 59
  agar-coated slides, 26
  capillary tubes, 217
  cellulose sponges, 230
  cheese cloth, 147
  chromatography beads, 72
  cotton, 217
  courtship pheromone, 289
  crystalline cellulose, 70
  dialysis bag, 48
  filter paper, 217, 238, 289
  gel-based, 30, 52, 57, 58, 242
  glass, 217
  glass microscope slide, 24, 30, 350
  krill eye, 63
  microbeads, 85
  microencapsulation, 85, 258
  muslin disks, 229
  normal-phase chromatography particles, 69
  oviposition-deterring pheromone, 237
  palatable prey
  phagostimulants, 239
  phytagel, 113, 119
  Plexiglas™ settlement plates, 113
  polyethylene vials, 217
  remote control, 49
  rubber septa, 217
  sodium-alginate-based foods, 59
  SPEX® beads, 154
  squid paste, 62
  syringe, 47
  thin-layer chromatography plates, 239
  tiles, 28, 343
Storeria dekayi, 299, 303
Strombus costatus, 86
Submandibular mental glands, 286
Surface pheromones, 227
Symbiodinium sp., 23
Synechococcus sp., 20

Tannins, 53
Taricha, 288
Taricha granulosa, 291, 293, 295
Taricha rivularis, 277
Tarichatoxin, 295
Terpenes, 188
Territorial behaviors, 352
  beaver castoreum, 354
  golden hamsters, 353
  mountain beaver, 374
  overmarking, 353
  rabbits, 355
  scent marks, 353
  urine-contaminated traps, 354
Territorial pheromone, 282
  chameleons, 301
  cloacal glands in alligators, 298
  cloacal odors, 283
  fecal pellets, 282
  postcloacal gland, 284
  salamanders, 282
Thalassia, 57
Thalassiosira pseudonana, 68
Thalassoma bifasciatum, 78
Thamnophis, 297, 307
Thamnophis elegans vagrans, 301
Thamnophis marcianus, 300
Thamnophis radix, 299, 308

Thamnophis sirtalis, 299, 302, 303, 305
Thigmotactic, 303
Thigmotropic response, 155
Time-lapse photography, 224
TLC, 10, 55
Toxins, 77
  amphibians, 294
  entomopathogenic fungus, 157
Trail pheromones, 225
  choice tests, 225
  rats, 359
Trailing, 302
Tribolium castaneum, 164
Tridentata marginata, 73
Trigeminal chemoreception, 328
Tripneustes gratilla, 87
Triturus cristatus, 289
Tropidurus delanonis, 306
Tube foot retraction assay, 64
Turbo undulata, 87
Turtles, 298
  inguinal musk glands, 298
  mental glands, 306
Tyrosine, 236

Ulva, 58, 65
Unaspis yanonensis, 237
Uric acid determination, 254
Uromyces, 155
Ustilago maydis, 147
Uta stansburiana, 306

Vaginal smear technique, 368
Varanus albigularis, 310
Vesicular-arbuscular mycorrhizal, 160
Vibrio alginolyticus, 18
Video microscopy, 18
Videotape, 48
Vipera aspis, 296
Vipera berus, 296, 303, 308
Vipera latastei, 310
Volatile chemical, 217
  moving air, 217
Vomeronasal chemoreception, 278, 328

Wind tunnel, 217, 220
  alarm pheromones, 226
  analyzing motion tracks, 223
  construction, 218, 220
  host odors, 223
  moving pattern, 222
  pheromones, 220
  smoke-tests, 222
  treadmills, 222
Wolves, 342
  urine, 342
Wrasses, 78

Yeast, 147

Zooplankters, 69
Zoospore behavior, 158
Zostera marina, 10, 25

MIX
Papier aus verantwortungsvollen Quellen
Paper from responsible sources
FSC® C105338

If you have any concerns about our products,
you can contact us on
ProductSafety@springernature.com

In case Publisher is established outside the EU,
the EU authorized representative is:
Springer Nature Customer Service Center GmbH
Europaplatz 3, 69115 Heidelberg, Germany

Printed by Libri Plureos GmbH
in Hamburg, Germany